D1227924

BIOGRAPHICAL NOTE

The senior author of the present work, George Frederick Kunz, was born in New York on 29 September 1856. His career as a mineralogist included a position as a gem expert with Tiffany and Company in New York, and as a special agent in charge of precious stones with the U.S. Geological Survey from 1883. In addition to the present work, Kunz was the author of *Ivory and the Elephant* (1915) and *Shakespeare and Precious Stones* (1916). He died in 1932.

Published in Canada by General Publishing Company, Ltd., 895 Don Mills Road, 400-2 Park Centre, Toronto, Ontario M3C 1W3.
Published in the United Kingdom by David & Charles, Brunel House, Forde Close, Newton Abbot, Devon TQ12 4PU.

Bibliographical Note

This Dover edition, first published in 2001, is an unabridged republication of the work originally published by The Century Co., New York, in 1908 and reprinted by Dover Publications, Inc., in 1993. The illustrations have been regrouped and these changes are reflected in the revised List of Illustrations and the Index. Those illustrations originally in color appear in black and white here. Biographical information about George Frederick Kunz has been added.

International Standard Book Number: 0-486-42276-3

Manufactured in the United States of America
Dover Publications, Inc., 31 East 2nd Street, Mineola, N.Y. 11501

CONTENTS

LIST OF ILLUSTRATIONS

FACING PAGE

MAPS

INTRODUCTION

THE preparation of this book has been a joint labor during the spare moments of the two authors, whose time has been occupied with subjects to which pearls are not wholly foreign—one as a gem expert, and the other in the fisheries branch of the American government. But for the views and expressions contained herein, they alone are personally responsible, and do not represent or speak for any interest whatever. For many years the writers have collected data on the subject of pearls, and have accumulated all the obtainable literature, not only the easily procurable books, but likewise manuscripts, copies of rare volumes, original edicts, and legislative enactments, thousands of newspaper clippings, and interesting illustrations, many of them unique, making probably the largest single collection of data in existence on this particular subject. It was deemed advisable to present the results of these studies and observations in one harmonious volume, rather than in two different publications. This publication is not a pioneer in an untrodden field. As may be seen from the appended bibliography, during the last two thousand years hundreds of persons have discussed pearls—mystically, historically, poetically, and learnedly. Among the older writers who stand out with special prominence in their respective periods are the encyclopedist Pliny, in the first century A.D.; Oviedo and Peter Martyr of the sixteenth century; the physician Anselmus De Boot, and that observant traveler and prince of jewelers, Tavernier, in the seventeenth century. It would be difficult to do justice to the many writers of the nineteenth century and of the present time; but probably most attention has been attracted by the writings of Hessling and Möbius of Germany; Kelaart, Streeter, Herdman, and Hornell of Great Britain; Filippi of Italy, and Seurat and Dubois of France. While the book is a joint work in the sense that each writer has contributed material to all of the chapters and has critically examined and approved the entire work, the senior author has more closely applied himself to the latter half of the text, covering antiquity values, commerce, wearing manipulation, treatment, famous collections, aboriginal use, and the illustrations, while the junior author has attended to the earlier half of the book, with reference to

history, origin, sources, fisheries, culture, mystical properties, and the
literature of the pearl.

The senior author has had exceptionally favorable opportunities to
examine the precious objects contained in the various imperial and
royal treasuries. Through the courtesy of the late Count Sipuigine,
Court Chamberlain, and of the late General Philamanoff, custodian of
the Ourejena Palata, he was permitted to critically examine the Russian crown jewels in the Summer Palace on the Neva, and in the
Palata in the Kremlin, at Moscow, he examined the crowns and
jewels of all the early czars. Through the courtesy of Baron von
Theile, he was permitted to inspect carefully and in detail the wonderful jewels of the Austrian crown, which are beautifully ordered
and arranged. The English and Saxon crown jewels were also seen
under favorable conditions which permitted detailed examination, and
the jewel collections of almost all the principal museums of Europe
and America were carefully studied. As regards the literature of the
subject, the senior author has gathered together the largest known
existing collection of works treating of pearls and precious stones.

In covering so comprehensive a subject, many obligations have been
incurred from individuals and officials, to whose courtesy and assistance is due much of the interest of this work. To list all of these is
impossible, yet it would be ungrateful not to note the following: her
Majesty Queen Margherita of Italy; his Royal Highness the Gaikwar
of Baroda; to H. R. H. le Prince Ruprecht of Bavaria, of Munich; to
the late Prince Sipuigine, then chamberlain of the Russian Imperial
Appanages; to Sir Edward Robert Pearce Edgcumbe for data relative
to fisheries of East Africa; Dr. H. C. Bumpus, director of the American Museum of Natural History, New York, for many courtesies
in regard to materials and illustrations; Sir Caspar Purdon Clarke,
director, Dr. Edward Robinson, assistant director, J. H. Buck,
curator of Metal-work, and A. G. St. M. D'Hervilly, assistant curator
of Paintings, all of the Metropolitan Museum of Art, for numerous
courtesies; Archer M. Huntington, founder of the Hispanic Society
and Museum in New York City; Dr. Bashford Dean, Prof. Friedrich
Hirth, Chinese professor, Dr. Berthold Laufer, Prof. A. V. Williams Jackson, professor of Indo-Iranian languages, and Prof. M. H.
Saville, all of Columbia University, New York City; J. Pierpont
Morgan, for the right to publish the illustration of Ashburnham
missal; Dr. W. Hayes Ward, Assyriologist; Dr. Charles S. Braddock,
formerly Chief of Medical Inspection for the King of Siam; Robert
Hoe, for the two plates of unique Persian illustrations from his manu-
scripts; Edmund Russell, for East Indian material; F. Cunliffe-
Owen, the author of diplomatic subjects; Ten Broeck Morse; Walter

Joslyn; Stansbury Hagar; Henri de Morgan, explorer; Dr. Nathaniel L. Britton, director New York Botanical Garden, J. H. Lawles, and Ludwig Stross, for many courtesies; Miss M. de Barril and Miss Belle da Costa Greene, all of New York; Dr. Stewart W. Culin, of the Brooklyn Institute of Arts and Sciences; the Contessa Casa Cortez, for Peruvian information, of Brooklyn; Dr. Charles B. Davenport, of the Carnegie Institution Laboratory, Cold Spring Harbor; Arthur C. Parker, archæologist, State Museum, Albany, N. Y.; A. S. Clark, antiquarian, Peekskill, N. Y.; Dr. Richard Rathbun, assistant secretary, Dr. Cyrus Adler, curator, Dr. Otis S. Mason, curator of Ethnology, all of the Smithsonian Institution; Dr. S. W. Stratton, chief of the Bureau of Standards; Miss E. R. Scidmore; Gilbert H. Grosvenor, editor, National Geographic Magazine; Hon. William Eleroy Curtis; his Excellency Enrique C. Creel, Embajador de Mexico, and James T. Archbold, war correspondent, all of Washington, D. C.; Prof. W. P. Wilson, director Philadelphia Commercial Museum, Clarence B. Moore, Academy of Natural Sciences, and T. Louis Comparette, curator Numismatic Collection, U. S. Mint, all of Philadelphia; Prof. Henry Montgomery, University of Toronto, Toronto, Canada; Dr. Warren K. Moorehead, archæologist, Andover, Mass.; H. D. Story, and Theo. M. Davis, curators of the Boston Museum of Fine Arts, Boston, Mass.; Miss Mathilde Laigle of Wellesley College; Prof. F. W. Putnam and Alfred M. Tozzer, Peabody Museum of Archæology, Cambridge, Mass.; Prof. Edward S. Morse, Salem, Mass.; Dr. Hiram Bingham, Yale University; W. E. Frost, Providence, R. I.; Dr. Edgar J. Banks, University of Chicago, Chicago, Ill.; Hon. F. J. V. Skiff, director, for several photographs of museum material, and Dr. George A. Dorsey, curator of Anthropology of the Field Columbian Museum; Dr. A. R. Crook, curator of the Museum of Natural History, Springfield, Ill.; Richard Hermann, director Hermann Museum, Dubuque, Ia.; Charles Russell Orcutt, San Diego, Cal.; David I. Bushnell, St. Louis, Mo.; Dr. J. H. Stanton, Prairie du Chien, Wis.; Joe Gassett, Clinton, Tenn.; Prof. Wm. C. Mills, University of Ohio, Columbus, O., for material covering the new Ohio mound discoveries; Mrs. Marie Robinson Wright, author and South American traveler, New York City; Miss Helen Woolley of Judson College, Alabama; Prof. Dr. Eugene Hussak, Rio Janeiro; Hon. George E. Anderson, Consul General of the United States, Rio de Janeiro, Brazil; Señor L. E. Bonilla, Consul General of Colombia; Madam Zelia Nuttall, Coyoacan, Mexico; Prof. Waldstein, University of Cambridge, Cambridge, England; Dr. O. F. Bell, assistant keeper Ashmolean Museum, Oxford; Dr. Stephen W. Bushell, Chinese authority; Lady Christopher Johnston, Dr. William F. Petrie, University College, Dr. Charles

Hercules Read, director of the department of Archæology, British Museum, for illustrations and data; Cyril Davenport, antiquarian writer of the British Museum, for the illustration of the English crown, and crown information; to Sir John Evans, late veteran archæologist and writer; Thomas Tyrer, chemist, W. Talbot Ready, A. W. Feaveryear, E. Alfred Jones, author on metal-work, Edwin W. Streeter, all of London, England; Prof. H. P. Blackmore, curator Blackmore Museum, Salisbury, England; Dr. Thos. Gann, Harrogate, England; Prof. Arthur E. Shipley, Cambridge, England; Dr. Wilfred Grenfell, Labrador; T. W. Lyster, librarian of the National Library of Ireland, Prof. R. F. Scharff, director of the National Museum of Ireland, Dublin, W. Forbes Hourie, all of Ireland; Mr. James Hornell, Dr. W. A. Herdmann, all on information concerning the Ceylon fisheries; Prof. James M. Milne, Belfast, Ireland; David MacGregor, Perth, Scotland; Joseph Baer & Co., Frankfurt, Germany; Herrn C. W. Kesseller, Idar, Germany; Prof. Dr. Carl Sapper, University of Tübingen, Germany; Geheimrath Prof. Dr. Max Bauer, University of Marburg, Germany; Herrn Prof. Dr. Hofer, director Biologische Versuchsstation, Munich; Herrn Ernst Gideon Bek, Pforzheim, Germany; Hon. Albert H. Michelsen, American Consul at Turin; Sabbatino De Angelis, of Naples, Italy; Mons. Alphonse Falco, of the Chambre Syndicale Pierres Précieuses of Paris; Prof. A. Lacroix, Musée Histoire Naturale, Paris; Mons. Georges Pellisier, Paris; Sr. Gaston J. Vives, La Paz, Mexico; Prof. R. Dubois, Facuelté des Sciences, University of Lyons, France; Prof. P. Candias, director of the National Museum, Athens, Greece; Prof. G. A. F. Molengraaff, University of Delft, Holland; the late Prof. Dr. Furtwängler of Munich; Dr. Otto Leiner, Custus Landes-Museum at Constanz, Baden; Herrn Dr. A. B. Meyer, Herrn Carl Marfels, Berlin; Prof. Dr. H. Schumacher, University of Bonn; Geheimrath C. F. Hintze, Breslau; Herrn R. Friedlaender & Sohn, Berlin; Herrn Reg.-Rath Dr. W. von Seidlitz, Dresden; Dr. R. Jacobi, director König Zoologichen Museum, Dresden, Germany; his Excellency Dr. Szalaz, director Hungarian National Museum; Dr. S. Radischi, director National Industrial Museum of Budapest; and to Herrn A. B. Bachrach, Budapest, Hungary; Frau Melanie Glazer, of Prague, and Herrn V. Fric, Prague, Bohemia; Herrn Prof. Dr. F. Heger, Custus Imperial Archæological Collection, Vienna; Herrn H. von Willer and Herrn Max Zirner, of Vienna; Herrn Leopold Weininger, the artisan goldsmith of Austria, for many courtesies; Prof. W. Vernadskij, University of Moscow; Mons. C. Faberje, Joaillier de la Cour, St. Petersburg, Russia; his Excellency Baron P. Meyerdorff, assistant director, Musée des Antiques, Ermitage Impériale, St. Petersburg, for important data and illustrations; his Excellency N. J. Moore, Premier, Western Australia; Dr. K. Van Dort, engineer of Bankok, Siam; Dr. J. Henry Burkill, of the India Museum, Calcutta, India; Alphaeus E. Williams, manager of the De Beers Mine, Kimberley; Capt. E. L. Steever, District Governor of Jolo, Philippine Islands; Dr. T. Nishikawa, Zoölogical Institute; K. Mikimoto, both of Tokio, Japan; Dr. S. M. Zwemer of Bahrein, Persian Gulf; Mr. Hugh Millman of Thursday Island, Australia; Julius D. Dreher, American Consul at Tahiti, Society Islands; and not least, by any means, the uniform promptness and completeness with which the officials of the British Colonial Service have responded to the many inquiries which the writers have addressed to them.

THE AUTHORS.

September, 1908.

I

PEARLS AMONG THE ANCIENTS

THE BOOK OF THE PEARL

I

PEARLS AMONG THE ANCIENTS

The richest merchandise of all, and the most soveraigne
commoditie throughout the whole world, are these pearles.
PLINY, *Historia naturalis.*
Lib. IX, c. 35.

ERFECTED by nature and requiring no art to enhance their
beauty, pearls were naturally the earliest gems known to
prehistoric man. Probably the members of some fish-eating
tribe—maybe of the coast of India or bordering an Asiatic
river—while opening mollusks for food, were attracted by their luster.
And as man's estimation of beauty developed, he found in them the
means of satisfying that fondness for personal decoration so characteristic of half-naked savages, which has its counterpart amid the
wealth and fashion of the present day.

Pearls seem to be peculiarly suggestive of oriental luxury and
magnificence. It is in the East that they have been especially loved,
enhancing the charms of Asiatic beauty and adding splendor to barbaric courts celebrated for their display of costume. From their possession of the rich pearl resources it is natural that the people of India
and of Persia should have early found beauty and value in these jewels,
and should have been among the first to collect them in large quantities. And no oriental divinity, no object of veneration has been without this ornament; no poetical production has lacked this symbol of
purity and chastity.

In a personal memorandum, Dr. A. V. Williams Jackson, professor
of Indo-Iranian languages in Columbia University, states that it is
generally supposed that the Vedas, the oldest sacred books of the Brahmans, contain several allusions to pearl decorations a millennium or

3

more before the Christian era, as the word *krisana* and its derivatives —which occur a half dozen times in the Rigveda, the oldest of the Vedas—are generally translated as signifying "pearl." Even if this interpretation of the term be called into question on the ground that the Hindus of the Panjab were not well acquainted with the sea, there can be little or no doubt that the Atharvaveda, at least five hundred years before the Christian era, alludes to an amulet made of pearls and used as a sort of talisman in a hymn[1] of magic formulas.

Those two great epics of ancient India, the Ramayana and the Mahabharata, refer to pearls. The Ramayana speaks of a necklace of twenty-seven pearls, and has pearl drillers to accompany a great military expedition.[2] An old myth recounts the offerings made by the elements as gifts worthy of the deity: the air offered the rainbow, the fire a meteor, the earth a ruby, and the sea a pearl. The rainbow formed a halo about the god, the meteor served as a lamp, the ruby decorated the forehead, and the pearl was worn upon the heart.

The literature of Hinduism frequently associates the pearl with Krishna, the eighth avatar or incarnation of Vishnu, the most popular god of Hindu worship. One legend credits its discovery to the adorable Krishna, who drew it from the depths of the sea to adorn his daughter Pandaïa on her nuptial day. Another version makes the pearl a trophy of the victory of Krishna over the monster Pankagna, and it was used by the victor as a decoration for his bride.

In the classic period of Sanskrit literature, about the first century of the Christian era, there were abundant references to pearls, generally called *mukta* (literally "the pure") ; and there are dozens of words for pearl necklaces, circlets, strings, and ornamental festoons, particularly in the dramas of Kalidasa—the Hindu Shakspere, who lived about the third century A.D.—and of his successors.

In the Mahavansa and the Dipavansa, the ancient chronicle histories of Ceylon in the Pali language, are several early Cingalese records of pearl production and estimation.[3] The Mahavansa lists pearls among the native products sent from Ceylon about 550 B.C., King Wijayo sending his father-in-law gifts of pearls and chanks to the value of two lacs of rupees; and notes that about 300 B.C., several varieties of Ceylon pearls were carried as presents by an embassy to India.

In the ancient civilization of China, pearls were likewise esteemed; this is evidenced by the frequent mention of them in traditional history, their employment in the veneration of idols, and as tribute by

[1] See pp. 301, 302.
[2] See Jacobi, "Das Ramayana," Bonn, 1893.
[3] Geiger, "Dipavansa und Mahavansa, die beiden Chroniken der Insel Ceylon," Erlangen, 1901.

Ancient Chinese crown with pearls   Ancient Chinese pearl rosary

Chinese priests keeping guard over the tombs of the kings,
in Mukden, where the crowns are preserved

GRECIAN PEARL AND GOLD NECKLACE

Of about third century B.C.

Now in Metropolitan Museum of Art, New York

foreign princes to the emperor. One of the very earliest of books, the Shu King (dating from about 2350–625 B.C.), notes that, in the twenty-third century B.C., Yü received as tribute oyster pearls from the river Hwai, and from the province of King Kau he received "strings of pearls that were not quite round."[1] That ancient Chinese dictionary, the Nh'ya, originating thirty centuries ago, speaks of them as precious jewels found in the province of Shen-si on the western frontier.

Many fantastic theories regarding pearls are to be found in ancient Chinese literature. Some writers credited them as originating in the brain of the fabled dragon; others noted that they were especially abundant during the reign of illustrious emperors, and they were used as amulets and charms against fire and other disasters. Curious allusions were made to pearls so brilliant that they were visible at a distance of nearly a thousand yards, or that rice could be cooked by the light from them. And one found about the beginning of the Christian era, near Yangchow-fu, in the province of Kiang-su, was reported so lustrous as to be visible in the dark at a distance of three miles.

In Persia, the popularity of pearls seems to date from a very early period. Professor Jackson states that if they are not mentioned in the extant fragments of the ancient Zoroastrian literature, the Avesta and the Pahlavi, or by the Middle Persian books from the seventh century B.C. to the ninth century A.D., it is probably a mere accident, due to the character of the work or to the fragmentary condition of the literature; for pearls were well known during that entire period, and seem to be indicated in extant sculptures. The coin and the gem portraits of Persian queens commonly show ear-pendants of these. The remains of a magnificent necklace of pearls and other gems were recently found by J. de Morgan in the sarcophagus of an Achæmenid princess exhumed at Susa or Shushan, the winter residence of the kings of Persia. This necklace, perhaps the most ancient pearl ornament still in existence, dates certainly from not later than the fourth century B.C., and is now preserved in the Persian Gallery of the Louvre.[2] Even if we had no other evidence, it would be natural to assume that the knowledge of pearls was as wide-spread among the Iranians in antiquity as it was among the Hindus, since the Persian Gulf, like the Indian Ocean, has been famous for its fisheries from ancient times.

In the ruins of Babylon no pearls have been found; indeed, it would be surprising if they could survive for so many ages in the relatively moist soil which contains much saltpeter. Inlays of mother-of-pearl and decorations of this material have been secured from the ruins of Bismaya, which Dr. Edgar J. Banks refers to about 4500 B.C.

[1] Legge, "The Shu King," Oxford, 1879, pp. 67, 69.   [2] See p. 404.

There is likewise little evidence that pearls were extensively employed by the ancient Assyrians, notwithstanding that excavations at Nineveh and Nimrud have furnished much information regarding their ornaments; and the collars, bracelets, sword-hilts, etc., wrought in gold and ornamented with gems, show that the jewelers' art had made much progress. This is not wholly trustworthy as determining the relative abundance; for being of organic or non-mineral origin, pearls would not have survived the burial for thousands of years so well as the crystal gems. An inscription on the Nineveh Obelisk, which states, according to Sir Henry Rawlinson, that in the ninth year of his reign Temenbar received, as "tribute of the kings of the Chaldees, gold, silver, gems, and pearls,"[1] shows that the sea-born gems were highly valued there.

The mother-of-pearl shell was in use as an ornament in ancient Egypt certainly as early as the sixth dynasty (*circa* 3200 B.C.), the period of the Tanis Sphinx. In a recent letter from Luxor, where he is studying the ruins of ancient Thebes, Dr. James T. Dennis states that he has found several of these shells bearing cartouches of that period; and in the "pan-bearing graves" of the twelfth dynasty (2500 B.C.), the shell occurred not only complete, but cut in roughly circular or oblong angular blocks and strung on chains with beads of carnelian, pottery, etc.

So far as can be determined from the representations of ancient Egyptian costumes, pearls do not seem to have been employed to any great extent in their decoration. The necklaces, earrings, and other jewels found in the tombs, which are composed largely of gold set with crystal gems, contain the remains of a few pearls, but give no indication that they were numerous. In fact, no evidence exists that they were used extensively before the Persian conquest in the fifth century B.C.; and probably it was not until the time of the Ptolemies that there began the lavish abundance which characterized the court of Alexandria at the height of her power.

The authorities differ in regard to the mention of pearls in ancient Hebrew literature; although in the Authorized Version of the Old Testament, this significance has been given to the word *gabîsh* in Job xxviii. 18, where the value of wisdom is contrasted with that of *gabîsh*. Some writers claim that this word refers to rock crystal. Other authorities are of the opinion that the word *peninim* in Lam. iv. 7, which has been translated as "rubies," actually signifies pearls. In Gen. ii. 12, Prof. Paul Haupt has proposed to render *shoham* stones by pearls, since the Hebrew word translated "onyx," if connected with the Assyrian *sându*, might mean "the gray gem." It does not

[1] Rawlinson, "Cuneiform Inscriptions of Babylonia and Assyria," London, 1850, p. 38.

appear that they entered into the decorations of the Tabernacle and the Temple, or were largely employed in the paraphernalia of the synagogue.

In the New Testament, however, there are numerous references to the estimation in which pearls were held. In his teachings, Christ repeatedly referred to them as typifying something most precious: "The kingdom of heaven is like unto a merchant man, seeking goodly pearls: who, when he had found one pearl of great price, went and sold all that he had, and bought it" (Matt. xiii. 45, 46); and in "casting pearls before swine," in that great Sermon on the Mount (Matt. vii. 6). In picturing the glories of the Heavenly City, St. John made the twelve gates of pearls (Rev. xxi. 21); and what could better serve as portals through the walls of precious stones?

In the Talmud, pearls are frequently mentioned, and usually as signifying something beautiful or very costly, as "a pearl that is worth thousands of zuzim" (Baba Batra, 146a); a "pearl that has no price" (Yerushalmi, ix. 12d); the coats which God made for Adam and Eve were "as beautiful as pearls" (Gen. R. xx. 12), and the manna was "as white as a pearl" (Yoma, 75a). Their purchase formed one of the exceptions to the law of Ona'ah (overcharge), for the reason that two matched pearls greatly exceeded the value of each one separately (Baba Mezi'a, iv. 8).

The high value attached to pearls by the ancient Hebrews is illustrated by a beautiful Rabbinical story in which only one object in nature is ranked above them. On approaching Egypt, Abraham hid Sarah in a chest, that foreign eyes might not behold her beauty. When he reached the place for paying custom dues, the collectors said, "Pay us the custom"; and he replied, "I will pay your custom." They said to him, "Thou carriest clothes"; and he stated, "I will pay for clothes." Then they said to him, "Thou carriest gold"; and he answered, "I will pay for gold." On this they said to him, "Surely thou bearest the finest silk"; and he replied, "I will pay custom for the finest silk." Then said they, "Truly it must be pearls that thou takest with thee"; and he answered, "I will pay for pearls." Seeing that they could name nothing of value for which the patriarch was not willing to pay custom, they said, "It cannot be but that thou open the box and let us see what is within." So the chest was opened, and the land was illumined by the luster of Sarah's beauty.[1]

The love which the early Arabs bore to pearls is evidenced by the references to them in the Koran, and especially the figurative description given of Paradise. The stones are pearls and jacinths; the fruits

[1] Gen. R. xl. 6. This story also exists somewhat altered in Arabic literature; see Weill's "Biblical Legends of the Mussulmans," New York, 1846.

of the trees are pearls and emeralds; and each person admitted to the delights of the celestial kingdom is provided with a tent of pearls, jacinths and emeralds; is crowned with pearls of incomparable luster, and is attended by beautiful maidens resembling hidden pearls.[1]

The estimation of pearls among the art-loving Greeks may be traced to the time of Homer, who appears to have alluded to them under the name τρίγληνα (triple drops or beads) in his description of Juno; in the Iliad, XIV, 183:

> In three bright drops,
> Her glittering gems suspended from her ears.

and in the Odyssey, XVIII, 298:

> Earrings bright
> With triple drops that cast a trembling light.

Classical designs of Juno usually show the three pear-shaped pearls pendent from her ears. The ancient Greeks probably obtained their pearls from the East through the medium of Phenician traders, and a survival of the word τρίγληνα seems to exist in the Welsh *glain* (bead), the name having been carried to Britain by the same traders, who exchanged textiles, glass beads, etc., for tin and salt.

The Persian wars in the fifth century B.C., doubtless extended the acquaintance which the Greeks had with pearls, as well as with other oriental products, and increased their popularity. One of the earliest of the Greek writers to mention pearls specifically appears to have been Theophrastus (372–287 B.C.), the disciple and successor of Aristotle, who referred to them under the name μαργαρίτης (*margarites*), probably derived from some oriental word like the Sanskrit *maracata* or the Persian *mirwareed*. He stated that pearls were produced by shell-fish resembling the *pinna,* only smaller, and were used in making necklaces of great value. In Pliny's "Historia naturalis," that great storehouse of classical learning, reference is made to many other writers—mostly Greeks—who treated of gems; but virtually all of these writings have disappeared, except fragments from Theophrastus, Chares of Mytilene, and Isidorus of Charace.

From Greece admiration for pearls quickly extended to Rome, where they were known under the Greek word *margaritæ*. However, a more common name for this gem in Rome was *unio,* which Pliny explained by saying that each pearl was unique and unlike any other one. The conclusion of the historian Ammianus Marcellinus (330–395 A.D.),

---

[1] Sale, "Preliminary Discourse to the Quran," London, 1882, Vol. I, pp. 153-159.

that it was because each one was found singly in a shell,[1] seems scarcely correct. Claude de Saumaise, the French classical scholar, thought that the common name for an onion was transferred to the pearl, owing to its laminated construction.[2] According to Pliny, the Romans used the word *unio* to distinguish a large perfect pearl from the smaller and less attractive ones, which were called *margaritæ*.[3]

It was not until the Mithridatic Wars (88–63 B.C.) and the conquests by Pompey that pearls were very abundant and popular in Rome, the great treasures of the East enriching the victorious army and through it the aristocracy of the republic. In those greatest spectacular functions the world has ever known—the triumphal processions of the conquering Romans—pearls had a prominent part. Pliny records that in great Pompey's triumphal procession in 61 B.C. were borne thirty-three crowns of pearls and numerous pearl ornaments, including a portrait of the victor, and a shrine dedicated to the muses, adorned with the same gems.[4]

The luxuries of Mithridates, the treasures of Alexandria, the riches of the Orient were poured into the lap of victory-fattened Rome. From that time the pearl reigned supreme, not only in the enormous prices given for single specimens, but also in the great abundance in possession of the degenerate descendants of the victorious Romans. The interior of the temple of Venus was decorated with pearls. The dress of the wealthy was so pearl-bedecked that Pliny exclaimed in irony: "It is not sufficient for them to wear pearls, but they must trample and walk over them";[5] and the women wore pearls even in the still hours of the night, so that in their sleep they might be conscious of possessing the beautiful gems.[6]

It is related that the voluptuous Caligula (12–41 A.D.)—he who raised his favorite horse Incitatus to the consulship—decorated that horse with a pearl necklace, and that he himself wore slippers embroidered with pearls; and the tyrannical Nero (37–68 A.D.), not content with having his scepter and throne of pearls, provided the actors in his theater with masks and scepters decorated with them. Thus wrote the observant Philo, the envoy of the Jews to the Emperor Caligula: "The couches upon which the Romans recline at their repasts shine with gold and pearls; they are splendid with purple coverings interwoven with pearls and gold."

Yet not all the men of Rome were enthusiastic over the beautiful "gems of the sea, which resemble milk and snow," as the poet Manlius

[1] *Lib.* XXIII, c. 6.
[2] "Plinianæ Exercitationes in Solinum," 1629, pp. 822-4.
[3] "Historia naturalis," *Lib.* IX, c. 59.
[4] *Ibid., Lib.* XXXVII, c. 2.
[5] *Ibid., Lib.* IX, c. 53.
[6] *Ibid., Lib.* XXXIII, c. 3. Also Böttiger, "Sabina oder Morgenscenen," Leipzig, 1803, Vol. I, p. 158.

called them. Even then, as now, there were some faultfinders. The immortal Cæsar interdicted their use by women beneath a certain rank; Martial and Tibullus inveighed against them; the witty Horace directed his stinging shafts of satire against the extravagance. Referring to a woman named Gellia, Martial wrote: "By no gods or goddesses does she swear, but by her pearls. These she embraces and kisses. These she calls her brothers and sisters. She loves them more dearly than her two sons. Should she by some chance lose them, the miserable woman would not survive an hour."[1]   Hear what stern old Seneca had to say: "Pearls offer themselves to my view. Simply one for each ear? No! The lobes of our ladies have attained a special capacity for supporting a great number. Two pearls alongside of each other, with a third suspended above, now form a single earring! The crazy fools seem to think that their husbands are not sufficiently tormented unless they wear the value of an inheritance in each ear!"[2]

The prices reported for some choice ones at that time seem fabulous. It is recorded by Suetonius, that the Roman general, Vitellius, paid the expenses of a military campaign with the proceeds of one pearl from his mother's ears: *"Atque ex aure matris detractum unionem pigneraverit ad itineris impensas."*   In his "Historia naturalis," Pliny says that in the first century A.D., they ranked first in value among all precious things,[3] and reports sixty million sestertii[4] as the value of the two famous pearls—"the singular and only jewels of the world and even nature's wonder"—which. Cleopatra wore at the celebrated banquet to Mark Antony. And Suetonius[5] places at six million sestertii the value of the one presented by Julius Cæsar as a tribute of love to Servilia, the mother of Brutus, who thus wore

> The spoils of nations in an ear,
> Changed to the treasure of a shell.

Or, as St. Jerome expressed it in his "Vita Pauli Eremitæ":

> *Uno filo villarum insunt pretia.*

We are told by Ælius Lampridius that an ambassador once brought to Alexander Severus two remarkably large and heavy pearls for the empress. The emperor offered them for sale, and as no purchaser was found, he had them hung in the ears of the statue of Venus, saying: "If the empress should have such pearls, she would give a bad example

[1] Martial, "Epigrammata," VIII, 81.
[2] Seneca, "De beneficiis," *Lib.* VII, c. 9.
[3] Pliny, "Historia naturalis," *Lib.* IX, c. 35.
[4] Equivalent to 1,875,000 ounces of silver, worth about $1,300,000 at the present time, but of far greater value in Roman days.
[5] "Divus Julius Cæsar," c. 50.

to the other women, by wearing an ornament of so much value that no one could pay for it."

The word "margarita" was used symbolically to designate the most cherished object; for instance, a favorite child. In an inscription published by Fabretti, p. 44, No. 253, the word *margaritio* has the same significance. (Sex. Bruttidio juveni margaritioni carissimo, vixit annis II mensibus VII, diebus XVIII.)[1]

While the ancient writers were familiar with the pearl itself, they knew little of the fisheries, and related many curious stories which had come to Athens and Rome. Pliny and Ælianus quoted from Megasthenes that the pearl-oysters lived in communities like swarms of bees, and were governed by one remarkable for its size and great age, and which was wonderfully expert in keeping its subjects out of danger, and that the fishermen endeavored first to catch this one, so that the others might easily be secured. Procopius, one of the most entertaining of the old Byzantine chroniclers, wrote of social relations between the pearl-oysters and the sharks, and of methods of inducing the growth of pearls.

The principal fisheries of antiquity were in the Persian Gulf, on the coasts of Ceylon and India, and in the Red Sea. The pearls referred to in ancient Chinese literature appear to have been taken from the rivers and ponds of that country, while those in Cochin China and Japan seem to have come from the adjoining seas. The pearls were distributed among the nations in control of the fisheries, and from them, other people received collections, either as presents, in conquest, or by way of trade. History makes no mention of pearls having been obtained elsewhere than in the Orient up to the time of Julius Cæsar, when small quantities of inexpensive ones were collected in Britain by the invading Romans. And in the first century A.D., Pliny states that small reddish pearls were found about Italy and in the Bosphorus Straits near Constantinople.

A number of specimens of pearls of the artistic Greeks and of the luxurious Romans are yet in existence, and some of these are in a fairly good state of preservation. A notable and interesting example is a superb Greek necklace of pearls and gold, referred to the third century B.C., and now in the Metropolitan Museum of Art in New York. Several earrings now in that museum, in the Hermitage at St. Petersburg, the British Museum, the Louvre in Paris, and in the Boston Museum of Fine Arts, are shown in this book. Some of these may have decorated ears that listened to the comedies of Aristophanes, the tragedies of Euripides, the philosophies of Plato, or the oratory of Demosthenes. A number of classic statues have the ears pierced

[1] "Dictionnaire des Antiquités Grecques et Romaines," Paris, 1904, Vol. III, pp. 1595–6.

for earrings, notably the Venus de Medici now in the Tribuna of the Uffizi, Florence; and a magnificent pair of half-pearls is said to have decorated the Venus of the Pantheon in Rome.[1]   Pearl grape earrings are shown on the artistic intaglio by Aspasios, representing the bust of the Athene Parthenos of Phidias, which has been in the Gemmen Münzen Cabinet at Vienna since 1669.

The beautiful Tyszkiewicz bronze statuette of Aphrodite was acquired in 1900 by the Boston Museum of Fine Arts, and has even yet a pearl in a fairly good state of preservation suspended from each ear by a spiral thread of gold which passes quite through the gem and also through the lobe of the ear.   This statuette has been described as "the most beautiful bronze Venus known."[2]   Professor Froehner considers that it belongs nearer to the period of Phidias (*circa* 500–430 B.C.) than to that of Praxiteles (*circa* 400–336 B.C.); but Dr. Edward Robinson does not concur in this opinion, and refers it to the Hellenic period (*circa* 330–146 B.C.).

However, considering the very large accumulations, relatively few pearls of antiquity now remain, and none of these is of great ornamental value.   Those in archæological collections and art museums are more or less decayed through the ravages of time and accident to which they have been subjected.   While coins, gold jewelry, crystal gems, etc., of ancient civilizations are relatively numerous, the less durable pearls have not survived the many centuries of pillage, waste, and burial in the earth.

A well-known instance of this decay is found in the Stilicho pearls, which owe their prominence to the incident of their long burial.   The daughters of this famous Roman general, who were successively betrothed to the Emperor Honorius, died in 407 A.D., and were buried with their pearls and ornaments.   In 1526, or more than eleven centuries afterward, in excavating for an extension of St. Peter's, the tomb was opened, and the ornaments were found in fair condition, except the pearls, which were as lusterless and dead as a wreath of last year's flowers.

[1] See p. 449.   [2] Froehner, "La Collection Tyszkiewicz," Munich, 1892.

II

MEDIEVAL AND MODERN HISTORY

II

MEDIEVAL AND MODERN HISTORY OF PEARLS

I 'll set thee in a shower of gold, and hail
Rich pearls upon thee.
Antony and Cleopatra, Act II, sc. 5.

THE popularity of pearls in Rome has its counterpart in the Empire of the East at Byzantium or Constantinople on its development in wealth and luxury after becoming the capital of that empire in 330 A.D. Owing to its control of the trade between Asia and Europe, and the influence of oriental taste and fashion, enormous collections were made; and for centuries after Rome had been pillaged, this capital was the focus of all the arts, and pearls were the favorite ornaments. The famous mosaic in the sanctuary of San Vitale at Ravenna, shows Justinian (483–565) with his head covered with a jeweled cap, and the Empress Theodora wearing a tiara encircled by three rows of pearls, and strings of pearls depend therefrom almost to the waist. In many instances the decorations of the emperors excelled even those of the most profligate of Roman rulers. An examination of the coins, from those of Arcadius in 395 to the last dribble of a long line of obscure rulers when the city was captured and pillaged by Venetian and Latin adventurers in 1204, shows in the form of diadems, collars, necklaces, etc., the great quantity of pearls worn by them. The oldest existing crown in use at the present time, the Hungarian crown of St. Stephen, which is radiant with pearls, is of Byzantine workmanship.

Outside of Constantinople, the demand and fashion for pearls did not cease with the downfall of the Roman Empire and the spoliation of Rome in the fifth century. The treasures accumulated there, and the gems and jewels, were carried away by the conquering Goths and scattered among the great territorial lords of western and northern Europe.

In the ancient cities of Gaul, in Toulouse and Narbonne, the Ostrogoth and the Visigoth kings collected enormous treasures. The citadel of Carcassonne held magnificent spoils brought from the sacking of Rome in 410 by Alaric, king of the Ostrogoths, consisting in part of jewels from the Temple, these having been carried to Rome after the spoliation of Jerusalem in 70 A.D. Several beautiful objects of this and somewhat later periods are yet in existence, notably the Visigothic crowns and crosses, in the Musée de l'Hôtel de Cluny, Paris, the most beautiful of which are probably the crown and the cross of Reccesvinthus.[1]

Even as the treasures of Rome were despoiled by the Ostrogoths and the Visigoths, so, later, their collections were depleted by the military operations of the Franks, when Narbonne was pillaged; when Toulouse was sacked by Clovis, or Chlodowig, in 507; when the churches of Barcelona and Toledo were despoiled by Childebert in 531 and 542; and by various expeditions in succeeding years.

The military triumphs of the Franks placed them in the highest rank among the peoples of Europe, in the sixth and seventh centuries, in the possession of treasures of jewels which enriched their palaces and great churches. And the taste which the triumphs of war had developed was maintained by the trade carried on by the Jewish and Syrian merchants. The inhabitants of Gaul were extremely fond of objects of art, of rich costumes, and of personal decorations; and the courts of some of the early kings rivaled in magnificence those of oriental monarchs. Especially was this true during the reign of King Dagobert (628–638), who competed in splendor with the rulers of Persia and India. His skilful jeweler, Eligius (588–659), was raised to the bishopric of Noyon, and eventually—under the name of St. Eloi —became one of the most popular saints in Gaul. Under direction of this artistic bishop, the ancient churches received shrines, vestments, and reliquaries superbly decorated with pearls and other gems. Indeed, for several centuries following the time of Eligius, the greatest treasures of jewels seem to have been collected in the churches.

The use of gems in enriching regalia, vestments, and reliquaries in Europe, advanced greatly during the reign of Charlemagne (768–814); and princes and bishops competed with each other in the magnificence of their gifts to the churches, sacrificing their laical jewels for the sacred treasures. Few of the great ornaments of Charlemagne's time are now in existence in the original form. Doubtless the most remarkable pieces are the sacred regalia of the great emperor, preserved among the imperial treasures in Vienna.

An artistic use for pearls at that time was in the rich and elegant bindings of the splendidly written missals and chronicles, finished in the highest degree of excellence and at vast expense. An artist might devote his whole life to completing a single manuscript, so great was

[1] See p. 415.

FRONT COVER OF ASHBURNHAM MANUSCRIPT OF THE FOUR GOSPELS

From the ninth century.   One quarter of the actual dimensions

Owned by J. Pierpont Morgan, Esq.

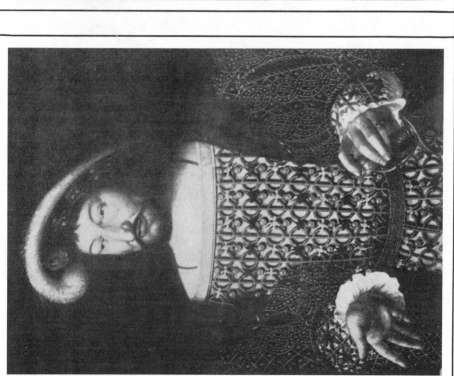

ISABELLE DE VALOIS

By Pantoia de la Cruz, Prado Museum, Madrid

FRANCIS I, KING OF FRANCE, 1494–1547

Louvre, Paris

the detail and so exquisite the finish. Vasari states that Julio Clovio devoted nine years to painting twenty-six miniatures in the Breviary of the Virgin now in the royal library at Naples. The library at Rouen has a large missal on which a monk of St. Andoen is said to have labored for thirty years. These books were among the most valued possessions of the churches, and their bindings were enriched with gold and pearls and colored stones. The wealthy churches had many such volumes; Gregory of Tours states that from Barcelona in 531 A.D. Childebert brought twenty *"evangeliorum capsas"* of pure gold set with gems. Several of these superbly bound volumes are yet in existence, in the Basilica of St. Mark in Venice; in the treasury of the cathedral at Milan; among the imperial Russian collections in the Ourejenaya Palata at Moscow, etc.; and they furnish probably the most reliable examples of artistic jewel work of the Dark Ages.

The most remarkable specimen of these books in America is doubtless the Ashburnham manuscript of the Four Gospels, now owned by J. Pierpont Morgan, Esq., which affords an interesting example of the jeweler's art. For many centuries it belonged to the Abbey of the Noble Canonesses, founded, in 834, at Lindau, on Lake Constance. After an extended examination, Mr. Alexander Nesbit concluded that the rich cover of the manuscript was probably made between 896 and 899 by order of Emperor Arnulf of the Carolingian dynasty. Most of the ninety-eight pearls appear to be from fresh water, and probably all of them were obtained from the rivers of Europe. This is one of the few remaining pieces of the magnificent ecclesiastical jeweling of that period.

After the death of Charlemagne, internal dissensions, separations and the division of the Empire into the nations of Europe, annihilated commerce, oppressed the people, and impoverished the arts. In the ninth century, the Normans pillaged many of the palaces and churches in Angoulême, Tours, Orléans, Rouen, and Paris, and destroyed or carried away large treasures. The tenth and the eleventh centuries were indeed the Dark Ages in respect to the cultivation of the arts; yet even during that period the churches of western Europe received many gems from penitent and fear-stricken subjects. The heart of man, filled with the love of God, laid its earthly treasure upon the altar in exchange for heavenly consolation. Pious faith dedicated pearls to the glorification of the ritual; altars, statues, and images of the saints, priestly vestments, and sacred vessels, were surcharged with them. The great museums and the imperial collections contain some beautiful and highly venerated objects of this nature.

In the meantime pearls of small size and of fair luster had been collected in the rivers of Scotland, Ireland, and France, the headwaters of the Danube, and in the countries north thereof. In England, as noted in the preceding chapter, they were obtained by Cæsar's invading legions, who carried many to Rome. Ancient coins indicate that pearls formed the principal ornament of the simple crowns worn by the early kings of Britain previous to Alfred the Great.

The river pearls were not so beautiful as oriental ones; but, owing to the ease with which they were obtained, they were employed more extensively and especially in ecclesiastical decorations, the principal use for pearls from the eighth to the eleventh century. Apparently authentic specimens of fresh-water pearls of an early period are the four now in the coronation spoon of the English regalia, which is attributed to the twelfth century.

From the most ancient times until the overthrow of the Roman Empire, practically the only use for pearls was ornamental; but after the eighth century there developed a new employment for these as well as for other gems. Natural history was little studied in Europe from the ninth to the fourteenth century, except for the effect which its subjects had in medicine and magic, which were closely allied. Largely through Arabic influence, the practice of medicine had developed into administering most whimsical remedies, among which gems, and especially pearls, played a prominent part, and belief in the influence of these was as strong as in that of the heavenly bodies. For this application, large demands had arisen for pearls, which seem to have been prescribed for nearly every ill to which the flesh was heir. On account of their cheapness, the small ones—seed-pearls—were used principally; though larger ones were preferred by persons who could afford them. While many of these so-called medicinal pearls were obtained from the Orient, most of them were secured from the home streams in the north of Europe and in the British Isles.

After the decadence of Roman power in the East, the rulers of India and Persia, through their control of the fisheries, again accumulated enormous quantities of pearls. All of the early travelers to those countries were astonished at the lavish display of these gems in decorative costume.

The manuscript of Renaudot's two Mohammedans, who visited India and China in the ninth century, notes that the kings of the Indies were rich in ornaments, "yet pearls are what they most esteem, and their value surpasses that of all other jewels; they hoard them up in their treasures with their most precious things. The grandees of the court, the great officers and captains, wear the like jewels in their collars."[1]

[1] Renaudot, "Ancient Accounts of India and China by Two Mohammedan Travelers," London, 1733, p. 98.

Inventories of some of the oriental collections of later times seem to be extravagant fiction rather than veritable history. In that interesting book dictated in a Genoese prison to Rusticiano da Pisa, accounts are given by Marco Polo of great treasures seen by the first Europeans to penetrate into China. He describes the king of Malabar as wearing suspended about his neck a string of 104 large pearls and rubies of great value, which he used as a rosary. Likewise on his legs were anklets and on his toes were rings, all thickly set with costly pearls, the whole "worth more than a city's ransom. And 't is no wonder he hath great store of such gear; for they are found in his kingdom. No one is permitted to remove therefrom a pearl weighing more than half a *saggio*. The king desires to reserve all such to himself, and so the quantity he has is almost incredible."[1]

Later travelers give wonderful descriptions of this excessive passion for pearls. Literature is full of this appreciation, and of the part which these gems played in the affairs of the Orientals. Who has not dwelt with delight upon those imperishable legends such as are embodied in the Arabian Nights, of the pearl voyages by Sindbad the Sailor, of the wonderful treasure chests, and of the superb necklaces adorning the beautiful black-eyed women!

The returning Crusaders in the twelfth and thirteenth centuries, and the development of the knightly orders, had much to do with spreading through Europe a fondness for pearls in personal decoration. Those who, like Chaucer's knight, had been with Peter, King of Cyprus, at the capture and plunder when "Alexandria was won," returned to their homes with riches of pearls and gold and precious stones. And learning much relative to decorative art from Moorish craftsmen, the jewelers of western Europe set these in designs not always crude and ineffective.

Although they were well known and valued, pearls do not seem to have been much used in England before the twelfth century, as the Anglo-Saxons were not an especially art-loving people. The word itself is of foreign derivation and occurs in a similar form in all modern languages, both Romance and Teutonic; *perle,* French and German; *perla,* Italian, Portuguese, Provençal, Spanish, and Swedish; *paarl,* Danish and Dutch. Its origin is doubtful. Some philologists consider it Teutonic and the diminutive of *beere,* a berry; Claude de Saumaise derives it from *pirula,* the diminutive of *pirum,* a sphere; while Diez and many others refer it to *pira* or to the medieval Latin *pirula,* in allusion to the pear shape frequently assumed by the pearl.[2]

[1] "The Book of Ser Marco Polo," London, 1871, Vol. II, p. 275.

[2] Analogous to the uniform European word for this gem, is the extension of the Sanskrit form, *mukta,* from Persia to the Sulu Islands. In Tamil, the word for pearl is *mootthoo;* in Hindustani, it is *mootie;* in Cingalese, *mootoo;* and in Malay, *mutya* or *mootara.* (Ainslie, "Materia Indica," London, 1826, Vol. I, pp. 292-297.)

The word pearl seems to have come into general use in the English language about the fourteenth century. In Wyclif's translation of the Scriptures (about 1360), he commonly used the word *margarite* or *margaritis,* whereas Tyndale's translation (1526) in similar places used the word *perle.* Tyndale translated Matt. xiii. 46: "When he had founde one precious pearle"; Wyclif used "oo preciouse margarite." Also in Matt. vii. 6, Tyndale wrote, "Nether caste ye youre pearles before swyne"; yet Wyclif used "margaritis," although twenty years later he expressed it "putten precious perlis to hoggis." Langland's Piers Plowman (1362), XI, 9, wrote this: *"Noli mittere Margeri perles Among hogges."* The oldest English version of Mandeville's Travels, written about 1400, contained the expression: "The fyn Perl congeles and wexes gret of the dew of hevene"; but in 1447, Bokenham's "Seyntys" stated: "A margerye perle aftyr the phylosophyr Growyth on a shelle of lytyl pryhs"; and Knight de la Tour (about 1450) stated: "The sowle is the precious marguarite unto God."

The word is given "perle" in the earliest manuscripts of those old epic poems of the fourteenth century, "Pearl" and "Cleanness," which have caused so much learned theological discussion and which testify to the great love and esteem in which the gem was held. The first stanza of "Pearl" we quote from Gollancz's rendition:

> Pearl! fair enow for princes' pleasance,
> so deftly set in gold so pure,—
> from orient lands I durst avouch,
> ne'er saw I a gem its peer,—
> so round, so comely-shaped withal,
> so small, with sides so smooth,—
> where'er I judged of radiant gems,
> I placed my pearl supreme.[1]

The fourteenth-century manuscript in the British Museum gives this as follows:

> Perle plesaunte to prynces paye,
> To clanly clos in gold so clere,
> Oute of oryent I hardyly saye,
> Ne proved I never her precios pere,—
> So rounde, so reken in uche a raye,
> So smal, so smothe her sydez were,—
> Queresoever I jugged gemmez gaye,
> I sette hyr sengeley in synglere.

[1] Gollancz, "Pearl, an English Poem of the Fourteenth Century," London, 1891.

And from a modern rendering of "Cleanness" we quote:

> The pearl is praised wherever gems are seen,
>   though it be not the dearest by way of merchandise.
> Why is the pearl so prized, save for its purity,
>   that wins praise for it above all white stones?
> It shineth so bright; it is so round of shape;
>   without fault or stain; if it be truly a pearl.

In the thirteenth and fourteenth centuries throughout Europe pearls were very fashionable as personal ornaments, and were worn in enormous quantities; the dresses of men as well as of women were decorated and embroidered with them, and they were noted in nearly every account of a festive occasion, whether it were a marriage, a brilliant tourney, the consecration of a bishop, or the celebration of a victory in battle.

The faceting of crystal gems was not known at that time, and those dependent on artifice for their beauty were not much sought after. Although the diamond had been known from the eighth century, it was not generally treasured as an ornament, and not until long after the invention of cutting in regular facets—about 1450—did it attain its great popularity.

In the Dark Ages, it was customary for princes and great nobles to carry their valuables about with them even on the battle-fields; first, in order to have them always in possession, and second, on account of the mysterious power they attributed to precious stones. Since jewels constituted a large portion of their portable wealth, nobles and knights went into battle superbly arrayed. In this manner the treasures were easily lost and destroyed; consequently, relatively few of the personal ornaments of that period are preserved to the present time.

Among the greatest lovers of pearls in the fourteenth and fifteenth centuries were the members of the ducal house of Burgundy, and especially Philip the Bold (1342–1404), Philip the Good (1396–1467), and Charles the Bold (1433–77), and some of the gems which they owned are even now treasured in Austria, Spain, and Italy. When Duke Charles the Bold, in the year 1473, attended the Diet of Treves, accompanied by his five thousand splendidly equipped horsemen, he was attired in cloth of gold garnished with pearls, which were valued at 200,000 golden florins.[1] We are told that "almost a sea of pearls" was on view at the marriage of George the Rich with Hedwig, the daughter of Casimir III of Poland, at Landshut, in 1475. Among the many ornaments was a pearl chaplet valued at 50,000 florins which

[1] Sachs, "Kaiserchronik," Vol. IV, p. 261.

Duke George wore on his hat, and also a clasp worth 6000 florins.[1] Members of the related houses of Anjou and Valois also held great collections. Nor in this account should we omit some of the English sovereigns, including especially Richard II (1366–1400), one of the greatest dandies of his day.

During the fifteenth century, enormous quantities of pearls were worn by persons of rank and fashion. A remarkable 1483 portrait of Margaret, wife of James III of Scotland, which is now preserved at Hampton Court, shows her wearing such wonderful pearl ornaments that she might well be called Margaret from her decorations. As this queen was praised for her beauty, we fear the artist has scarcely done justice to her appearance; or possibly since that period tastes have changed as to what on a throne passes for beauty. Her head-dress is undoubtedly the most remarkable pearl decoration which we have seen of that century.

The uxorious and sumptuous Henry VIII of England (1491–1547) spent much of the great wealth accumulated by his penurious father, Henry VII, in enriching the appearance of his semi-barbaric court. In this reign, the spoliation of the Catholic cathedrals and churches contributed many pearls to the royal treasury; and onward from that time, they were prominently displayed among the ornaments of the women of rank in England. Most of the portraits of Henry's wives show great quantities of these gems; many of them with settings doubtless designed by artistic Hans Holbein the Younger (1497–1543); and during the succeeding reigns the women near the throne were commonly depicted with elaborate pearl decorations.

The cold, unflattering portraits by Holbein of the court celebrities of that period, not only of the gracious women and of the dandified men, but of the clergy as well, show the prominence of pearls. Note his portrait of Jane Seymour, of Anne of Cleves, of Christina of Denmark, and the pearl-incrusted miter of Archbishop Warham of Canterbury.

An interesting story is told of Sir Thomas More, the learned chancellor of Henry VIII, showing his view of the great display of jewels which distinguished the period in which he lived:

His sonne John's wife often had requested her father-in-law, Sir Thomas, to buy her a billiment sett with pearles. He had often put her off with many pretty slights; but at last, for her importunity, he provided her one. Instead of pearles, he caused white peaze to be sett, so that at his next coming home, his daughter-in-law demanded her jewel. "Ay, marry, daughter, I have not forgotten thee!" So out of his studie he sent for a box, and solemnlie deliv-

---

[1] Staudenraus, "Chronik der Stadt Landshut," 1832, Vol. I, p. 172.

ered it to her. When she, with great joy, lookt for her billiment, she found, far from her expectation, a billiment of peaze; and so she almost wept for verie griefe.[1]

Meanwhile, in the yet unknown America, pearls were highly prized, and their magic charm had taken an irresistible hold on aborigines and on the more highly civilized inhabitants of Mexico and Peru. In Mexico the palaces of Montezuma were studded with pearls and emeralds, and the Aztec kings possessed pearls of inestimable value. That they had been collected elsewhere for a long time is evidenced by the large quantities in the recently opened mounds of the Ohio Valley, which rank among the ancient works of man in America. As in the Old World, so in the New, they had been used as decoration for the gods and for the temples, as well as for men and women.

The principal immediate effect of Columbus's discovery and of the commercial intercourse with the New World, was the great wealth of pearls which enriched the Spanish traders. The natives were found in possession of rich fisheries on the coast of Venezuela, and somewhat later on the Pacific coast of Panama and Mexico, whence Eldorado adventurers returned to Spain with such large collections that—using an old chronicler's expression—"they were to every man like chaff." For many years America was best known in Seville, Cadiz, and some other ports of Europe, as the land whence the pearls came. Until the development of the mines in Mexico and Peru, the value of the pearls exceeded that of all other exports combined. Humboldt states that till 1530 these averaged in value more than·800,000 piastres yearly.[2] And throughout the sixteenth century the American fisheries—prosecuted by the Spaniards with the help of native labor—furnished Europe with large quantities, the records for one year showing imports of "697 pounds' weight" into Seville alone.

For two centuries following the discovery of America, extravagance in personal decoration was almost unlimited at the European courts, and the pearls exceeded in quantity that of all other gems. Enormous numbers were worn by persons of rank and fortune. This is apparent, not only from the antiquarian records and the historical accounts, but also in the paintings and engravings of that time; portraits of the Hapsburgs, the Valois, the Medicis, the Borgias, the Tudors, and the Stuarts show great quantities of pearls, and relatively few other gems.

Probably the largest treasures were in possession of the Hapsburg

[1] Jones, "History and Mystery of Precious Stones," London, 1880, p. 135.
[2] Humboldt, "Personal Narrative of Travels to the New Continent," London, 1822, Vol. II, p. 273.

family, which furnished so many sovereigns to the Holy Roman Empire, to Austria, and to Spain, and which, by descent through Maria Theresa, continued to rule the Holy Roman Empire until its abolition in 1806, and has since ruled Austria and Hungary.

A number of superb pieces of jewelry owned centuries ago by members of this illustrious family are yet in existence; notably the buckle of Charles V, and especially the imperial crown of Austria, made in 1602 by order of Rudolph II.[1]

Two great women of that period are noted for their passion for pearls, Catharine de' Medici (1519-89), and Elizabeth of England (1533-1603). It requires but a glance at almost any of their portraits, wherein they are represented wearing elaborate pearl ornaments, to see to what an extent they carried this fondness. And many other women were not far behind them, among whom were Mary Stuart, Marie de' Medici, and Henrietta Maria. And not only by the women, but by the men also, pearls were worn to what now seems an extravagant extent. Nearly all the portraits of Francis I (1494–1547), Henry II (1519–59), Charles IX (1550–74), and Henry III (1551–89) of France; of James I (1566–1625), and of Charles I (1600–49) of England, and likewise of other celebrities, show a great pear-shaped pearl in one ear. Many portraits also show pearls on the hats, cloaks, gloves, etc.

When the Duke of Buckingham went to Paris in 1625, to bring over Henrietta Maria to be queen to Charles I, he had, according to an account in the "Antiquarian Repertory," in addition to twenty-six other suits, "a rich suit of purple satin, embroidered all over with rich orient pearls, the cloak made after the Spanish mode, with all things suitable, the value whereof will be twenty thousand pounds, and this, it is thought, shall be for the wedding day at Paris."

In the rich and prosperous cities of southern Europe, pearls were no less popular. From its share of the spoils of the Byzantine Empire, after its partition in 1204, pearls and other riches were plentiful in Venice, and they were increased by the rapidly developing trade with the Orient. In the rival maritime cities, Genoa and Pisa, the gem was equally popular; and likewise in Florence "the Beautiful." When Hercule d'Este sought Lucrezia Borgia (1480–1519) in marriage for his son, her father, Pope Alexander VI, plunging both hands in a box filled with pearls, said: "All these are for her! I desire that in all Italy she shall be the princess with the most beautiful pearls and with the greatest number."[2]

Separated by three centuries of time and by the intervening simplicities of puritanism and democracy, it is difficult for us to appreciate

[1] See p. 473.   [2] Yriarte, "Autour des Borgia," Paris, 1891, pp. 136, 137.

MARIA THERESA (1717-1780), QUEEN OF HUNGARY

By Martin de Mytens, 1742

LADY ABINGER   MRS. ADAIR

LADY WIMBORNE   HON. MRS. RENARD GRÉVILLE

MARCHIONESS OF LANSDOWNE

LADY LONDONDERRY

BARONESS DE FOREST

the passion for pearls in Europe at that period, which may well be called the Pearl Age.

The sumptuary laws which prevailed at different times in France, England, Germany, and other countries, did not overlook this extravagance; and an entire volume might be devoted to the efforts to curb the excessive use. In France they were probably most stringent during the reign of Philip IV (1285–1314), of Louis XI (1461–83), of Charles IX (1560–74), of Henry III (1574–89), and of Louis XIII (1610–43). In Germany almost every city had its special restrictions. A sumptuary law of Ulm, in 1345, provided that no married woman or maiden, either among the patricians or the artisans, should wear pearls on her dresses; and another, in 1411, restricted them to "one pearl chaplet," and this should not exceed twelve *loth* (half ounce) in weight. A Frankish sumptuary law of 1479 provided that ordinary nobles serving a knight at a tourney should not wear any pearl ornaments, embroidered or otherwise, excepting one string around the cap or hat. The regulations decreed by the Diet of Worms, in 1495, set forth that the citizens who were not of noble birth, and nobles who were not knights, must withhold from the use of gold and pearls. A similar provision was enacted by the Diet of Freiburg in 1498, and likewise by the Diet of Augsburg in 1530, which permitted the wives of nobles four silk dresses, but without pearls. In the sumptuary law of Duke John George of Saxony, April 23, 1612, we read: "the nobility are not allowed to wear any dresses of gold or silver, or garnished with pearls; neither shall the professors and doctors of the universities, nor their wives, wear any gold, silver or pearls for fringes, or any chains of pearls, or caps, neck-ornaments, shoes, slippers, shawls, pins, etc., with gold or silver or with pearls." Beadles, burgomasters, and those connected with the law-courts were forbidden to wear chains of pearls and ornaments of precious stones on their dresses, caps, etc., or slippers or chaplets with pearls.

Probably in no place were these laws more stringent than in the art-loving republic of Venice from the fourteenth to the sixteenth century. This seems remarkable in view of the fact that this city was largely dependent for its wealth and prominence on commerce with the East, of which pearls constituted a prominent item.

The earliest Venetian restriction that we have found regarding pearls was made in 1299; when, in a decree determining the maximum number of guests at a marriage ceremony and the extent of the bridal trousseau, the grand council of the republic provided that no one but the bride should wear pearl decorations, and she should be permitted only one girdle of them on her wedding dress. This enactment was modified in 1306, but numerous other restrictions were substituted, notably in 1334, 1340, 1360, 1497, and 1562. These differed in many particulars: some forbade ornaments or trimmings of pearls, gold, or silver on the dresses of any women except a member of the Doge's family; and other enactments required that, after a definite period of married life, no woman should be permitted to wear pearls of any kind. But an examination of the documents and of the paintings of that period shows that these decrees had little effect, and the luxury of the "Queen of the Adriatic" in the use of pearls at the most brilliant epoch in her history is aptly reproduced in the portraits by Giovanni Bellini, Lorenzo Lotto, the great Titian, Tintoretto, Paul Veronese, and other artists of the highest rank. In the engraving by Hendrik Goltzius of a marriage at Venice in 1584, not one of the many women present seems to be without her necklace and earrings of pearls, and some of them have several necklaces.[1] And the same appears true of the principal female figures in Jost Amman's noted engraving, "The Espousal of the Sea," executed in 1565.[2]

As preservation of the republic became more difficult with declining resources and with the continued growth of dazzling splendor, a resolution in the Senate, dated July 8, 1599, set forth that "the use and price of pearls has become so excessive and increases to such an extent from day to day, that if some remedy is not provided, it will cause injury, disorders, and notable inconvenience to public and private well-being, as each one of this council in his wisdom can very easily appreciate." And then it was enacted: "That, without repealing the other regulations which absolutely prohibit the wearing of pearls, it shall be expressly enjoined that any woman, whether of noble birth or a simple citizen, or of any other condition, who shall reside in this our city for one year (except her Serenity the Dogaressa and her daughters and her daughters-in-law who live in the palace), after the expiration of fifteen years from the day of her first marriage, shall lay aside the string of pearls around her neck and shall not wear or use, either upon her neck or upon any other part of her person, this string or any other kind of pearls or anything which imitates pearls, neither in this city nor in any other city or place within our dominion, under the irremissible penalty of two hundred ducats."

And yet ten years later, on May 5, 1609, another law enacted in the Senate stated:

Although in the year 1599 this council decided with great wisdom that married women should be permitted to wear pearls for only fifteen years after their first marriage, nevertheless it is very evident that the desired end has not been attained, and the extravagance has continued up to the present time

---

[1] See Yriarte, "Venice," Paris, 1878, p. 236.   [2] *Ibid.*, pp. 252, 253.

and still continues with the gravest injury to private persons. Therefore, as it is necessary to remedy, by a new provision, not only this considerable incommodity, but also to prevent in the future the introduction into the city of a greater quantity of pearls than are found here at present, it is enacted, that married women as well as those who shall marry in the future (except the Serene Dogaressa and her daughters and her daughters-in-law living in the palace) of whatever grade and condition they may be, who have resided in this city for one year, cannot wear pearls of any kind except for ten years immediately following the day of their first marriage; and after that period they must lay aside these pearls which they are forbidden to wear on any part of their persons, at home or abroad, and as well in this as in the other cities, lands, and other places of our dominion, under the penalty of two hundred ducats. And if the husband of the offending wife is a noble, he shall be proclaimed in the greater council and declared a debtor to the office of the governors of the revenue in the sum of twenty-five ducats for each fine; and if he is a citizen or of any other condition, besides the penalty of two hundred ducats and the fine of twenty-five ducats above mentioned, he shall be banished for three years from Venice and the Duchy, and the same for each offence. And pearls or anything which imitates pearls, shall be forbidden to all other women, men and boys or girls of every age and condition at all times and in all places, under the same penalty of two hundred ducats. In the future no one shall in any manner bring pearls to this city as merchandise, under the penalty of their seizure and forfeiture. And the merchant shall be imprisoned for five consecutive years; and if he flees, he shall be banished from the city and district of Venice and from all other cities, lands, and places of our dominion for eight consecutive years. . . . And all who at present have pearls to sell are required to deposit a list of them with the sumptuary office, so as to avoid all fraud which could be practiced in this matter.

A copy of the title-page of this enactment is presented above.

The decrees and edicts were not confined to Venice, or to Italy, France, or Germany; they made their appearance quite generally throughout western and northern Europe and the interdictions of the civil authorities were strengthened by the voice of the bishops and other clergy, especially in the imperial cities of southern Germany. Yet the united authority of church and state was ineffectual in stemming the tide of fashion and personal fancy, and whether or not pearls should be worn became one of the much discussed questions of that period.

To the question, "Whether the statute and regulation of Bishop Tudertinus, who had excommunicated all women who wore pearls, was binding," Joannes Guidius replied that many denied that this was so, and made the subtle defense that "the women had not accepted it and all had worn pearls, and it was considered that such a law was binding only when it was accepted by those for whom it was intended."[1]

And as to the validity of the statutes requiring that women should not wear more than a definite number of pearls, he decided that "such a statute is valid and in itself good. And if the question is put whether every woman who infringes incurs the penalty, an answer may be gathered from the sayings of the doctors, who distinguish between married and unmarried women. They consider that an unmarried woman is obliged to obey the statute and regulation or to incur the penalty. But as to a married woman, if her husband approves, she should obey the statute; if, however, the husband objects, then the wife ought to wish to obey the statute, but in effect she should rather obey her husband, for she is most immediately and strongly bound to do this."[2] Aided by such ingenious opinions as these, the women continued to follow their own inclinations notwithstanding the opposition of church and state.

Other fine distinctions were drawn by the lawyers of that day regarding ownership of gems under certain conditions. For instance, it was decided that pearls given by a father to his unmarried daughter remained her property after marriage because "they are given for a reason, namely to induce a marriage"; yet "pearls handed to a wife by her husband are not considered as her property, but must be given to his heirs, since it is supposed that they were given only for her adornment. The same holds good as respects pearls handed to a daughter-in-law by her father-in-law."[3]

However, the greed of fashion, which law-makers and bishops could not arrest, was gradually satiated; and, influenced probably by the horrors of the Thirty Years' War, more simple taste prevailed in the latter part of the seventeenth century.

In the meantime, improvements in cutting and polishing had greatly increased the beauty and popularity of diamonds and other crystal

[1] Guidius, "De Mineralibus," Frankfort, 1627, p. 74.   [2] *Ibid.*, p. 73.   [3] *Ibid.*, pp. 75-77.

gems, and this adversely affected the demand for pearls. Furthermore, cleverly fashioned imitations manufactured at a low cost also served to decrease the relative rank and fashion of the sea-born gems. In the eighteenth century, pearls were relatively scarce; the resources of the American seas were largely exhausted, likewise the Ceylon and Red Sea fisheries were not to be depended on, and practically the entire supply came from the Persian Gulf, with a few from European rivers and the waters of China. As a result, although they continued to be prized by connoisseurs, pearls were not so extensively sought after by the rank and file of jewel purchasers.

It should be noted, however, that from the most ancient times, the princes of India and of Persia have had their pick and choice of the output from Ceylon and the Persian Gulf; and the largest single collections of the Western world have never equaled the possessions of some of those rulers. Some Indian princes have loaded themselves with thousands of pearls, and individual ornaments have been valued not only by oriental, but by European experts, at several millions of dollars.

The great diamond resources of Brazil were discovered in 1727, and after a few years these came on the market at the rate of 140,000 carats annually. At that time ladies of rank did not esteem diamonds so highly as pearls. This distinction was accentuated by Lord Hervey in his account of the coronation, in 1727, of George II and his consort Caroline, who wore not only the great pearl necklace inherited from Queen Anne, but "had on her head and shoulders all the pearls she could borrow of the ladies of quality at one end of the town, and on her petticoat all the diamonds she could hire of the Jews and jewelers at the other; so that the appearance and the truth of her finery was a mixture of magnificence and meanness not unlike the *éclat* of royalty in many other particulars, when it comes to be nicely considered and its source traced to what money hires or flattery lends." [1] In a portrait of Charlotte (1744–1818), wife of George III, the pearls and diamonds appear equally popular.

On the entry of the British into possession of Ceylon in 1796, the fisheries of that country were resumed with great success after thirty years of idleness, resulting in very large outputs for several seasons. But owing to exhaustion of the areas, they were soon reduced, and the yield became small and uncertain.

About 1845, pearls came on the market from the Tuamotu Archipelago and other South Sea islands, and the industry was revived on the Mexican coast. The pearls from these localities are noted for their range of coloration, and particularly for the very dark shades,

[1] Croker, "Lord Hervey's Memoirs," London, 1848, Vol. I, pp. 88, 89.

black or greenish black being especially prominent. But the fashion, and thus, necessarily, the demand, had always been for white and yellow pearls; consequently, these black ones were of little value in the markets until about ten years later, when they became fashionable in Europe largely through their popularity with Empress Eugénie of France, then at the height of her power. To this queen, pearls owe much of their high rank in fashion in the nineteenth century; and on her head they were royal gems royally worn, as appears from Winterhalter's portrait of her, showing her magnificent necklace.

The discovery of the resources on the Australian coast about 1865, and the development of the fishery there for mother-of-pearl, resulted in many large white pearls coming from that coast. The search was confined to the relatively shoal waters, until the introduction of diving-suits about 1880. The use of these facilitated a considerable extension of the fisheries not only on the Australian coast, but also in Mexico, the Malay Archipelago, several of the South Sea islands, and some minor localities.

In America, few jewels were worn previous to the Civil War, owing to the absence of great wealth and to the simplicity of taste in personal decorations. The rapid increase in wealth and luxury, on the termination of that war, resulted in a great demand for gems, and the most brilliant and showy ones were selected, especially diamonds. This demand was the more readily supplied by the discovery of the South African mines, with their great yield from 1870 to the present time. So popular did that gem become that many a young man invested his first earnings in a "brilliant," and an enormous diamond in the shirt-front became the caricatured emblem of a prosperous hotel clerk.

But in the last quarter of the nineteenth century, in Europe, as well as in America and elsewhere where gems are worn, luxury found in pearls a refinement, associated with richness and beauty, exceeding that of diamonds and other crystal gems, and in the last few years they have taken the highest rank among jewels. This change in fashion and the increase in wealth among the people developed vastly greater demands and consequently very much higher prices. These have resulted in greatly extending the field of search, and during the last two or three decades many new territories have been brought into production.

By far the most important of these new regions is the Mississippi Valley in America, the pearl resources of which were made known about a score of years ago. As the exploitation developed, the gems from these streams added very largely to the supply, especially of the baroque or irregular pearls, which have increased greatly in fashion in the last ten years.

Notwithstanding the popular idea that pearls are scarce owing to depletion of the fisheries, they are doubtless produced in greater quantities at present than ever before in the history of the world. True, they were more plentiful in Rome after the Persian conquest, and in Spain immediately following the exploitation of tropical America; but it is highly probable that in no equal period have the entire fisheries of the world yielded greater quantities than in the five years from 1903 to 1907 inclusive. Certain individual fisheries are now less productive than at the height of their prosperity; those in the Red Sea do not compare favorably with their condition in ancient times, the European resources are nearly exhausted, the supplies from the Venezuelan coast do not equal those obtained early in the sixteenth century, the yield from Mexico is not so extensive as twenty-five years ago, and the same is true of some other regions. On the other hand, the great fisheries of Persia and Ceylon are yet very prosperous, the Ceylon fishery of 1905 surpassing all records, and the number of minor pearling regions has largely increased.

The present value of pearls — which has advanced enormously since 1893 — is due to the extended markets and the increased wealth and fashion in Western countries, rather than to diminished fisheries. The oriental demand still consumes the bulk of the Persian and Indian output, and the vast increase in wealth among the middle classes in America, Europe, and elsewhere, has increased the demand tenfold over that of a century ago. While women no longer appear ornamented from head to foot as in the sixteenth century, pearls are in the highest fashion, and the woman of rank and wealth usually prizes first among her jewels her necklace of pearls.

III

ORIGIN OF PEARLS

III

ORIGIN OF PEARLS

Heaven-born and cradled in the deep blue sea, it is the purest
of gems and the most precious.

S. M. ZWEMER.

THE origin of pearls has been a fruitful subject of speculation and discussion among naturalists of all ages, and has provoked many curious explanations. Most of the early views—universally accepted during those centuries when tradition had more influence than observation and experiment—have no standing among naturalists at the present time. And although much information has been gained as to the conditions accompanying their growth, and many theories are entertained, each with some basis in observed fact, science does not yet speak with conclusive and unquestioned authority as to the precise manner of their origin and development.

Owing to the chaste and subdued beauty of pearls, it is not strange that poets of many countries have founded their origin in tears—tears of angels, of water-nymphs, of the lovely and devoted. Sir Walter Scott in "The Bridal of Triermain" refers to—

> The pearls that long have slept,
> These were tears by Naiads wept.

In one of his most lovely and consoling thoughts, Shakspere says:

> The liquid drops of tears that you have shed,
> Shall come again, transform'd to orient pearl,
> Advantaging their loan with interest
> Of ten times double gain of happiness.

And we quote from Rückert's "Edelstein und Perlen":

> I was the Angel, who of old bowed down
> From Heaven to earth and shed that tear, O Pearl,
> From which thou wert first-fashioned in thy shell.

To thee I gave that longing in thy shell,
Which guided thee and caused thee to escape,
O Pearl, from the bewitching sirens' song.

In luster they so closely resemble the limpid, sparkling dewdrop as it first receives the sun's rays, that the ancients very naturally conceived that pearls are formed from drops of dew or rain. The usual legend is, that at certain seasons of the year, the pearl-oysters rise to the surface of the water in the morning, and there open their shells and imbibe the dewdrops; these, aided by the breath of the air and the warmth of the sunlight, are, in the course of time, transformed into lustrous pearls; but if the air and the sunlight are not received in sufficient quantities, the pearls do not attain perfection and are faulty in form, color, and luster. However remarkable and even absurd this may seem at present, it appears to have been universally accepted for centuries by the most learned men of Europe as well as by primitive people who delight in the mystical and fantastic. This opinion was recorded in the Sanskrit books of the Brahmans and in other oriental literature. The classical and medieval writings of Europe contain numerous references to it; and it is found even yet in the traditions and folk-lore of some peoples.

In the first century A.D., Pliny wrote in his "Historia naturalis," according to Dr. Philemon Holland's quaint translation:

The fruit of these shell fishes are the Pearles, better or worse, great or small, according to the qualitie and quantitie of the dew which they received. For if the dew were pure and cleare which went into them, then are the Pearles white, faire, and Orient; but if grosse and troubled, the Pearles likewise are dimme, foule, and duskish; pale they are, if the weather were close, darke and threatening raine in the time of their conception. Whereby (no doubt) it is apparent and plaine, that they participate more of the aire and sky, than of the water and the sea; for according as the morning is faire, so are they cleere: but otherwise, if it were misty and cloudy, they also will be thicke and muddy in colour. If they may have their full time and season to feed, the Pearles likewise will thrive and grow bigge; but if in the time it chance to lighten, then they close their shells together, and for want of nourishment are kept hungrie and fasting, and so the pearles keepe at a stay and prosper not accordingly. But if it thunder withall, then suddenly they shut hard at once, and breed only those excrescences which be called *Physemata*, like unto bladders puft up and hooved with wind, no corporal substance at all: and these are the abortive & untimely fruits of these shell fishes.[1]

Pliny's views were probably derived from the ancient authorities of his time, particularly from Megasthenes, Chares of Mytilene, and Isi-

[1] "The Naturall Historie of C. Plinius Secundus," London, 1601, Book IX, ch. 35.

VENEZUELA SHELL

(*Margaritifera radiata*)

Showing growth of pearls

PANAMA SHELL

(*Margaritifera margaritifera mazatlanica*)

With pearls attached

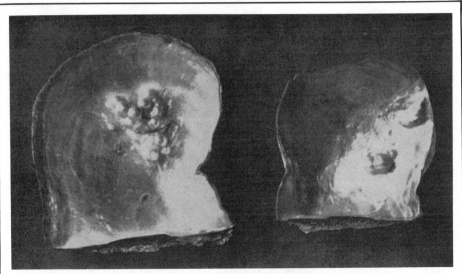

Shells from Venezuela (*Margaritifera radiata*) with attached pearls

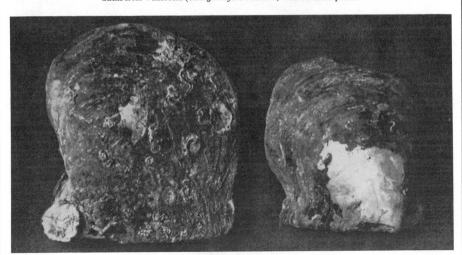

Exterior view of same

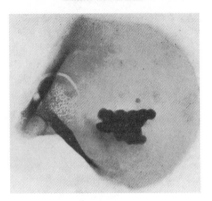

X-ray photograph of shell, printed through exterior
of shell and showing encysted pearls

dorus of Charace; and these curious fictions were incorporated by subsequent writers and influenced popular opinion for many centuries. With scarcely a single exception, every recorded theory from the first century B.C. to the fifteenth century evidences a belief in dew-formed pearls.

This theory is referred to by Thomas Moore in his well-known lines:

> And precious the tear as that rain from the sky,
> Which turns into pearls as it falls in the sea.

The Spanish-Hebrew traveler Benjamin of Tudela, in his "Masaoth" in Persia (from 1160 to 1173), wrote: "In these places pearls are found, made by the wonderful artifice of nature: for on the four and twentieth day of the month Nisan, a certain dew falleth into the waters, which being sucked in by the oysters, they immediately sink to the bottom of the sea; afterwards, about the middle of the month Tisri, men descend to the bottom of the sea, and, by the help of cords, these men bringing up the oysters in great quantities from thence, open and take out of them the pearls."[1]

From the "Bustan," one of the most popular works of Sadi, the Persian poet (1190–1291 A.D.), Davie quotes:

> From the cloud there descended a droplet of rain;
> 'T was ashamed when it saw the expanse of the main,
> Saying: "Who may I be, where the sea has its run?
> If the sea has existence, I, truly, have none!"
> Since in its own eyes the drop humble appeared,
> In its bosom, a shell with its life the drop reared;
> The sky brought the work with success to a close,
> And a famed royal pearl from the rain-drop arose.
> Because it was humble it excellence gained;
> Patiently waiting till success was attained.

Even the usually well-informed William Camden (1551–1623), in whose honor the Camden Historical Society of England was named, accepted the theory of dew-formed pearls. He stated that the river Conway in Wales "breeds a kind of shells, which being pregnated with dew, produce pearl."[2] Also, speaking of the Irt in county Cumberland, England, he said: "In this brook, the shell-fish, eagerly sucking in the dew, conceive and bring forth pearls, or (to use the poet's word) shell berries (*Baccas concheas*)."[3]

A recent letter from the American consul at Aden indicates that this

---

[1] "Travels of Rabbi Benjamin of Tudela," Gerrans's edition, London, 1783, p. 23.   [2] "Camden Britannia," 2d edition, London, 1722, Vol. II, p. 801.
[3] *Ibid.*, Vol. II, p. 1003.

view is held even yet by the Arabs of that region. In giving their explanation for the present scarcity in the Red Sea, he states: "There is a belief among them that a pearl is formed from a drop of rain caught in the mouth of the pearl-oyster, which by some chemical process after a time turns into a pearl; and as there has been very little rain in that region for several years past, there are few pearls."

So firmly established throughout Europe was the belief in dew-formed pearls, that its non-acceptance by the native Indians of America excited the commiseration of the Italian historian Peter Martyr, in his "De Orbe Novo," one of the very first books on America, published in 1517. He states: "But that they [pearls of Margarita Island on the present coast of Venezuela] become white by the clearnesse of the morning dewe, or waxe yelowe in troubled weather, or otherwise that they seeme to rejoice in fayre weather and dear ayre, or contrary-wise, to be as it were astonished and dymme in thunder and tempests, with such other, the perfect knowledge hereof is not to be looked for at the hands of these unlearned men, which handle the matter but grossly and enquire no further than occasion serveth."[1] Peter Martyr was distinguished for his learning, was an instructor at the court of Spain at the height of its power, and came in contact with the most enlightened men of Europe, consequently it may be assumed that he reflected the best opinions of his time.

It was not long before the aborigines of America were not alone in discrediting the views which had prevailed in Europe for more than fifteen hundred years. That practical old sailor Sir Richard Hawkins concluded that this must be "some old philosopher's conceit, for it can not be made probable how the dew should come into the oyster."[2] A similar view is expressed by Urbain Chauveton in his edition of Giro-lamo Benzoni's "Historia del Mondo Nuovo," published at Geneva in 1578. From his reference to pearl-oysters on the Venezuelan coast, we translate:

Around the island of Cubagua and elsewhere on the eastern coast, are sandy places where the pearl-oysters grow. They produce their eggs in very large quantities and likewise pearls at the same time. But it is necessary to have patience to let them grow and mature to perfection. They are soft at the beginning like the roe of fish; and as the mollusk gradually grows, they grow also and slowly harden. Sometimes many are found in one shell, which are hard and small, like gravel. Persons who have seen them while fishing say that they are soft as long as they are in the sea, and that the hardness comes to them only when they are out of the water. Pliny says as much, speaking of the Orientals in Book IX, of his Natural History, ch. 35.

[1] Richard Eden edition, London, 1577, 10th ch. of 3rd Decade, fol. 148a.   [2] Hawkins, "Voyage to the South Sea in 1593," London, 1847, p. 133.

But as to that author and Albert the Great and other writers upon the generation of pearls, who have said that the oysters conceive them by means of the
dew which they suck in, and that according as the dew is clear or cloudy the
pearls also are translucent or dark, etc., etc.,—all this is a little difficult to
believe; for daily observation shows that all the pearls found in the same
shell are not of the same excellence, nor of the same form, the same perfection
of color, nor the same size, as they would or must be if they were conceived
by the dew all at one time. Besides this, in many of the islands the Indians
go fishing for them in ten or twelve fathoms depth, and in some cases they
are so firmly attached to the rocks in the sea that they can be wrenched off
only by main strength. Would it not be difficult for them to inhale the
quintessence of the air there? It seems then that it is the germ and the
most noble part of the eggs of the oyster which are converted into pearls
rather than any other thing; and the diversities of size, color, and other
qualities, proceed from the fact that some are more advanced than others,
as we see eggs in the body of the hen.[1]

The old theory of dew-formed pearls was illustrated even as late as
1684 on a medal struck in honor of Elena Piscopia of the Corraro
family of Venice. This bore an oyster-shell open and receiving drops
of dew, and underneath was engraved the motto *"Rore divino"* (By
divine dew). Even yet one hears occasionally from out-of-the-way
places—as in the instance reported by the American consul at Aden—
of pearls formed from rain or dew, notwithstanding that there seems
to exist absolutely no justification for it in scientific zoölogy.

Probably the most popular theory entertained from the fifteenth to
the seventeenth century was that pearls were formed from the eggs
of the oyster. This was intimated by Chauveton in the quotation above
given, and it was also referred to by many naturalists.

In an interesting letter, dated Dec. 1, 1673, and giving as his authority the testimony of an eye-witness, "Henricus Arnoldi, an ingenious
and veracious Dane," Christopher Sandius wrote: "Pearl shells in
Norway do breed in sweet waters; their shells are like mussels, but
larger; the fish is like an oyster, it produces clusters of eggs; these,
when ripe, are cast out and become like those that cast them; but sometimes it appears that one or two of these eggs stick fast to the side of
the matrix, and are not voided with the rest. These are fed by the
oyster against her will and they do grow, according to the length of
time, into pearls of different bigness."[2] This possibly hit the mark
with greater accuracy than the observations of the "ingenious and
veracious Dane" warranted, for he seems to have had quite a different
idea as to the manner in which the pearls are "fed by the oyster against

[1] Benzoni, "Novæ Novi Orbis Historiæ," Geneva, 1578, pp. 161–163. [2] "Philosophical Transactions," 1674, No. 101, p. 11.

her will" from those generally entertained by naturalists at the present time.

However, Oliver Goldsmith settled the matter by declaring briefly: "Whether pearls be a disease or an accident in the animal is scarce worth enquiry." [1] Thus it seems that notwithstanding all that had been written and the extended attention given to the subject, theory prevailed to the almost complete exclusion of practical investigation, with little intelligent advance over Topsy's " 'spect they just growed."

Owing, doubtless, to the scarcity of pearl-bearing mollusks in their vicinities, naturalists of Europe were somewhat slow in giving attention to the origin of pearls. This is further accounted for by the fact that the gems occur more frequently in old and diseased shells than in the choice specimens which have naturally attracted the notice of conchologists.

One of the first of the original observations made on this subject was that by Rondelet, who, in 1554, advanced the idea that pearls are diseased concretions occurring in the mollusca, similar to the morbid calculi in the mammalia.[2]

The first writer to intimate the similarity in structural material or substance between pearls and the interior of the shell in which they are formed, appears to have been Anselmus de Boot (*circa* 1600), who wrote that the pearls "are generated in the body of the creature of the same humour of which the shell is formed; . . . for whenever the little creature is ill and hath not strength enough to belch up or expel this humour which sticketh in the body, it becometh the rudiments of the pearl; to which new humour, being added and assimilated into the same nature, begets a new skin, the continued addition of which generates a pearl." [3] The Portuguese traveler, Pedro Teixeira (1608), stated: "I hold it for certain that pearls are born of and formed of the very matter of the shell and of nothing else. This is supported by the great resemblance of the pearl and the oyster-shell in substance and color. Further, whatever oyster contains pearls has the flesh unsound and almost rotten in the parts where the pearls are produced, and those oysters that have no pearls are sound and clean fleshed." [4]

Somewhat more than one hundred years later, this theory was confirmed by investigations made by the famous physicist Réaumur (1683-1757). Microscopic examination of cross sections of pearls show that they are built up of concentric laminæ similar, except in curvature, to those forming the nacreous portion of the shell. In a

[1] Goldsmith, "History of the Earth and Animated Nature," 1774, Vol. VI, p. 54.
[2] Rondelet, "Universæ Aquitilium Historiæ Pars Altera," Lugduni, 1554.

[3] "Gemmarum et Lapidum Historia," Hanoviæ, 1609.
[4] "The Travels of Pedro Teixeira," Hakluyt Society, p. 180.

paper published by the French Academy of Science in 1717,[1] Réaumur noted this condition, and suggested that pearls are misplaced pieces of organized shell, and are formed from a secretion which overflows from the shell-forming organ or from a ruptured vessel connected therewith, and that the rupture or overflow is ordinarily produced by the intrusion of some foreign or irritating substance.

Sir Edwin Arnold calls attention to this theory in his beautiful lines:

> Know you, perchance, how that poor formless wretch—
> The Oyster—gems his shallow moonlit chalice?
> Where the shell irks him, or the sea-sand frets,
> He sheds this lovely lustre on his grief.

In pursuance of this idea, we find, in 1761, the Swedish naturalist Linnæus, "the father of natural history," experimenting in the artificial production of pearls by the introduction of foreign bodies in the shell, and meeting with some degree of success. His discovery was rated so highly that it has been announced by some writers as the reason why the great naturalist received the patent of nobility, which is generally supposed to have been the reward for his services to science.

It seems that Linnæus's discovery but verified the old saying that there is nothing new under the sun, for later it was announced[2] that in China—where so many inventions have originated—this idea had been put to practical account for centuries preceding, and the crafty Chinaman had succeeded in producing not only small pearly objects, but even images of Buddha, with which to awe the disciples of that deified teacher.

The method consisted in slightly opening or boring through the shell of the living mollusk and introducing against the soft body a small piece of nacre, molded metal, or other foreign matter. The irritation causes the formation of pearly layers about the foreign body, resulting, in the course of months or of years, in a pearl-like growth. While these have some value as objects of curiosity or of slight beauty, they are not choice pearls, nor for that matter were those produced by Linnæus.

It will be observed that the theory of Réaumur, and also that of Linnæus, required the intrusion of some hard substance, such as a grain of sand, a particle of shell, etc., to constitute a nucleus of the pearl; and this is the accepted explanation at the present time as to the origin of many of the baroque or irregular pearls, and likewise the pearly "blisters" and excrescences attached to the shell. But not so as to the

[1] "Mémoires de l'Académie des Sciences," 1717, pp. 177–194.
[2] Grill, in "Abhandlungen der Königlichen Schwed. Akademie der Wissenschaften," Vol. XXXIV, p. 88, 1772.

choice or gem pearls, those beautiful symmetrical objects of great luster which are usually referred to in speaking of pearls.

Examinations of many of these have failed, except in rare instances, · to reveal a foreign nucleus of sand or similar inorganic substance. In searching many fresh-water mussels, Sir Everard Home frequently met with small pearls in the ovarium, and he further noticed that these, as well as oriental pearls, when split into halves, often showed a brilliant cell in the center, about equal in size to the ova of the same mollusk. From these observations, in 1826 he deduced his "abortive ova" theory, and announced:

A pearl is formed upon the external surface of an ovum, which, having been blighted, does not pass with the others into the oviduct, but remains attached to its pedicle in the ovarium, and in the following season receives a coat of nacre at the same time that the internal surface of the shell receives its annual supply. This conclusion is verified by some pearls being spherical, others having a pyramidal form, from the pedicle having received a coat of nacre as well as the ovum.[1]

Naturalists generally accepted these conclusions, that pearls originate in pathological secretions formed, either as the result of the intrusion of hard substances, or by the encysting or covering of ova or other objects of internal origin; and there was no important cleavage of opinion until the development of the parasitic theory, as a result of the researches of the Italian naturalist Filippi, and those following his line of investigations. This theory is not severely in conflict with those of Réaumur, Linnæus, Home, etc., but relates principally to the identity of the irritating or stimulating substance which forms the nucleus of the pearl.

In examining a species of fresh-water mussel, the *Anodonta cygnea,* occurring in ponds near Turin, and especially the many small pearly formations therein, Filippi observed that these were associated with the presence of a trematode or parasitic worm, which he named *Distomum duplicatum,* and which appears to be closely allied to the parasite which causes the fatal "rot" or distemper in sheep. Under the microscope, the smallest and presumably the youngest of these pearls showed organic nuclei which appeared undoubtedly to be the remnants of the trematode. In Anodonta from other regions, which were not infested with the distoma, pearls were very rarely found by Filippi. In a paper,[2] published in 1852, containing a summary of his observations, he concluded that a leading, if not the principal, cause of pearl-forma-

[1] "Philosophical Transactions," 1826, Pt. III, pp. 338–341.

[2] "Sull'origine delle Perle. Il Cimento, revista di Scienze," Torino, 1852, Vol. I, pp. 429–439.

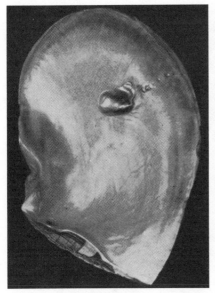

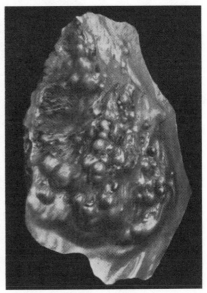

Mexican pearl-oyster (*Margaritifera margaritifera mazatlanica*) with adherent pearl

Group of encysted pearls in shell of Australian pearl-oyster (*Margaritifera maxima*)

American Museum of Natural History

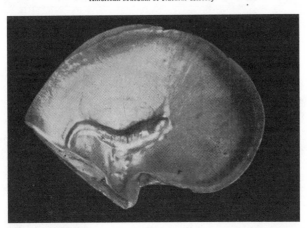

Mexican pearl-oyster (*Margaritifera margaritifera mazatlanica*) with encysted fish

American Museum of Natural History

Group of encysted pearls
(Oriental)

Reverse of same group, showing outline of the individual pearls

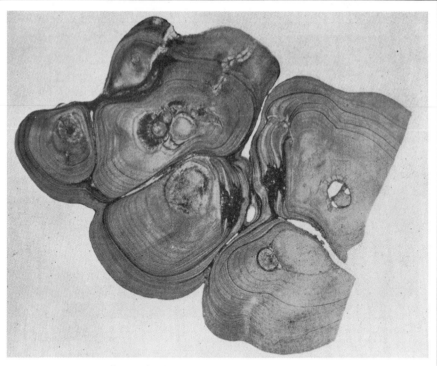

Cross section of an irregular pearl, magnified 80 diameters

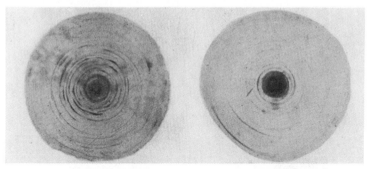

Cross sections of pearls, magnified 30 diameters

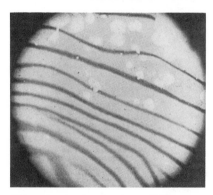

Thin section of mother-of-pearl, magnified, showing
sponge borings which traversed the pearl shell

Structure of conch pearl produced by fracturing,
magnified 80 diameters

tion in those mussels was the parasite above noted; and in later papers [1] he included such other forms as *Atax ypsilophorus* within the list of parasitic agencies which might excite the pearl-forming secretions, comparing their action to that of the formation of plant-galls.

The discovery of the parasitic origin of pearls was extended to the pearl-oysters and to other parasites by Küchenmeister [2] in 1856, by Möbius [3] in 1857, and by several other investigators. Prominent among these were E. F. Kelaart and his assistant Humbert, who, in 1859 [4] disclosed the important relation which the presence of vermean parasites bears to the origin of pearls in the Ceylon oysters. These naturalists found "in addition to the Filaria and Cercaria, three other parasitical worms infesting the viscera and other parts of the pearl-oyster. We both agree that these worms play an important part in the formation of pearls." Dr. Kelaart likewise found eggs from the ovarium of the oyster coated with nacre and forming pearls, and also suggested that the silicious internal skeletons of microscopic diatoms might possibly permeate the mantle and become the nuclei of pearls. Unfortunately, Dr. Kelaart's investigations were terminated by his death a few months thereafter.

In 1871, Garner ascribed the occurrence of pearls in the common English mussel (*Mytilus edulis*) to the presence of distomid larvæ.[5] Giard,[6] and other French zoölogists, made similar discoveries in the case of Donax and some other bivalves. In 1901, Raphael Dubois confirmed the observations of Garner, associating the production of pearls in the edible mussels on the French coasts with the presence of larvæ of a parasite, to which he gave the name of *Distomum margaritarum,* and boldly announced: "La plus belle perle n'est donc, en définitive, que le brillant sarcophage d'un ver." [7]

Prof. H. L. Jameson, in 1902, disclosed the relation which exists between pearls in English mussels (Mytilus) and the larvæ of *Distomum somateriæ.*[8] The life history of this trematode, as revealed by Dr. Jameson, is especially interesting from a biological standpoint, since it is entertained by three hosts at different times: the first host is a member of the duck family; the second is the Tapes clam (*Tapes decussatus*), or perhaps the common cockle (*Cardium edule*), which incloses

[1] "Memorie della Reale Academia delle Scienze di Torino," 1855, Vol. XV, pp. 331–358; 1857, Vol. XVI, pp. 419–442, and 1859, Vol. XVIII, pp. 201–232.

[2] Müller's "Archiv für Anatomie," 1856, pp. 269–281.

[3] "Die echten Perlen," Hamburg, 1858. Dr. Möbius died in Berlin, on April 26, 1908. He was born at Eilenburg, in Saxony, in 1825.

[4] "Report on the Natural History of the Pearl Oyster of Ceylon," Trincomali, 1859.

[5] "Journal of the Linnean Society," Vol. XI, pp. 426–428.

[6] "Société de Biologie, Séance du 29 décembre, 1903."

[7] "Comptes Rendus de l'Académie des Sciences," Vol. 133, pp. 603–605, Oct. 14, 1901.

[8] "Proceedings of the Zoölogical Society of London," Vol. I, pp. 140–166.

the first larval stage, and the third is the edible mussel, in which the second larval stage of the parasite stimulates the formation of pearls. At the Brighton Aquarium and the Fish Hatchery at Kiel, Dr. Jameson claims to have succeeded in artificially inoculating perfectly healthy mussels with these parasites by associating them with infested mollusks, and thereby producing small pearls.

From Dr. Jameson's interesting paper we abridge the following account of the manner in which the pearls are developed. The trematode enters *Mytilus edulis* as a tailless cercaria, and at first may often be found between the mantle and the shell. The larvæ, after a while, enter the connective tissue of the mantle, where they come to rest, assuming a spherical form, visible to the naked eye as little yellowish spots about one half millimeter in diameter. At first the worm occupies only a space lined by connective-tissue fibrils, but soon the tissues of the host give rise to an epithelial layer, which lines the space and ultimately becomes the pearl-sac. If the trematode larva completes its maximum possible term of life, it dies, and the tissues of the body break down to form a structureless mass which retains the form of the parasite, owing to the rigid cuticle. In this mass arise one or more centers of calcification, and the precipitation of carbonate of lime goes on until the whole larva is converted into a nodule with calcospheritic structure. The granular matter surrounding the worm, if present, also undergoes calcification. The epithelium of the sac then begins to shed a cuticle of conchiolin, and from this point the growth of the pearl probably takes place on the same lines and at the same rate as the thickening of the shell.[1]

Fully as remarkable as the observations of Dr. Jameson are the results claimed by Professor Dubois in experimenting with a species of pearl-oyster (*M. vulgaris*) from the Gulf of Gabes on the coast of Tunis, where they are almost devoid of pearls, a thousand or more shells yielding on an average only one pearl. Conveying these to the coast of France in 1903, he there associated them with a species of trematode-infested mussel (*Mytilus gallo-provincialis*), and after a short period they became so infested that every three oysters yielded an average of two pearls.[2] This claim has not been without criticism; but who ever knew scientists to agree?

In the pearl-oyster of the Gambier Islands (*M. margaritifera cumingi*), Dr. L. G. Seurat found that the origin of pearls was due to irritation caused by the embryo of a worm of the genus *Tylocephalum*, the life of which is completed in the eagle-ray, a fish which feeds on the pearl-oyster.[3]

[1] "Proceedings of the Zoölogical Society of London," 1902, pp. 148-150.
[2] "Comptes Rendus de l'Académie des Sciences," Paris, 1903, Vol. CXXXVII, pp. 611-613.
[3] Seurat, "Observation sur l'évolution de l'Huître perlière des Tuamotu et des Gambier," 1904.

In 1903, Prof. W. A. Herdman, who, at the instance of the colonial government, and with the assistance of Mr. James Hornell, examined the pearl-oyster resources of Ceylon, announced: "We have found, as Kelaart did, that in the Ceylon pearl-oyster there are several different kinds of worms commonly occurring as parasites, and we shall, I think, be able to show that Cestodes, Trematodes, and Nematodes may all be concerned in pearl-formation. Unlike the case of the European mussels, however, we find that in Ceylon the most important cause is a larval Cestode of the Tetrarhynchus form."[1]

In his investigation of the Placuna oyster in 1905, Mr. James Hornell found that the origin of pearls was due to minute larva of the same stage and species as that which causes the pearls in the Gulf of Manar oyster.[2]

The spherical larvæ of this tapeworm sometimes occur in great abundance, and there is evidence of forty having been found in a single pearl-oyster. Mr. Hornell states that the living worm does not induce pearl-formation, this occurring only when death overtakes it while in certain parts of the oyster. As a consequence, pearls are more numerous in oysters which have been long infected, where the worms are older and more liable to die. This parasitic worm has been traced from the pearl-oyster to the trigger-fishes, which eat the pearl-oysters, and thence into certain large fish-eating rays, where it becomes sexually mature and produces embryos which enter the pearl-oyster and begin a new cycle of life-phases.

It seems, therefore, that the latest conclusions of science appear entirely favorable to the parasitic theory as explaining at least one, and probably the most important, of the causes for the formation of pearls; and that some truth exists in the statement that the most beautiful pearl is only the brilliant sarcophagus of a worm. This morphological change is not peculiar to mollusks, for in most animal bodies a cyst is formed about in-wandering larvæ. Fortunately for lovers of the beautiful, in the pearl-oysters the character of the cyst-wall follows that of the interior lining of the shell, and not only simulates, but far surpasses it in luster.

While the theory that pearls are caused by the intrusion of some unusual substance has the evidence of actual demonstration in many instances, and is unquestionably true to a large extent, yet microscopic examination of some pearls suggests the theory that a foreign substance is not always essential to their formation, and that they may originate in calcareous concretions of minute size, termed "calcosphe-

[1] "Pearl Oyster Fisheries of the Gulf of Manaar," London, 1903, Vol. I, p. 11.
[2] Hornell, "Report on the *Placuna pla-centa* Pearl Fishery of Lake Tampalakamam," Colombo, 1906.

rules." As regards their origin, Professor Herdman classifies pearls into three sorts: (1) "Ampullar pearls," which are not formed within closed sacs of the shell-secreting epithelium like the others, but lie in pockets or ampullæ of the epidermis. The nuclei may be sand-grains or any other foreign particles introduced through breaking or perforation of the shell. (2) "Muscle pearls," which are analogous to gallstones, formed around calcospherules at or near the insertion of the muscles. And (3) "Cyst pearls," in which concentric layers of nacre are deposited on cysts containing parasitic worms in the connective tissue of the mantle and within the soft tissues of the body.[1]

Even a particle of earth, clay, or mud may form the nucleus of a pearl. This was illustrated a few years ago in a fine button-shaped pearl, which was accidentally broken under normal usage and was found to consist of a hard lump of white clay surrounded by a relatively thin coating of nacre. More remarkable yet are the cases in which a minute fish, a crayfish, or the frustule of a diatom has formed the nucleus.

Several instances have been described by Woodward, Gunther, Putnam, Stearns, and others, where small fish have penetrated between the mantle and the shell of the mollusk, and the latter has resented the intrusion by covering the intruder with a pearly coating. In two or three instances the secretion occurred in so short a time that the fish suffered no appreciable decomposition, and its species is readily identified by observation through the nacreous layer. Among the remarkable specimens of this nature which have come under our observation are two very curious shells received in March, 1907, from the Mexican fisheries. One of these specimens shows an encysted fish, so quickly covered and so perfectly preserved that even the scales and small bones are in evidence; indeed, one can almost detect the gloss on the scales of the fish; and in the other—with a remarkable comet-like appearance—a piece of ribbed seaweed is apparently the object covered.

From the foregoing, it appears that the pearl is not a product of health associated with undisturbed conditions, but results from a derangement in the normal state of the mollusk. Unable to resist, to rid itself of the opposing evil, it exercises the powers given to it by a beneficent Creator and converts the pain into perfection, the grief into glory. Nature has many instances of the humble and lowly raised to high degree, but none more strikingly beautiful than this. One of the lowest of earth's creatures, suffering a misfortune, furnishes a wonderful lesson upon the uses of pain and adversity by converting its affliction into a precious gem symbolical of all that is pure and

[1] "Pearl Oyster Fisheries of the Gulf of Manaar," London, 1903, Vol. I, p. 10.

beautiful. As written by a forgotten poet: "Forasmuch as the pearl is a product of life, which from an inward trouble and from a fault produces purity and perfection, it is preferred; for in nothing does God so much delight as in tenderness and lustre born of trouble and repentance." As the great Persian poet Hafiz says:

Learn from yon orient shell to love thy foe,
And store with pearls the wound that brings thee woe.

IV

STRUCTURE AND FORMS

IV

STRUCTURE AND FORMS OF PEARLS

"This maskellez perle that boght is dere,
  The joueler gef fore alle hys gold,
Is lyke the reme of hevenes clere";
  So sayde the fader of folde and flode,
"For hit is wermlez, clene and clere,
  And endelez rounde and blythe of mode,
And commune to all that ryghtwys were."
FOURTEENTH-CENTURY MSS. OF "PEARL,"
IN THE BRITISH MUSEUM.

AS Kadir Munshi says, "pearls have no pedigree"; their beauty is not to be traced to their origin, but exists wholly in the excellence of the surroundings in which they develop.

The pearl-bearing mollusks are luxurious creatures, and for the purpose of protecting their delicate bodies they cover the interior of their shells with a smooth lustrous material, dyed with rainbow hues, and possessing a beautiful but subdued opalescence. No matter how foul, how coral-covered, or overgrown with sponges or seaweeds the exterior may be, all is clean and beautiful within. This material is nacre or mother-of-pearl. It consists ordinarily of an accumulation of extremely thin semi-transparent films or laminæ of a granular organic substance called conchiolin, with the interstices filled with calcareous matter. The nacre decreases in thickness from the hinge toward the lip of the shell, and terminates a short distance from the extreme edge.

Next to the nacre is the middle layer or the shell proper. In species of *Margaritifera,* this stratum is commonly formed of layers of calcareous prisms arranged vertically to the shell surface. External to this middle or prismatic layer is the epidermis or periostracum, the rough outer coating of varying shades, usually yellow or brown. Where the waves are rough, and the bottom hard and rocky, this covering is thick and heavy, to afford greater protection; but where the waters are smooth and gentle, and the bottom free from rocks, Nature—never working in vain—furnishes only thin sides and slight defense. As is the case with the nacre, the prismatic layer and the periostracum decrease in thickness from the hinge to the edge, and the inside lip of the shell shows the gradual union of the three superimposed layers. The two outer layers are formed by the thick edge of the mantle, the remaining portion—or nearly the entire surface—of this organ secretes the nacral layer.

Not only is the interior of the shell made lustrous and beautiful, but this tendency is exerted toward all objects that come in contact with the soft body of the mollusk, either by intrusion simply within the shell, or deeply within the organs and tissues of the animal itself. All foreign bodies—such as small parasites, diatoms, minute pebbles, etc., —irritate the tender tissues of the mollusk, and stimulate the pearly formation which in course of time covers them. At first the nacreous covering is very thin; but with added layer after layer the thickness is enhanced, and the size of the object increases as long as it remains undisturbed and the mollusk is in healthful growth.

Chemically considered, aside from the nucleus, the structure of pearls is identical in composition with that of the nacre of the shell in which they are formed. Analyses have shown that those from the fresh-water mussels of England and Scotland, and from the pearl-oysters of Australia and of Ceylon, have nearly identical composition in the proportion of about 5.94 per cent. of organic matter, 2.34 of water, and 91.72 per cent. of carbonate of lime.[1] The specific gravity ranges from nearly 2 to about 2.75, increasing with the deposit of the nacreous coatings. The following summary by Von Hessling[2] shows the results of certain determinations of specific gravity:

Authority	Specific Gravity	Note
Muschenbroet	2.750	at moderate temperature
Brisson	2.684	at 14° Réaumur
Möbius	2.686	4 fine pearls, weighing 2.396 gms.
"	2.650	24 pearls, weighing 6.221 gms.
"	2.336	63 brown pearls from Mazatlan, weighing 4.849 gms.
Voit	2.722	Bavarian pearls, $3\frac{3}{16}$ carats, medium quality
"	2.616	" " $3\frac{5}{8}$ carats, finer quality
"	2.724	" " $1\frac{3}{4}$ carats, very fine
"	2.578	" " gray, with some luster
"	2.765	" " brown, ranking between good & black
"	2.238	" " poor black pearls, impure

The distinctive characteristic, the great beauty of a true pearl, is its luster or orient, which is a subdued iridescence, rather than the glittering brilliance of the diamond; and unless the shelly growth be lustrous it does not rank as a gem pearl, no matter how perfect its

[1] Harley, "Proceedings of the Royal Society," Vol. XLIII, p. 461. [2] "Die Perlenmuscheln," Leipzig, 1859, pp. 294, 295.

form or beautiful its color. This luster is due to the structural arrangement of the surface as well as to the quality of the material. The nacreous material forming true pearls, and likewise mother-of-pearl, is commonly deposited in irregular tenuous layers, very thin and very small in area compared with the surface of the pearl. These laminæ overlap one another, the surfaces are microscopically crumpled and corrugated, and the edges form serrated outlines. The greater the angle which the laminæ form with the surface, the closer will be these serrated outlines, and where the plane of the exterior lamina is parallel with the plane of the surface the lines are not present. This arrangement causes the waves of light to be reflected from different levels on the surface, just as in a soap bubble, and the minute prisms split the rays up into their colored constituents, producing the chromatic or iridescent effect.

The cause is wholly mechanical, and an impression of the surface made in very fine wax shows a similar iridescence. Also, if a piece of mother-of-pearl be immersed in acid until the surface lime or shelly matter is dissolved, the pellucid membrane shows the iridescence until it is so compressed that the corrugations are reduced. About two score years ago an Englishman invented steel buttons with similar minute corrugations producing pearly effect, but the manufacture was unprofitable, owing, principally, to their liability to tarnish.

In the shells of some mollusks—as the edible oysters (*Ostrea*) or the giant clam (*Tridacna*),—there is almost a total absence of the crumpled corrugated laminæ, and, consequently, there is little luster. In others the nacre is of better quality, resulting in superior orient, and it probably reaches its highest degree of perfection in the pearl-oyster (*Margaritifera*).

As the curvature of the surface of pearls is greater, and the minute striæ are more numerous, than in ordinary mother-of-pearl, it follows that the iridescence is likewise greater.

Superior nacre is more or less translucent, depending on its quality; and to the iridescence of the outer laminæ is added that of many interior ones, so that the luster is vastly increased. The position of the pearl within the shell may greatly affect the quality of the material and, consequently, the orient. The choicest are commonly found within the soft parts of the animal, and those of poorer quality are at the edges of the mantle, or within the fibers of the adductor muscle of bivalves.

The structure of pearls may be studied by examining thin cross sections under the microscope, or by transmitted polarized light. It appears that ordinarily a pearl is made up of many independent laminæ superimposed one upon another "like the layers of an onion," or, rather, resembling the leaves near the upper part of a well formed cabbage. When subjected to sufficient heat, the laminæ separate from each other, as do shells of edible oysters and similar mollusks under like conditions. When broken by a hammer, a pearl may exhibit this laminated formation. If not split directly through the center, the central section may retain the spherical form; and as this commonly remains attached to one of the parts, its concave impression appears in the other portion of the broken pearl. The outer laminæ of many pearls may be removed with a fair prospect of finding a good subjacent surface, and this may be continued until the size is greatly reduced. These laminæ are not always similar in color or luster.

However, not all pearls are laminated in this manner. Instead of superimposed layers, some of them exhibit a crystalline form, composed of beautiful prismatic crystals radiating from the center to the circumference. In at least one oriental pearl examined, these crystals were in well defined arcs, and were further separated into concentric rings of different degrees of thickness, depth of color, and distance apart. Another specimen—a Scotch pearl—combined in separate layers both the laminated form and the crystalline structure.

Dr. Harley points out that some crystalline pearls apparently originate in mere coalescences of mineral particles, rather than in well defined nuclei.[1] Microscopic sections of crystalline pearls convey the idea that the prisms branch and interlace with one another, and also that in some instances they are of fusiform shape. However, these appearances seem to be due simply to the cross sections having cut the prisms at different angles.

Pearls showing these types were exhibited at a meeting of the Royal Society of London, June 8, 1887. That exhibit also contained a section of a west Australian pearl of curiously complex crystalline formation; instead of one central starting-point, it had more than a dozen scattered about, from which the crystalline prisms radiated in all directions.

Since the three superimposed layers of the shell are secreted by separate parts of the mantle, viz., the nacre by the general surface, the prismatic layer by the inner edge, and the epidermis by the outer edge, it follows that if a pearl in course of formation is moved from one of these distinctive portions of the palial organ to another, the nature of its laminæ changes. Thus, if a pearl formed on the broad surface of the mantle is moved in some way to the inner edge of that organ, it may be covered with a prismatic layer; if then moved to the outer edge it may receive a lamina of epidermis, and then by changing again to the broad surface of the mantle it receives further coats of nacre.

---

[1] Harley, "Proceedings of the Royal Society of London," Vol. XLV, p. 612.

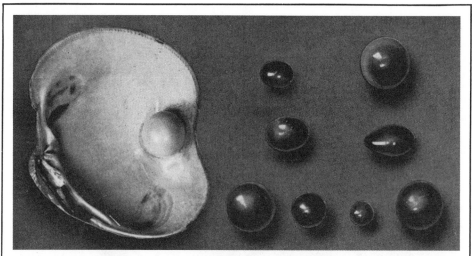

Pearls from common clam (*Venus mercenaria*) of eastern coast of America

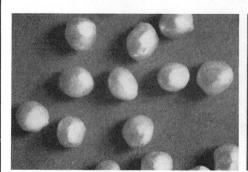

Pearl "nuggets" from the Mississippi Valley

Wing pearls from the Mississippi Valley

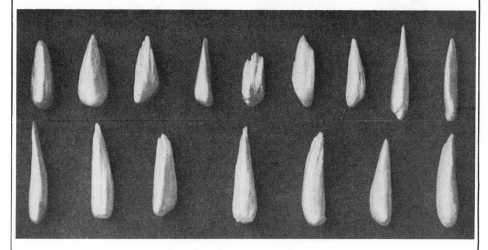

Dog-tooth pearls from the Mississippi Valley

ACTUAL SIZES OF PEARLS FROM ⅛ GRAIN TO 160 GRAINS

The structure of pearls from univalve mollusks, such as the conch, the abalone, etc., as well as those from some bivalves, as the Pinna, for instance, differs from that of the true pearls formed in species of Margaritifera. Instead of the alternate layers of conchiolin and of carbonate of lime, many of these have an alveolar structure. When greatly magnified, the surface of a Pinna pearl appears to be formed of very small polygones, which, as decalcification shows, are the bases of small pyramids radiating from the nucleus. The walls of these pyramids are formed of conchiolin, and they are filled with carbonate of lime of a prismatic crystalline structure. This is simply a modification of the parallel laminæ in the Margaritifera pearls, for, as Dubois points out, in some sections we can see portions where the alveolar formation has proceeded for a time coincidentally with the lamellar form.

Pearls are affected by acids and fetid gases, and may be calcined on exposure to heat. Their solubility in vinegar was referred to by the Roman architect Vitruvius ("De Architectura," L. viii.. c. 3) and also by Pausanias, a Greek geographer in the second century ("Hellados Periegesis," L. viii, c. 18); but it seems that there could be little foundation for Pliny's well-known anecdote in which Cleopatra is credited with dissolving a magnificent pearl in vinegar and drinking it —"the ransom of a kingdom at a draught"—to the health of her lover Antony.[1] It is no more easy to dissolve a pearl in vinegar than it is to dissolve a pearl button—for the composition is similar, and one may easily experiment for himself as to the difficulty in doing this. Not only does it take many days to dissolve in cold vinegar the mineral elements of a pearl of fair size, but even with boiling vinegar it requires several hours to extract the mineral matter from one four or five grains in weight, the acid penetrating to the interior very slowly. And in neither case can the pearl be made to disappear, for even after the carbonate of lime has dissolved, the organic matrix of animal matter—which is insoluble in vinegar—retains almost the identical shape, size, and appearance as before. If the pearl is first pulverized, it becomes readily soluble in vinegar, and might be thus drunk as a lover's potion, but it would scarcely prove a *bonne bouche*.

Pearls assume an almost infinite variety of forms, due largely to the shapes of the nuclei, and also to their positions within the mollusk. The most usual—and, fortunately, also the most valuable—is the spherical, resulting from a very minute or a round body as a nucleus and the uniform addition of nacre on all sides. Of course, spherical pearls can result only where they are quite free from other hard sub-

---

[1] "Historia Naturalis," *Lib*. IX, c. 35. This is also referred to by Macrobius in Saturnaliorum conviviorum *Lib*. II, c. 13.

stances;.consequently they originate only in the soft parts of the mollusk and not by the fixation of some nucleus to the interior surface of the shell.

The perfectly spherical pearls range in weight from a small fraction of a grain to three hundred grains or more, but it is very, very rare that one of choice luster weighs more than one hundred grains. The largest of which we have any specific information was that among the French crown jewels as early as the time of Napoleon, an egg-shaped pearl, weighing 337 grains. The largest pearl known to Pliny in the first century A.D. weighed "half a Roman ounce and one scruple over," or 234½ grains Troy. These very large ones, weighing in excess of one hundred grains, are called "paragons." The small pearls—weighing less than half a grain each—are known as "seed-pearls." The very small ones, weighing less than ½5 of a grain, are called "dust-pearls." These are too small to be of economic value as ornaments.

Slight departures from the perfect sphere, result in egg shapes, pear shapes, drop shapes, pendeloque, button shapes, etc. Some of these are valued quite as highly at the present time as the spherical pearls, and many of the most highly prized pearls in the world are of other than spherical form. Indeed, pearls of this kind are found of larger size than the perfectly round pearls. The egg-shaped pearl,[1] called "la Régente,"—one of the French crown jewels sold in May, 1887—weighed, as stated above, 337 grains. The great pear pearl described by Tavernier—"the largest ever discovered"—weighed about 500 grains. A button pearl received from Panama in 1906 weighed 216 grains.

Wider departures from the spherical form result in cylindrical, conical, top-shaped, etc. Some pearls present the appearance of having been turned in a lathe with intricate tooling. Remarkable examples of these "turned pearls" have been found, competing in their circular perfection with the best work of a jeweler's lathe.

Many standard varieties of non-spherical, but normally shaped pearls, are recognized by the fishermen and the jewelers. For instance, in the nomenclature of the American fishermen, *bouton,* or button pearls are divided into "haystacks" and "turtle-backs," according to the height of the projection. Also, certain imperfections result in distinguishing names: "bird's-eye" refers to a pearl having a little imperfection on the best surface; "ring-arounds" have a dark or discolored ring about them; and "strawberries" have numerous minute projections on the surface.

During its growth, a spherical pearl may come in contact with a foreign body, such as grit or a vegetable film, and the additional nacral layers envelop the adjacent matter until it is entirely concealed within

[1] Now in the French crown brooch in the possession of the Princess Youssoupoff of Russia.

the pearl, its position being recognized only by the excrescence on one side, and, with continued increase in size, even this may be almost overcome.

Sometimes double, triple, or multiple pearls are formed; each of these may have a separate nucleus and grow independently for a time until they adjoin each other; continuing to grow, they become so united as to form a connected mass. The "Southern Cross" is a remarkable example of this. It appears to consist of seven nearly spherical pearls attached to one another in a straight line, and one projecting from each side of the second in the row, thus forming a Roman cross.[1]

A few years ago, near Sharks Bay, on the coast of western Australia, a cluster was found containing about 150 pearls closely compacted. This cluster measured about one and a half inches in length, three quarters of an inch in breadth, and half an inch in thickness.

When a growing pearl is very near to the nacreous lining of the shell, the pressure between the two hard substances results in a rupture of the pearl-forming sac and the epithelial layer of the shell, and the pearl comes in actual contact with the nacre. The pearl gradually becomes attached to the shell, and the under portion is prevented from growing further; the upper or exposed surface receives other layers, resulting in the formation of a *bouton*. As the shell around the pearl continues to grow, it gradually closes about, and almost wholly conceals the pearl. Since it is constantly wasting away on the exterior surface as it grows on the interior, it follows that in time the shell passes the pearl quite through to the outside, where it rapidly decays. Thus the oyster virtually forces the annoying intruder directly through the wall of its house instead of by way of the open door, and magically closes the breach with its marvelous masonry.

These embedded pearls are generally faulty and of diminished luster, but in the aggregate, large quantities of imperfect ones, and especially half and quarter pearls, are secured in this manner. Sometimes—particularly in the Australian fisheries—large pearls are thus found, weighing twenty, forty, sixty, and even eighty grains; and when the faulty outside layers of nacre are removed, a subjacent surface of fine luster may possibly be revealed. In bivalves, these adherent pearls are commonly in the deep or lower valve, except in those unusual cases where the mollusks have been lying in a reverse position. At the fisheries, the surfaces of the shells are carefully inspected for evidence of pearly nodules, and these are broken open in search for encysted objects. Cutters of mother-of-pearl occasionally find embedded pearls of this kind which have escaped the vigilant eyes of the fishermen.

[1] See p. 466.

We read of an instance in an important paper treating of the jeweling trade of Birmingham: "A few years since [the paper was written in 1866] a small lot of shells was brought to Birmingham, which either from ignorance or mistake had not been cleared of the pearls at the fishery. A considerable number were found and sold, and one especially was sold by the man who had bought the shell for working into buttons, for £40. The purchaser, we believe, resold the same for a profit of £160; and we have heard that it was afterward held in Paris for sale at £800."

A choice gem which was found in New York, in October, 1905, in an Australian shell, sold finally for $1200.

The intrusion and continued presence of grains of sand or similar material between the mantle and the shell causes the formation of nacre over the foreign body, resulting in a *chicot* (blister pearl), or possibly a quarter or a half-pearl. The growth of a *chicot* sometimes results from the mollusk covering a choice pearl which has become loosened from the soft tissues and adheres to the shell, as above cited. Hence, it is sometimes desirable to break a *chicot* to secure its more valuable inclosure. In the account of his interesting pearling experiences on the Australian coast, Henry Taunton states: "During the first season's shelling at Roebuck Bay, we came across an old worm-eaten shell containing a large blister, which was removed in the usual manner by punching a ring of minute holes around its base; a slight tap was then sufficient to detach it. For many weeks it was untouched, no one caring to risk opening it, for if filled with black ooze, which is frequently the case, it would be of little value. At last, baffled in his attempt to solve the problem, and emboldened by an overdose of 'square face,' the skipper gave it a smart blow with a hammer, which cracked it open, and out rolled a huge pearl, nearly perfect, and weighing eighty grains. A few specks and discolorations were removed by a skilful 'pearl-faker,' and it was sold in London for £1500."[1]

Blister pearls are also caused by the defensive or protective action of the mollusk in resisting the intrusion of some animal, as a boring sponge or a burrowing worm, which has begun to penetrate the outer layers of the shell. This stimulation causes the mollusk to pile nacreous material upon the spot, thus making a substantial mound closely resembling a segment of a large pearl. This walling-out of intruders is not the result of intelligent forethought or of instinct, analogous to the repairing of a damaged web by a spider, or the retunneling of a collapsed gallery by ants; it is a pathological rather than an intelligent action.

When the nucleus of a pearl is large and very irregular, it necessarily

[1] Taunton, "Australind," London, 1903, p. 224.

BROOCHES MADE OF PETAL, DOG-TOOTH, AND WING PEARLS
From the Upper Mississippi Valley

GRAY PEARLS IN THE POSSESSION OF AN AMERICAN LADY AND BROOCH
FROM TIFFANY & CO.'S EXHIBIT, PARIS EXPOSITION, 1900

follows that the deposited nacre roughly assumes the irregular out-line of the inclosed object. This is strikingly shown in pearls covering a minute fish, a crayfish, or a small crab. Several specimens have been found in which the species could be identified by examination through the nacreous coating.

In the American Unios there is a strong tendency to produce elongated pearls near the hinge of the shell, which are consequently known as "hinge pearls." The occurrence and form of these suggest that their origin may not be due to nuclei, but that they result from an excess of carbonate of lime in the water, and that the animal stores a surplus of nacre in this convenient form. There are several standard forms of these hinge pearls. Many are elongated or dog-toothed, some are hammer-shaped, others resemble the wings of birds, the petals of flowers, the bodies of fish, and various other objects. A large percentage of the pearls found in Unios of the Mississippi Valley are of these types.

Some irregular pearls or baroques are very large, weighing an ounce or more. A well-known example is the Hope pearl, described on page 463, which weighs three ounces. These monster pearls sometimes as-sume odd shapes, such as clasped hands, the body of a man, lion, or other animal, etc.

Although baroques may have a pearly luster, they are not highly prized unless unusually attractive, and they have little permanent value, apart from their estimation in the eyes of admirers of the curious and unique. They are used largely in *l'art nouveau,* and in forming odd and fanciful objects of jewelry, the designer taking ad-vantage of the resemblance which they bear to common objects of every-day life, and by additions of gold and other ornaments complet-ing the form which nature had merely suggested.

Some remarkable examples of baroque mountings have been pro-duced, and a few are to be found in most of the large pearl collections. In a single case in the Imperial Treasury at Vienna are baroques forming the principal parts or figures of a horse, stag, lamb, tortoise, lizard, cock, dragon, butterfly, gondola, hippopotamus, female bust, and three mermaids. Other well-known collections are those of the royal family of Saxony in the Grüne Gewölbe at Dresden; those in the Palace of Rosenberg at Copenhagen; in the Waddesden (Rothschild) collection of the British Museum; among the jewels in the Louvre in Paris; with the treasures of the Basilica of St. Mark in Venice; and in the museum of the University of Moscow.

A remarkable pearl-like ornament more common in Asia than in the Occident, is the *coque de perle,* which is an oval section of the globose whorl of the Indian nautilus. The exterior or convex surface is highly lustrous, but the material is very thin. It is commonly provided with a suitable filling or backing of putty or cement to impart solidity, and is used like a blister pearl. Sometimes two perfectly matched *coques de perle* are filled and cemented together, giving the appearance of an abnormally large oblong or nearly spherical pearl.

The color of pearls has no connection with the luster. In general it is the same as that of the shell in which they are formed. Black pearls are found in the black shells of Mexico, and pink pearls in the pink-hued *Strombus* of the Bahamas. Ceylon pearls are seldom of any other color than white, and Sharks Bays are almost invariably quite yellow or straw-colored, while those of Venezuela are commonly yellowish tinged. But from other localities, pearls simulate every tint of the rainbow, as well as white and black. The most common, as well as the most desirable ordinarily, is white, or rather, silvery or moonlight glint,—*"la gran Margherita,"* as Dante calls it; but yellow, pink, and black are numerous. They may also be piebald—a portion white and the rest pink or brown or black. Some years ago there was on the market a large bean-shaped pearl of great luster, one half of which was white and the other quite black, the dividing-line being sharply defined in the plane of the greatest circumference. The pearls from Mexico, the South Sea islands, and the American rivers are especially noted for their great variety of coloration, covering every known tint and shade, and requiring such a master as Théophile Gautier to do justice to them.

Many theories have been advanced to explain the coloration of pearls. When the old idea of dew formation prevailed, it was considered that white pearls were formed in fair weather, and the dark ones when the weather was cloudy. It was further considered that the color was influenced by the depth of the water in which they grew: that in deep water they were white, but where it was so shallow that the sunlight easily penetrated, the pearls were more likely to be dark in color. Tavernier curiously explained that the black pearls of Panama and Mexico owed their color to the black mud in which the pearl-oysters of those localities lived, and that Persian Gulf pearls were more inclined to yellow than those of Ceylon, owing to the greater putrefaction of the flesh before they were removed therefrom.[1] Two centuries ago the color of a pearl was attributed to that of the central nucleus, and it was concluded that if the nucleus was dark, the pearl would be of a similar hue.[2] This theory has also been upset, for pearls are found white on the exterior and quite dark within, and also with these conditions reversed.

[1] "Tavernier's Travels," London, 1889, Vol. II, p. 115. See p. 97.

[2] See "Report of the Royal Society," Oct. 13, 1688.

The color of a pearl is determined by that of the conchiolin, as appears from its remaining unchanged after decalcification. While generally it is the same as that of the mother-of-pearl at the corresponding point of the shell in which it is formed, there are many exceptions to this, and the reasons for the varying tints and colors are probably to be found in the changes in position of the pearl, the ingredients of the water, the health of the mollusk, accidents of various kinds, etc. These factors will be referred to later in discussing the pearls from different mollusks and regions; but in general it is no more easy to explain the colors of pearls than it is to say why one rose is white and another is yellow.

Medieval writers had much to say regarding unripe or immature pearls, likening them to eggs in the body of a hen, which follow a uniform rate of growth; and this idea is not entirely absent even in contemporaneous writings. However, it is an interesting fact that the humble mollusks, like the five wise virgins with prepared lamps, keep their gems perfect in beauty and luster at all times. It matters not whether the pearl be removed when it is only the size of a pin-head or not until it reaches that of a marble, it is at all times a complete, a ripe, a perfect pearl, and the largest surpasses the smallest only in the characteristics and properties which are incidental to size. Imparting perfection and completion every day, every moment, the mollusk utilizes the added time simply in enlarging its beautiful work.

Although art has made wonderful progress in that direction, the pearl, like truth, is not easily imitated. There is as much difference between the ubiquitous imitations and the perfect gem as there is between a chromolithograph and a silvery Corot, or between the effects of cosmetics and the freshness of youth. While to the unskilled, or under superficial inspection, the false has some of the properties of the genuine, it is only necessary to place them side by side to make the difference apparent. However clever the imitation may be in color, in form, and in density, it always lacks in richness, in sweetness, and in blended iridescence.

V

SOURCES OF PEARLS

V

SOURCES OF PEARLS

Rich honesty dwells like a miser, sir, in a poor house,
as your pearl in your foul oyster.

As You Like It, Act V, sc. 4.

IN geographic range, the sources of pearls are widely distributed, each one of the six continents yielding its quota; but the places where profitable fisheries are prosecuted are restricted in area. First in point of value, and possibly of antiquity also, are the fisheries of the Persian Gulf, giving employment ordinarily to thirty thousand or more divers. The yield in the likewise ancient fisheries of the Gulf of Manaar is uncertain, but sometimes remarkably large. The Red Sea resources are now of slight importance compared with their extent in the time of the Ptolemies. Other Asiatic fisheries are in the Gulf of Aden, about Mergui Archipelago, on the coast of China, Japan, Korea, and Siam, and also in the rivers of China, Manchuria, and Siberia.

Aside from those produced in the Red Sea and the Gulf of Aden, the pearl fisheries of Africa are of small extent. Some reefs exist on the lower coast of the German East African territory and also in Portuguese East Africa, but they have not been thoroughly exploited.

In most of the inshore waters of Australasia pearls may be secured; the fisheries are most extensive on the northern coast of Australia, in the Sulu Archipelago, and about the Dutch East Indies. Tuamotu Archipelago, Gambier, Fiji. and Penrhyn are prominent in the South Pacific Ocean.

In the seas of Europe few pearls have been found, but the rivers have yielded many; and although the resources have been greatly impaired, many beautiful gems are yet found there.

South America contributes the important reefs on the coast of Venezuela—the land of unrest and revolutions, whose fisheries were first exploited by Columbus. Other South American countries in which pearls are collected are Panama, Ecuador, Peru, etc. In

North America, pearls are found in the pearl-oyster of the Gulf of California, the abalone of the Pacific coast, the queen conch of the Gulf of Mexico, and in the Unios of most of the rivers, especially those of the Mississippi Valley.

Since pearly concretions partake of the characteristics of the shell within which they are formed, it follows that practically all species of mollusks whose shells have a well-developed nacreous lining yield pearls to a greater or less extent. But the number of these species is relatively small. They belong chiefly to the *Margaritiferæ,* or pearl-oyster family of the sea, and to the *Unionidæ,* or family of fresh-water mussels. Pearls occur also in some univalves, but not so abundantly as in bivalves of the families mentioned. Broadly stated, we may hope to find pearls within any mollusk whose shell possesses a nacreous surface; and it is useless to search for them in shells whose interior is dull and opaque, such as the edible oyster for instance.

The great bulk of the pearls on the market, and likewise those of the highest quality, are from the *Margaritiferæ,* which are widely distributed about tropical waters. Although these mollusks are spoken of as pearl-oysters, they are not related in any way to the edible oysters (*Ostrea*) of America and Europe.[1] The flesh is fat and glutinous, and so rank in flavor as to be almost unfit for food, although eaten at times by the poorer fishermen in lieu of better fare. The origin of the name is doubtless due to the fact that in the somewhat circular form of the shell they resemble oysters rather than the elongated mussels of Europe, to which they are more nearly related in anatomy. Also in that—like their namesakes—they are monomyarian, having only one adductor muscle.

The two valves or sides of the pearl-oyster shell are nearly similar in shape and almost equal in size; whereas in the edible oysters one valve is thin and somewhat flat, while the other is thicker, larger, and highly convex. In the latter, also, the hinge, or umbo, is an angular beak; but in the pearl-oysters the umbo is prolonged by so-called ears or wings into a straight line the length of which is nearly equal to the breadth of the shell.

The byssus, or bunch of fibers, by which pearl-oysters attach themselves to the bottom indicates their relationship to the mussels. The possession of a small foot and somewhat extended migratory powers

---

[1] Neither is there any special significance in the popular terms "clams," "mussels," etc., as applied to the pearl-bearing species of the rivers. The "clams," or Unios of the Mississippi Valley, resemble neither the long clams (*Mya*) nor the round clams (*Venus*) of the Atlantic coast; the mussels of the fresh-water lakes are quite distinct from the edible ones of brackish waters, and the Pinna oyster and the giant clam (*Tridacna*) have little resemblance to the mollusks with which these terminal names are commonly associated.

—at least in the first years of growth—also distinguish them from the sedentary edible oysters. But from an economic point of view, the principal difference is the possession of a thick, nacreous, interior lining in the shells of pearl-oysters, which is wholly lacking in the edible species. Like their namesakes, the pearl-oysters are exceedingly fertile, a single specimen numbering its annual increase by millions.

Commercially considered, the pearl-oysters are roughly divisible into two groups, (1) those fished exclusively for the pearls which they contain, and (2) those whose shells are so thick as to give them sufficient value to warrant their capture independently of the yield of pearls. The best examples of the first group are the pearl-oysters of Ceylon and of Venezuela, and to a less extent those of the Persian Gulf, the coast of Japan, and of Sharks Bay, on the Australian coast. Of the second group, the pearl-oysters of Torres Straits and of the Malay Archipelago are the most prominent members. Between these two groups are the many species and varieties whose shells and pearls are more evenly divided with respect to value, including those of Mexico, Panama, the Red Sea, the South Sea islands, etc.

Some conchologists recognize a large number of species of *Margaritiferæ,* while other authorities consider many of these as local variations of the same species. There is much difference in the size, color, and markings of the shells in different localities, owing to varying geographical and physical conditions. The distinction of species and the nomenclature herein adopted are those of Dr. H. L. Jameson, who has recently revised and rearranged the collection of shells belonging to this family in the British Museum of Natural History,[1] and to whom we are indebted for descriptive notes relative to several of the species.

The greatest pearl-producer in the family of pearl-oysters is the *Margaritifera vulgaris* of the Gulf of Manaar and the Persian Gulf, and to a much less extent of the Red Sea. It occurs in various other inshore waters of the Indian Ocean, and about the Malay Archipelago and the coast of Australia and New Guinea, although it is not the principal pearl-oyster of those waters. An interesting account of its immigration into the Mediterranean Sea through the Suez Canal was given by Vassel in 1896.[2]

This species is quite small, averaging two and a half inches in diameter in Ceylon waters, and somewhat more in the Persian Gulf,

---

[1] Jameson, "On the identity and distribution of the mother-of-pearl oysters; with a revision of the subgenus *Margaritifera."* Proceedings of the Zoölogical Society of London, Vol. L, 1901, pp. 372–394.

[2] Vassel, "Sur la Pintadine du Golfe de Gabes, Comptes Rendus Assoc. Franç.," 1896, pp. 458–466.

whence large quantities of the shell are exported under the name of "Lingah shell." The Ceylon variety has the nacreous lining almost uniformly white over the entire surface, only the lip having a slightly pinkish ground color. The exterior is marked by seven or eight reddish brown radial bands on a pale yellow ground. In addition to its greater size, the Persian variety is darker, and the lip of the shell has a reddish tinge.

For centuries the *Margaritifera vulgaris* has sustained the great pearl fisheries of Ceylon, India, and Persia, and at present yields the bulk of pearls on the market, especially the seed-pearls and also those of medium size. It produces relatively few large ones, rarely exceeding twelve grains in weight. These pearls are commonly silvery white, and for their size command the highest prices, because of their beautiful form and superior luster. Excepting the Venezuelan species, this is the only pearl-oyster which at present supports extensive fisheries exclusively for pearls; in the fisheries for all other species the value of the shells furnishes considerable revenue, and in some localities this represents several times as much as the income from the pearls.

Ranking next to *Margaritifera vulgaris* in extent of pearl-production is the *Margaritifera margaritifera,* which is widely distributed about the tropical inshore waters of the Pacific and Indian oceans. It is very much larger than the Lingah oyster, good specimens measuring seven or eight inches in diameter, and the nacreous interior is usually of a darker color. In addition to its yield of pearls, the shell of this species is of value in the mother-of-pearl trade, and contributes largely to the economic results of the fisheries. Indeed, in several regions the shell is of more value than the pearls, which represent only an incidental yield. As Jameson notes, the color and markings of the shell, though extremely variable, generally suffice to distinguish this species. The ground color of the exterior ranges through various shades from yellowish brown to very dark brown. Its characteristic markings consist of from ten to eighteen radial rows of white and yellow spots, running from the umbo, or hinge, to the margin.

Several varieties of *Margaritifera margaritifera* are recognized. The type species occurs along the north coast of Australia, from Brisbane on the east to Sharks Bay on the west; on the New Guinea coast; at Formosa; and about many of the islands of the Pacific. The well-known "black-lip shell" of Australian waters is of this species; it shows a greenish black on the margin of the nacre. The yield of this is very small compared with that of the large pearl-oyster of Australia.

The *Margaritifera margaritifera* occurs on the eastern coast of Arabia in two varieties, which differ somewhat from the type species.

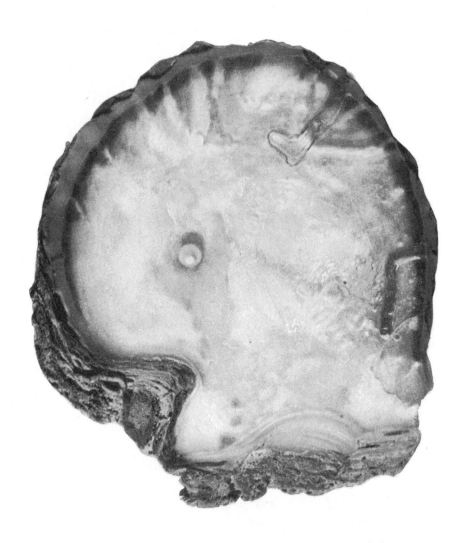

SHELL OF PEARL-OYSTER WITH ATTACHED PEARL

(Margaritifera margaritifera mazatlanica)

From Costa Rica

PEARL-OYSTER OF CEYLON (*Margaritifera vulgaris*)
Natural size

PINNA OR WING SHELL (*Pinna seminuda*)
One third natural size

These have been designated by Jameson as *M. margaritifera persica* and *M. margaritifera erythræensis*. These are much larger than the Lingah shell of the Persian Gulf, but are smaller than the Australian species. The percentage of pearls in them is less than in the Lingah species, but from a commercial point of view this is to some extent offset by the greater value of the shell. The *M. m. persica* is more numerous in the gulf than the *M. m. erythræensis,* and large quantities of the shell are marketed in Europe. Formerly the shipments were made principally by way of Bombay, hence the shell is known in the mother-of-pearl trade as "Bombay shell." The exterior is of a light grayish or greenish brown color, with yellowish white radial bands. The nacre has a slightly roseate tint, and the margin is greenish yellow. The pearls found herein are more yellowish in color and attain a larger size than those from the Lingah oyster.

The *M. m. erythræensis* occurs also in the Red Sea and along the shores of the Arabian Sea. Among mother-of-pearl dealers it is known as "Egyptian shell" or "Alexandria shell," owing to the fact that prior to the opening of the Suez Canal shipments were commonly made by way of Alexandria. The color of the nacre is darker than that of its related variety in the Persian Gulf. In the trade, three grades of this shell are recognized, classed according to the shade of color. The lightest comes from Massowah and near the southern end of the Red Sea, and the darkest from farther north, in the vicinity of Jiddah and Suakim.

The islands of the southern Pacific, and of eastern Polynesia especially, yield another variety of *M. margaritifera,* to which the name *M. m. cumingi* has been given. The nacre is of a dark metallic green, and in the mother-of-pearl trade the shell is designated as "black-edged." It attains a large size, only slightly smaller than the large Australian species; many individual specimens measure ten inches in diameter, and weigh six or seven pounds for the two valves. Belonging to this variety are those oysters whose shells are known in the markets of Europe and America as "Tahiti," "Gambier," and "Auckland" shells, the name designating the port of shipment.

Yet another subspecies, the *M. m. mazatlanica,* occurs on the coasts of Panama and Mexico, and especially in the Gulf of California. This is likewise green-edged, and the exterior color is yellow or light brown. This shell has been marketed in quantities since 1850, and is known in the mother-of-pearl trade as "Panama shell." It is smaller than the Australian species, specimens rarely exceeding eight inches in diameter. It yields a large percentage of the black pearls that have been so fashionable in the last fifty years.

Since 1870, the largest pearls have been found mainly in a very large species of pearl-oyster, *Margaritifera maxima,* obtained off the north and west coasts of Australia, among the Sulu Islands, and elsewhere in the Malay Archipelago. In the fisheries for this species, the mother-of-pearl is the principal object sought, and the pearls are obtained incidentally. It is the largest of all the members of this family, reaching in exceptional cases twelve or thirteen inches in diameter, and weighing upward of twelve pounds; while the Ceylon oyster rarely exceeds four ounces in weight. So marked is this difference, that the Australian species is often designated the "mother-of-pearl oyster," and the Ceylon species the "pearl-oyster." Jameson notes that it differs from the *Margaritifera margaritifera,* its nearest competitor in size, in its much longer hinge, its shape, its lesser convexity, and in its color and markings. As described by him, the color ranges from pale yellowish brown to deep brown, with traces of radial markings of dark brown, green, or red in the umbonal area. In its marginal region, the shell is marked by a series of circumferential lines about one third of a millimeter apart.

Several geographical varieties of this species are recognized in the mother-of-pearl trade, differing principally in the coloring of the interior surface. The chief commercial varieties are "Sydney" or "Queensland," "Port Darwin," "West Australian," "New Guinea," "Manila," "Macassar," and "Mergui." The nacre of those from the Australian coast is almost uniformly silvery white. That of the "Manila shell" is characterized by a broad golden border surrounding the silvery white nacre. The "Macassar shell" lacks the golden border of the "Manila shell," and is similar in its uniform whiteness to the "Sydney shell," but its iridescence is much greater.

The *Margaritifera carcharium,* from Sharks Bay, on the coast of Australia, yields yellow pearls and small quantities of mother-of-pearl. This species is small—three or four inches in diameter. The color is grayish or greenish yellow, with several somewhat indistinct radial bands of brownish green. The nacre has a yellowish green tint, with a margin of pale yellow, with brown markings.

In the West Indies and on the Atlantic coast of tropical America, especially the coast of Venezuela, occurs the *Margaritifera radiata.* This species is quite small, and seems to be closely allied to the Ceylon oyster. Like the latter, the nacreous interior is rich and brilliant, but owing to its small size, the shell is wholly valueless as mother-of-pearl. The principal and almost the only fishery for this species is on the Venezuelan coast, in the vicinity of Margarita Island, the islands of Cubagua, and Coche.

The coast of Japan yields the *Margaritifera martensi,* which occurs among the numerous islands in the southern part of the empire, but does not extend beyond 40° north latitude. This species is likewise small, and closely resembles the pearl-oyster of Ceylon, from which it differs principally in coloration. As noted by Jameson, brown and white predominate in the exterior coloring, and the interior of the lip is marbled with yellow ocher and chocolate brown, instead of pink, as in the Ceylon shell.

There are numerous other species of pearl-oysters, but they are of slight economic importance, and do not support fisheries of value.

As only a small percentage of the individual mollusks contain pearls, it follows that vast quantities are destroyed without any return whatever, and handling them merely adds to the expense of the industry, as well as reduces the resources of the reefs. This could be obviated if it were possible, without opening them, to determine the individual mollusks containing pearls.

Among the several methods proposed for this purpose, especially interesting is the use of X-rays, which was suggested by Raphael Dubois of Lyons, France, in 1901.[1] The shells of some pearl-oysters—those of Ceylon and of Venezuela for instance—are relatively thin, and it was thought that by the means of the rays the presence of pearls could be ascertained, and non-pearl-bearers could be saved from opening, and be returned to the reefs without injury. Although the calcareous shell partly interrupts the radiations, it is not difficult to recognize the presence of large pearls.

The theory has never been found practical in application, owing largely to the rough and irregular exterior of the shell and the small size of the pearls. The presence of the larger pearls may be ascertained by this method; but it is exceedingly probable that a very large percentage of the small ones, and especially the seed-pearls, would be overlooked. Furthermore, if in their sixth year oysters contain no pearls, the probability of appearance therein later is very small, and little benefit would result from their return to the water. As to saving the trouble of opening the non-pearl-bearing mollusks, labor in the pearling regions is usually inexpensive, and this cost is far more than offset by the reasonable certainty of securing practically all the small as well as the large pearls by the present method of operation. Owing to the greater thickness and the economic value of the large pearl-oysters—as those of Australia or of Mexico, for instance—the application of X-rays to them is obviously impractical. However, when pearl-oyster culture becomes a highly developed industry, with personal ownership in those mollusks returned to the water, some method such as this might be of great value.

Pearls are yielded by various species of *Unionidæ* or *Naiades* occur-

[1] See "Comptes Rendus de l'Académie des Sciences," Paris, 1904, Vol. CXXXVIII, pp. 301, 302.

ring in the rivers of America, Scotland, Saxony, Bavaria, Norway, Sweden, Russia, France, China, etc. These mollusks exist exclusively in the fresh-water streams, lakes, and ponds, and quickly die when submerged in salt water. The *Unionidæ* are of particular interest in America, as it is here that this group is most abundant, and nearly every stream east of the Rocky Mountains contains more or less of them. The Mississippi basin abounds in Unios, or "clams," as they are known to the fishermen of that region, and furnishes about 400 of the 1000 recognized species of this important family.

The Unios are most abundant in clear, running water, where the bottom is gravelly or sandy. The interiors of the shells are iridescent, and vary greatly in tint, exhibiting many delicate shades of color from silvery white to straw color, pink, purple, brown, etc.

About five hundred species of American fresh-water mussels have been recognized by conchologists. Many of these differ from one another so very slightly that they are scarcely distinguishable from an examination of the shells themselves, or even from the descriptions, and a detailed index to the complete list is of little economic importance. The professional fishermen and the shell-buyers take the trouble to name only the species with which they deal, which includes only about twenty-five species, all of which are margaritiferous, though some to a greater extent than others. In the pearling regions a popular nomenclature exists, the names given by the fishermen having reference to the shape, color, etc.

The niggerhead (*Quadrula ebena*) is the most numerous in the Mississippi, and it is extensively used in button manufacture. The thick shell of this species is almost round, with a black outer surface and a pearly white interior. At maturity it averages about four inches in diameter and four ounces in weight. Owing to its uniform whiteness and the flatness of its surface, it is well adapted to button manufacture, and for this purpose more than twenty thousand tons are taken in the Mississippi Valley every year. When the fishery originated, the niggerhead was very abundant in some places, and especially between La Crosse and Burlington. From a single bed near New Boston, Illinois, measuring about 200 acres in area, 7500 tons, or about 70,-000,000 individual shells, were removed in three years. In 1897, a bed of 320 acres near Muscatine furnished 500 tons, or about 4,750,000 shells. This species occasionally yields valuable pearls.

Two species of Unios, *Quadrula undulata* and *Q. plicata,* are known among the fishermen as "three-ridges." The former is also known as the "blue-point" from the fact that the sharp edge is usually tinged faint blue on the inside. Although not the best for button manufacture, the shells yield the greatest number of pearls.

A species somewhat similar to the niggerhead is the bullhead (*Pleurobema œsopus*). This shell is thick and opaque, the nacre is not so iridescent as that of the niggerhead, nor does it yield pearls of such good quality. These two species are not evenly distributed over the bottom of the streams, but occur in great patches or beds, sometimes several feet in thickness and covering many hundreds of acres. Some of the beds are several miles in length, and they may be separated by twenty or thirty miles in which the mollusks are so scarce that profitable fishing can not be made; but usually the reefs are smaller and more closely situated.

The sand shells (*Lampsilis*)—of which there are several species— do not occur in large beds, but are scattered over the sandy beaches and sloping mud-banks. In shape they are narrow and long, adults measuring five or six inches in length. Owing to the small waste in cutting, due to uniformity in thickness, these shells are sold to button manufacturers for more than the niggerhead, which in turn is more valuable than the bullhead.

The buckhorn (*Tritigonia verrucosa*) is very long and narrow; on the dark brown exterior it is rough, as is the horn from which it takes its name, while the interior shows a beautiful display of colors. This is not found in beds, but lies scattered among other species. It sells at a relatively high price—usually in excess of $20 per ton—for button manufacture.

Another species is the butterfly (*Plagiola securis*), which is very prettily marked on the outside with faintly colored dotted stripes of varying length. Over a background of dark yellow run black stripes to the outer edge of the shell, with dark dots between the stripes. The shell is small and thick, and like the sand shell and the buckhorn, is found in small quantities. Owing to the beauty and permanency of its luster, this shell is in demand for button manufacture, and its pearls are often very beautiful.

Other well-known species are the pancake (*Lampsilis alatus*), the maple-leaf (*Quadrula wardi*), and hackle-back (*Symphynota complanata*). On the Atlantic seaboard, the principal species in which pearls have been found are *Unio complanata;* the *Alasmodon arcuata,* which has hinge teeth, and a species of Anodon. Pearls from the *Unio complanata* are usually smaller but more lustrous than those from either of the other species.

Among the many fresh-water mussels are found some remarkable conditions of animal life. Probably the most curious is the parasitic stage of certain species. When hatched from the egg, each one of these is provided with hooks or spines, by means of which it attaches itself to the gills or fins of a swimming fish and becomes embedded therein. After confinement in this cyst for a period of two months or more, the small mollusk works its way out and falls to the bottom of the river or pond, where its development continues along lines more conventional to molluscan life.

In most of the species of Unios the sexes are separate; but it has been determined that in some the individuals are provided with both sets of sexual organs. It is claimed by some naturalists that certain species may change from one sex to another; yet this does not seem to have been positively established.

Not the least interesting of the habits of the Unios is the manner in which they "walk," bushels of them changing their habitation in a few hours. The shell opens slightly and the muscular tongue-like "foot" is thrust out, and by pressure of this on the bottom, the mollusk is propelled in a jerky, jumpy movement with more speed than one would suppose possible for the apparently inert creature.

The number of eggs produced by an individual in one season ranges from a few hundred in some species to many millions in others, as in the *Quadrula heros,* for instance. Most of the fresh-water mollusks are of slow growth, reaching maturity in six or eight years, and it is believed that if undisturbed they live to be from fifteen to fifty years old; indeed, some writers credit them with attaining an age of one hundred years.

While outwardly there is no positive indication of the existence of pearls, they are relatively scarce in young mollusks, and likewise in those having a normal, healthy appearance, with smooth exterior free from blemishes, and they are found generally in the older, irregular, and deformed shells, which bear excrescences and the marks of having parasites. However, some of the choicest pearls have come from shells relatively young and apparently in perfect condition.

It has been pointed out that with the fresh-water Unios there are three indications on which the fishermen to some extent rely for determining the presence of pearls from the outward aspects of the shell. There are, first, the thread or elevated ridge extending from the vertex to the edge; second, the kidney-shape of the shell, and third, the contortion of both valves toward the middle plane of the mollusk.

A single mollusk may contain several small pearls,—more than one hundred have been found,—but in such cases usually none has commercial value. Ordinarily only one is found in the examination of very many shells. Of these objects it may be truthfully said that "many are found, but few are chosen," few that are of first quality or are worthy of a fine necklace. In many instances, several pounds of cheap pearls would be gladly exchanged for a choice gem weighing an equal number of grains.

On the Atlantic seaboard of America, the Anodontas, or "mussels," as they are known locally, are more numerous than the Unios. They prefer the still waters of the ponds and lakes, rather than the swift currents of the streams. The shell is much thinner than that of the Unios, and it is usually not so brilliant in color and iridescence; consequently the pearly concretions obtained from them are less lustrous.

The rivers of Europe, and of Asia also, contain numbers of pearl-bearing mussels. In many localities the yield of pearls has at times attracted attention and produced much profit, though probably never equaling the present extent of the Mississippi River finds. The principal pearl-bearer of Europe is the *Unio margaritifera,* the shell of which has been of some local importance in the manufacture of pearl buttons. In Great Britain it is known as the pearl-mussel; in France as the *moule* or *huître perlière;* in Germany as *perlenmuschel;* in Belgium as *paarl mossel de rivieren;* in Denmark as *perle-skiael;* in Sweden as *perlmussla;* in Russia as *schemtschuschuaja rakavina,* and in Finland as *simpsuckan cuosi.* The *Unio margaritifera* likewise exists in Siberia, and possibly elsewhere in Asia. Other species of Unio exist there and in Mongolia, Manchuria, etc., as, for instance, *U. mongolicus, U. dahuricus,* etc. A leading species in eastern China, the *Dipsas plicatus,* has long been extensively employed in the artificial production of pearly objects or culture-pearls.[1] Unio pearls show less uniformity of tints than those derived from the pearl-oysters. They present an extended series of shades, corresponding to those on the interior of the shells, from almost perfect white through various tints of cream, pink, yellow, bright red, blue, green, russet, and brown. The metallic shades are numerous, especially the steels and the coppers.

Most of the members of the *Mytilidæ* family, which includes the marine mussels, are of slight luster; and the pearly concretions found in them are of the grade known as "druggists' pearls," so-called because, formerly, they were used in a powdered form in astringent and other medicines. However, some of these mussels on the European coast yield pearls that are fairly lustrous. The white and the pink are most numerous, but purple, red, bronze, and yellow are by no means uncommon.

A few pearls are also obtained from the sea-wings or wing-shells (*Pinna*), the silkworms of the sea, found in the Red Sea, the Mediterranean Sea, the Indian Ocean, the southern coast of America, and elsewhere. These shells are narrow at the umbo, or hinge, long, and fan-shaped; they are generally brittle, and present a horn-like appearance. The interior is commonly of a silvery reddish or orange-colored hue, and this tint is imparted to the pearls. The most characteristic feature of the *Pinna* is the thick rope of silky fibers, from four to ten,

[1] See p. 288.

and sometimes twenty or more inches in length, constituting the byssus, a remarkable provision by means of which it anchors itself to the bottom and thus outrides the storm. Formerly the byssus was gathered in Sicily, washed in soap and water, dried, corded, and fabricated into gloves and similar articles of a fine texture. The finished garments were of a beautiful golden brown color, resembling the burnished gold on the backs of some splendid flies or beetles.

The yield of *Pinna* pearls is very small. A few are obtained from the Mediterranean, especially on the Adriatic coast. These are usually rose-tinted or reddish in color, but of diminished orient, and inferior in size. *Pinna* pearls are also reported from the Isle of Pines and from New Caledonia, where they are commonly very dark, almost black in color.

The window-glass shell (*Placuna placenta*), the *vitre chinoise* of some writers, yields a few small, irregularly shaped pearls of a dull leaden color. It occurs in the inshore waters of the Indian and the southwestern Pacific oceans; fisheries are prosecuted in Tablegram Lake, near Trincomali, on the northeast coast of Ceylon; on the coast of Borneo, especially at Pados Bay, and to a less extent in some other localities. This mollusk is quite distinct from the true pearl-oyster, and in adult life is devoid of the byssus, living on the muddy bottom of the shallow waters. The shell is almost circular, the right valve is quite flat, and the left only slightly convex. It is remarkable for its transparency, especially in the first year of growth, when the beating of the heart of the mollusk is visible through it. Reaching maturity in about two years, the shell becomes white and translucent, resembling pressed isinglass somewhat in its texture. It then measures about six or seven inches in length, and nearly the same in width. The outside is rough; the interior is glazed over and has a subdued pearly luster. It is so thin and transparent that with a strong light very coarse print can be read through it. It is commonly used in the East Indies as a substitute for glass in windows, admitting a soft mellow light into the room. For this purpose it is usually cut into small rectangular or diamond-shaped pieces, about five or six square inches in area, and these are inserted into sash frames. It forms a good economical substitute for glass, not only in windows of native residences, but also in lanterns and the like.

The giant clam (*Tridacna gigas*) of tropical waters yields a few large opal-white symmetrical pearls, with faint luster and of little value. The transversely oval shell of the *Tridacna,* with its great squamous ribs, is probably the largest and heaviest in existence, single pairs weighing upward of 500 pounds. It is found in tropical seas, and especially in the Indian Ocean. It is much used for ornament,

SHELL AND PEARLS OF THE COMMON CONCH

(Strombus gigas)

Of Florida and the West Indies

particularly for fountain-basins, and for *bénitiers,* or holy-water fonts. A beautiful pair used as *bénitiers* in the Church of St. Sulpice in Paris is said to have been a gift of the Republic of Venice to Francis I. There seems to be no established fishery for this mollusk, and the pearls very rarely come on the market. About four years ago in New York City an effort was made to market one weighing about 200 grains. The owner represented that it was a "cocoanut pearl," and offered to sell it for $2000; whereas its actual value was probably not over $10 or $20, and that only for a museum collection.

Pearls of slight luster also occur in the quahog, or hard clam (*Venus mercenaria*), of the Atlantic coast of the United States. Although these are rare, they are generally of good form, and some weigh upward of eighty grains each. They are commonly of dark color, purplish, ordinarily, but they may be white, pale lilac, brown, and even purplish black, or black. The white ones—which so nearly resemble ivory buttons as readily to pass for them at a casual glance— are of little value; but fine dark ones have retailed at from $10 to $100 each. There is little demand for them, for unless the color is very good, they possess slight beauty, lacking the orient peculiar to choice pearls. Pearls have also been reported from the edible clam of the Pacific coast of America.

Shelly concretions are found in the edible oyster of America (*Ostrea virginica*), as well as in that of Europe (*O. edulis*); but these are commonly objects of personal interest or of local curiosity, rather than of artistic or commercial value, as they are lacking in luster and iridescence. Most of them are dull or opal-white, some are purple, and a few are white on one side and purple on the other. As many as fifty of these formations have been found in a single oyster. Sometimes they are of odd appearance, suggesting the human eye or face, and recently one was found which bore a striking resemblance to a human skull. Notwithstanding many news items to the contrary, it is doubtful whether the choicest pearl from an edible oyster would sell for as high as $20 on its own merits; professional shuckers have opened thousands of bushels of oysters without finding one which would sell for ten cents.

Among univalves, the most prominent pearl-producer is probably the common conch or great conch (*Strombus gigas*) of the West Indies and the Florida coast, which secretes beautiful pink pearls of considerable value. This is one of the largest of the univalve shells, some individuals measuring twelve inches in length, and weighing five or six pounds. The graceful curves and the delicate tints of lovely pink color make it exceedingly attractive. The conch abounds in the waters of the West Indies, especially in the Bahamas, where many thousands are annually taken for the shell, which forms quite an article of commerce. The flesh is esteemed as food and is also used for bait; and it is particularly in preparing for these purposes that the pearls are found, as no established fisheries exist for the pearls alone.

The ear-shells or abalones (Haliotidæ) found on the coasts of California, Japan, New Zealand, and other localities in the Pacific, secrete pearly concretions, sometimes with fine luster, but usually of small value. These shells resemble in general outline the form of the human ear. Distinguishing characteristics are the flatly-spiral bowl-like shape, and the regular series of holes in the back near the distal margin, for the admission of water to the respiratory organs. The holes are on the left side and parallel with the columellar lip, and those nearest the apex close up as the shell increases in size. The shells are rough externally, but beautifully nacreous within. In variety and intensity of coloring, the nacre is superior to that of the pearl-oysters, but it is not so harmonious, and it does not form so thick and flat a layer.

Abalone pearls are especially interesting on account of their brilliant and unusual colors. Green predominates, but blue and yellow also occur. Although commonly very small, some of the well-formed ones exceed seventy-five grains in weight, and those of irregular shape may be very much larger. The ear-shells also produce many irregular pearly masses. Although these are without an established commercial value, their beautiful greenish or bluish tints adapt them for artistic jeweled objects, such as the body of a fly or of a beetle.

Similar concretions are found in species of turbos and turbinella, especially the Indian chank (*Turbinella rapa*), which yields pink and pale red pearls. The pearly nautilus (*Nautilus pompilius*) yields a few yellowish pearls, especially those taken in Australian waters; but from the paper nautilus—"the sea-born sailor of his shell canoe"— no pearls are obtained, owing to the non-lustrous nature of the shell.

In bygone days, especially in Asia, and also to some extent in Europe, pearls were credited as coming from many non-molluscan sources. The Rabbis had the idea that they came also from fish, as noted in the story of a tailor who was rewarded by finding a pearl in one which he bought (Gen. R. xi. 5). The Raganighantu of Narahari, a Kashmir physician of about 1240 A.D., reported them as coming from bamboos, cocoanuts, heads of elephants, bears, serpents, whales, fish, etc.;[1] although it conceded that these were deficient in luster, which is recognized as the characteristic feature of pearls. We understand, therefore, that this use of the word signifies only

[1] Garbe, "Die Indischen Mineralien," Leipzig, 1882.

hard concretions of a spherical form. In the apology for his book, prison-bound Bunyan wrote:

> A pearl may in a toad's head dwell,
> And may be found in an oyster shell.

The crystal gems—the diamonds, rubies, etc.—are practically unlimited in their longevity, existing thousands of years unchanged in condition. Except those which have been discovered by man, the earth contains about as many as it ever did, and it is not unreasonable to suppose that in course of time a considerable percentage of the total will be discovered. But in the seas as well as in the rivers, the longevity of pearls is greatly restricted, and

> Full many a gem of purest ray serene
> The dark, unfathomed caves of ocean bear [1]

to run their course of existence and decay unseen and unknown. Perishable while in the seas, almost as cereals and fruits on land, the harvest must be gathered with promptness or it is wasted. And it seems probable that only a small percentage of the beautiful gems produced in the waters have gladdened the sight of man.

With considerable hesitancy we have attempted to estimate the number of persons employed in the pearl fisheries of the world, and the aggregate local value of their catch. For two or three regions, this is not a matter of great difficulty. For instance, the divers employed in the Ceylon fishery are numbered each season, and the auction sales of their catch furnish a reasonably satisfactory basis for determining the value of the output. Likewise in Australia, Venezuela, and some minor localities, the fishermen are numbered; but the reports are less satisfactory as to the value of the pearls. In the Persian Gulf, the Red Sea, the Gulf of California, and the islands of the Pacific, where pearl-diving is a profession and a regular source of livelihood, the number of employees is fairly constant. But in the rivers and ponds of America, as well as of Europe and of Asia, where neither experience nor costly equipment is required for the industry, and pearls to the value of very many thousands of dollars are obtained by men, women, and even children, on pleasure bent, as well as in the widely fluctuating professional fisheries, the problem is far more difficult.

Contending with these many difficulties, we venture to present the following estimate of the number of persons employed in the pearl fisheries of the world, and the value of the output in 1906.

[1] Gray's Elegy.

Localities.	Fishermen No.	Pearls Local Values.	Shells Local Values.
Asia:			
Persian Gulf	35,000	$4,000,000	$110,000
Ceylon [1]	18,500	1,200,000	40,000
India	1,250	100,000	95,000
Red Sea, Gulf of Aden, etc. [2]	3,000	200,000	150,000
China, Japan, Siberia, etc.	20,000	400,000	50,000
Total	77,750	$5,900,000	$445,000
Europe:			
British Isles	200	15,000	
Continent of Europe	1,000	100,000	3,000
Total	1,200	$115,000	$3,000
Islands of the Pacific:			
South Sea islands	4,500	125,000	500,000
Australian coast [3]	6,250	450,000	1,200,000
Malay Archipelago	5,000	300,000	800,000
Total	15,750	$875,000	$2,500,000
America:			
United States rivers	8,500	650,000	350,000
Venezuela	1,900	275,000	10,000
Mexico	1,250	210,000	200,000
Panama	400	40,000	75,000
Miscellaneous	1,000	75,000	25,000
Total	13,050	$1,250,000	$660,000
Grand total	107,750	$8,140,000	$3,608,000

Our returns do not represent the annual output of pearls in the values best known to gem buyers. The difference in price between pearls in the fisherman's hands in the Persian Gulf or at the Pacific islands, and that for which they are exchanged over the counters in New York or Paris, is nearly as great as the difference in value of wool on the sheep's back and of the same material woven into fashionable fabrics. For each dollar received by the fisherman, the retail buyer probably pays three; and it is not unreasonable to suppose that the pearls herein represented probably sold ultimately for an aggregate of $24,420,000.

This summary falls far short in giving a correct idea of the importance of the pearl fisheries in furnishing a livelihood to humanity; for it takes no consideration of that great body of men who contribute incidentally to the prosecution of the fisheries, such as shell-openers,

[1] In 1905, the Ceylon pearl yield approximated $2,000,000 in value.

[2] Including African coast.
[3] Including Sharks Bay.

pearl-washers, watchmen, cooks, laborers, etc. In the Ceylon pearl fishery of 1906, for instance, our estimate shows 18,500 fishermen; but there were 40,000 persons engaged at the pearl camp alone, and many others were given employment in boat-building, supplying provisions, selling the pearls, etc., and this does not include the wives and children depending on the industry for sustenance. Indeed, it seems not unreasonable to estimate that instead of only the 18,500 fishermen, 85,000 persons were in a large measure dependent for their livelihood on the Ceylon fishery in 1906.

Estimated on the same basis, we have a total of 500,000 persons depending largely on the pearl fisheries of the world for their support. Thus we see that pearl buyers and pearl wearers not only gratify a commendable admiration for the beautiful, but contribute largely to the economic balance whereby one class of humanity either sustains or is dependent upon another, even though these classes be so widely separated as the crown of Russia from the half-starved diver of the tropical seas. How strange is the providence of God, who, by granting the pearl to the poor Arab, the Tamil of India, the South Sea Islander, and the forgotten Selang of Mergui, makes the greatest and wealthiest in the world contribute to their support.

VI

PEARLS FROM ASIA

THE PERSIAN GULF, FISHERIES OF INDIA, CEYLON
PEARL FISHERIES, RED SEA AND ARABIAN
SEA, CHINA, JAPAN, SIBERIA, ETC.

THE PEARL FISHERIES OF THE PERSIAN GULF

Dear as the wet diver to the eyes
Of his pale wife, who waits and weeps on shore,
By sands of Bahrein in the Persian Gulf ;
Plunging all day in the blue waves ; at night,
Having made up his toll of precious pearls,
Rejoins her in their hut upon the shore.
SIR EDWIN ARNOLD.

THE pearl fisheries of the Persian Gulf are the most famous and valuable in the world, and have been prosecuted for more than two thousand years. A translation by that eminent Assyriologist, Jules Oppert, of a cuneiform inscription on a broken obelisk, erected presumably by a king of Nineveh, seems to indicate a very early origin for these fisheries.[1] Professor Oppert's translation is :

In the sea of the changeable winds (*i.e.,* the Persian Gulf),
his merchants fished for pearls;
In the sea where the North Star culminates,
they fished for yellow amber.

The earliest writing of Europeans on the East refer to these fisheries. An account of them was given by the Greek writer Megasthenes, who accompanied Seleucus Nicator, the Macedonian general, in his Asiatic conquests, about 307 B.C. Shortly afterward they were noted by the Greek historian, Isidorus of Charace, in his account of the Parthian Empire. Extracts from Nearchus preserved by Arrian also mention them. Ptolemy speaks of the pearl fisheries which existed from time immemorial at Tylos, the Roman name for the present Island of Bahrein. These resources were well known in the days of Pliny. In his "Historia Naturalis," Book IX, ch. 35, he says: "But the most perfect and exquisite [pearls] of all others be they that are gotten about Arabia, within the Persian Gulf." [2] Pliny states also

---

[1] Oppert, " L 'Ambre jaune chez les Assyriens."   [2] Holland's edition of 1601, p. 254.

(Book VI, ch. 25) that Catifa (El Katiff), on the Arabian coast opposite Bahrein, was the center of an important fishery.

In the ninth century these fisheries were noted by Massoudi, one of
the earliest Arabian geographers.[1]   In the latter part of the twelfth
century they were visited and described by the Spanish-Hebrew traveler, Rabbi Benjamin of Tudela.[2]   The Arabian traveler, Ibn Batuta,
wrote of them about 1336.[3]   In 1508 they were noted in the account
of Lodovico Barthema's expedition to the Island of Ormus.   According to him:

At three days' journey from this island they fished the largest pearls
which are found in the world; and whoever wishes to know about it, behold!
There are certain fishermen who go there in small boats and cast into the
water two large stones attached to ropes, one at the bow, the other at the stern
of each boat to stay it in place.   Then one of the fishermen hangs a sack from
his neck, attaches a large stone to his feet, and descends to the bottom—about
fifteen paces under water, where he remains as long as he can, searching for
oysters which bear pearls, and puts as many as he finds into his sack.   When
he can remain no longer, he casts off the stone attached to his feet, and ascends
by one of the ropes fastened to the boat.   There are so many connected with
the business that you will often see 300 of these little boats which come from
many countries.[4]

Shortly following the visit of Barthema, the Portuguese under
Albuquerque took possession of the principal ports of the Persian
Gulf, and they imposed heavy taxes on the pearl fishery throughout
the century of their retention.   While under their jurisdiction, the
fisheries were visited and described by J. H. van Linschoten in 1596,
who wrote:

The principall and the best that are found in all the Orientall Countries, and
the right Orientall pearles, are between Ormus and Bassora in the straights, or
Sinus Persicus, in the places called Bareyn, Catiffa, Julfar, Camaron, and
other places in the said Sinus Persicus, from whence they are brought into
Ormus.   The king of Portingale hath also his factor in Bareyn, that stayeth
there onlie for the fishing of pearles.   There is great trafficke used with them,
as well in Ormus as in Goa.[5]

This was the Ormus where the treasures of the Orient were gathered in abundance, the half-way house between the East and the West,
making it one of the greatest emporia of the world.   So renowned
was its wealth and commerce that it was a saying among the Portu-

[1] Reinaud, "Mémoire sur l'Inde," Paris,
1849.

[2] "Travels of Rabbi Benjamin of Tudela,"
London, 1783.

[3] Lee, "Ibn Batuta," 1829, p. 65.

[4] "The Travels of Lodovico di Barthema,
1503 to 1508," London, 1863, p. 95.

[5] "Discours of Voyages into ye Easte and
West Indies," London, 1598, folio, ch. 84.

Cargo boat in pearl fishery of the Persian Gulf

Huts of mats and palm leaves, the homes of the pearl fishermen at Menamah, Bahrein Islands, Persian Gulf

ARAB PEARL-DIVERS AT WORK IN THE PERSIAN GULF

guese, were the whole world a golden ring, Ormus would be the jeweled signet. It was built on an island, supported a population of 40,000 persons, and was particularly well situated as a distributing point for the pearls, which enriched the argosies of Portugal, and contributed so largely to

> the wealth of Ormus and of Ind,
> Or where the gorgeous East with richest hand
> Show'rs on her kings barbaric pearl and gold,

which Milton celebrates in "Paradise Lost." This wonderful Ormus, in the sixteenth and seventeenth centuries one of the wealthiest places in the world, is now only a fishing village of less than a hundred huts.

It was at Ormus, nearly a century later, in 1670, that the shrewd old jewel merchant, Tavernier, whose acquaintance with gems doubtless equaled that of any man of his time, saw what he called "the most beautiful pearl in the world"; not so much for its size, for it weighed only 48¼ grains, nor for its regularity in form, but because of its most wonderful luster.[1]

In describing the fisheries, which had been retaken by the Persians in 1622, Tavernier wrote in 1670, according to Ball's translation:

There is a pearl fishery round the island of Bahren, in the Persian Gulf. It belongs to the King of Persia, and there is a good fortress there, where a garrison of 300 men is kept. . . . When the Portuguese held Hormuz [Ormus] and Muscat, each boat which went to fish was obliged to take out a license from them, which cost fifteen *abassis* [$5.45], and many brigantines were maintained there, to sink those who were unwilling to take out licenses. But since the Arabs have retaken Muscat, and the Portuguese are no longer supreme in the Gulf, every man who fishes pays to the King of Persia only five *abassis,* whether his fishing is successful or not. The merchant also pays the king something small for every 1,000 oysters. The second pearl-fishery is opposite Bahren, on the coast of Arabia-Felix, close to the town of El Katif, which, with all the neighboring country, belongs to an Arab prince.[2]

During the century following Tavernier's time, the fisheries were vigorously prosecuted, owing to the impoverished condition of the reefs in India and America, and to the large demand for pearls, not only by the Oriental courts, but by the wealth and fashion of Europe. Except for the last four years, when the Ceylon fishery was very productive, throughout the eighteenth century the Persian Gulf was almost the only important source of supply for pearls. For several years following the reopening of the Ceylon fishery in 1796, that region

---

[1] See p. 457, for Tavernier's description of this gem.

[2] Tavernier, "Travels in India," London, 1889, Vol. II, p. 108.

diverted some of the attention which the Persian waters had been receiving, but it was not long before these regained their ascendancy.

In 1838, Lieutenant J. R. Wellsted, an officer in the British India service, reported that the fisheries of the gulf employed 4300 boats, manned by somewhat more than 30,000 men.[1] Of these boats, 3500 were from the Island of Bahrein, 100 from the Persian coast, and the remaining 700 from the Pirate Coast situated between Bahrein and the entrance to the Gulf of Oman. Lieutenant Wellsted estimated the value of the pearls secured annually as approximately £400,000, which is somewhat less than the average value of the output in recent years.

Twenty-seven years later, according to Sir Lewis Pelly,[2] who was in the Indian service from 1851 to 1877, there were 1500 boats at Bahrein, and the annual return from the whole fishery was £400,000, the same as previously reported by Wellsted. In 1879, the value of the output was estimated at £600,000 by the British Resident, Colonel Ross, and at £800,000 by Captain L. E. Durand, of the British Protectorate of the Persian Gulf. Owing to the increased market value, the average output in the last five years has amounted to approximately four million dollars annually. This refers to the local value only, which is greatly increased by the time the pearls leave the markets in Bombay and Bagdad.

The Persian Gulf is nearly 600 miles long, with an average width of somewhat more than 100 miles. The Strait of Ormus—thirty to sixty miles wide—connects it with the Gulf of Oman, which opens directly into the Arabian Sea. The depth of water rarely exceeds thirty fathoms. Oyster-reefs are well distributed throughout the gulf, and are in greatest abundance on the Arab side between the 24th and 27th degrees of north latitude and the 50th and 54th degrees of east longitude, at a distance of from a few hundred yards to sixty miles from the shore, and especially in the vicinity of the Bahrein Islands. The oysters are scattered over level areas of coral rock and sand, with depths ranging from two to eighteen fathoms.[3] The divers rarely descend in deeper water than twelve fathoms, notwithstanding that valuable pearls are apparently obtainable at greater depths.

Although the British Protectorate extends over the Persian Gulf, insuring the peaceful prosecution of the fisheries and the settlement of intertribal contentions by the government resident, the fisheries are under the regulations of the maritime Arab sheiks. The restrictions imposed by these, however, are principally with a view to collecting a revenue from each boat employed. The total amount realized thereby

[1] Wellsted, "Travels in Arabia," London, 1838, Vol. I, ch. 17, pp. 264, 265.

[2] In report to the Government of Bombay, dated December 15, 1865.

[3] Schlagintweit, "Nachrichtsblatt der deutschen Gesellschaft," Frankfurt-am-Main, 1883, pp. 153-156.

is unknown, but there is good reason for supposing that it is considerable.

The fisheries are carried on during the greater part of every year, presenting a strong contrast to the Ceylon fishery, which is prosecuted usually less than forty days, and in only about one year in three on an average. This is especially remarkable when it is considered that no particular care is taken of the Persian reefs and, except for certain tribal restrictions, the fishermen may work whenever and wherever they choose. Owing to the extended area over which the fishing is prosecuted and the existence of undisturbed breeding-oysters in the deeper waters, the reefs are not readily exhausted, notwithstanding the tens of millions of mollusks annually removed therefrom.

The fisheries are at their height from June to September, when nearly every person on the coast is interested in some capacity, if not in fishing, at least in furnishing supplies, cleaning shells, buying pearls, etc. In April and May the water on the deep banks is so cold that the fishermen confine their efforts to the more shallow areas. During the winter months, the cold weather and the northwesterly gales interfere with the work, except such as is prosecuted in the smaller bays and inlets.

The pearling operations are financed mostly by Indian *bunnias,* or traders, principally from Bombay, who furnish capital for equipment, supplies of food, etc., and who purchase the pearls in gross lots. These men bear very hard on the fishermen, furnishing the supplies and buying the pearls almost at their own prices; and the poor divers who explore the depths and secure the pearls derive from their exertions little more than the crudest necessaries of life, and are usually in debt to the traders.

The actual fishing operations are carried on mainly by the maritime tribes of Hasa and Oman, including those on the Pirate Coast. The inhabitants of the Bahrein Islands and the adjacent shores have been devoted to pearling from time immemorial; but the Wahabis of the Pirate Coast—the *Ichthyophagi* of Ptolemy's time—have more recently, under the persuasive influence of British gunboats and magazine-rifles, substituted pearling for their two-century inherited life of fanatical piracy. Referring to these people in his quaint sketches of Persia eighty years ago, Sir John Malcolm wrote: "Their occupation is piracy, and their delight murder, and to make it worse they give you the most pious reasons for every villainy they commit. They abide by the letter of the sacred volume, rejecting all commentaries and traditions. If you are their captive and offer all to save your life, they say, "No! It is written in the Koran that it is not lawful to plunder the living; but we are not prohibited from stripping the dead.' So saying they knock you on the head."[1] Most of the Wahabi pearlers congregate in the mat-hut settlements of Dobai, Abu Thubi, and Ras-el-Kheima, located at the mouths of creeks which form fairly good harbors for the small boats. The Batina coast also furnishes some pearl fishermen, these coming principally from Fujaira, Shenas, Sohar, Suaik, and Sib.

The headquarters for the pearling fleet are at Bahrein Island, the largest of the insular group bearing the same name, the islets of Moharrek, Sitrah, and Nissan completing the group. This is the early home of Chaldean civilization, and one of the traditional sources of the Phenicians, and whence came that fish-god who—according to the Babylonian myth—bore the ark over the deluge. This island, the center of the greatest pearl fishery in the world, is half-way down on the southern side of the Persian Gulf, and twenty miles from the mainland of "Araby the blest." It is about twenty-eight miles in length, and ten in width at the widest part. The population approximates 60,000, all Moslems, except about 100 Banyan traders from Sindh, India. The northern half of the island is described as of great beauty, being a garden of pomegranate, lemon, citron, and quince-trees, and especially the magnificent date-palms, with numerous springs furnishing an abundance of excellent fresh water. The principal settlement, Manama, with about 10,000 inhabitants, is poorly built, the houses consisting mostly of huts of mats and palm-leaves; yet it presents a better appearance than any other settlement along this coast.

The one great industry, and the center of all interest throughout this region, is the pearl fishery. The present conditions are precisely as Palgrave wrote in 1863: "It is from the sea, not from the land, that the natives subsist; and it is also mainly on the sea that they dwell, passing amid its waters the one half of the year in search of pearls, the other half in fishery or trade. Hence their real homes are the countless boats which stud the placid pool, or stand drawn up in long black lines on the shore, while little care is taken to ornament their land houses, the abodes of their wives and children at most, and the unsightly strong boxes of their treasures. 'We are all, from the highest to the lowest, slaves of one master—Pearl,' said Mohammed bin Thanee to me one evening; nor was the expression out of place. All thought, all conversation, all employment, turns on that one subject; everything else is merely by-game, and below even secondary consideration."[2]

According to recent returns, the Persian Gulf fisheries employ about

[1] Malcolm, "Sketches of Persia," London, 1827, p. 27.

[2] "Personal Narrative of Journey through Arabia," London, 1865, p. 100.

SHAH SULAIMAN (1647-1694)

From a Persian manuscript in the library of Robert Hoe, Esq.

AGHA MOHAMMED (1666-1725)

Founder of the present Persian dynasty

From a Persian manuscript in the library of Robert Hoe, Esq.

HIS IMPERIAL MAJESTY, MOHAMMED ALI, SHAH OF PERSIA
Wearing the Kajar crown

3500 boats,[1] large and small, of which 1200 of the best are owned at Bahrein, 700 on the coast of El Hassa from El Katar to Kuweit, and the remaining 1600 are from various parts of the gulf, and especially from the Pirate Coast east of El Katar. They measure from one to fifty tons. The smaller ones, with three to fifteen men each, work near the shores; the larger, carrying fifteen to thirty men, fish over the whole gulf, remaining out for weeks at a time. These craft are very picturesque with their artistic rigs and spoon-shaped sails, and when the fishery is at its height the scene is one of rare interest. The boats from Bahrein are of excellent construction made by native workmen using local materials, with home-woven sailcloth and rigging of twisted date-fiber. Each of the larger ones usually evidences a lingering trace of Semitic influence in its *kubait,* or figurehead, covered with skin of the sheep or goat sacrificed in the launching ceremonies.[2] The boats from El Hassa and the Pirate Coast are usually smaller and less substantial than those from Bahrein, the fishermen from the latter place far surpassing those of the mainland in civilization and industrial wealth.

The fleet is manned by approximately 35,000 fishermen. In addition to the *nakhoda,* or captain, who is often the owner of the boat, the crew consists of *ghoas* or divers, who are mainly Arabs and Sedees, and *sebs,* or rope-tenders, who are usually Bedouins or Persians and attend the divers and perform other duties. Many Hindus from India, and flat-nosed, sable-hued Negroes from the east coast of Africa find employment here. On each of the larger boats is a general utility man, known as *el musully,* literally the "prayer-man," who, in addition to various other duties, relieves those *sebs* who stop to pray.

Among the fishermen are all types and classes to be met with in this part of the world, with the usual contingent of the lame, the halt, and the blind. There are a number of fishermen who have been maimed and mutilated by shark bites. A surprisingly large number of men who have become totally blind engage in diving, and they usually do fairly well where the oysters are abundant on the reefs. And one or two unfortunate divers are reported who continue the work even though handicapped by the loss both of a leg and of eyesight, this interfering less with their diving than with their movements on land.

The fishery in this region owes absolutely nothing to modern civilization in the method of securing the pearls from the depth of the sea; it is carried on to-day practically as it was six hundred years ago, and probably has been without important variation for two thousand years.

[1] Lord Curzon reports 4500 boats, and some other authorities state 5000, but this probably includes a number of tenders.

[2] For this and some other data on the pearl-ing fleet of Bahrein we are indebted to the kindness of Dr. S. M. Zwemer, who has spent many years at the Bahrein Islands as a missionary.

Aside from a loin-cloth, the diver is devoid of clothing except that rarely, early in the season when *polypi* abound, he is enveloped in a cotton overall as a protection. Over each finger and thumb he wears a shield or stall (*khubaat,* or finger-hat), about two inches long, made of flexible leather, to protect the fingers from the sharp shells and coral-growths. As each fisherman usually wears out at least two sets of these shields each season, it will be seen that a very large quantity of them is required to supply the entire fleet.

The divers use stones on which they descend feet foremost. Although this is less spectacular than the method of diving practised by the natives of the South Sea islands, it enables the fisherman to reach the bottom more speedily and with far less effort. The diving-stones range in weight from thirty to fifty pounds each, depending largely on the depth of water and the weight of the fisherman. They are somewhat oval in shape, and have one end perforated to admit a rope. Immediately above the attachment is formed a loop, resembling a stirrup, to receive the diver's foot. When prepared for the day's work, each stone is suspended by a stout rope over outriggers projecting from the side of the boat, and by a slip-knot is temporarily held four and a half or five feet below the surface of the water. A very stout diver may have a stone affixed to his waist to overcome his greater buoyancy. Usually two divers use one stone together and descend alternately. Each one has an attendant in the boat who assists him in ascending, and looks after the ropes, baskets of shells, etc.

In preparing for descent, the fisherman takes hold of the rope from which the diving-stone is suspended, puts one foot in the loop just above the stone and places the other foot in the rim of a net basket, eighteen inches wide, made of coir rope. When ready, he signals his attendant, inhales several good breaths, closes his nostrils with a *fitaam* or nostril-clasp of flexible horn attached to a cord around his neck, raises his body somewhat above the surface to give force to the descent, releases the slip-knot retaining the stone, and sinks rapidly to the bottom. Immediately disengaging his foot from the stone, he throws himself in a stooping position on the ground and collects as many oysters as possible during the fifty seconds or more in which he is able to remain under water. When near his limit of endurance, he hastily gives a signal jerk to the rope attached to the basket, and the watchful attendant hauls him up as speedily as possible, the diver frequently quickening the ascent by hand over hand movement up the rope. When near the surface, he lets go of the rope and with his arms close to his body pops above the surface puffing and blowing. The contents of the net bag are emptied into a large basket by the attendant, and the dead shells and other refuse are separated from the live oysters and thrown back into the sea, the diver having worked too rapidly at the bottom to discriminate closely as to what he gathered.

In the meantime, the stone has been drawn up and suspended by the slip-knot in its customary position and the diving partner is resting at the surface preparatory to descending. Thus, diving alternately at intervals of five or six minutes, each fisherman descends thirty or forty times in an ordinary day's work. The number of oysters gathered at each descent depends on such conditions as their abundance, the depth and clearness of the water, etc. It ranges from none to fifty or more, but ordinarily ten or twelve is a good average. As the men commonly work on shares, the shells brought up by each diver or by each pair of divers are kept separate.

The best type of Arab divers are very careful of themselves, drying the body thoroughly with towels on coming out of the water, taking intervals of rest during the day's work; and even while in the water between dives they may enjoy the luxury of a cheroot or pipe, or possibly a cigarette may pass from mouth to mouth of several men.

When pursuing their work, the divers are abstemious. After devotions at sunrise and a light breakfast of perhaps dates or rice and coffee, they begin fishing. About noon they knock off for coffee, prayers, and an hour's siesta, and then resume work for several hours. When the day's work is over and they have faced Meccaward with the customary prayers, they rest and eat a substantial meal, commonly of dates and fish roasted over a charcoal fire.

In equal depths the Arab fishermen remain under water longer than those of India who resort to the Ceylon fishery, but this is partly counterbalanced by the latter descending somewhat more frequently. When preparing for a lengthy dive, the fisherman imbibes large quantities of air, opening his mouth and inhaling large volumes.

The length of time a diver remains submerged in the average depth of seven or eight fathoms rarely exceeds sixty seconds, although some may remain seventy, eighty, and even ninety seconds on special occasion. A fully substantiated instance is reported from Manaar of an Arab diver having remained 109 seconds in seven fathoms of water. This occurred April 13, 1887, and was witnessed and reported[1] by Captain James Donnan, the inspector of the fishery. Wellsted reports[2] a diving contest in the Persian Gulf in which only one man, of the hundreds who competed, remained down 110 seconds; the depth, however, is not noted.

There are numerous reports of much longer stays than these; in-

[1] "Reports by the Superintendent of the Fishery and the Inspector of the Pearl Banks," Colombo, 1887.

[2] Wellsted, "Travels in Arabia," London, 1838, Vol. I, p. 266.

deed, a study of the published evidence bearing upon the subject furnishes surprising results. Ribeiro wrote, in 1685, that a diver could remain below while two *credos* were repeated: *"Il s'y tient l'espace de deux credo."*[1] In his interesting account of the Ceylon fishery, Percival stated that the usual length of time for divers to remain under water "does not much exceed two minutes, yet there are instances known of divers who could remain four or even five minutes, which was the case with a Caffre boy the last year I visited the fishery. The longest instance ever known was of a diver who came from Anjango in 1797, and who absolutely remained under water full six minutes."[2] Le Beck says, that in 1797, he saw a diver from Karikal remain down for the space of seven minutes.[3] The merchant traveler, Jean Chardin, reported in 1711 that the divers remain up to seven and a half minutes under water: *"Les plongeurs qui pêchent les perles sont quelquefois jusqu'à demi-quart-d'heure sous l'eau."*[4]

In 1667, the Royal Society of London addressed an inquiry on this subject to Sir Philiberto Vernatti, the British Resident at Batavia in the East Indies. Vernatti's reply gave certain details regarding the Ceylon fishery, but did not touch upon the length of diving because, as he stated, he could not "meet with any one that can satisfy me, and being unsatisfied myself, I cannot nor will obtrude anything upon you which may hereafter prove fabulous; but shall still serve you with truth."[5] Two years later, and presumably after investigation, Vernatti reported: "The greatest length of time that pearl-divers in these parts can continue under water is about a quarter of an hour; and that by no other means than custom; for pearl-diving lasts not above six weeks, and the divers stay a great while longer at the end of the season than at the beginning."[6]

The anatomist Diemerbroeck relates[7] the case of a pearl diver who, under his own observation, remained half an hour at a time under water while pursuing his work; and this was seriously adopted without comment by John Mason Goode in his "Study of Medicine."[8] Ibn Batuta, "the Doctor of Tangier," wrote about 1336 that "some remain down an hour, others two hours, others less."[9] A still earlier writer,

[1] "Histoire de l'Isle de Ceylon," Amsterdam, 1701, ch. 22, p. 169.
[2] "An Account of the Island of Ceylon," London, 1803, ch. 3, p. 91.
[3] "Asiatic Researches," London, 1798, Vol. V, p. 402.
[4] Chardin, "Voyages en Perse," Paris, 1811, Vol. III, p. 363.
[5] Sprat, "History of the Royal Society," London, 1667, p. 169.

[6] Philosophical Transactions for 1669, No. 43, p. 863.
[7] Diemerbroeck, "Anatome Corporis Humani," Ultrajecti, 1672.
[8] Sixth American Edition, New York, 1835, Vol. I, p. 239.
[9] Reinaud, "Fragments Arabes," Paris, 1845, p. 126. Lee, "Ibn Batuta," London, 1829, p. 65.

Jouchanan ibn Masouiah,[1] in his book on stones, states that "the diver, when he dives, places upon his nose a *masfâṣa* lest water should enter into him, and breathes through the fissure, and remains under water for half an hour." According to Sebaldus Rau[2] this *masfâṣa* was an article resembling a hood or cap, which the diver placed over his nose. It was made of some impervious material and had a projection so long that it reached to the surface of the water. The same writer believes that this object was alluded to by Aristotle ("De part. animal.," *Lib.* II, c. 16), where he likens the trunk of the elephant to the instrument used by certain divers for aiding their respiration, so that they could remain longer in the water and draw in air from above the surface.[3] And here we cease pursuit of further records, lest our faith in recorded testimony be too severely tested.[4]

A superficial inspection of the above evidence, from the one or two hours noted by Ibn Batuta about the year 1336, to the half an hour of Diemerbroeck in 1672, the quarter of an hour of Vernatti in 1669, the seven and one half minutes of Chardin in 1711, the six minutes of Percival in 1803, to the 110 seconds of the present time, seems to indicate very clearly a gradual but somewhat remarkable decrease in the ability of the Asiatic divers, and that the pearl fishermen of the present day are very different creatures from their ancestors. And especially is this so when it is considered that the above records are not isolated reports selected for the particular purpose of showing a decrease in the length of diving; on the contrary they are authoritative and representative publications of their respective periods. We do not recall having seen in any report issued previous to 1675, an intimation that the limit of time was less than ten minutes.

However, a careful consideration of the subject leads to the belief that there has been no serious decrease in the length of time that the Arab and Indian divers remain under water, and that either the writers were misinformed or that the individual cases reported were extremely exceptional. Ibn Batuta's instance of one to two hours

---

[1] A Christian physician who lived in the time of the Khalif Wathek Billa, about 842 A.D.

[2] "Specimen Arabicum," Traiecti ad Rhenum, 1784, p. 64.

[3] *Ibid.*, p. 65.

[4] Writers describing the early pearl fisheries on the American coast, and especially at Cubagua on the present coast of Venezuela, also reported very lengthy stays. In 1526, Gonzalo Fernández de Oviedo y Valdés wrote: "The thing that causeth men most to marvel is to consider how many of them can remain at the bottom for the space of one whole hour, and some more or less, according to expertness." ("Natural Historia de las Indias," Toledo, 1526.) About 1588, the Jesuit priest José de Acosta wrote: "I did see them make their fishing, the which is done with great charge and labor of the poor slaves, which dive 6, 7, yea 12 fathoms into the sea . . . ; but yet the labor and toil is greatest in holding their breath, sometimes a quarter, yea, half an hour together under water." (Acosta, "Natural and Moral History of the Indies," Hakluyt Society, 1880, p. 227.)

could easily be caused by a mistake in copying Arabic manuscript, or in the translation. The case related by Diemerbroeck in which a pearl diver remained submerged half an hour, is more perplexing, especially as the physician reports that this was done under his own observation. The numerous reports of five or six minutes may have been based on a very exceptional case.

These statements are viewed as highly incredible by men who have spent scores of years at the fisheries. A man may remain submerged for several minutes, but the conditions are vastly different from the activities of pearl-gathering at a depth of ten fathoms, where the pressure of the water is nearly thirty pounds to the square inch, and the slightest exercise is fatiguing. Unless the time is taken by a watch, it is easy to overestimate the stay; the seconds pass very slowly when one is waiting momentarily for the appearance of the diver's head above the water, and certainly to the nearly exhausted fisherman with straining chest and palpitating heart, the last few seconds must seem extremely long indeed. An instance is noted in which an Arab diver remained submerged seventy-one seconds, and on his reappearance, naïvely inquired if he had not been down ten minutes. It seems doubtful whether the 110 seconds herein noted has been greatly exceeded, in recent years at least, by Arab or Indian divers, who do not appear to equal the semi-amphibious natives of the South Sea islands in their exploits.

One of the most curious features of the pearling industry is the manner in which the fishermen secure supplies of drinking water. In the vicinity of Bahrein, numerous fresh-water springs exist at the bottom of the gulf in depths of two or three fathoms, and the fishermen dive into the depth of the salt water down to where the fresh water is springing forth and there fill a skin or other suitable receptacle which they bring to the surface. By running a pipe down near the bottom in the vicinity of one of these springs, an abundance of fresh water may be pumped into the boat.

Three species—or at least three varieties—of pearl-bearing oysters are obtained in the Persian Gulf. These are known locally as *mahar,* *sudaifee,* and *zinni.* Of these, the *mahar* or Lingah oyster, which corresponds to the Ceylon pearl-oyster, yields the greatest quantity of pearls, and those of the finest quality. It measures three or three and a half inches in diameter, and is found in deeper water than the others. The *sudaifee* and the *zinni,* which are larger, yield pearls in much smaller quantities than the *mahar.*

On large boats, which remain out for two or three weeks at a time, the oysters are left on deck overnight, and the following morning they are opened by means of a curved knife (*miflaket*), four or five inches in length. The smaller boats working near shore convey the catch to the land for the opening and searching for pearls.

The Persian Gulf pearls are commonly not so white as those from Ceylon, but they are found of larger size, and it is believed in Asia that they retain their luster for a greater length of time. Many of the Persian Gulf pearls, especially those from *sudaifee* and *zinni* shells, have a distinctly yellow color. Tavernier made a curious explanation of this. He stated:

As for the pearls tending to yellow, the color is due to the fact that the fishermen sell the oysters in heaps, and the merchants awaiting sometimes up to 14 or 15 days till the shells open of themselves, in order to extract the pearls, some of these oysters lose their water during this time, decay, and become putrid, and the pearls become yellow by contact. This is so true that in all oysters which have retained their water, the pearls are always white. They are allowed to open of themselves, because if they are opened by force, as we open our oysters in the shell, the pearls may be damaged and broken. The oysters of the Manar Strait open of themselves, 5 or 6 days sooner than those of the Gulf of Persia, because the heat is much greater at Manar, which is at the tenth degree of North latitude, while the island of Bahrein is at about the twenty-seventh. And consequently among the pearls which come from Manar there are few yellow ones found.[1]

Tavernier was more familiar with the pearls themselves than with the methods of the fishery. The yellow color is not due to contact with the putrefactive flesh, and is independent of the manner of opening. In fact, if putrefaction caused the yellow color, this shade would be far more prevalent in the Manaar or Ceylon pearls than in those from Bahrein, for practically all of the Ceylon oysters are permitted to putrefy, whereas only a portion of those in the Persian Gulf are opened in this manner. Furthermore, notwithstanding that it is nearer the equator, the heat at Manaar during the pearling season is not to be compared with that at Bahrein when the season is at its height, for the Persian Gulf during July and August is notorious as one of the hottest places on the globe.

While the great bulk of the pearls are either white or yellowish, these fisheries yield a few pink, bluish, gray, and occasionally even black pearls. These unusual colors are not especially prized. A curious and remarkably detailed story has gone the rounds in which the qualities of Persian and Ceylon pearls are compared, to the disparagement of the latter, and during the last hundred years few accounts have been published of this fishery without recording it. We notice it first in Morier's "Journey through Persia in 1808 and 1809,"[2] but

[1] Tavernier, "Travels in India," Ball edition, Vol. II, pp. 114, 115. [2] London, 1812, p. 55.

possibly it antedated that report. The statement is that the pearls of Ceylon peel off, while those of Persia are as "firm as the rock on which they grow"; and though they lose in color and luster one per cent. annually for fifty years, they still lose less than those of Ceylon, and at the expiration of the fifty years they cease to diminish in appearance.

The pearl output in the Persian Gulf at the present time appears from the official returns to exceed four million dollars annually at local valuation. The exports in 1903 were reported at £827,447, and in 1904, £1,077,241. It is generally understood that all of the pearls are not entered in the official figures, and the valuations in the markets of Asia and Europe are greatly in excess of these amounts. The profits of the fishery are divided among a great number of persons. A large percentage goes to the shrewd *bunnias* from India, who finance the fishery operations, and who, by all sorts of tricks connected with advances of supplies, valuation of the catch, etc., manage to make a very good thing out of the business. It is nothing unusual for the valuation of a lot of pearls to double and even treble after leaving the hands of the fishermen.

While many of the gulf pearls—and especially of the small seed-pearls—go to Bagdad, the great bulk of them are sold to representatives of Hindu and Arab merchants of Bombay for shipment to that city, which to the Bahrein fisherman is the heart of the outside world. Few of the pearls go directly into Arabia or Persia, as the certain sale in the larger Bombay market is preferable to a sometimes higher but less regular price in other markets. Indeed, pearls may usually be purchased at a less cost in India than a stranger would be obliged to pay at Bahrein. The Bombay merchants "sow the earth with Orient pearl," dealing direct with London, Paris and Berlin, and with the oriental jewelers. Most of the yellow pearls find oriental purchasers, with whose dark complexions they harmonize better than the silvery white ones. They are also more popular because of a belief existing throughout the East that they are less likely to lose their luster with the lapse of years.

The shell of the pearl-oysters is not used locally, but large quantities are exported to Europe for manufacture. Although it is the smallest and cheapest produced in the gulf, yet, owing to the enormous quantity taken for their pearls, the shell of the *mahar* (*Margaritifera vulgaris*) constitutes the bulk of the exports. Formerly most of the shipments were made from the harbor of Lingah, hence it is known in the markets of Europe as "Lingah shell." But in the last three or four years, much of it has been transported to Europe via Bander Abbas and Bushire. A German firm at Bahrein is extensively employed in exporting this shell, and several Indian merchants are also engaged in the trade. The total exports in 1906 amounted to 3262 tons, valued at $26,408 according to the port returns, but worth about $135,000 in Europe. Very large quantities are received in London, and over 2500 tons have been offered at auction in a single year. This shell is very small, averaging about three inches in diameter and about one and a half ounces in weight. It is the cheapest of all mother-of-pearl. The best quality sells in London for ten to twenty shillings per hundredweight, but the ordinary grade is worth usually less than nine shillings, and sometimes as low as three shillings per hundredweight. America formerly imported it, but few lots have been received since the exploitation of the Mississippi shell about fifteen years ago.

The shell of the larger species of pearl-oysters in the Persian Gulf is worth considerably more than the "Lingah shell," selling in Europe for £12 to £60 per ton, yet manufacturers consider it as furnishing only poor qualities of mother-of-pearl. Several hundred tons are exported annually. It measures six or seven inches in diameter and is used principally in making cheap grades of buttons.

THE PEARL FISHERIES OF CEYLON

Errors, like straws, upon the surface flow ;
He who would search for pearls must dive below.
DRYDEN, *All for Love*, Prologue.

SECOND in extent to those of Persia only, are the intermittent and uncertain pearl fisheries of the Gulf of Manaar. This is an arm of the Indian Ocean, from 65 to 150 miles in width, separating the island of Ceylon from the southernmost part of India. The pearl-oyster banks —known locally as paars—are situated off the northwest coast of Ceylon and also in the vicinity of Tuticorin on the Madras coast of the mainland. The Ceylon fisheries are under the control of the colonial government of the British Empire, and those of the mainland are monopolized by the Madras government. Notwithstanding the fact that they are outside of the three-mile limit established as the bound of national jurisdiction, exclusive privileges are exercised over these fisheries by the respective governments,[1] and poaching vessels are liable to seizure and punishment.

Though possibly not so ancient as those of Persia, the Ceylon pearl fisheries are of great antiquity. References to them occur in Cingalese

[1] See *infra.*, p. 125.

records dating from 550 B.C. Pliny, Ptolemy, Strabo, and other ancient writers speak of their importance.

The "Periplus of the Erythræan"—written about the end of the second century A.D.—refers to these fisheries, and states that, owing to the dangers involved, it was customary to employ convicts therein. In the days of the "Arabian Nights," under the name "Serendib," this was the scene of the pearling adventures of Sindbad the Sailor, and the reputation of the valuable pearl resources is reflected in those wonderful tales.

The first extensive description we have of the Gulf of Manaar fisheries was given by the Venetian traveler, Marco Polo, who visited the region about 1294. He wrote:

The pearl-fishers take their vessels, great and small, and proceed into the gulf where they stop from the beginning of April till the middle of May. They go first to a place called Bettelar, and then go 60 miles into the Gulf. Here they cast anchor and shift from their large vessels into small boats. You must know that the many merchants who go divide into various companies, and each of these must engage a number of men on wages, hiring them for April and half of May. Of all the produce they have first to pay the king, as his royalty, the tenth part. And they must also pay those men who charm the great fishes to prevent them from injuring the divers whilst engaged in seeking pearls under water, one-twentieth of all that they take. These fish-charmers are termined *Abraiaman;* and their charm holds good for that day only, for at night they dissolve the charm so that the fishes can work mischief at their will. These *Abraiaman* know also how to charm beasts and birds and every living thing. When the men have got into the small boats they jump into the water and dive to the bottom, which may be at a depth of from 4 to 12 fathoms, and there they remain as long as they are able. And there they find the shells that contain the pearls, and those they put into a net bag tied round the waist, and mount up to the surface with them, and then dive anew. When they can't hold their breath any longer they come up again, and after a little down they go once more, and so they go on all day. These shells are in fashion like oysters or sea-hoods. And in these shells are found pearls, great and small, of every kind, sticking in the flesh of the shell-fish. In this manner pearls are fished in great quantities, for thence in fact come the pearls which are spread all over the world. And I can tell you the King of that State hath a very great receipt and treasure from his dues upon those pearls.[1]

That quaint old missionary bishop, Friar Jordanus, in his "Mirabilia Descripta, or the Wonders of the East" (*circa* 1330), reports that "more than 8000 boats" were sometimes employed for three months continually in these fisheries, which were then prosecuted under the

[1] "The Book of Ser Marco Polo," London, 1871, Vol. II, pp. 267, 268.

THE "PRINCE OF PEARLS"; THE LATE RANA OF DHOLPUR IN HIS PEARL REGALIA

THE LATE MAHARAJAH OF PATIALA

jurisdiction of the Cingalese kings of Kandy, and that the quantity of pearls taken was "astounding and almost incredible."[1]

This number of boats seems entirely too large, especially in view of the fact that Jordanus secured his information at second hand; but it leaves the impression that the fisheries of that period were of great importance.

When the Portuguese, attracted by the wealth of its resources, obtained control of this region about 1510, they exacted from the local rulers an annual tribute in pearls and spices. Later they conducted the fisheries on their own account, permitting the native fishermen to retain one fourth of the catch as compensation for their work, and dividing the remainder into three equal portions, for the king, the church, and the soldiers, respectively.

Linschoten, who visited India about 1590, leaves this interesting account of the fishery at that time:

"There are also other fishings for pearle, as between the Iland of Seylon, and the Cape de Comoriin, where great numbers are yearlie found, for that the King of Portingale hath a captaine there with soldiers that looketh unto it; they have yearlie at the least above 3 or 4 thousand duckers [divers], yt live onlie by fishing for pearles, and so maintaine themselves." He describes the methods of fishing, which appear to be similar to those of the present time, and adds: "When they have made an end of the day's fishing, all the fishers with the captaine, soldiers, laborers and watchmen for the king, goe together, and taking all the pearls [pearl-oysters] that are caught that day they divide them into certaine heaps, that is, one part for the king, another part for the captaine and soldiers, the third part for the Jesuits, because they have their Cloyster in that place, and brought the countrie first into the Christian faith, and the last part for the Fishers, which is done with Justice and Equalitie. This fishing is done in the Summer tyme, and there passeth not any yeare but that divers Fishers are drowned by the Cape de Comoriin (which is called the King's fishing) and manie devoured by fishes, so that when the fishing is done there is great and pitiful noyse and cry of women and children heard. Yet the next yeare they must do the same work againe, for that they have no other means to live, as also for that they are partlie compelled thereunto by the Portingales, but most part because of the gaine."[2]

The best description we have seen of the Ceylon fisheries at the time of the Portuguese occupation, is that of Caesar Frederick, a Venetian trader, who referred to the period from 1563 to 1581. Frederick reported, according to Hickocke's translation in the Hakluyt edition:

[1] Jordanus, " Mirabilia Descripta," Hakluyt Society, 1863, p. 28.

[2] "The Voyage of John Huyghen van Linschoten to the East Indies," Hakluyt Society, 1884, Vol. II, pp. 133-135.

The sea that lieth between the coast which descendeth from Cao Comori, to the lowe land of Chilao, and from Island Zeilan, they call the fishing of Pearles, which fishing they make every yeare, beginning in March or April, and it lasteth fiftie dayes, but they doe not fishe every yeere in one place, but one yeere in one place, and another yeere in another place of the same sea. When the time of this fishing draweth neere, they send very good Divers, that goe to discover where the greatest heapes of Oisters bee under water, and right agaynst that place where greatest store of Oisters bee, there they make or plant a village with houses and a Bazaro, which standeth as long as the fishing time lasteth, and it is furnished with all things necessarie, and nowe and then it is neere unto places that are inhabited, and other times farre off, according to the place where they fishe. The fishermen are all Christians of the countrey, and who so will may goe to fishing, paying a certain dutie to the king of Portugall, and to the Churches of the Friers of Saint Paule, which are in that coast. All the while that they are fishing, there are three or foure Fustes armed to defend the Fishermen from Rovers. It was my chance to bee there one time in my passage, and I saw the order that they used in fishing, which is this. There are three or foure Barkes that make consort together, which are like to our litle Pilot boates, and a little lesse, there goe seven or eight men in a boate: and I have seene in a morning a great number of them goe out, and anker in fifteene or eighteene fadome of water, which is the ordinarie depth of all that coast. When they are at anker, they cast a rope into the sea, and at the end of the rope, they make fast a great stone, and then there is readie a man that hath his nose and his eares well stopped, and annointed with oyle, and a basket about his necke, or under his left arme, then he goeth downe by the rope to the bottome of the Sea, and as fast as he can hee filleth the basket, and when it is full, he shaketh the rope, and his fellows that are in the Barke hale him up with the basket: and in such wise they go one by one untill they have laden their barke with oysters, and at evening they come to the village, and then every company maketh their mountaine or heape of oysters one distant from another, in such wise that you shall see a great long rowe of mountaines or heapes of oysters, and they are not touched until such time as the fishing bee ended, and at the ende of the fishing every companie sitteth round about their mountaine or heape of oysters, and fall to opening of them, which they may easilie doe because they bee dead, drie and brittle: and if every oyster had pearles in them, it would be a very good purchase, but there are very many that have no pearles in them: when the fishing is ended, then they see whether it bee a good gathering or a badde: there are certaine expert in the pearles whom they call Chitini, which set and make the price of pearles according to their carracts [carats or weight], beautie, and goodnesse, making foure sorts of them. The first sort bee the round pearles, and they bee called Aia of Portugale, because the Portugales doe buy them. The second sorte which are not round, are called Aia of Bengala. The third sort which are not so good as the second, they call Aia of Canara, that is to say, the kingdome of Bezeneger. The fourth and last sort, which are the least and worst sort, are called Aia of Cambaia. Thus the price being set, there are merchants of every countrey which are readie with their money in their handes, so that in a fewe dayes all is brought up at the prises set according to the goodnesse and caracts of the pearles.[1]

A remarkable instance of the immutability of custom in the Orient is found in the fact that, except in a few minor particulars, Frederick's account, written more than three centuries ago, could serve as a description of the methods of the fisheries in recent years. The industry was then very extensive, as appears from an account shortly afterward (about 1608) by Pedro Teixeira, who reported[2] that from 400 to 500 boats were employed, and from 50,000 to 60,000 persons resorted to the fishery.

In 1658, possession of Ceylon and India passed from the Portuguese to the Dutch, who for a time continued the pearl fisheries after the manner practised by their predecessors; but owing to contentions as to the details of management, they soon resorted to leasing them each year to the highest bidder, or to several bidders, for a definite money payment. The successful bidders prosecuted the industry in the same manner as the government had previously done, employing the same native fishermen and compensating them with one fourth of the oysters secured. Under the Dutch rule the fisheries were very unprofitable, and particularly so during the last seventy years of their authority. There was practically no fishing from 1732 to 1746, and there was also a suspension—but not entirely from lack of oysters or of pearls—from 1768 until the territory passed into the control of the British in 1796.

The colonial government of the British Empire continued the Dutch policy of leasing, only restricting the limits of territory and season for fishing. Many objections were found to this method. It was difficult to regulate the business properly, and there were no reliable means of determining its proceeds and conditions. At length in 1835, the government began to operate the fishery on its own account, as the Portuguese had done two hundred years before, allowing the fishermen one fourth of the oysters taken by them and selling the remaining three fourths for the benefit of the treasury. In this way the full value of the resources was realized without mystery, deception, or concealment, and the plan worked satisfactorily for all concerned.

Owing, presumably, to the long period in which they had lain undisturbed, the Ceylon oyster reefs were in excellent condition at the beginning of British rule. In 1796 the government derived a revenue

[1] "Hakluyt's Voyages," Vol. V, Glasgow, 1904, pp. 395-397. Benjamin Franklin states that the Mediterranean divers, finding the light below obscured by the surface waves, used to let a little oil out of their mouths at intervals, which, rising to the surface, smoothed the waters. This might be a suggestion to modern marine and fresh-water pearl fishers.

[2] "The Travels of Pedro Teixeira," Hakluyt Society, 1902, pp. 174-181.

of Rs.1,100,000 therefrom, and in 1797 the revenue was Rs.1,400,000; these two years were by far the most productive during the first century of British occupation.

Several very interesting reports on the industry were prepared about that time. Especially to be noted among these were the accounts by Henry J. LeBeck in 1798;[1] by Robert Percival in 1803;[2] and by James Cordiner in 1807,[3] to which reference is made for detailed accounts of the fisheries of that period.

The Ceylon fishery was prosecuted about every other year from 1799 to 1809, and the annual returns ranged from £15,022 in 1801 to £84,257 in 1808. From 1810 to 1813, inclusive, there was a blank so far as receipts were concerned. In 1814 the fishery was very good, bringing in a revenue of £105,187. With the exception of very slight returns in 1815, 1816, and 1820, no oysters were then obtained until 1828. Excepting 1832 and 1834, the industry was prosecuted each year from 1828 to 1837, the revenue to the government averaging about £30,000 annually. Then came a long blank of seventeen years, for there was no fishing from 1838 to 1854, and likewise from 1864 to 1873. Indeed, so depleted had the beds throughout the Gulf of Manaar become in 1866, that serious consideration was given to the possibilities of securing seed oysters from the Persian Gulf for restocking the reefs; but fortunately this was rendered unnecessary by the discovery soon afterward of a few oysters on several reefs on both the Ceylon and the Malabar coasts.

From 1855 to 1863, and also from 1874 to 1881, the returns were only ordinary, the highest being £51,017 in 1863, and £59,868 in 1881, —the best year since 1814; and during these two periods fishing was entirely omitted in nearly one half the seasons. There were five lean years from 1882 to 1886, and the 1887 fishery was only fair, with a yield of £39,609. But the returns for 1888 were large, amounting to £80,424; and those for 1891 were even greater, being £96,370, representing a yield of 44,311,441 oysters. No oysters were caught from 1892 to 1902, inclusive. In 1903, the fishery was profitable, yielding 41,180,137 oysters, and the share of the government amounted to £55,303; and in 1904 the yield was almost the same, being 41,039,085 oysters and a revenue of £71,050 to the government.

In 1905 occurred the greatest fishery in the modern history of Ceylon. The season extended from February 20 until April 21, giving forty-seven working days, exclusive of Sundays and five days of bad

[1] "Asiatic Researches," London, 1798, pp. 393, et seq.
[2] "The Island of Ceylon," 1803, ch. 3.
[3] "Description of Ceylon," 1807, Vol. II, pp. 36-78.

weather, the longest period in over half a century.[1]   The boats employed numbered 318, with 4991 divers and 4894 attendant *manduks*. The yield of oysters exceeded all records, amounting to 81,580,716 in number, or nearly twice as many as in any previous year within the period of British occupation.   The prices at which these sold ranged from Rs.24 to Rs.124 per thousand, with an average of Rs.48.89 for the entire season.   The government received Rs.2,510,727 as its share of the revenue, which was twice as much as in any previous year since the British have been in control, and doubtless the largest received by any government in the history of the industry.   The oysters falling to the share of the divers must have sold for at least Rs.1,255,363 (since 1881 the divers have received one third of the catch as their compensation, instead of one fourth).   The profits of the merchants, who purchased and opened the government oysters as well as those of the divers, doubtless amounted to fully as much, making a total of Rs.5,021,453, or nearly $2,000,000 as a low estimate of the local value of the pearls secured at Ceylon in 1905.

Owing to the great success in 1905, an enormous number of persons flocked to the camp at the beginning of the season in 1906.   Employment was given to 473 boats, the largest number on record, and over 8600 divers were engaged, with an equal number of attendants. Owing to unfavorable weather and the great quantity of oysters removed in 1905, the catch in 1906 was less than in that record year, amounting to 67,150,641 in number, from the sale of which Rs.1,376,-746 was realized.   The prices covered a wide range.   For the large Cheval oysters, even Rs.276, Rs.291, and Rs.309 per 1000 were received.   The inferior, stunted oysters from the Muttuvaratu paar ranged from Rs.20 to Rs.41 per 1000, and even at these prices many buyers sustained losses.   On the other hand considerable money was made by the buyers of those from Cheval, in which some very large and beautiful pearls were found.

The results of the 1907 fishery were surprisingly good, excellent prices being obtained.   The proceeds from the sale of two thirds of the 21,000,000 oysters amounted to Rs.1,040,000, or just under $350,-000.   The fishery lasted thirty-six working days.   Only 173 boats were used, as it was considered that a fleet of this size is fully as large as can be employed advantageously to the greatest satisfaction of all interested.

According to the compilations of the colonial secretary's office, the gross revenue to the government from 1796 to 1907, inclusive,

---

[1] In 1881, the number of days was the same—47, the season extending from March 4 to April 27.   In 1891 there were 40 working days, in 1904 there were 33, in 1903 there were 36, and in 1906 there were 36 days of actual fishing.

amounted to £2,098,830. If to this be added the fishermen's share and the merchants' compensation, we have a total of about £4,200,000 or $21,000,000 as the local value of the pearls produced in Ceylon during the period of British occupation. The value of these in the markets of Asia and Europe was undoubtedly very much greater.

In many respects the Ceylon pearl fisheries are the most interesting in the world. Owing to their ready accessibility and thorough organization, they are far better known than any others. Reliable data exist as to the number of oysters taken during each season since 1854, and it is possible to estimate roughly the pearls obtained therefrom. Throughout the 112 years of British occupation, and previously to some extent under the successive rule of the Cingalese kings, of the Portuguese, and of the Dutch, for centuries, the reefs were annually examined by official inspectors, and fishing was permitted only in those years when they appeared in satisfactory condition.

A noticeable feature of these fisheries is their uncertainty, a prosperous season being followed by an absence of fishing sometimes extending over ten years or more. This is not of recent development. Over eight hundred years ago a total cessation of yield for a considerable period was recorded[1] by Albyrouni, who served under Mahmud of Ghazni. He stated that, in the eleventh century, the oysters which formerly existed in the Gulf of Serendib (Ceylon) disappeared simultaneously with the appearance of a fishery at Sofala in the country of the Zends, where previously the existence of pearls had been unknown; hence it was conjectured that the pearl-oysters of Serendib had migrated to Sofala.

In the 249 years since Ceylon passed from the dominion of the Portuguese in 1658, there have been only sixty-nine years in which the pearl fisheries were prosecuted. During the last century there were only thirty-six regularly authorized fisheries. Enormous quantities of oysters have appeared on the reefs, giving rise to hopes of great results, only to end in disappointment, owing to their complete disappearance. In the fall of 1887, for instance, examination of one of the reefs revealed an enormous quantity of oysters, covering an area five miles in length by one and a half miles in width, with "600 to 700 oysters to the square yard" in places. It was estimated by the inspection officials that there were 164,000,000 oysters, which exceeded the total number taken in the preceding sixty years, and which should have yielded several million dollars' worth of pearls in the following season, according to the usual returns. But some months later not an oyster was to be found on this large reef, the great host presumably having been destroyed by action of the sea. Numerous reasons are

[1] See Reinaud's "Fragments Arabes," Paris, 1845, p. 125.

assigned for the failure of promising reefs. Those most frequently heard are that the currents sweep the oysters away, that they are devoured by predaceous enemies, that they are covered by the shifting bottom, or that they voluntarily move to new grounds.

The oysters are found in well-known and permanently located banks or paars in the upper end of the Gulf of Manaar, in the wide shallow plateau off the northwest end of the island and directly south of Adams Bridge. The hard calcrete bottom is formed mostly of sand combined with organic remains in a compact mass and with more or less coral and shell deposits. The density of the water, as determined by Professor Herdman (to whose important and valuable report[1] we are indebted for much information), is fairly constant at 1.023, and the temperature has a normal range of from 82° to 86° F. during the greater part of the year. The charts and records refer to about twenty paars, but most of these have never yielded extensively, either to the English or to the Dutch. In the aggregate, they cover an area fifty miles in length and twenty miles in width. Most of them are from five to twenty miles from the shore, and at a depth of five to ten fathoms. The principal paars are Cheval, Madaragam, Periya, Muttuvaratu, Karativu, Vankalai, Chilaw, and Condatchy. Only three have afforded profitable fisheries in recent years, i. e.: Cheval, Madaragam and Muttuvaratu.

The other paars are of practically no economic value at the present time. They become populated with tens of millions of oysters, which mysteriously disappear before they are old enough for gathering. Especially is this true of the Periya paar, which is about fifteen miles from the shore, and runs eleven miles north and south, varying from one to two miles in width. Frequently this is found covered with young oysters, which almost invariably disappear before the next inspection, owing, probably, to their being covered by the shifting bottom caused by the southwest monsoon. The natives call this the "Mother paar," under the impression that these oysters migrate to the other paars.

The Ceylon government has given very careful attention to all matters affecting the prosperity of the pearl resources. It has maintained a "Pearl Fishery Establishment," consisting of a superintendent, an inspector and numerous divers, attendants, and sailors. The inspector examines the paars, determines when and to what extent they should be fished, and directs the operations. The superintendent conducts the work on shore, divides and sells the oysters, etc. The expense of this establishment has approximated $40,000 per annum when there has been a fishery, and about $22,500 without fishery expenses.

[1] "Pearl Oyster Fisheries of the Gulf of Manaar," 5 vols., London, 1903-1906.

It has been decided by naturalists that Ceylon oysters less than four years old produce very few marketable pearls; in the fifth, and again in the sixth year the value of the yield doubles, and in the seventh it is supposed to increase fourfold. Beyond that age there appears to be little increase, and there is the risk of the oysters dying, and of the pearls deteriorating or becoming lost. Eight years seems to be the natural limit of life. While experience has shown that the most profitable period for taking the pearl-oysters is when they are from five to seven years old, the mollusks are liable to disappear, especially after the fifth year, and the danger of waiting too long is as great as that of beginning too early. The fishing on any particular bank is determined by various circumstances and conditions, and is permitted only after careful examination.

The different beds are inspected from time to time, and no fishing is permitted until the condition of the pearl-oysters on the particular reef thrown open seems to warrant the most valuable returns. In the examination of a bed apparently in suitable condition, several thousand oysters—usually eight or ten thousand—are taken up and the pearls found therein are examined and valued. If they average Rs.25 or Rs.30 per thousand oysters, profitable results may be expected, provided there is a sufficient quantity of oysters on the bed. This method of determining the fishery is very ancient. Tavernier wrote, about 1650, "before they fish, they try whether it will turn to any account by sending seven or eight boats to bring 1000 oysters each, which they open, and if the oysters per 1000 yield five *fanos* or above, they then know the fishing will turn to account." [1] And much the same method was described by Ribeiro in 1685.

When it has been decided to hold a fishery, public notice is given by advertisement, stating which of the many paars or reefs will be open, and the estimated quantity of oysters to be removed, the number of boats that will be given employment, and the date for beginning the season and the length of time it will probably last. This notice is usually given in December preceding the fishery, and it is the signal for preparation by tens of thousands of persons in this part of Asia, and especially on the Madras and the Malabar coasts of India, and on the coast of Arabia. The fishermen, the merchants, and the multitude of artisans, mechanics, and laborers who contribute to the industry, set their homes and business in order so that they may attend. We give the notice issued in 1907, both in Cingalese and in English. [2]

Early in February the area to be gleaned is again examined, the limits of the oysters are charted and buoyed off, the number that may be obtained is estimated as accurately as possible, and valuation

[1] Tavernier, "Travels in India," Vol. II, ch. 21.   [2] See pp. 110, 111.

samples are collected. Several thousand oysters are taken up, the pearls are removed, examined, and valued by uninterested experts, and the results are published, so that prospective buyers may have a reliable idea as to their value. Otherwise this would not be possible until the merchants had washed some of their own purchases, which ordinarily would not be for a week or ten days after the opening of the season.

The fishery usually begins late in February or early in March, as the sea is then relatively calm, the currents least perceptible, and there is less danger of storms. It is prosecuted from a temporary settlement or camp on the sandy shore at a place conveniently near the reefs. The important fisheries of the five years ending in 1907, were centered at the improvised settlement known as Marichchikadde. Although prosecuted from the coast of Ceylon, relatively few Cingalese attend compared with the large numbers who assemble from India, Arabia, and elsewhere.

A week or two before the opening of the season, the boats begin to arrive, sometimes fifty or more in a single day, laden with men, women and children, and in many cases with the materials for their huts. In a short time the erstwhile desolate beach becomes populated with thousands of persons from all over the Indian littoral, and there is the noisy traffic of congregated humanity, and a confusion of tongues where before only the sound of the ocean waves was heard. Beside the eight or ten thousand fishermen, most of whom are Moormen, Tamils, and Arabs, there are pearl merchants—mainly Chetties and Moormen, boat repairers and other mechanics, provision dealers, priests, pawnbrokers, government officials, koddu-counters, clerks, boat guards, a police force of 200 officials, coolies, domestic servants, with numbers of women and children. And for the entertainment of these, and to obtain a share of the wealth from the sea, there are jugglers, fakirs, gamblers, beggars, female dancers, loose characters, with every allurement that appeals to the sons of Brahma, Buddha or Mohammed. Natives from the seaport towns of India are there in thousands; the slender-limbed and delicate-featured Cingalese with their scant attire and unique head-dress; energetic Arabs from the Persian Gulf; burly Moormen, sturdy Kandyans, outcast Veddahs, Chinese, Jews, Portuguese, Dutch, half-castes, the scum of the East and the riffraff of the Asiatic littoral, the whole making up a temporary city of forty thousand or more inhabitants.[1]

[1] The report of the Chief of Police at the 1905 fishery states: "In the camp there were 40,000 to 50,000 persons, of whom it may be said that not less than a tenth were gamblers, vagrants, and rogues, who, without occupation in their own country, made their way to Marichchukkadi with the hope of making money to gamble in oysters." ("Reports on the Pearl Fisheries for 1905," Colombo, p. 17.)

THE

Ceylon Company of Pearl Fishers,

LIMITED.

NOTICE

Is hereby given that a Pearl Fishery will take place at Marichchukkaddi, in the Island of Ceylon, on or about February 20, 1907.

The Banks to be fished are—

The Karativu, Dutch Moderagam and Alanturai Pars, estimated to contain 21,000,000 oysters, sufficient to employ 100 boats for twenty-one days with average loads of 10,000 each per day.

The North-West and Mid-West Cheval, estimated to contain 2,000,000 oysters, sufficient to employ 100 boats for two days with average loads of 10,000 oysters.

The Muttuvaratu Par, estimated to contain 8,000,000 oysters, sufficient to employ 100 boats for eight days with average loads as before stated : each boat being fully manned with divers.

2. It is notified that fishing will begin on the first favourable day after February 19. Conditions governing the employment of divers will be issued separately.

3. Marichchukkaddi is on the mainland, eight miles by sea south of Sillavaturai, and supplies of good water and provisions can be obtained there.

4. The Fishery will be conducted on account of the Ceylon Company of Pearl Fishers, Ltd., and the oysters put up to sale in such lots as may be deemed expedient.

இலங்கை முத்துக்குளிப்பு சங்கத்தார் லிமிட்டெட்.

————

1907-ம் ஹு பெப்ரவரி மீ 20-ந் திகதியிலாவது அதற்
கு சற்றே முன் பின்னைவது இலங்கைத்தீவிலுள்ள மரிச்சிக்
கடையில் முத்துக்குளிப்பு நடக்குமென்று இதனை யாவருக்
கும் அறிவித்தல் செய்யப்படுகிறது.

முத்துக்குளிப்பு நடக்கும் பார்களாவன:— காரத்தீவு,
டச்சுமொடராகம், ஆலந்தூல பார்கள். இப்பார்களில் நாளொ
ன்றுக்கு தோணி ஒன்றுக்குச் சராசரி 10,000 சிப்பிகள் வீதம்
100 தோணிகள் 21 நாள் முத்துக்குளிக்கப் போதுமான
21,000,000 சிப்பிகளிருக்கிறதென்று மதிப்பிடப் பட்டிருக்கி
றது

வடமேல், மேல்மத்திய செவல்பார். இப்பார்களில் தோ
ணி ஒன்றுக்குச் சராசரி 10,000 சிப்பிகள் வீதம் 100 தோ
ணிகள் 2 நாள் குளிக்கப்போதுமான 2,000,000 சிப்பிகளிருக்
கிறதென்று மதிப்பிடப்பட்டிருக்கிறது

முத்துவரற்றுப்பார். இப்பாரில் மேற் சொல்லப்பட்ட
விகிதப்படி 8 நாளுக்கு 100 தோணிகள் குளிக்கப்போதுமான
8,000,000 சிப்பிகளிருக்கிறதென்று மதிப்பிடப்பட்டிருக்கி
றது

2. பெப்ரவரி மீ 19-ந் திகதிக்குப்பின் குளிப்புக்கு அ
னுகூலமென்று காணப்படும் முதல் நாளிலேயே முத்துக்கு
ளிப்பு ஆரம்பமாகுமென்று இதனுல் யாவருக்கும் அறிவித்த
ல செய்யப்படுகிறது. முத்துக்குளிகாரர் சம்பந்தமாய் அனுச
ரிக்கப்படும் ஏற்பாடுகள் பிரத்தியேகமாய் விளம்பரமாகும்.

3. மரிச்சிக்கடை சலாபத்துறைக்குத் தெற்கே கடல்
மார்க்கமாய் 8 மைல் தூரத்துக்கப்பால் கண்டத்தில் கடற்கரை
போரமாயிருக்கிறது, அங்கு நல்லதண்ணீரும் உணவுக்குவே
ண்டிய பதார்த்தங்களும் கிடைக்கும்.

4. முத்துக்குளிப்பு இலங்கை முத்துக்குளிப்பு சங்கத்
தாரின் கணக்கில் நடத்தப்படும். உசிதமென்று தோன்றுகிற
தொகைகளாக சிப்பிகள் விற்கப்படும்.

A populous town springs up with well-planned and lighted streets and vast numbers of temporary abodes of all sorts, according to the means and the caste of the occupants, some of them just large enough for two or three persons to creep into. Although made mostly of poles, mats, *cajans* or plaited fronds of the cocoanut tree, they furnish ample shelter for the locality and season, the uncertainty of the fishery from year to year being sufficient argument against expensive and substantial buildings. Numerous wells and cisterns yield water for the use of all. Sanitary measures are strictly enforced, with a liberal use of disinfectants. At a considerable distance southward from the settlement are constructed the private *toddis,* or inclosures, for decomposing the oysters and washing the pearls therefrom. Nearer the camp or settlement itself are the police court, the jail, the bank, the post and telegraph offices, the auction room, the hospital and the cemetery—all to endure through a strenuous six weeks of toil and labor, of money-getting and gambling, and then the inhabitants "fold their tents like the Arabs, and silently steal away," leaving the debris to the shore-birds and the jackals.

The fishing fleet consists of several hundred boats[1] of various rigs and sizes. These are interesting on account of their picturesque appearance and also their remarkable diversity of types in hull and rigging: there is the broad and roomy Jaffna dhoney, commonly painted black; the lugger-like Paumben boat; the very narrow and speedy canoes,—not unlike the single masted bugeyes of the Chesapeake region—from Kilakarai and neighboring villages, most noticeable owing to their great number and their bright colors—red, green, or yellow; the clumsy looking, single masted Tuticorin lighters, sharp sterned and copper bottomed, the largest boats in the fleet, ranging in capacity from twenty to forty tons each; and, most singular of all, the three masted great canoes from Adirampatnam and Muttupat on the Tanjore coast, pale blue in color and with curved prow. In addition to these standard types, added novelty is imparted by a few boats of design so odd and fantastic as would be conceived only by the mind of an oriental builder.

Reaching the camp at the beginning of the season, these boats are examined by the officials as to condition and equipment and, if found satisfactory, are registered and numbered. When the quantity of oysters to be removed is small, many more boats may arrive than is necessary or than can find profitable employment. Formerly when this occurred a lottery was held to determine those to be employed. More recently the officials have endeavored to engage all boats passing

---

[1] In 1906 there were 473 boats employed; in 1905, 318; in 1857, 1858, 1859, and 1863, over 400 boats reported for employment.

the inspection, although to do so might necessitate arranging the fleet into two divisions, each fishing on alternate days. In 1874, the boats were arranged in three divisions, the red, blue and green, with fifty boats in each; in 1879, and again in 1881, there were two divisions, the red and the blue; and likewise in 1880, in 1903 and in 1906 there were two, the red and the white divisions. Of the 318 boats employed in the 1905 fishery, 143 were from Kilakarai, seventy-four from Jaffna, thirty-five from Tuticorin, thirty-four from Paumben, nine from Manaar, six from Negapatam, five from Colombo, four each from Tondi and Kayalpatam, and one each from Devipatam, Adrapatam, Ammopatam, and Koddaipatam.

The number of persons on each boat ranges from about twelve to sixty-five, with an average for the entire fleet of about thirty-five men per boat. This includes the *sammatti,* or master, who represents the owner; the *tindal,* or pilot; the *todai,* or water-bailer, who is very necessary on these leaky craft, and who also takes charge of the food and drinking water; at times a government inspector or "boat guard"; and from five to thirty divers, with an equal number of *manducks,* or attendants.[1] The *sammattis, tindals,* and *todais* are nearly all from the coast of southern India. The "boat guards" or inspectors are natives of Ceylon, and are employed by the government to prevent the fishermen from opening the oysters. Most of the *manducks* are from the Indian coast.

Of the 4991 divers employed in 1905, 2649 were Moormen or Lubbais from Kilakarai, Tondi, etc., on the Madura coast; 923 were Arabs; 424 were Erukkalampiddi Moormen from Ceylon, and the remaining 995 were Tamils from Tuticorin, Rameswaram and elsewhere on the Madras coast, Malayalans from the Malabar coast, with small numbers from other localities on the Asiatic coasts.

Among the 8600 divers in 1906, were 4090 Arabs, the largest number of those people employed in recent years. In 1905 there were only 923 Arab divers, in 1904 only 238, and previously the number was much less. Some have worked on the Ceylon coast since 1887, but most of them are newly arrived from Bahrein and Kuweit, where they received their training as pearl-divers. They are very energetic and skilful fishermen, far surpassing the Tamils, coming early in the season and staying late, and working on many days when rough seas deter the Indian divers from venturing out.

The Erukkalampiddi divers of Ceylon are by no means so energetic or steady in work as the Arabs, and commonly desert the fishery be-

---

[1] Some years ago, notably in the early sixties, each and every boat was required to have ten divers, thus making a total of 23 persons in each boat. (See Vane's "Report on Ceylon Pearl Fisheries," 1863.)

fore the close. The Tamil divers belong to the Parawa and Kadeiyar castes.

The season in the Ceylon fishery is very short, only about six or eight weeks at the most; and the holidays and storms usually reduce the number of actual working days to less than thirty. In no other pearl fishery of importance is the season less than four months in length, and in most of them it extends through more than half of the year. Owing to this restricted time, there is greater activity in the Ceylon fishery compared with the value of the output than in any other pearl fishery in the world.

Although the season is short, it is strenuous. Arising shortly after midnight, the thousands of fishermen breakfast, perform their devotions and prepare to get under way so as to reach the reefs about sunrise. There each boat takes its position on the ground allotted for the day's work, and which has been marked in advance by buoys topped with flags; and shortly afterward, on a signal from the guard vessel, the diving commences. This is carried on in the same manner as already described for the Persian Gulf, except that the Indian divers do not use nose-clips, only compressing the nostrils with the fingers during the descent. Rarely do they descend to a greater depth than ten fathoms.

The divers work in pairs, each pair using a single diving stone in common, and descending alternately, precisely as in the Persian Gulf. It is remarkable what few changes have occurred in the methods of the fishery in the last six centuries; the description[1] of Marco Polo, who visited the region about 1294, and of writers somewhat more recent, indicating that, in the main features, it was then conducted in the same manner as at the present time.

An exception to the usual mode of diving is practised by the Malay-alam fishermen, who, in some seasons—as in 1903, for instance—attend in large numbers from Travancore and northward on the Malabar coast. These men are rather low in skill and physical endurance.[2] They dive head foremost from a spring-board, and even with this assistance,—or possibly we should say, handicapped by this method,— they find the average depth of eight fathoms too great for them to work in with much comfort, rarely remaining under water longer than forty-five seconds.

The number of oysters secured on each visit to the bottom ranges from nothing to seventy-five or more, averaging between fifteen and fifty. This depends not only on the ability of the fishermen, but also on the abundance of oysters and the ease with which they may be collected. Sometimes they are held together in loose bunches of

[1] *Supra.*, p. 100.　　[2] Hornell, "Reports on the Pearl Fisheries of 1904," Colombo, p. 31.

Unloading oysters from the vessels into the kottus, at Marichchikadde, Ceylon

The pearling fleet on the shore at Marichchikadde, Ceylon

Hindu workmen preparing to drill pearls, Marichchikadde, Ceylon

Indian pearl merchants ready for business

Children of Persian pearl dealers

five to ten in each, and a diver can easily gather one hundred in the short length of time he remains submerged. In other localities they may be somewhat firmly attached individually to the bottom, so that some force is necessary to release them, thus reducing the possible quantity. Ordinarily one dive clears a space of several square yards.

Since 1904, a steamer has been employed each season by the government for dredging oysters in connection with experiments in oyster-culture. The officer in charge of this work concludes that "dredging is economically a more sound method of fishing than is diving."[1] This view is disputed by the superintendent of the fishery, who points out that the average catch by the steamer when dredging mature oysters only slightly exceeds that of an ordinary diving boat, and the cost of maintenance and operation is vastly greater.[2] A remarkable tribute to the skill of the nude divers, brought out by this discussion, is that, during some days when they were at work, the sea was too rough for dredging by the steamer, notwithstanding that she was a typical Grimsby or North Sea trawler of 150 tons measurement, built in 1896.[3]

A rough comparison of the Ceylon method of catching pearl-oysters with that practised by the American oyster-growers may not be un-interesting. On a basis of 400 to the bushel, the total Ceylon catch of 81,580,716 pearl-oysters in 1905 represents a trifle more than 200,000 bushels, or about the quantity annually produced by each of the half dozen leading oyster-growers of this country. Each one of these growers requires only about three steamers, at a total cost, maybe, of $25,000, and manned by twenty-five men; instead of one steamer at a cost of $25,000 and 318 diving boats manned by 10,000 men, which was the equipment in Ceylon. To be sure, the conditions under which the work is prosecuted are different—however, not so entirely unlike as might be supposed—and the American season is about six months long instead of the two months in Ceylon; but the comparison is presented simply as a suggestion of the possibilities of dredging on the Ceylon reefs.

Until 1885, one of the most novel features of the fishery was the employment of shark-charmers or "binders of sharks" (*kadal-kotti* in the Tamil language, *hai-banda* in Hindustani), whose presence was rendered necessary by the superstition of the Indian divers. The fishermen placed implicit reliance upon the alleged supernatural pow-ers of these impostors, resembling in some respects that reposed in the "medicine men" by the American Indians, and would not dive without their supervision. It is unknown at what period the influence of these

[1] "Reports on the Pearl Fishery for 1904," p. 7. [2] "Reports on the Pearl Fishery for 1905," p. 23.
[3] *Ibid.*, p. 22.

semi-priests developed, but at the time of Marco Polo's visit about 1294, they were in the full bloom of their authority, receiving one twentieth of the total catch of oysters,[1] which amounted to a very considerable sum. It is probable that the number of shark-charmers was then quite large, some writers more recently referring to one for each boat. During the Portuguese occupation the number was reduced to twelve, and at the beginning of the British influence, it was further reduced to two.

Interesting descriptions have been given of the methods by which these men exercised their alleged powers. In 1807, Cordiner stated:

One goes out regularly in the head pilot's boat. The other performs certain ceremonies on shore. He is stripped naked, and shut up in a room, where no person sees him from the period of the sailing of the boats until their return. He has before him a brass basin full of water, containing one male and one female fish made of silver. If any accident should happen from a shark at sea, it is believed that one of these fishes is seen to bite the other. The divers likewise believe that, if the conjurer should be dissatisfied, he has the power of making the sharks attack them, on which account he is sure of receiving liberal presents from all quarters.[2]

Amusing stories are told of the shrewdness displayed by these fellows in inventing explanations to redeem their credit when a fisherman became a victim of the sharks. These accounts are by men who evidently bore no good-will toward the shark-charmers, and it would be of interest to hear from the other side; but we have been unable to find any one who has appeared in print in their defense.

The British government, in its policy of noninterference with the superstitions or semi-religious customs of the natives, tolerated these seeming impostors, owing, probably, in a measure, to the fact that the superstitious belief in their necessity was favorable to the preservation of the resources, since it restricted poaching on the reefs. However, the government endeavored to prevent an extravagant misuse of the influence, and restricted the compensation of the shark-charmers to one oyster per day from each diver. Later, they were remunerated by the government, and were not allowed, under any pretense whatever, to demand, exact, or receive oysters or any other compensation from the boatmen, divers, or any other persons. And, finally, in 1885, the shark-charmers were done away with entirely, after having exacted their toll for upward of six centuries at least.

The dangers to which the Ceylon divers are exposed have been greatly exaggerated, and especially the risks from sharks. Poets tell

[1] "The Book of Ser Marco Polo," London, 1871, Vol. II, p. 267.

[2] Cordiner, "Description of Ceylon," Vol. II, p. 52.

how "the Ceylon pearler went all naked to the hungry shark," and the struggle of the diver has been a favorite theme with sensational writers. As a matter of fact, the trouble from this source is very slight, and the occupation is less dangerous than that of most of the deep-water fisheries, not to be compared, for instance, with that of the winter haddock-fishery off the New England coast. Even in 1905, when 4991 divers and an equal number of assistants were employed in pearling, not a single fatal accident was reported, and although much rough weather prevailed, not a fishing boat was lost. In the important fishery of 1904, with 3049 divers, only one fatal accident occurred, this was an elderly Moorman, whose death at the bottom was apparently due either to apoplexy or to exhaustion from remaining under water too long.

The superintendent of the fishery reported that not a single shark was seen during the 1904 season.[1] According to the statement of Sir William Twynam, whose Ceylon pearl fishing experience and observation equal those of any European, he has never known of a diver being carried off by a shark, and has heard of only one case—"which was a very doubtful one."[2] Prof. James Hornell, the inspector of pearl banks, reported in 1904: "During all the months I have spent upon the pearl banks during the last two years and a half I have never had a glimpse of a shark dangerous to man. Several times the boatmen have caught basking sharks of considerable size, but all were of a species that lives almost entirely upon small crustaceans."[3] The late Mr. A. M. Ferguson wrote in 1887: "I think it is pretty certain that in the whole course of the Ceylon fisheries only two human beings have fallen victims to these fierce fishes."[4]

The diving continues until a signal is given from the guard vessel about twelve or one o'clock, this time depending largely on the beginning of the sea breeze which roughens the water and interferes with the work, and likewise serves to speed the passage of the sail vessels to the shore. Occasionally the breeze is unfavorable, and the boatmen are obliged to row for miles, delaying their return in some instances until nightfall. Then the shore is lighted up to guide them to the landings, and extra precautions are maintained to prevent them from getting away with some of the oysters in the darkness.

It is claimed—and doubtless with much truth—that it is not unusual for the boatmen to take advantage of the time spent in reaching the shore to surreptitiously open many of the oysters and extract the pearls therefrom, throwing the refuse back into the sea. It would ap-

---

[1] "Reports on the Pearl Fisheries of 1904," p. 17.

[2] *Ibid.*, p. 17.

[3] *Ibid.*, p. 34.

[4] "Royal Asiatic Society Proceedings," 1887-1888, p. 100.

pear from some authorities that this is a general practice. One official
—and probably the one in the best position to know—reported in 1905
that more than 15,000,000 oysters, or nearly one fifth of the enormous
catch during that season, were illicitly opened.[1] However, this statement is strongly disputed by the superintendent of the fishery, who
states:

As a matter of fact the opening of oysters that goes on in the boats is of
a much more casual description than this. The divers occasionally pick out
some of the best looking oysters that happen to be conspicuous, or some that
open, and look inside them. It is quite possible that a valuable pearl might be
found in this way, but the chances are against it. It is hardly likely that the
divers would throw into the sea an enormous quantity of perfunctorily examined oysters in which they have a share and which contain pearls, while
they were aware that immediately on landing they could get good prices for
their shares.[2]

The government officials have endeavored to put a stop to whatever
looting may exist, searching boats and occupants at the shore, revoking
the license of any boat showing evidence of oysters having been opened
or carrying knives or other appliances for that purpose. The fishermen are alleged to resort to all sorts of devices to secrete their illicit
find of pearls, concealing them in the nose, ears, eyes, and other parts
of the body, and even hiding them in parcels in the furled sails or attached to the embedded anchor. In some seasons—as in 1904 and
1905—the government employed a guard for each boat. But serious
criticism has been made of the integrity of these guards, who, with
compensation of only one rupee per diem, could scarcely be expected
to resist the action of thirty or forty fishermen and report their doings,
when by silence they would have much to gain, and "the guards simply
add to the number of thieves on board" was reported by one superintendent.

Doubtless the most interesting sight in the Ceylon fishery is afforded by the return, about mid-afternoon, of the hundreds of novel,
sail-spreading boats running before the wind and crowded with tur-
baned fishermen dressed in their few brilliant rags, and each anxious
to be the first at the wave-washed beach, where they are welcomed by
an equal if not greater number of officials, merchants, laborers, and
camp followers, gathered on the shore to learn the result of the fishery.
The fantastic appearance of the boats, the diversified costumes of the
people, the general scene of animation, afford a view which for novelty
is rarely equaled even in the picturesque Orient.

The average number of oysters brought in daily by each boat is

[1] " Reports on the Pearl Fisheries of 1905," p. 40.   [2] *Ibid*., p. 24.

about 10,000. Some days when the weather is unfavorable many of the boats return empty; on other days they may have 25,000 or more. In 1905 the maximum catch in one day for one boat was 29,990, while in 1904 a single boat brought in 37,675 oysters. The catch by the entire fleet one day in 1905 was 4,978,686 oysters, or an average of 16,485 for each of the 302 boats out on that occasion.

Each person taking part in the fishery receives as his compensation a definite portion of the oysters. By government regulations, published in 1855 and yet operative, each *sammatti, tindal,* and *todai* receives daily one dive of oysters from each diver in the boat to which they are respectively attached. In some instances the hire of the boat is paid for in cash—about Rs.1.50 per day from each diver,—but in most cases either one fifth or one sixth of each diver's portion is devoted to this purpose. After these provisions have been made, each diver gives one third of his remaining portion to his *manduck,* retaining the balance for himself. The Moormen divers from Kilakarai commonly contribute one dive daily to the mosque of their native town,[1] in addition to the portions given to the *sammatti, tindal,* and *todai.* Previous to 1855, the Hindu temples of the Madras Presidency were allowed to operate a certain number of boats on their own account, but this led to so many abuses that it was abolished.

After the boats are run up on the firm, hard beach, all the oysters are removed by the crews of the boats into the government *koddu* or palisade, a large wattle-walled and palm-thatched inclosure with square pens, each bearing a number corresponding to that of each boat. This is done under close supervision to prevent a diversion of the oysters from the regular channels, which otherwise would be relatively easy among the animation and excitement caused by the thousands of persons about the landing-place.

Within the government inclosure, the oysters taken by each boat are divided by the fishermen themselves into three portions as nearly equal as possible. This applies not only to the oysters falling to the share of the divers and *manducks,* but also to those set apart for the *sammatti, tindals,* and *todais,* for hire of the boat and even for the Kilakarai mosque. An official indicates one of these as the share of the fishermen, who at once remove their portion from the inclosure through a narrow gate on the landward side. By this arrangement a satisfactory division of the oysters is secured and all cause for complaint or unfairness is removed. Previous to 1881, the fishermen received only one fourth of the catch as compensation for their work; but in that year their portion was increased to one third, at which it has since remained.

[1] "Reports on the Pearl Fishery for 1904," Colombo, p. 6.

As soon as the fishermen pass out of the government *koddu* with their quota, they are met by a crowd of natives eager to buy the oysters in small lots, and frequently at so many per rupee—ranging from eight to twelve ordinarily. This "outside market" is one of the many interesting features of the camp, for there are few persons on the shore who do not risk small sums in testing their fortunes in this lottery. And a wonderful lottery it is too, in which a man may risk a few coppers and win a prize worth hundreds of dollars. A poor Tamil once bought five oysters for half a rupee, and in one of them he found the largest pearl of the season. Any not sold among this eager, animated throng are at once marketed with a native buyer. The diver then hastens to immerse himself in one of the bathing tanks provided for the purpose. It is claimed that if this bath is omitted after immersion all the morning in the salt water of the gulf, the diver is liable to fall ill; and a sufficient supply of fresh water for this purpose is an important factor in the arrangement of the camp.

Owing to their sale in much smaller lots, or as we may say, at retail, the fishermen succeed in getting relatively high prices for their oysters, and their earnings exceed one half of the government's share. In 1905 this amounted to probably £86,000, or an average of about $1350 for each of the 318 boats. However, some crews made very much more than this, with a corresponding decrease for the others. Although 1905 was a record year for large returns, even in an ordinary season pearl fishing is relatively profitable, as a skilled diver earns five or six times as much as a common laborer in Ceylon. The regulations particularly forbid the employment of divers for a monetary consideration instead of for a share of the oysters according to the established custom.

The remaining two thirds of the oysters in the *koddu* are the property of the government. These are combined and counted. At nine o'clock each evening they are sold at auction, and by noon of the following day all have been removed, and the inclosure is ready for the incoming catch.

At the auction the number of oysters to be sold that evening is announced, and bids are invited. Some one starts the bidding at, maybe, Rs.20 or 25, and this is advanced by successive bids until the limit appears to be reached, which may possibly be Rs.50 or 60. The successful bidder is permitted to take as many oysters in multiples of 1000 as he chooses; and after he is supplied, other merchants desiring them at that particular price are accommodated. If there is no further demand for them at that price, the bidding on the remaining oysters is begun precisely as at first, and when the maximum bid is reached, all merchants willing to give that amount are furnished with as many as they wish in multiples of 1000 as before. If this does not exhaust the oysters, the bidding on the remainder is started up again, and so on until all are sold.[1] No one knows at the time whether he is buying a fortune in gems or only worthless shells.

The prices at which the oysters are sold at auction may differ greatly from the estimated valuation of the samples secured in the February examination. For instance, in 1905 the valuation of the South Mada-ragam oysters was Rs.17.86 per 1000, yet the auction sales on the first day began at Rs.53 and went up to Rs.61 per 1000, or three times the valuation; and about the same general proportion of increase prevailed for the oysters from the remaining banks, a result of great advances in the market for pearls.

The auction prices for the different lots and from day to day are fairly constant. But the shrewd Indian merchants know their busi-ness well and keep in close touch with the yield, so that there are many variations in the selling price that are puzzling to the uninitiated. A somewhat higher estimation is placed on the oysters from certain banks, and also on those from rocky portions of a particular reef, owing to their reputation for yielding a larger percentage of pearls. The estimation of particular oysters varies to some extent according to the amount of adhering rock and coral growth. As already shown, the prices in 1906 covered the remarkable range of from Rs.20 to 309 per 1000. Superstitious belief in luck also has its influence, and a buyer may consider a certain day as unfavorable for him and abstain from bidding on that occasion; or considering a particular day as lucky, he may bid very high to secure a considerable portion of the sales.

The prices in different seasons vary greatly. In 1860, the average was Rs.134.23 per 1000, which was unprecedentedly large; the nearest to this was Rs.79.07 in 1874 and Rs.49 in 1905. In 1880, the average price per 1000 was only Rs.11, which was the lowest ever recorded. The records for individual days greatly exceed these limits. The highest figures at which oysters have sold on any one day was Rs.309 per 1000 in 1906, the equivalent for each oyster of 10½ cents in American money. In 1874, the price reached Rs.210 per 1000, and in 1905, the maximum price was Rs.124, or about 4¼ cents for each oyster.

The oyster-buyers are principally wealthy Chetties from Madura, Ramnad, Trichinopoli, Parambakudi, Tevakoddai, Paumben, Kumbha-konam, and other towns of southern India. These are quite differ-ent from the scantily clothed Naddukoddai Chetties so common in Ceylon. Many of them are fashionably dressed in semi-European

[1] "Colonial Sessional Papers," 1904, Colombo, p. 653.

costume, with walking-stick, patent leather boots, and other evidences of contact with Europe. Smaller quantities of oysters are purchased by Moormen of Kilakarai, Ramnad, Bombay, Adrampatam, Tondi, etc. A few oysters are also purchased by the Nadans or Chánár caste people of Perunali, Kamuti, and Karakal. Over 99 per cent. of the 50,346,601 oysters sold by the government in 1905 were secured by Indian buyers, and less than one per cent. by Cingalese. A few of the oysters—from two to five per cent.—are sent to Indian and Ceylon ports, but most of them are opened at the fishing camp.

The purchaser of only a small number of oysters may open them at once by means of a knife, and with his fingers and eyes search for the pearls. By this method very small pearls may be easily overlooked, and it is scarcely practicable in handling large quantities of oysters. These are removed to private inclosures known as *toddis* or *tottis,* situated some distance from the inhabited portions of the camp; where, exposed to the solar heat, they are permitted to putrefy, and the fleshy parts to be eaten by the swarms of big red-eyed bluebottle flies, and the residue is then repeatedly washed.

Shakspere may have had in view some such scene as this when he spoke of the "pearl in your foul oyster." The lady who cherishes and adorns herself with a necklace of Ceylon pearls would be horrified were she to see and especially to smell the putrid mass from which her lustrous gems are evolved. The great quantity of repulsive bluebottle flies are so essential to success in releasing the pearls from the flesh, that a scarcity of them is looked upon as a misfortune to the merchants. However, except it may be at the beginning of a fishery, there is rarely ever a cause for complaint on this score, for commonly they are so numerous as to be a great plague to persons unaccustomed to them, covering everything, and rendering eating and drinking a difficult and unpleasant necessity, until darkness puts a stop to their activities. But the intolerable stench, impossible of description, the quintessence of millions of rotting oysters, fills the place, and makes existence a burden to those who have not acquired odor-proof nostrils. This animal decomposition seems almost harmless to health; indeed, the natives evidently thrive on it, and eat and sleep without apparent notice of the nauseous conditions. And yet vegetable decomposition in this region is usually followed by fatal results. Notwithstanding sanitary precautions and the usual quarantine camp and hospitals, cholera occasionally becomes epidemic and puts a stop to the fishery, as was the case in 1889; but this probably was due more to the violation of ordinary sanitary laws than to the decaying oysters.

In a large *toddi* the oysters are placed in a *ballam,* or a dug-out tank or trough, fifteen or twenty feet long and two or three feet deep, smooth on the inside so that pearls may not lodge in the crevices. This tank is covered with matting, and the *toddi* is closed up, sealed, and guarded for a week or ten days, when the fly maggots will have consumed practically all of the flesh tissues, leaving little else than the shells and pearls. The tank is then filled with sea water to float out the myriads of maggots. Several nude coolies squat along the sides to wash and remove the shells. The valves of each shell are separated, the outsides rubbed together to remove all lodgments for pearls, and the interior examined for attached or encysted pearls. The washers are kept under constant supervision by inspectors to prevent concealment of pearls; they are not permitted to remove their hands from the water except to take out the shells, and under no circumstances are they allowed to carry the hands to the mouth or to any other place in which pearls could be concealed.

After the shells have been removed, fresh supplies of water are added to wash the debris, which is turned over and over repeatedly, the dirty water being bailed out through sieves to prevent the loss of pearls. After thorough washings, every particle of the *sarraku,* or material at the bottom of the *ballam,* consisting of sand, broken pieces of shell, pearls, etc., is gathered up in a cotton cloth. Later the *sarraku* is spread out on cloths in the sun to dry, and the most conspicuous pearls are removed. When dry, the material is critically examined over and over again, and winnowed and rewinnowed, and after it seems that everything of value has been secured, the refuse is turned over to women and children, whose keen eyes and deft fingers pick out many *masi-tul* or dust-pearls; and even after the skill of these has been exhausted, the apparently worthless refuse has a market value among persons whose patience and skill meets with some reward. It is due largely to the extreme care in the search that so many seed-pearls are found in Ceylon.

And this leads to a discussion of what is commonly known in Ceylon as the "Dixon washing machine." This is an invention of Mr. G. G. Dixon who constructed it at Marichchikadde in 1904 and 1905, at a total cost to the government of about Rs.162,000,[1] including all expenses incidental to the experiment. The machine involves two separate processes; the first consists in separating the shells from the soft portion of the oysters, and the second in recovering the pearls from the resultant *sarraku* after it has been dried. In 1905, about 5,000,000 oysters were put through this machine,[2] but with what result has not been announced.

The shells having pearls attached to the interior surface are turned

[1] "Colonial Sessional Papers," 1906, Colombo, p. 330.

[2] "Reports on the Pearl Fisheries for 1905," Colombo, p. 25.

over to skilled natives, who remove the valuable objects by breaking the shell with hammers, and then with files and other implements remove the irregular pieces of attached shell and otherwise improve the appearance.

In no fishery in the world is the average size of the pearls secured smaller, nor is the relative number greater than in that of Ceylon. It is rare that one is found weighing over ten grains, and the number weighing less than two grains is remarkable. For roundness and orient they are unsurpassed by those of any region. However, Ceylon pearls worth locally Rs.1000 ($400) are by no means abundant. The most valuable one found in the important fishery of 1904, is said to have been sold in the camp for Rs.2500. The fishery of 1905 yielded one weighing 76½ *chevu,* and valued at Rs.12,000.

The quantity of seed-pearls obtained in the Ceylon fishery exceeds that of any other—probably all other parts of the world. The very smallest—the *masi-tul,*—for which there is no use whatever in Europe, have an established value in India, being powdered for making *chunam* for chewing with betel. Those slightly larger,—*tul* pearls—for which also there is no market in Europe, are placed in the mouth of deceased Hindus of wealth, instead of the rice which is used by poorer people.

The great bulk of the Ceylon pearls are silvery white in color, but occasionally yellowish, pinkish, and even "black" pearls are found, although the so-called "black" pearls are really brown or slate-colored. In some seasons these are relatively numerous, as in 1887, for instance.

Notwithstanding the large product at the fishery camp, it is difficult to purchase single pearls or small quantities there at a reasonable price, the merchants objecting to breaking a *mudichchu,* or the lot resulting from washing a definite number of oysters.

The shells obtained in the Ceylon fisheries do not possess sufficient thickness of lustrous nacre for use as mother-of-pearl, and are mostly used for camp-filling. A few are burned and converted into *chunam, i.e.:* prepared lime for building purposes, or to be used by natives for chewing with the betel-nut. Forty or fifty years ago, before the large receipts of mother-of-pearl from Australia and the southern Pacific, there was a good market for the shell for button manufacture and the like, but since 1875 only the choicest have been used for this purpose, and these are worth only about $25 per ton delivered in Europe.

It will be observed that up to the close of the season of 1906, the Ceylon fisheries were operated by the colonial government as a state monopoly. In 1904, proposals were made to the British colonial office by a London syndicate with a view to leasing the fisheries for a term of years. The original suggestion was that they should be leased for thirty years in consideration of an annual rental of £13,000 or Rs.195,-000, together with a share of the net profits after payment of a reasonable rate of interest on the investment; and later it was suggested that the rental be Rs.100,000 a year and twenty per cent. of the profits after seven per cent. on capital had been paid to the shareholders. But the government preferred a definite money payment without any rights to share in the profits realized; and after lengthy negotiations this was fixed at Rs.310,000 annually, with certain preliminary payments. Accordingly, on November 30, 1905, a preliminary agreement was executed between the crown agents for the colonies, acting on behalf of the government of Ceylon, and representatives of the Ceylon Company of Pearl Fishers, Limited. On February 27, 1906, this agreement was confirmed and made effective by special ordinance[1] of the governor and legislative council of Ceylon, and the crown agents were authorized to execute the lease as of January 1, 1906.

The principal financial terms of this lease required the company to purchase the expensive Dixon pearl-washing machine at a cost of Rs.120,000, which was Rs.42,000 less than it cost the government during the preceding two years; to purchase at a cost of Rs.62,501 the steamship *Violet*, which the government had used in its experimental oyster-culture; to reimburse the government each year the amount spent in policing, sanitation and hospital services at the fishery camp, which had in some individual seasons amounted to more than Rs.200,000; to expend each year from Rs.50,000 to Rs.150,000 in the development of pearl-oyster culture; and to pay an annual rental of Rs.315,000, a rate based roughly on the average return of the preceding twenty years, including the record year of 1905.

The company was authorized to take up the pearl-oysters by means of divers, or by steam dredges, or by such other mechanical means as might appear most advantageous, and to carry on such experiments with the immature oysters as appeared most conducive to the profitable working of the fisheries, provided they do nothing to make the resources less valuable at the expiration of the lease.

One of the most interesting features of the lease is that relating to the power of the colonial government to grant an exclusive right of fishing on the banks outside the three-mile limit. The question of this exclusive right arose in 1890, but was not conclusively determined. Fearing lest this authority did not exist, the terms in which the right of fishing was conveyed were carefully chosen by the attorney general to protect the government from liability "should any inter-

[1] Ordinance No. 8 of 1906.

national question arise";[1] and the government leased to the company "all the right or privilege which the lessors have hereto exercised and enjoyed of fishing for and taking pearl-oysters on the coasts of Ceylon between Talaimannar and Dutch Bay Point, to the intent that the company *so far as the lessors can secure the same* may have the exclusive right, liberty and authority to fish for, take and carry away pearl-oysters within the said limits. . . . But nothing in this lease shall be taken to make the lessors answerable in damages if *owing to any cause beyond the control of the lessors* the company is prevented from fully exercising and enjoying such exclusive right and privilege."[2]

In the meantime, while the negotiations were in progress, there occurred the very profitable fishery of 1905, from which the colonial government derived a revenue of Rs.2,510,727, or approximately eight times the proposed annual rental; and before the lease was finally concluded occurred the fishery of 1906, with its revenue of Rs.1,376,-746. While it is true that a succession of barren seasons prevailed from 1892 to 1902, yet, as the revenue in 1903 was Rs.829,548, and in 1904 it was Rs.1,065,751, there was, in the four years ending in 1906, a revenue to the government of Rs.5,782,772, or nearly as much as the total amount to be derived from the lease during the twenty years it was to run. These figures seemed to furnish strong reasons for retaining such a valuable source of revenue, with its possibilities of still greater expansion under the supervision and direction of specialists in the employ of the government.

Many of the inhabitants of Ceylon saw in this a decided objection to the lease, and there was a general feeling of indignation in the colony, with public meetings in protest, and the like. In reply to a memorial prepared at one of these meetings held in Colombo, Lord Elgin, the British secretary of state for the colonies, wrote under date of May 9, 1906:

The memorialists have protested against the lease on the double ground that a lease on any terms is contrary to the best interests of Ceylon, and that the rent agreed upon is "under existing circumstances wholly inadequate." There must always be in cases of this kind a difference of opinion as to whether a fixed annual sum, with immunity from all expense and sundry other advantages, is or is not preferable to continuing to face all the risks for the sake of all the profits. In the present instance the lease appears to me to have been drafted with a sincere desire to safeguard to the utmost the property and interests of the Colony.

It may be true that the development of the fishery upon a scientific system affords good prospect of a greater return in the future than has been obtained in the past, and affords at least the hope that the barren cycles which have

---

[1] "Ceylon Sessional Papers," 1906, p. 328.   [2] *Ibid.*, pp. 333, 335.

Street scene in Marichchikadde, the pearling camp of Ceylon

Return of the fleet from the pearl reefs to Marichchikadde, Ceylon

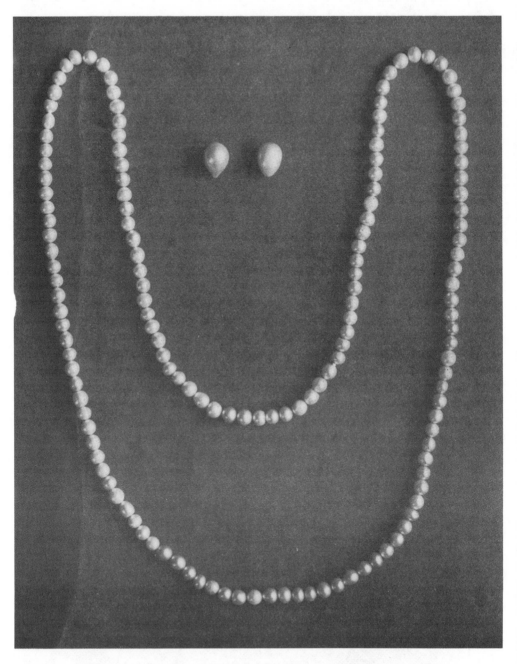

PEARLS PRESENTED BY THE IMAM OF MUSCAT TO PRESIDENT VAN BUREN
Now in the United States National Museum, Washington, D. C.

been so common in the past will not recur to the same extent. But the operations necessary to that end are of a highly technical and experimental character, and I am very doubtful whether any machinery which could be set in motion by the Government would be suited to develop processes at once so doubtful and so delicate. In twenty years' time the Colonial Government will receive back the fishery, not only intact, but in the most perfect state to which commercial enterprise and scientific methods can raise it, and, in the meanwhile, a regular and substantial payment is assured. Twenty years are no doubt a considerable period in the lifetime of individuals; but if within that time all the resources that science can contribute toward systematic development of the fisheries have been applied and thoroughly tested, the period will not, I think, be regarded as excessive or unfortunate in the history of a fishery which has lasted for more than two thousand years.[1]

The Ceylon Company of Pearl Fishers, Limited, with a paid up capital of £165,000, has just entered into possession of its lease, and it is uncertain what changes will be made in the methods of the fishery or what measure of success will follow the attempts at pearl oyster-culture and the growth of pearls. The attention of the pearling interests of the world is now directed to the work of this company in the development of its magnificent leasehold, and it seems not unlikely that greater changes will be made in the methods of the industry during the ensuing decade than have occurred in the whole of the last ten centuries.[2]

A curious fishery, with the *Placuna placenta* for its object, exists in Tablegram Lake, a small bay in northeastern Ceylon adjacent to the magnificent harbor of Trincomali, which Nelson declared to be "the finest in the world." At intervals during the nineteenth century, the Ceylon government leased the Tablegram Lake fishery to native bidders for a period of three consecutive years. In 1857, Dr. Kelaart visited the place and calculated that in the three years preceding, eighteen million oysters had been removed.[3] Owing to scarcity of the mollusk, no fisheries have existed since 1890, but from 1882 to 1890 they were regularly leased at an average of Rs.5000 for each term of three years. Prof. James Hornell, who made a careful examination in 1905, reported that if the business were carried on providently and systematically, "it should become the source of a fairly regular annual revenue to Government of from Rs.10,000 to Rs.12,000, possibly even more."[4]

[1] "Ceylon Sessional Papers," 1906, p. 650.

[2] The Government Commission has interdicted the fishing for this year (1908), as experts have reported the pearl-oysters were not plentiful enough and were also immature, being only five years old. The next fishery will be in 1909.

[3] Kelaart, "Report on the Tablegram Pearl-Oysters," Trincomali, 1857, 6 pp.

[4] Hornell, "Report on the *Placuna placenta* Pearl Fishery of Lake Tampalakamam," Colombo, 1906.

The *Placuna* oysters are caught by Moormen divers, who are scarcely equal physically to the pearl fishery in the sea. They rarely descend more than four fathoms, and most of the work in Tablegram Bay is in less than two fathoms. Each diver returns with from one to five or more oysters, depending on their abundance, and receives one half of the catch as his share of the proceeds. Unlike the method in the pearl-oyster fishery of Ceylon, the *Placuna* oysters are opened while fresh, this work being performed by coolies, who are compensated at the rate of about Rs.3 per 1000.

THE PEARL FISHERIES OF INDIA

There are two moments in a diver's life:
One, when a beggar, he prepares to plunge;
Then, when a prince, he rises with his prize.`
ROBERT BROWNING.

NOTWITHSTANDING the great fame of the pearl fisheries of India, those prosecuted within the limits of British India proper are of small extent. The only pearl resources within the empire are the rarely productive reefs on the Madras coast in the vicinity of Tuticorin, the relatively modern fisheries of Mergui Archipelago, and some small reefs of only local importance on the Malabar coast and in the Bombay presidency.

The celebrity of India in connection with the pearl fisheries has never rested on the extent of those within the territorial limits or under the control of this government. It originated in the fact that it is largely Indian capital which finances the fisheries of Ceylon and of the Persian Gulf; nearly all of the divers and others employed in Ceylon are from the coast of this empire, and most of the pearls are purchased by merchants of Bombay, Madura, Trichinopoli, and other large towns. Thus, from an economic and industrial point of view, the pearl fisheries of Ceylon, and to a less extent those of the Persian Gulf, have contributed to the fame and to the wealth of the Empire of India.

The pearl fisheries off Tuticorin in the Madras presidency have been referred to incidentally in the account of the fisheries of Ceylon. They are separated by only a few miles of water, and are prosecuted by the same fishermen and in precisely the same manner. Consequently, it is difficult to discuss them separately, especially in their early history and during the time that this part of the world was under the rule of the Portuguese and later of the Dutch.

The fisheries of the Madras coast compete in antiquity with those of Ceylon. Indeed, from the time of Ptolemy to the seventeenth century, the industry seems to have been prosecuted largely from the Mad-

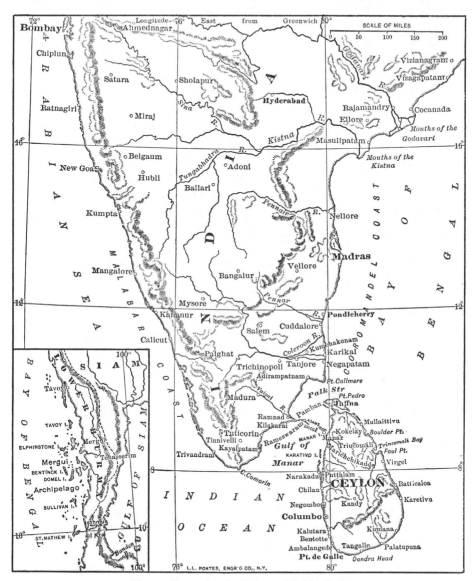

The pearling regions in Ceylon and British India

ras side of the gulf, centering at Chayl or Coil on the sandy promontory of Ramnad. This place appears to be the Κόλχοι of Ptolemy, the Ramana Koil of the natives, as well as the Cael of the travelers of the Middle Ages. But during the last three hundred years, the

Ceylon side has been the scene of the greatest pearling operations; and from the Madras coast, the fisheries have not been prosecuted except at long intervals, averaging once in fifteen or twenty years.

Owing to the scarcity of oysters and to other causes, the fishery was prosecuted on the Madras coast in only eight years of the whole period from 1768 to 1907. These years of productivity were 1822, 1830, 1860, 1861, 1882, 1889, 1890, and 1900; and even then the yield was relatively small. The largest was 15,874,500 oysters in 1860, from which the Madras government derived a revenue of Rs.250,276; and about half as many oysters were obtained in 1861 with a revenue of Rs.129,003. Numerous and prolonged experiments in conserving the reefs and in cultivating the oysters have been made without success. The reason usually given for the greater wealth of oysters on the Ceylon side is, that it is more sheltered from the strong currents which sweep down the Bay of Bengal into the Gulf of Manaar and impinge directly on the coast of the mainland.

The headquarters of the fishery are at Tuticorin, near to Madura, the Benares of the south, the holy "City of Sweetness" which the gods have delighted to honor from time immemorial. But the camp is commonly erected of palmyra and bamboo on the barren shore several miles distant from Tuticorin. The 1890 fishery was at Salápatturai, and that of 1900 at a place which received the mouth-filling name of Veerapandianpatanam.

The preparations for pearling at Tuticorin are similar to those on the Ceylon coast. In the autumn the reefs are examined by government inspectors, and if the conditions seem to warrant a fishery in the following spring, arrangements are made therefor and the proper notification issued. The announcement follows the general plan of that in Ceylon. The following, from the Fort St. George "Gazette," Madras, January 16, 1900, is a copy of the notification preceding the last fishery which has occurred:

Notice is hereby given that a pearl fishery will take place at Veerapandian-patanam on or about the 12th March, 1900.

1. The bank to be fished is the Theradipulipudithapar, estimated to employ 100 boats for twenty days with average loads of 7,000 oysters per day.

2. It is therefore recommended that such boat owners and divers as may wish to be employed shall be at Tuticorin on or before the 1st of March next and anchor their boats abreast of the government flagstaff; the first day's fishing will take place on the 12th of March, weather permitting.

3. The fishery will be conducted on account of Government, and the oysters put up for sale in such lots as may be deemed expedient.

4. The arrangements of the fishery will be the same as have been usual on similar occasions.

5. Payments to be made in ready money in rupees or in Government of India notes. Checks on the Bank of Madras or Bank Agencies will be received on letters of credit being produced to warrant the drawing of such checks.

6. All particulars can be obtained on application to the Superintendent of Pearl Fisheries, Tuticorin.

Tinnevelly Collector's Office,   Sd/—J. P. BEDFORD,
    16th November 1899.   *Collector.*

On the long sweep of desolate shore at a place convenient to the reefs, a temporary camp is erected, just as is done on the Ceylon coast. However, this camp is not nearly so large, only about one fourth or one fifth the size of that on the eastern side of the gulf. It resembles the larger one in the quarters for divers and merchants, the bazaars, the bungalows for the officials, the hospital, the sale and washing inclosures, etc.; in addition to these is the temporary Roman Catholic chapel.

The divers are mainly of the Parawa caste from Tuticorin, Pinnacoil, Pamban, etc. on the Madras coast. Although influenced by many Hindu superstitions, they are nominally Roman Catholics, as evidenced by the scapulars suspended from the neck, their ancestors having been converted and baptized through the zealous work of that prince of missionaries, St. Francis Xavier, in the sixteenth century. Even yet a chapel at Pinnacoil is held in special reverence by these people as a place where the saintly father preached. Professor Hornell writes that the present hereditary head of this caste is Don Gabriel de Croos Lazarus Motha Vaz, known officially as the Jati Talaiva More, or Jati Talaivan. He resides at Tuticorin, and is largely the intermediary between the government and the Parawa fishermen.

In the details of its prosecution, the Madras fishery differs in no important particular from that of Ceylon. The boats are manned and operated in precisely the same way; they fish in the morning only, taking advantage of the prevailing favorable winds; the divers carry the oysters into the government inclosure, and divide them into three equal lots, of which they receive one; the share of the government is auctioned daily, the divers disposing of theirs as they choose; and the oysters are rotted and washed in the same manner as in Ceylon.

In addition to the fishery for pearl-oysters at Tuticorin, two other species of pearl-producing mollusks are collected in the Madras presidency; one of these is a species of mussel (*Mytilus smaragdinus*, according to Dr. Edgar Thurston of the Madras Museum), which is collected from the estuary of the Sonnapore River near Berhampore;

and the other is the *Placuna placenta,* found in many places in this presidency, and especially in Pulicat Lake and in the vicinity of Tuticorin.

The Sonnapore mussels, which are small and bright green in color, are found adhering to the masses of edible oysters in depths of ten or twelve feet of water. They are caught in a novel manner, as described in a letter from the acting collector of customs at Ganjam. Thrusting a long bamboo pole deep into the bottom of the reef, the fisherman dives down, and holding on to this bamboo, breaks off as large a mass of the oysters as he can bring to the surface in one hand, helping himself up the bamboo pole with the other. Removing the mussels from the mass, he opens them with a suitable knife and by running his thumbs and fingers over the flesh tissues, detects the pearls therein. These pearls are of very inferior quality and of little ornamental value. They are sold mostly for chunám and for placing in the mouth of deceased Hindus.

Along the west coast of India, in the Bombay presidency, a few pearls are found at various places, but the output is of slight value. The most important of these is off the coast of Nawanagar, on the south side of the Gulf of Cutch, where the true pearl-oyster is found.

According to the "Jamnagar Diwan," the yearly value of the Nawanagar fisheries is about Rs.4000. This is smaller than formerly, as the reefs are in a depleted state; to give them a chance to recuperate, a close season was established in 1905. The oysters are found along a coast-line eighty miles in length extending from Mangra, near Jodya Bunder, to Pindera in the Gulf of Cutch, and also about the islands of Ajad, Chauk, Kalumbar, and Nora, which are also situated in the Cutch Gulf. They are not procured by diving, but are gathered off the rocks when the tide is out. During the monsoon, the collection is limited to eight days in the month; *i.e.,* from the twelfth to the fifteenth of each half according to the Hindu calendar.

The fisheries are by law restricted exclusively to the *waghers* of ten villages, which are Varinar, Sashana, Sika, Balachedi, Jhakher, Sarmat, Bharana, Salaya, Chudesar, and Bedi. The collection of the pearls is left entirely to these men, who at Divala—the Hindu new year—bring all the pearls gathered by them to the durbar. There an estimate is made of their value, one fourth of which is paid to the *waghers,* and the pearls are turned over to the representatives of the state treasury for sale. This method of conducting the industry has been long established. In recent years the government experimented in farming out the revenue, but the old custom has been resumed in order to placate the native fishermen.

A few pearl-oysters are also found on the Ratnagiri coast below

Bombay, and likewise at Kananur in the Malabar district. In 1901–1902, there was some local excitement about pearls found at Belapur and quantities were reported as collected; but since then little has been heard of the industry in that region.

Elsewhere on the west coast of India, pearls are obtained from the so-called "window-glass" shell, of the genus *Placuna*. The individual shells are flat, thin, and transparent, and are still used in Goa and vicinity as a substitute for glass in windows. This mollusk is abundant from Karachi, near the Baluchistan border, to the Kanara district south of Bombay; and wherever it occurs in any abundance it is collected for the sake of the small pearls found therein.

Of the fishery at Karachi, Mr. E. H. Aitken writes: "It is farmed out by Government for a good sum. In 1901, the amount realized was Rs.3650 for a period of three years; but the lessee lost heavily, and in 1904 the highest offer for a similar period of three years was Rs.1851. Pearls may be found in as many as ten to twenty per cent. of the mature mollusks." Pearls are far more numerous in the *Placuna* than in the pearl-oysters, but few of them are of sufficient size or luster to be used as ornaments, ranking with the so-called medicinal pearls of Europe. They are much softer in texture than the pearls of the Margaritiferæ. The largest are commonly of irregular form, with the surface slightly botryoidal or like the "strawberry" pearls of the Mississippi. While not often used as ornaments, they are highly valued by the Hindus in calcined or powdered form for medicinal purposes, and especially to be chewed with the betel-nut, and are also used in the original form in funeral rites, a small quantity being placed in the mouth of a deceased person.

In the Mergui Archipelago, which is within the territory of lower Burma and under the jurisdiction of the government of British India, patches of pearl-oyster reefs are scattered over an area roughly computed at 11,000 square miles, taking 97° 40′ as the western boundary. They occur principally in the strong tidal passages among the islands. The bottom is formed largely of porphyritic granite interspersed with sand and thinly covered with corals, coral cups, the long whip-like black coral (*Antipathes arborea*), and other submarine animal and vegetable growths.[1] These constitute a home most favorable to the growth and development of molluscan life.

Of the several species of pearl-bearing mollusks occurring in the Mergui Archipelago, by far the most important is the "mok," or large Australian pearl-oyster (*Margaritifera maxima*). The shell attains a maximum size of about thirteen inches in diameter, and the nacre is of a milky or silvery color. This species occurs in its two varieties of

[1] Jardine, "Report Relating to the Mergui Pearl Fisheries," Rangoon, 1894, p. 6.

"golden lip" and "silver edge," the former being in greater abundance. The "silver edge" shell is the more valuable owing to its uniformity of coloring, and the pearls found therein are of superior luster and orient.

The "pate goung," or Lingah pearl-oyster (*Margaritifera vulgaris*), is similar to that of the Gulf of Manaar. It is circular in shape and measures about two and one-half inches in diameter. The nacre is silvery, with slight yellowish tinge. Many of the pearls from this species are of a silvery color, but most of them are yellowish or golden. The fishery for this mollusk is of little importance compared with that for the larger pearl-oyster, which is the species referred to in Mergui when not otherwise mentioned.

The pearl fisheries of Mergui originated with the Selangs or Salangs, a nomadic race of maritime gipsies, the last remnants of whom live among the three thousand islands of this group. They are supposed to be of Malay descent; but their early history is unknown, and they are rapidly passing away in the conflict of existence with the neighboring peoples. Probably in no part of the world are the pearl fisheries prosecuted by a more primitive class of men. With their women and children, they live mainly in roomy dug-out boats; but during the southwest monsoon they erect temporary shelters on the shore, these consisting of a few frail sticks, supporting coverings of braided mats, and floors of bamboo strips.

They have few wants and derive a livelihood principally from gathering and bartering shells, pearls, cured *thadecon,* and nests of the sea-swallow (Collocalia). Within depths of six or eight fathoms they are fairly good divers, both the men and the women, but their physical endurance is slight. Their trade is mostly with Chinese merchants who visit them in small vessels. No information exists as to when the Selangs first found profit in searching for pearls; but it was probably many centuries ago, and for a long time they made contributions of them to the Buddhist rulers of Burma.

Shortly after the acquisition of Mergui Archipelago in 1826, representatives of the British government brought experienced divers from southern India to examine more fully the resources which the Selangs had made known; but as only seed-pearls were secured, the government concluded that they would yield an insignificant revenue, and the attempt to develop these resources was given up.[1]

However, the Selangs continued to fish in their primitive fashion; and as the market for the shell developed, the profits increased. But their wants were easily appeased, and the increased profits were counterbalanced by decreased activities. Old traders among the islands

[1] Nisbet, "Burma Under British Rule and Before," Westminster, 1901, Vol. I, p. 362.

tell of the opportunities of those days when choice pearls could be obtained for a pinch of opium or for a few ounces of tobacco.

Far from the highways of the world, the Selangs remained undisturbed in their beautiful seas until nearly twenty years ago. Meanwhile, 800 miles distant, Singapore had arisen from a desert shore to the rank of a great seaport, and the headquarters for the pearl fishery of the Malay Archipelago and of the northwestern coast of Australia. In this fishery the vessels were well equipped and depended on the use of diving apparatus rather than on nude divers.

Beginning about 1888, some of these vessels made occasional visits to the Mergui pearl-oyster reefs, and usually with very profitable results. This was the first instance in which diving apparatus was successfully introduced on any part of the Asiatic coast from the Red Sea to Malacca Strait. So great was the profit that nearly every one on the lower coast of Burma with sufficient capital or credit hastened to obtain a boat and diving equipment. The success of some of these early ventures was remarkable, single pearls worth $3000, $5000, and even $10,000 each being secured. The reefs in the shoal waters were rapidly depleted, to the great disadvantage of the nude Selangs, who can do little in deep water.

With a view to deriving a revenue from these well-equipped vessels, the government of Burma in 1898 divided the 11,000 square miles of pearling territory into five definite areas known as "blocks." The area within each of these blocks was surveyed, marked, and charted; and the financial commissioner from time to time determined as to each block whether licenses for pearl fishing should be issued, or whether the exclusive right therein should be leased. These leases were disposed of either by inviting tenders and granting the lease to any of the persons who might tender, or by public auction, as the financial commissioner might direct. By the terms of the lease, the lessee was obliged to register at the office of the deputy commissioner of finance the number of boats and pumps employed by him; to declare by letter, at the end of each month, the number, weight, and estimated value of all mother-of-pearl shell and pearls collected during the month, and to refrain from taking any mother-of-pearl measuring less than six inches from lip to hinge.

Outside the limits of blocks in which the exclusive pearl fishing was leased, licenses to use diving implements were granted in such number and on payment of such fees, not exceeding Rs.1000 per apparatus, as might from time to time be fixed, every such license expiring on June 30 next following the date on which it was granted, and no license was transferable.

The five blocks in which the Mergui pearling rights were leased are of large area, averaging somewhat over 2000 square miles each. The lessees customarily granted permits to subsidiary fishermen to operate in their respective blocks, on payment of a royalty, this ranging in amount from 12½ to 25 per cent. of the mother-of-pearl secured, and the pearls found were the absolute property of the fishermen.

Until 1900 the pearling rights were leased by blocks as above noted. Rights to catch trochus, green snail shells, and sea-slugs, were included in the lease. It was noticed that European pearlers always sublet the trochus and green snail rights, and it was decided to auction these separately; while as regards pearling proper the auction system was abolished in that year in favor of a system of licensing individual vessels for a fee of Rs.400 each. The right to collect pearls by nude diving was thought for some time to have been left free; but subsequently it was auctioned along with the rights to collect green snails, trochus and sea-slugs.

The following summary, compiled from data furnished by Mr. I. H. Burkill of the Indian Civil Service, shows the extent of the pearl and shell fisheries of Mergui for a series of years.

Year.	No. of Pumps.	Revenue from Pumps. Rs.	Revenue from Auction Rights. Rs.	Reported Value of Yield. Rs.
1904	70	28,000	22,500	149,239
1905	77	30,800	14,200	131,921
1906	80	32,000	15,300	124,798
1907	76	30,400	19,700	

The local headquarters of the industry are at Mergui, but most of the supplies are drawn from Maulmain and Rangoon, or from the more distant Singapore, where the industry is financed. The season extends from October to April or May, when the southwest monsoon begins and puts a stop to the fishery on this exposed coast.

The boats used are mostly of Burmese build. They measure from 25 to 35 feet in length, and 7 or 8 feet in width, and have 18 to 24 inches of draft, with curved or half-moon shaped keels, and with high square sterns. Owing to the very light draft and the amount of free-board, they are deficient in weatherly qualities; but are fast sailors before the wind and are easily rowed from place to place. For this reason they are especially suited to the industry in Mergui, because during the pearling season calms and light winds prevail and oars form the principal motive power, especially in the channels and passageways between the islands where the tides are frequently very swift.

If a number of boats are of the same ownership, a schooner of thirty to one hundred tons' capacity is commonly provided as a floating station and base of supplies for them; the gathering of such a fleet

FROM THE TREASURY OF THE EMIR OF BOKHARA

Necklace and earrings. Property of an American lady

CARVED "JERUSALEM SHELL" FROM THE RED SEA

In the collection of Dr. Bashford Dean

presents an interesting sight, like a great white hen among her brood of chickens.

Most of the boats are from Mergui, and are chartered at a monthly rate of from Rs.105 to Rs.120 each, including a crew of four or five Burmans with their subsistence, consisting principally of rice and salted fish; the charterer is further required to pay each member of the crew four annas, or one rupee, for each day actually employed in operating the diving pump. In addition to these men, each boat carries one diver and an attendant, commonly known as "tender." The boat is sailed or rowed by the crew, as directed by the diver; and while the latter is submerged, the boat and crew are under the supervision of the attendant.

The divers are the most important men in the fleet, for on their ability and efficiency depends the success of the enterprise. A very considerable portion of them are natives of the Philippine Islands, although many Japanese have been employed recently, and the number is increasing. The compensation is at the rate of £2 to £4 per month, and £20 for each ton of mother-of-pearl secured. The attendants are likewise mostly Manilamen, but many Malayans and Burmans are employed; the wages range from Rs.50 to Rs.80 per month, including provisions. The peculiar duties of the attendant are to help the diver into his dress, place the shoulder leads into position, screw on the helmet, and especially to receive and respond to signals and to direct the movements of the vessel in accordance therewith.

The scaphander, or diving-dress, is composed of solid sheet rubber, covered on both sides with canvas. The head-piece is made of tinned copper, and is fitted with three glasses, one at the front and one on each side, so as to afford the diver as wide a view as is consistent with strength of construction. It has a valve by which he can regulate the pressure of the atmosphere. The dress has a double collar, the inner portion coming up around the neck, and the other hermetically fastened to the breastplate. The breastplate is likewise made of copper. The suit is connected with the air-pump by means of a stout rubber tube which enters the helmet, and through which air is supplied to the diver incased therein. This air-tube consists of three or four lengths—each of fifty feet—of light hose, commonly called "pipe." This is buoyant so that it may be easily pulled along, and may not readily foul among the rocks. However, when working on very rough bottom with sharp-edged stones, the lower length is of stouter material in order to resist the chafing on the bottom. Before descending, the air-line is loosely coiled around the diver's arm to prevent a sudden strain on it when it is tightened, and a signal-line is attached to his waist to enable him to communicate with the men above.

In fishing, if the current is slight, the boat is permitted to drift therewith, and if there is little or no current, it is propelled by oars as may be required. The diver—fully dressed in the rubber suit with helmet, etc.,—goes overboard easily by means of a Jacob's ladder of five or six rungs on the port side of the boat, and is lowered by an attendant, who gives close attention to the lines, the crew having manned the pump in the meantime. On reaching bottom, the diver walks along, following the course of the moving boat and swinging his shoulders from side to side to take in a wide vision in his search for oysters. In clear water he can discover them at a distance of twenty-five or thirty feet, even when fifteen fathoms below the surface; but sometimes the water is so clouded that it may be necessary for him to go almost on hands and knees to see them, and when the seaweeds are thick and high, he may locate them almost as much by feeling as by sight. Owing to this difficulty in seeing the oysters, the work is suspended in rough weather and for many days following. The catch is placed in a sack or basket of quarter-inch rope, which is raised when filled, emptied, and returned to the bottom by means of a rope.

Finding the shell is by no means an easy matter, and much natural hunter-craft is necessary. Of a neutral color, it is not at all conspicuous as it lies on a gray coral bed, itself covered with coral or sponge or hidden in dense masses of gorgeous seaweeds. Still less visible is the shell on a muddy bottom, for there it embeds itself and exposes only half an inch or so of the "lip." As the boat is impelled by the tide, the diver may have to walk rapidly in a swinging gait; and if he should stumble or fall while stooping to pick up the shell, recovery of balance may be difficult. He must be constantly on the alert and has many dangers to avoid. Sharks are numerous in these clear tropical waters; but although disaster sometimes results, they are timid, a stream of air bubbles from the sleeve of the dress sending them away in fright. More fruitful sources of danger are fouled air-pipes, broken pumps, falling into holes, and especially paralysis from recklessly deep diving.

When the diver wishes to come up, he closes the escape valve in his helmet; his dress fills and distends with air, causing a speedy return to the surface, and the tender hauls him alongside by means of the life-line. After "blowing" for a few minutes with the helmet removed, and usually enjoying the indispensable cigarette, he returns to the bottom.

When the Mergui reefs were first exploited by diving apparatus, the bulk of the shells were secured from depths of ten to twelve fathoms. These shallow reefs have been exhausted, temporarily, at least, and the divers now work in deeper water, fifteen, twenty, and even twenty-five fathoms, if the bottom is very uneven and rocky. Many shells are found in the depressions between the large boulders, which may be twenty or thirty feet deeper than the surrounding areas.

The oysters are opened by means of the long-bladed working-knife of the country, known as *dah-she*. The flesh is thrown into a large basket or washtub, where it is searched by the proprietor of the boat, who takes each piece between the hands and squeezes and feels through every part of it. After the flesh has been carefully examined, the sediment at the bottom of the tub is washed and panned to obtain those pearls which have fallen through the flesh tissues. The Mergui pearls are commonly of good color and luster, and compare favorably with those from the Sulu Archipelago or the Dutch East Indies.

The sea-green shell of the snail (*Turbo marmoratus*) is gathered in large quantities by the nude diving Selangs, who barter it to Chinese traders at the equivalent of Rs.8 or 10 per 100 in number. The flesh is also dried and disposed of to these traders under the name of *thadecon*, at about Rs.3 per *viss* of 3.33 pounds. This mollusk yields a few greenish yellow pearls.

In 1895, three pearl reefs were discovered off the Bassein coast in the district of Irawadi.[1] These proved fairly remunerative for one season and a portion of another, when they were abandoned.

THE PEARL FISHERIES OF THE RED SEA, GULF OF ADEN, ETC.

> Under the Ptolemies, and even long after—under the Califs—these were islands whose merchants were princes; but their bustle and glory have since departed from them, and they are now thinly inhabited by a race of miserable fishermen.
>
> JAMES. BRUCE (1790).

THE Red Sea was one of the most ancient sources of pearls, furnishing these gems for centuries before the Christian era, and particularly during the reign of the Ptolemies. These pearls were alluded to by Strabo, Ælianus, and other classical writers. Although the prominence of the fisheries has suffered by comparison with those of Persia and Ceylon, the yield has been more or less extensive from the days of Solomon up to the present time.

Of the several pearl-yielding mollusks in the Red Sea and on the

[1] Nisbet, "Burma Under British Rule and Before," Vol. I, p. 363.

southeast coast of Arabia, the largest and best known is that called "sadof" by the Arabs, and which has been identified by Jameson as *Margaritifera m. erythræensis.* This is closely related to the large species in the Persian Gulf. It is commonly four or five inches in diameter, and in exceptional instances attains a diameter of eight inches and a weight of three pounds or more. In addition to its

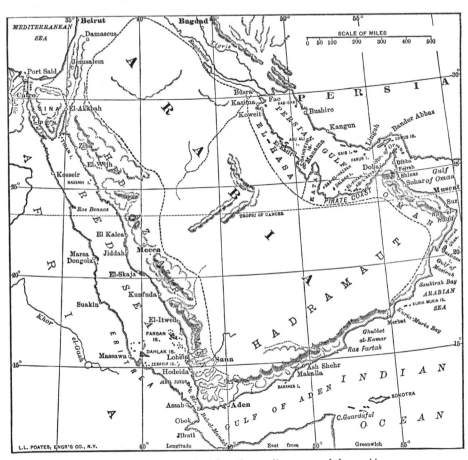

Persian Gulf and the Red Sea, the pearling center of the world

size, it is distinguished by a dark green coloring about the edges, and a more or less greenish tint over the nacreous interior surface; this color is darker in the vicinity of Jiddah and Suakin than at the southern end of the sea, or in the Gulf of Aden. This species occurs singly rather than congregated in beds or reefs. Although it is found in depths of fifteen fathoms or more, most of the fishing is in less than five fathoms of water.

The "sadof" yields pearls only rarely, and is sought principally for the shells, which afford good qualities of mother-of-pearl, the pearls furnishing an additional but always looked-for profit to the regular source of income. As in other regions, there is no constant relation between the value of the pearls and the quantity of mollusks taken. The oysters of some reefs are comparatively rich in certain years; while in other seasons or on other reefs the mollusks may be numerous but yield very few pearls.

The second species of importance in the Red Sea is similar to the Lingah oyster and is known to the Arabs as "bulbul." This is much smaller than the "sadof," averaging less than three inches in diameter. It is collected for the pearls exclusively, the shells being too small for industrial use; but only 3 or 4 per cent. of the individuals yield pearls.

It is claimed by writers of authority that it is the red Pinna pearl from this sea that is referred to in the Scriptures under the name *peninim* as the most precious product, and which has been translated as rubies.[1] The shell is extremely fragile, and the nacreous interior is white tinged with a beautiful red. It is of little importance in the commercial fisheries of the Red Sea at the present time.

The "sadof" is more scattered and less numerous than the "bulbul"; and in order to save much useless diving, it is customary to inspect the bottom before descending. Therefore, operations are largely restricted to calm weather, when the water is sufficiently clear to enable the divers to sight the individual oysters on the bottom. In recent years, water-telescopes have been used to assist in locating them. The most popular form consists of a tin can with a sheet of glass inserted in the bottom. The glazed end of the tin is submerged several inches below the surface, affording a far-reaching and much clearer vision. In this fishery the divers work from small canoes (*uri*), each manned by two men, one of whom rows while the other leans over the bow and searches for the oysters. When one is sighted, he dives into the water for it, and then returns to the boat to resume the search.

The pearling season begins commonly in March or April, and continues until about the end of May; it is renewed in the autumn, continuing through September and October.[2] The vessels employed are of two varieties: *dhows* carrying from twenty to eighty men each, and the much smaller *sambuks* or sail-boats without decks, each with from six to twenty-five men, most of whom are Negro slaves. Many of the large vessels are from the Persian Gulf. The *sambuks* are owned principally by Zobeid Bedouins inhabiting the coast between Jiddah and Yambo, and also the islands near the southern end of the

[1] See Proverbs xxxi, 1.   [2] Hesse, "Der Zoologische Garten," Dec. 1, 1898.

sea, which are very hot-beds of pearls, shells, religious frenzy and half famished Arabs and Negro slaves.

The "bulbul" oysters are taken in nearly the same manner as in the Persian Gulf. When the vessel is located over the reef, each diver descends, commonly with a short stick of iron or hard wood, with which he releases the oysters within reach; placing them in a sack, he is pulled up by an attendant when his breath is nearly exhausted.

The fisheries are prosecuted along both sides of the Red Sea and in the channels among the islands, from the Gulf of Akabah to Bab-el-Mandeb. They are especially extensive among the Dahlak Islands on the coast of the Italian colony Eritrea, where the population is largely supported by them. This was the center of the industry during the time of the Ptolemies and in the early Christian era. The fisheries are also important in the vicinity of Jiddah, the port of entrance for Mecca and Medina, holiest places of Islam. They likewise exist near Kosseir at the northern end of the sea, and at Suakin, Massawa, the Farsan Islands, and Loheia, near the lower end. They are carried on by Arabs, who succeed in evading efforts at control on the part of the local governments. Even on the African side, the Arab fishermen predominate, for the native Egyptian has never evinced much fondness for venturing on the sea.

On the southeast coast of Arabia, pearl fishermen are to be found at the various harbors from Aden to Muscat. Their fantastic dhows are met with in the harbor of Makalla, and also in that of Shehr. On the Oman coast, the ports of Sur and of sun-scorched Muscat do a considerable pearling business, not only locally but to the Sokotra Islands, and even on the coast of East Africa and Zanzibar, the trading baggalas adding pearling and illicit slave-trading to their many sources of income. A number of these traders, each with an instinct for pearls equal to that of a trained hound for game, visit the fishing centers at intervals, and exchange needful commodities for pearls and shells.

The Arab pearl divers of the Red Sea have been noted for the depths to which they can descend. Lieutenant J. R. Wellsted, of the Indian Navy, who had unusual facilities for acquaintance with their exploits, reported that in the Persian Gulf the fishermen rarely descended beyond eleven or twelve fathoms, and even then they exhibited signs of exhaustion; but that in the Red Sea they go down twice that depth. Among the most noted of these divers of the last century was old Serúr, who attracted the notice of many travelers. Lieutenant Wellsted states that he saw him descend repeatedly to twenty-five fathoms without the slightest evidence of distress; that he frequently dived in thirty fathoms, and is reported to have brought up mud from the bottom at a depth of thirty-five fathoms, which is about the record, the pressure of the water being nearly 90 pounds to the square inch. His sons were also remarkably expert; one of them when scarcely thirteen years of age would descend to a depth of twenty-five fathoms.[1]

An interesting story of an Arab's diving ability is told by Lieutenant Wellsted: "In 1827, we were cruising in the sloop *Ternate* on the pearl banks. Whilst becalmed and drifting slowly along with the current, several of the officers and men were looking over the side at our Arab pilot, who had been amusing himself in diving for oysters. After several attempts, his search proved unsuccessful. 'Since I cannot get oysters I will now,' said he, 'dive for and catch fish.' All ridiculed the idea. He went down again, and great was our astonishment to see him, after a short time, rise to the surface with a small rock-fish in each hand. His own explanation of the feat was, that as he seated himself at the bottom, the fish came around and nibbled at his skin. Watching his opportunity, he seized and secured his prey by thrusting his thumb and forefinger into their expanded gills."[2]

Owing to the character of the fishery and the lack of government supervision, it is extremely difficult to determine accurately the extent of this industry in the Red Sea and the Arabian Gulf. All over this coast extends the influence of the Hindu traders, who finance the fisheries and purchase most of the catch. The pearls are sent mostly to Bombay, and are not reported in the official returns of the Red Sea ports. The fishermen are suspicious of outside inquiries, and are far from anxious to impart reliable information. Probably the best estimates of the catch are to be obtained from Bombay merchants, from whom A. Perazzoli learned in 1898 that pearls to the value of 2,000,000 lire ($400,000) were carried from the Red Sea to Bombay each year.[3] In the last four or five years the output has been smaller than usual, owing to disturbed political conditions.

The annual product of "Egyptian" and "Bombay" shells in these fisheries is usually upward of 1000 tons, worth from $100 to $600 per ton, according to quality. Most of these go to Austria and France, only about 200 tons reaching London each year. Owing to the dark color and the lack of thickness in the nacreous layer, they are scarcely suitable for anything else than button-manufacture. Many of them are sent to Bethlehem and Jerusalem, where they are cut into various shapes for crosses, crucifixes, wafer-boxes, beads, and nearly every conceivable article in which mother-of-pearl is

[1] Wellsted, "Travels in Arabia," London, 1838, Vol. II, p. 238.
[2] *Ibid.*, Vol. I, pp. 268, 269.
[3] "Bolletino della Società d'esplorazione commerciale in Africa," Milan, June, 1898.

manufactured. Many of the choicest shells are incised with scriptural or allegorical designs for sale to tourists as well as for export. The best of the engraved shells sell for $10 to $50, and the cheaper ones for less than $1 each. This industry is of great importance in Bethlehem, giving employment to a considerable percentage of the eight thousand inhabitants of the village.

Doubtless in no pearl fishery in the world are greater hardships endured than in the Red Sea and along the coast of the Arabian Gulf. In practically every other region, the industry is carried on under government supervision, and there is little opportunity for ill-treatment of the humbler fishermen. But the fanatics who control the fishery on the Arabian coast—untrammeled by authorities and responsible to none—show little consideration for the poor divers, and particularly for the unfortunate black slaves brought from the coast of Africa.

These pearl fishermen lead a very eventful life, the divers especially. They see some wonderful sights down below the surface—plant life and creeping things and enemies innumerable. Dropping from the sun-scorched surface down into the deep cool waters, everything shows "a sea change, into something rich and strange," just as the eyes of the drowned man in Ariel's song are turned into pearls and his bones into coral.

And there are enemies innumerable. The terrible sharks, prowling about near the bottom, prove a source of perpetual uneasiness, and in the aggregate many fishermen are eaten by these blood-thirsty tigers of the sea. There are horrible conflicts with devil-fish equaling that in Hugo's "Toilers of the Sea." The saw-fish is also a source of danger, particularly in the Arabian Gulf, and instances are reported in which divers have been cut in two by these animals, which sometimes attain a length of twelve or fifteen feet, and possess a saw five feet long and three inches broad, armed on each edge with teeth two inches in length. Another menacing peril is the giant clam (*Tridacna gigas*), a monster bivalve, whose shell measures two or three feet in diameter, and is firmly anchored to the bottom. This mollusk occurs on many of the Asiatic pearling grounds. Lying with the scalloped edges a foot or more apart, a foot or a hand of the diver may be accidentally inserted. When such a fate befalls a fisherman, the only escape is for him to amputate the member immediately. Once in a while on the pearling shores a native may be found who has been maimed in this manner, but usually the unfortunate man does not escape with his life.

CAP OF STATE, FROM LOOTING OF SUMMER PALACE, PEKIN, IN 1860

Now in South Kensington Museum

FISHING FOR THE AWABI (ABALONE) SHELLS AT WADA-NO-HARA, JAPAN

THE PEARL FISHERIES OF CHINA, JAPAN, SIBERIA, ETC.

Do churls
Know the worth of Orient pearls?
Give the gem which dims the moon
To the noblest or to none.
EMERSON, *Friendship.*

IT appears from ancient Chinese literature, noted in the first chapter of this book, that pearl fisheries have existed in the rivers of China for several thousand years. The Chinese also derived pearls from the sea, and especially from the coast of the province of Che-kiang. Little is known of the early fisheries, but the fragmentary literature contains so many allusions to pearls as to lead us to believe that they were of considerable extent and importance.

It is related that about 200 B.C., a pearl dealer at Shao-hing, an ancient city between Hang-chau and Ning-po, on the shore of Hang-chau Bay, furnished to the empress a pearl one inch in diameter, for which he received five hundred pieces of silver; and to an envious princess the same dealer sold a "four-inch pearl." A hundred years later, the reigning emperor sent an agent to the coast to purchase "moon pearls," the largest of which were two thirds of an inch in diameter.

In the tenth century A.D., Mingti, one of the most extravagant of the early monarchs, used so many pearls—not only in his personal decoration but on his equipage and retinue,—that after a formal procession the way would be rich in the jewels which dropped from the gorgeous cortège. About 1000 A.D., an embassy to the emperor brought as tribute an ornament composed of strings of pearls, and also 105 liang (8¾ lbs.) of the same gems unmounted.

An interesting story is told of "pearl-scattering" by an embassy to the Chinese court from a Malayan state about 1060. Following the customs of their country, the ambassadors knelt at the threshold of the audience chamber, and then advanced toward the throne, bearing a golden goblet filled with choice pearls and water-lilies wrought of gold. These they scattered upon the floor at the feet of the emperor; and the courtiers, hastening to pick them up, secured ten liang (15 oz.) of pearls.[1]

The Keh Chi King Yuen, a Chinese encyclopedia, describes a pearl fishery in the southern part of Kwang-tung province, in the depart-

[1] Von Hessling, "Die Perlenmuscheln," Leipzig, 1859, p. 6.

ment of Lien-chau and near the city of Hóhpú. Fishing began in the spring, and was preceded by conciliating the gods through certain sacrifices, in order that the weather might be propitious and that no disaster might be suffered through sharks and other agencies. The five sacrificial animals,—horses, cattle, sheep, swine, and fowls,—were presented; but ordinarily paper images of these were economically substituted, as equally acceptable to the Chinese rulers of destiny. In the details of the diving, the fishery resembled somewhat that prosecuted about the same period in the Gulf of Manaar. The diver was let down by a rope, and after collecting the mollusks and placing them in a basket, he was drawn up at a given signal. Much complaint was made that the divers would open the mollusks, extract the pearls and conceal them in the mouth before returning to the surface.

The business became so perilous and the loss so great, that about the beginning of the sixteenth century, according to the same encyclopedia, dredges were adopted. These at first were simple rakes; later large dredges were trailed along between two boats, by means of which great quantities of shells were gathered. So important was the industry that an officer was designated by the viceroy of Canton to collect a revenue therefrom. It does not appear that pearls have been collected in considerable numbers on this part of the Chinese coast for very many years, probably not since the advent of Europeans.

Pearls are yet found in the river mussels in all parts of eastern Asia, from Siberia to the Indian Ocean, and from the Himalayas to the Pacific. It is represented that they are not from the *Unio margaritifera,* the common river-mussel of Europe, but from other species, such as *Unio mongolicus, U. dahuricus, Dipsas plicatus,* etc. It is quite impossible to obtain a reliable estimate of the total number of persons employed, or the output of pearls in China, but these items are certainly very much larger than the average Occidental believes.

In the vicinity of Canton the *Dipsas plicatus* has been used for centuries by the Chinese in the production of artificial pearls, this industry giving employment to thousands of persons.[1]

The pearl-mussel fishery is of importance in Manchuria, where it has been carried on for hundreds of years, not only by the citizens, but by the military department on account of the government, and especially in the streams which flow into the Songari, a tributary of the Amur. Jacinth relates that in case of a deficit, the officers and subalterns were punished by a deduction from their pay, and also by corporal chastisement.[2] Witsen speaks of the pearls from the River Gan, a tributary of the Amur, and also from the islands of the Amur, the

---

[1] See p. 288 for an account of the methods.

[2] Statist, "Beschreibung des chines. Reiches," 1842, Vol. II, p. 11.

boundary river of Manchuria. Pearl fisheries were established at these places by the Russians nearly two centuries ago.[1]

Pearls become finer and more plentiful the further we penetrate into Manchuria; and they are numerous in the lake of Heikow or Hing-chou-men, "Black Lake" or "Gate of Precious Gems," where they have long been exploited for the account of the emperor of China.

The occurrence of pearls in many parts of Asiatic Russia was noted by Von Hessling. In northern Siberia, according to Witsen, writing in 1705,[2] pearls were found in the waters about the town of Mangasea on the Turuchan; and Von Middendorf notes that they were found in the Tunguska River, which flows into the Yenisei. Whether, however, they come from the *Unio margaritifera* is considered doubtful by Von Middendorf. Witsen referred to their occurrence in the rivers and streams of Irkutsk and Onon, and this is confirmed by several writers of more recent times. Pallas says that the mussels found there are quite large, and speaks of the Ilim, which flows into the Angara, as another river where they occur.

Ancient books relating to Japan repeatedly allude to the occurrence of pearls on the coasts of that country. They are mentioned in the Nihonki, of the eighth century, the oldest Japanese history.

Tavernier wrote about 1670: "It is possible that of those who have written before me concerning pearls, none have recorded that some years back a fishery was discovered in a certain part of the coasts of Japan, and I have seen some of the pearls which the Dutch brought from thence. They were of very beautiful water, and some of them of large size, but all baroques. The Japanese do not esteem pearls. If they cared about them it is possible that by their means some banks might be discovered where finer ones would be obtained."[3]

In 1727, Kaempfer wrote that pearls, called by the Japanese *kainotamma* or shell jewels, were found in oysters and other mollusks almost everywhere about Saikokf. Every person was at liberty to fish for them. Formerly the natives had little or no value for them, till they learned of their estimation by the Chinese, who were ready to pay good prices for them, their women being very proud of wearing necklaces and other ornaments of pearls. "The largest and finest pearls are found in the small sort of oysters, called *akoja*, which is not unlike the Persian pearl-oyster. These are found only in the seas about Satzuma and in the Gulf of Omura (Kiusiu). Some of the pearls weigh from four to five candareens[4] and these are sold for a hundred

[1] Ranft, "Vollständige Beschreibung des russischen Reiches," Leipzig, 1767, p. 415.

[2] Witsen, "Nord en Oost Tartarye," 1705, p. 762.

[3] Tavernier, "Travels in India," Ball edition, London, 1889, Vol. II, pp. 113, 114.

[4] One candareen equals 5.72 grains.

kobans each.[1] The inhabitants of the Loochoo Islands buy most of those about Satzuma, since they trade to that province. Those found on the Gulf of Omura are sold chiefly to the Chinese and Tunquinese, and it is computed that they buy for about 3000 taels[2] a year. This great profit occasioned the strict orders, which were made not long ago by the princes both of Satzuma and Omura, that for the future there should be no more of these oysters sold in the market with other oysters, as had been done formerly."[3]

Kaempfer also noted that the Japanese obtained pearls from the yellow snail shell and from the *taira gai* (Placuna) in the Gulf of Arima, and especially from the awabi or abalone (Haliotis). This mollusk was much sought after for food, being taken in large quantities by the fishermen's wives, "they being the best divers of the country."[4]

Of the several species of pearl-oysters which occur in the coastal waters of Japan, the only one of importance at present is the *Margaritifera martensi*. While this occurs in very many localities, it is most numerous among the southern islands, where some fine pearls have been secured. The fishery for this species was quite extensive thirty or forty years ago, and the reefs were largely depleted. For nearly a score of years it has been used in growing culture-pearls, an account of which is given on pages 292, 293.

A few pearls are obtained from several other bivalve mollusks in Japan. Among the collections of the present writers are pearls from *Margaritifera martensi,* collected at Bay Agu; from *M. panasisæ,* about the Liu-kiu Islands; from *Pecten yezocusis,* in Sokhaido; from *Mytilus crasitesta,* in the Inland Sea, and from North Japan, and from a species of Dipsas found in Lake Biwa.

While the pearl fisheries of Japan are not of great importance in any single locality, the distribution of the reefs is so extensive that the aggregate yield is considerable.

The awabi or ear-shell (*Haliotis gigantea*), found on the coast of Japan, Korea, etc., yields many pearly forms. This species is much smaller than the California abalones. It has a fairly smooth, nacreous surface, but its value is depreciated by the great size of the marginal perforations, which render useless for commercial purposes all of the shell external to the line of perforations. While its opalescent tints make it desirable for manufacturing into certain styles of buttons and buckles, its principal use is for inlaying work or marquetry, for which it is especially adapted, owing to its fineness of texture and beauty of coloring even when reduced to thin sheets.

---

[1] One koban equals 66 cents.
[2] $4200.

[3] Kaempfer, "History of Japan," 1728, Vol. I, pp. 110, 111.
[4] *Ibid.,* Vol. I, p. 139.

Probably the most interesting of the abalone fisheries is that on the shores of Quelpaerd Island, about sixty miles south of the Korean coast, which is prosecuted largely by the women. Dressed only in a scanty garment, these women swim out to the fishing grounds, distant several hundred yards in some cases, carrying with them a stout knife and a small sack suspended from a gourd. On reaching the reefs, they dive to the bottom—sometimes to a depth of six or eight fathoms —and by means of the knife, remove the abalones from the bottom and place them in the sack. They may remain out an hour, diving repeatedly until the sack is filled, when they swim back to the shore. Pearls are found only rarely; in one lot of one hundred shells, only five were found bearing pearls; two with three pearls each, two with two pearls each, and one with a single pearl. The flesh of this mollusk after it has been cleaned and dried, is quite popular as an article of diet. Although white when fresh, the color changes to a dark red. The pieces of dried flesh, in the form of flat reddish disks four or five inches in diameter, are fastened on slender sticks—about ten to each stick—and displayed in the grocery shops in Seul and other cities.

In the Gulf of Siam on the Asiatic coast, pearls are obtained from a small oyster with a thin shell, presumably a variety of the Lingah oyster. The beds have not yet been thoroughly exploited, as the Siamese do not especially value pearls, attributing some superstitious sentiments of ill luck to them. However, from time to time Chinese traders have bought them from the Malay divers and sold them at great profit in the Singapore market. The known beds occur chiefly in the northern part of the gulf, on the west coast, and extend in a narrow belt for a distance of about one hundred miles. The fishing is prosecuted by nude divers in shallow water. A recent letter from Dr. K. Van Dort, a mining engineer of Bangkok, Siam, states that in 1906 in six weeks, with the aid of half a dozen divers he was able to collect 720 grains' weight of pearls, mostly small ones, but including one of 20 grains, one of 14 grains, two of 12 grains each, and seven over 9 grains in weight. He reports that the total value of the large ones in Bangkok was $1500, but the small ones could not be sold to any advantage, as they are little prized by the Siamese. The shells are of no commercial value, as they are too thin for industrial use other than for inlaid work. Some fine old specimens of marquetry in which these shells were used exist in the Buddhist temples at Bangkok. This art of inlaying is almost lost among the Siamese, and there is said to be only one man in the king's palace who can lay any claims to proficiency in working mother-of-pearl shell.

VII

PEARLS FROM THE AFRICAN COASTS

EAST AFRICAN PEARL FISHERIES

The Islanders with fleecy curls,
Whose homes are compass'd by the Arabian waves;
By whom those shells which breed the orient pearls
Are dived and fish'd for in their green sea caves.
TASSO, *Jerusalem Delivered.*

THE principal pearl fisheries of the coasts of Africa are those prosecuted in the Red Sea, between this continent and Asia. These have already been described in the preceding chapter, among the Asiatic fisheries; for, although situated between the two continents, they are prosecuted largely by Arabs rather than by natives of the western shores of the sea.

Other than those in the Red Sea, the only pearl resources in Africa which have received attention are on the eastern coast, south of the Gulf of Aden. Little information exists as to the origin of these fisheries. In a paper published by the Lisbon Geographical Society, January, 1903, Señor Ivens Ferranz states that, according to tradition, in remote times the Ibo Archipelago, on the northeast coast of Portuguese East Africa, was inhabited by a Semitic colony, which located there to fish for pearls, and these were carried through the Red Sea to King Solomon. He adds that there is little doubt that, after the great emigration which started from the Persian Gulf in 982 and founded Zanzibar, Kilwa, and Sofala on this coast, some Arabs engaged in fishing for pearls about the islands near Sofala.

In 1609 Joao dos Santos wrote that on the sandy sea-bottom about the Bazaruto Islands, which are about 150 miles south of Sofala, there were many large oysters which bore pearls, and the natives fished for them by diving in practically the same manner as in the Persian Gulf.[1]

In a personal memorandum, Sir Robert Edgcumbe states that in the very early times of Portuguese exploitation on the eastern coast of Africa, pearl fishing was carried on in these waters. For a long period the tenure of power exerted by the Portuguese was of a feeble character; they practically occupied no position of importance on the

[1] Joao dos Santos, "Ethopia Oriental," Lisbon, 1609, Vol. 1, c. 27.

mainland, but seized upon stations on the islands which offered decent harbors. Thus their chief settlements, such as Mozambique and Ibo, were on islands lying off the coast, and until recent years they made no serious attempt to occupy the mainland.

Arabs and Banyans carried on the commercial traffic of the country, as they still do, and they were more truly the masters of this coast than were the Portuguese, who were little more than nominal rulers. Trading to and from India in their small dhows, the Arabs and Banyans had full knowledge of the value of pearls, and undoubtedly secured all that were obtainable. But they observed no restrictions, and without doubt—for a time, at any rate—greatly impaired the productive power of the fisheries.

The principal pearl reefs of East Africa, so far as known at present, extend along the coast of the German East African territory from the Province of Uzaramo to the Rovuma River, the southern limit of that territory, and also into Portuguese East Africa as far south as Pemba Bay, a total distance of about 300 miles. Along much of this coast, there are islands lying from one to two miles off shore, and between these islands are barriers of reefs, which create a series of lagoons. In these lagoons, protected by the islands and the reefs from the influence of the surf formed by the Indian monsoons, there are large patches of coral rock and groups of living coral, which form excellent attachments for the pearl-oysters.

It is only recently that serious attention has been paid to these pearl resources, although year by year a considerable number of pearls have been collected by the natives and sold to Arabs and Banyans, who have sent them chiefly to India by way of Zanzibar. The natives of these parts are not very expert in diving, and they collect the oysters principally by wading out as far as they can at low tide. They do not wait for the mollusks to attain a proper age, and as a result they find few pearls of large size. Many parcels of pearls fished in this very elementary way pass through the custom-house, where they are subject to a small duty for export, and others are smuggled out of the country. Quantities of seed-pearls are sent to India, where they are used principally as a medicine and in cosmetics; and occasionally there are rumors that some choice pearls have been discovered.

In the German territory a concession of the fisheries was granted a few years ago to Dr. Aurel Schulz; and, although we are not in position to say what success he has met with, it is reported that he has secured a considerable number of pearls under four grains in weight, of fair shape and quality and of good marketable value.

A concession of the pearl fishery on the Portuguese coast north of Ibo has been granted to the East African Pearl Company. For this company an examination of the resources is now being made by Mr. James J. Simpson, acting under direction of Prof. W. A. Herdman, of the Liverpool University, the technical advisor of the Ceylon Company of Pearl Fishers.

At least four species of pearl-bearing mollusks exist here; these are *Margaritifera vulgaris, M. margaritifera, Pinna nobilis,* and a species of Perna, named in the order of their importance. A preliminary report of Mr. Simpson (supplied through the courtesy of Sir Robert Edgcumbe), states that among the Ibo Islands about one half of the bottom is sandy and the other half is covered with detached pieces of coral rock, groups of living corals, masses of nullipore, and expanses of fixed seaweed. On all of these in the shoal waters, there is such an abundance of pearl-oysters (*M. vulgaris*) that a single diver, by simply descending and bringing up a few in his hands each time, can secure about 200 in fifteen minutes. Oysters also occur singly on the sandy bottom, but not so abundantly. Within the three-year-old oysters there are many seed-pearls. It is evident that there has been an extensive removal of large oysters in recent years and that large pearls were then found; but the depredations of the natives now prevent the mollusks from attaining an age and size which render them useful as pearl-bearers.

Said Mr. Simpson in his report: "The women here play great havoc on the reefs by going out daily and collecting the pearl-oysters at low tide. All along the coast from Muliga Point to Arimba the shores are covered with shells. At one place we came across a heap of freshly-opened oysters which consisted of thirty or forty thousand at the lowest estimate; while an older heap contained between forty to sixty millions. Four women who were fishing on the reefs while we were there had over two thousand oysters in their baskets. Thus it is evident that immense quantities are annually destroyed. And the worst feature is that out of those destroyed, not one per cent. were over two years old."

It is the intention of the East African Pearl Company, as soon as the investigation of the resources is completed, to police the fishing grounds so as to put an end to the removal of immature oysters, which yield only seed-pearls, and to permit them to attain maturity. In addition to this, it is their purpose to utilize the extensive beds of oysters lying in comparatively deep water, which are now inaccessible to the natives owing to their lack of diving skill.

Sir Robert Edgcumbe writes that it is impossible to say more at present than that these fisheries at one time bore a high repute, and that the oysters have continued to exist in multitudes though fished by the natives in the immature state; and there is every indication that if properly policed and worked in a scientific way these fisheries should once again become of much importance. The fact that the pearl-bearing oysters are found there in large quantities, notwithstanding that they have been poached without restriction by the natives, indicates that only proper management and policing are necessary to make them valuable and productive.

On the lower coast of Portuguese East Africa, pearl fishing has been of some local importance. The reefs are most extensive about the Bazaruto Islands, previously referred to as about 150 miles south of Sofala. In 1888, when famine prevailed on this coast, the inhabitants of this archipelago, of both sexes and of all ages, fished for the large pearl-oysters (known locally as *mapalo*), selling their catch at Chiloane to Asiatic traders, who gave them a handful of rice for a large basket of the mollusks. It was estimated that during two months of that year, pearls to the value of eighty *contos* ($83,500) were taken.[1] In 1889 three British subjects attempted to renew the fishery by using dredges, but without success, owing, it is said, to the great weight of the implements.

The Kafirs of Bazaruto continue to fish irregularly, but their catch is not of importance. These pearls are carried by traders to Zanzibar, Muscat, and Bombay. The American consul writes that some years ago the Portuguese government granted a concession to a company of American fishermen to exploit the Bazaruto reefs, but the attempt to work the concession failed through "bad management, lack of funds, heavy expenses, and political difficulties," a combination apparently sufficient to wreck a similar attempt in the most favorable locality.

The American consul at Tamatave states that in 1907 the government of Madagascar awarded two grants for pearl fisheries, covering the entire western coast, a distance of one thousand miles, excepting two hundred miles, for which two grants were given in 1906. Apparently no effort was made to develop the earlier grants; the later ones may be operated, perhaps jointly. These concessions are personal, and may not be sold or transferred without the governor-general's consent. The use of divers, machinery, dredges, and other apparatus, and the building of necessary stations are allowed, if there be no interference with navigation, fishing, or coast travel. An annual tax is to be paid, with a stated increase each year, and revised according to the success of the enterprise. A report must be sent yearly to the governor-general. The grants may be revoked if work is not begun within a stated period; if the work is needlessly abandoned during one year, or if the tax is not paid. Whenever the interests of the colony or of the public service require it, the privilege may be withdrawn without indemnity.

[1] Lisbon Geographical Society Report, January, 1903.

VIII

EUROPEAN PEARL FISHERIES

THE BRITISH ISLES, THE CONTINENT OF EUROPE

VIII

PEARL FISHERIES OF THE BRITISH ISLES

And Britain's ancient shores great pearls produce.
MARBODUS (*circa* 1070).

THE occurrence of pearls in the British Isles was known two thousand years ago, and frequent references to them were made in Roman writings of the first and second centuries of the Christian era.

In his "Lives of the Cæsars," the biographer Suetonius, after speaking of the admiration which Julius Cæsar had for pearls, states that their occurrence in Britain was an important factor in inducing the first Roman invasion of that country in 55 B.C.[1] If this be true, the English-speaking people owe a vast debt of gratitude to these pearls in bringing their Briton ancestors in contact with Roman civilization; and the influence which they have thus exercised on the world's history has been greater than that of the pearls from all other regions or, we might add, than all other jewels.

The naturalist Pliny (23–79) stated: "In Brittaine it is certain that some do grow; but they bee small, dim of colour, and nothing orient. For Julius Cæsar (late Emperor of famous memorie) doth not dissimble, that the cuirace or breast-plate which he dedicated to Venus mother within her temple was made of English pearles."[2]

This decoration of pearls was a very proper offering to the goddess who arose from the sea.

The historian Tacitus noted in "Vita Agricolæ" that the pearls from Britain were dusky or brownish (*subfusca ac liventia*).[3] In his commentaries on the Gospel of Matthew, Origen (185–253), one of the Greek fathers of the church, described the British pearls as next in value to the Indian. Their surface, he stated, was of a golden color, but they were cloudy and less transparent than those from India.

---

[1] "Britanniam petiisse spe margaritarum, quarum amplitudinem conferentem, interdum sua manu exegisse pondus." "Divus Julius Caesar," c. 47.

[2] "Naturall Historie," Holland edition, London, 1601, *Lib*. IX, c. 35.

"Vita Agricolæ," c. 12.

We have no certain information whether the pearls secured by the Romans were from the edible mussel (*Mytilus edulis*) of the sea-coast or from the Unios of the fresh-water streams. Tacitus's statement that they were collected "as the sea throws them up," seems to locate them on the sea-coast; but conditions in modern times make it appear more probable that they were from the fresh waters.

Some of the very early coins of the country indicate that pearls were used to ornament the imperial diadem of the sovereigns of ancient Britain. In "Historia ecclesiastica gentis Anglorum," the celebrated English monk, Bede (673–735) surnamed "The Venerable," enumerated among other things for which Britain was famous in his day, "many sorts of shell-fish, among which are mussels, in which are often found excellent pearls of all colours; red, purple, violet and green, but mostly white."[1] And Marbodus, Bishop of Rennes, in his lapidarium, written about 1070, refers to the British pearls as equaling those of Persia and India. About 1094 a present of an Irish pearl was made to Anselm, Archbishop of Canterbury, by Gilbert, Bishop of Limerick.[2]

In the twelfth century there was a market for Scotch pearls in Europe, but they were less valued than those from the Orient.[3] An ordinance of John II, King of France, in August, 1355, which confirmed the old statutes and privileges of goldsmiths and jewelers, expressly forbade mounting Scotch and oriental pearls together in the same article, except in ecclesiastical jewelry (Orfèvre ne peut mettre en œuvre d'or ne argent parles d'Ecosse avec parles d'orient se ce n'est en grands joyaulx d'église).[4]

Writing in the sixteenth century, the historian. William Camden (1551–1623) stated in his "Britannia":

The British and Irish Pearls are found in a large black Muscle. . . . They are peculiar to rapid and stony rivers; and are common in Wales, and in the North of England, and in Scotland, and some parts of Ireland. In this country they are called by the vulgar *Kregin Diliw,* i.e. Deluge shells; as if Nature had not intended the shells for the rivers; but being brought thither by the Universal Deluge, they had continued there, and so propagated their kind ever since. Those who fish here for Pearls, know partly by the outside of these Muscles, whether they contain any; for generally such as have them are a little contracted or distorted from their usual shape. A curious and accomplished Gentleman, lately of these parts, showed me a valuable Collection of the Pearls of the Conway River in Wales; amongst which I noted a stool-pearl [button-

---

[1] Giles's edition, London, 1840, p. 6.
[2] Joyce, "Social History of Ancient Ireland," New York, 1903, Vol. II, p. 227.
[3] Nicolai, "Anglia Sacra," Vol. II, p. 236.

Also "Alberti Magni Opera Omnia," ed. Augusti Borgnet, Paris, 1890, Vol. V, p. 41.
[4] "Histoire de l'Orfèvrerie-Joaillerie," Paris, 1850, p. 46. De Laborde, "Emaux," Paris, 1852, Vol. II, p. 437.

OLD PRINT SHOWING FOUR METHODS OF CATCHING PEARL-BEARING MOLLUSKS

Reproduced from "Margaritologia, sive Dissertatio de Margaritis," by Malachias Geiger, Monachii, 1637

MADAME NORISCHKINE
NÉE STRAUDMAN

DUCHESSE ELIZABETH
(CONSTANTIN)

DAUGHTER OF GENERAL SOBELIEFF
FIRST COUNTESS BEAUHARNAIS

pearl], weighing seventeen grains, and distinguished on the convex side with a fair round spot of a Cornelian colour, exactly in the center.[1]

In 1560 "large handsome pearls" were sent from Scotland to Antwerp.[2] In 1620 a great pearl was found in the Kellie Burn, in Aberdeenshire. This was carried to King James by the provost, who was rewarded with "twelve to fourdeen chalder of victuals about Dunfermline, and the Customs of Merchants' goods in Aberdeen during his life." No record appears of the reward paid to the finder; possibly it was not worth recording.

In 1621 the Privy Council of Scotland issued a proclamation that pearls found within the realm belonged to the Crown; and conservators of the pearl fisheries were appointed in several of the counties, including Aberdeen, Ross, and Sutherland. It was the duty of the conservators, among other things, to nominate experts to fish for pearls during July and August, "when they are at chief perfection." The conservators and fishermen were compensated by selling those pearls of ordinary quality, but "the best for bignesse and colour" were to be remitted to the king. It was reported to the Privy Council that the conservator in Aberdeenshire did very well in the first year. "He hath not only taken divers pearls of good value, but hath found some in waters where none were expected." The first parliament of Charles I abolished these privileges.

Robert Sibbald, physician to Charles II, wrote that he had seen a necklace of Scotch pearls which was valued at two thousand crowns; they were "larger than peas, perfectly round, and of a brilliant whiteness."[3]

It is said that Sir Richard Wynne of Gwydir presented to Catherine of Braganza, queen of Charles II of England, a pearl from the Conway in Wales, which is said to be even yet retained in the royal crown. In his "Faerie Queene" (1590), Spenser speaks of the

> Conway, which out of his streame doth send
> Plenty of pearles to deck his dames withal.

The White Cart River in Scotland, on which the city of Paisley is situated, was distinguished, according to Camden, "for the largeness and the fineness of the Pearls that are frequently found hereabouts and three miles above."[4] And the pearls from Irton in Cumberland, England, were so noted at that time that "fair as Irton pearls" became

---

[1] Camden. "Britannia," 2nd edition, London, 1722, Vol. II, p. 802.

[2] Macpherson, "Annals of Commerce," Vol. II, p. 131.

[3] Sibbald, "Hist. Nat. Scotiae," 1684, Vol. III, p. 27.

[4] Camden, "Britannia," London, 1695, p. 924.

a byword in the north country.   In their history of Westmoreland and Cumberland,[1] Nicolson and Burn state that "Mr. Thomas Patrickson, late of How of this county (Cumberland), having employed divers poor inhabitants to gather these pearls, obtained such a quantity as he sold to the jewellers in London for above £800."   But in 1794 Hutchinson[2] stated that none had been seen for many years past.

Pearl fishing in Ireland was of some consequence in the seventeenth century.   Speaking of the Slaney River, Solomon Richards, in a description of Wexford about the year 1656, said: "It ought to precede all the rivers in Ireland for its pearle fishing, which though not abundant are yet excellent, for muscles are daily taken out of it about fowre, five and six inches long, in which are often found pearles, for lustre, magnitude and rotundity not inferior to oriental or any other in the world."[3]   In 1693 Sir Robert Redding wrote that there were four rivers in the county of Tyrone in northern Ireland which abounded in pearl mussels, all four emptying into Lough Foyle and thence into the sea.   They were also to be found in several rivers in the adjacent Donegal County.   Redding gave an interesting description of the fishery:

In the warm months before harvest is ripe, whilst the rivers are low and clear, the poor people go into the water and take them up, some with their toes, some with wooden tongs, and some by putting a sharpened stick into the opening of the shell; and although by common estimate not above one shell in a hundred may have a pearl, and of these pearls not above one in a hundred be tolerably clear, yet a vast number of fair merchantable pearls, and too good for the apothecary, are offered to sale by those people every summer assize. Some gentlemen of the country make good advantage thereof, and I myself, whilst there, saw a pearl bought for £2, 10s. that weighed 36 carats, and was valued at £40, and had it been as clear as some others produced therewith it would certainly have been very valuable.   Everybody abounds with stories of the good pennyworths of the country, but I will add but one more.   A miller took a pearl, which he sold for £4, 10s. to a man that sold it for £10 to another, who sold it to the late Lady Glenanly for £30, with whom I saw it in a necklace; she refused £80 for it from the late Duchess of Ormond.

The young muscles never have any pearl in them.   The shells that have the best pearls are wrinkled, twisted, or bunched, and not smooth and equal, as those that have none.   And the crafty fellows will guess so well by the shell, that though you watch them never so carefully, they will open such shells under the water, and put the pearls in their mouths, or otherwise conceal them. Yet sometimes when they have been taking up shells, and believing by such signs as I have mentioned, that they were sure of good purchase, and refused good sums for their shares, they found no pearl at all in them.   Upon discourse

[1] London, 1777, Vol. II, p. 24.
[2] "History of Cumberland," London, 1794, Vol. I, p. 573.
[3] Joyce, "Social History of Ancient Ireland," New York, 1903, Vol. II, p. 227.

with an old man that had been long at this trade, he advised me to seek not only when the waters were low, but on a dusky, gloomy day also, lest, said he, the fish see you, for then he will shed his pearl in the sand; of which I believe no more than that some muscles have voided their pearls, and such are often found in the sands.[1]

For several years following 1760, the Scotch pearl fisheries were of considerable local value. The zoölogist, Thomas Pennant, wrote of them several times in his "Tour of Scotland." Referring to the Tay and Isla rivers, then as now the center of the Scotch pearling, he states: "There has been in these parts a very great fishery of pearl, got out of the fresh-water muscles. From the year 1761 to 1764, £10,000 worth were sent to London, and sold from 10s. to £1 6s. per ounce. I was told that a pearl had been taken there that weighed 33 grains. But this fishery is at present exhausted, from the avarice of the undertakers. It once extended as far as Loch Tay."[2] And he adds later that, some years preceding, a pearl fishery was prosecuted in Loch Dochart with great success and the pearls were esteemed the fairest and largest of any.

From 1770 to 1860 the pearl resources of Scotland remained almost dormant, and Scotch pearls were rarely met with in the markets. In 1861 a German merchant, who was acquainted with the beauty· of these gems, traveled through the districts of Tay, Doon and Don, obtaining a great number which the poor people kept for their own pleasure, not esteeming them of any market value, and interested the fishermen in searching for the mussels. The seemingly high prices which he paid and the abundance of the pearls sent hundreds of persons to the rivers and small brooks. Those who were otherwise employed during the day devoted hours of the long summer nights to diligent search after the coveted shells; while boys and old persons, who were without regular avocations, waded day after day where there was a probability of reward. In the course of a short time pearls, good, bad and indifferent, reached the originator of the movement at Edinburgh, from Ayrshire, from Perthshire, and from the Highland regions far beyond the Grampians. He was soon the possessor of a collection which, for richness and variety, had seldom been surpassed. A trade in these gems was developed, the patronage of royalty was obtained, and once more Scotch pearls became fashionable, and their vogue was enhanced by the fondness which Queen Victoria entertained for them.

In addition to the rivers named, pearls were found in the Forth,

---

[1] "Transactions of the Royal Society of London, for 1693," Vol. XVIII, No. 198, pp. 659-663.

[2] Pennant, "Tour in Scotland," Chester, 1771.

the Teith, the Ythan, and the Spey in eastern Scotland. The summer of 1862 was most favorable for pearling, owing to the dryness of the season and the low water, and unusually large quantities of pearls were found, the prices ranging ordinarily from 10s. to £2 6s. Queen Victoria is said to have purchased one for forty guineas; others were bought by Empress Eugénie and by the Duchess of Hamilton. A necklace of them was sold for £350 in 1863.[1] The value of the entire catch in Scotland in 1864 was estimated at £12,000 to the fishermen, the yield being unusually large in that season owing to the unprecedented drought which permitted access to the deep beds of the rivers. In some of the streams the resources were quickly depleted, but in others the fisheries yielded profitable returns for many years. While most of the pearls were small, some of them were choice and of considerable individual value, ranging from £5 to £150, and £500 is said to have been paid for one fine specimen.

The pearl-mussel of the British Isles (*Unio margaritifera*) has a thick, coarse and unsightly shell, from 3 to 7 inches in width and 1½ to 2½ inches in length from the umbo to the lip. The rough exterior is dark brown, and it is sometimes twisted, distorted and barnacled.

It generally lies scattered and detached over the pebbly bottoms, but it also exists in reefs or beds which are sometimes of considerable extent. These occur usually where a stretch of water is still and deep, and oftentimes where the depth places the mussels beyond the reach of the fishermen. Apart from the pearls it contains, the mussel is of no economic value except that in some localities the mollusk is used for bait in cod-fishing.

In recent years the pearl-mussel has been numerous in several of the rivers of Scotland, such as the Tay, Earn, and Teith in Perthshire; the Dee, the Don, and the Ythan in Aberdeenshire; the Spey and Findhorn in Inverness-shire, and also the classic Doon of Burns, the Nith, the Annan and others in southern Scotland; however, it is rare in the Clyde and the Tweed.

The Teith has long been famed for pearl-bearing, though like other rivers it has become nearly fished out. The Tay produces many pearls, yet as a rule they are not of the best class. Some of its tributaries, as the Tummel and the Isla, also bear pearls; those in the Isla are usually fine and rank higher than those from the Tay. The Earn is also famous for the fine quality of its pearls, but the whole river was robbed of its wealth some years ago by a body of professional fishermen, and it has not yet recovered from the raid; few pearls now exist there save in the deeper pools, where doubtless may still be hid "full many a gem of purest ray serene."

[1] London "Times," December 24, 1863.

In Ireland pearls have been found principally in the rivers of counties Kerry, Donegal, Tyrone, Antrum, etc. In an article in "The Field," December 10, 1864, Mr. F. T. Buckland stated that they abound near Oughterard, and that a man called "Jemmy the Pearl-catcher," who lived there, told him that he knew when a mussel had a pearl in it even without opening the shell, because "she [the mussel] sits upright with her mouth in the mud, and her back is crooked," that is, corrugated like a ram's horn. Pearls are yet found in several localities in the Emerald Isle, notably in the river Bann in the northeastern part and in the beautiful Connemara district in western Ireland. In 1892 the Bann yielded one of the choicest pearls that ever came from Ireland. Within the last twelve months Lady Dudley, wife of the Viceroy of Ireland, presented to Queen Alexandra a number of pearls from the Connemara. These were mounted in a green enameled brooch, and excited so much admiration that an active demand for similar gems quickly developed in County Galway.

Mr. D. MacGregor, a well-known jeweler of Perth, to whom we are indebted for much information relative to pearls in Scotland, states that no attention whatever is given to conserving the mussel; on the contrary, the waters are unscrupulously despoiled by the greedy pearl fisherman who destroys all that he finds, since, by chance, they may yield the coveted gem. Immense numbers are thus wantonly destroyed, which if allowed to grow and propagate would be more likely to contribute to the pearl yield, as it is well known that it is the aged mussels in which a pearl is most likely to be found. There is no close time, and so extensive have been the raids upon the mussels in recent years that they have been rapidly exterminated in places accessible to the fishermen; and should the spoliation continue and extend to the deep waters, the pearl-mussel may soon become extinct.

Pearl fishing is not prosecuted throughout the year, as it can be carried on only in the dry season when the waters are low. There are a number of professional fishermen who search in their favorite streams, and sometimes very profitably, as much as £200 having been gained in a single season by one fisherman. One of the most noted of these was "Pearl Johnnie," who a few years ago hailed from Compar-Angus, in Perthshire, and who styled himself "Pearl Fisher to the Prince of Wales," by reason of some dealings he once had with his Royal Highness. He was very successful in his experience of more than thirty years. There is little mystery in the search; skill does not always avail, and men, women and children are rewarded or disappointed indiscriminately. The bed of the stream is searched until the patches of mussels are discovered, and this is usually the most tedious part of the work. These may be in very shoal water, where a small boy has only to wade with water above his knees and pick up the mussels by stooping; but more frequently the water covers a man's hips, and at times he is immersed almost to the shoulders.

The equipment of a pearl fisherman is simple. If he wades, he commonly wears long boots with tops reaching to his breast. Provided with a pole five or six feet long having a cleft at the lower end, and with a tube several inches in diameter with the lower end closed by a glass, he invades the home of the pearl-mussel. Thrusting the tube or water glass beneath the surface, he scans the bed of the stream, and when a mussel is sighted, the cleft pole is brought into use and it is picked up by means of these primitive tongs. Owing to   close resemblance which the pearl-mussel bears to the stones in the river-bed, good eyesight is required to avoid overlooking it. A bag by the fisher's side receives the catch; and when this is well filled, he goes to the bank of the stream and opens his lottery, in the great majority of cases to find that he has drawn a blank.

A boat is seldom used, simply because it is not available, but in the tidal waters it is indispensable. The "box" is a risky device for fishing in the deeper waters. It is a small contrivance, somewhat like the ancient British coracle, in which the fisherman sits or lies over on his chest; venturing out in the deeper parts which can not be waded, he carefully peers through the tube and draws up his find with the long cleft stick. This is a tiresome method, but some places can not be readily fished in any other manner.

In Aberdeenshire, Perthshire, etc., there are a few men who regularly spend the season "at the pearls." The knowing ones dispose of their best finds to wealthy residents or to strangers and tourists who frequent the vicinity. In addition to these experienced fishermen, many of the idlers and unemployed about the riverside towns, and also the farm servants in the country, search the waters in their neighborhood in the hope of picking up some gems. But very often it is severe and disappointing labor, for the pearl-seeker may travel far and endure privation and hardships for days, and yet, after destroying hundreds and even thousands of mussels, he may be rewarded with only a little almost worthless dross; but again and again he returns to the elusive game, inspired by the "hope which springs eternal in the human breast."

The British pearls are in great variety of colors, but most of them are practically valueless on account of the absence of orient or luster; for one possessing the white pearly luster, fifty may be found of a dull color and devoid of value. Many of these opaque pearls are dark,

The Valley of the Tay

The River Earn

Photographs by The Raeburn Portrait Studio, Perth, Scotland

SCOTCH PEARL RIVERS

GREAT CAMEO PEARL, ACTUAL SIZE 22 INCHES

Sold at auction in Amsterdam in 1776 for 180,000 florins. Note great baroque pearl
forming body of the swan at the base, diameter 1.37 inches.

lusterless brown, and handfuls of them sell for only a few shillings. A large percentage are of a grayish or milky color, or of a bluish white tinge; these seldom attain much value unless aided by excellence of shape and purity of skin. A few are of a dark, fiery tint and of great luster. Sometimes the pearl is of a beautiful pink tint, sometimes of a light violet, or other exquisite shade. The fine pink ones are very rare and are highly prized. The best are those having the sweet, pure white light which constitutes the inimitable loveliness of a pearl; but few of them are found even in the most favorable seasons, and usually these are from the streams in the northeastern counties and some of the streams in the southwest. Very few combine the qualities of perfection in shape and luster; and the product of many seasons might be examined in vain to furnish enough pearls to make a well-matched necklace of gems weighing from five to ten grains each. But occasionally beautiful specimens are discovered, weighing fifteen or twenty grains or more. One found in Aberdeenshire a few years ago, perfect in shape and luster, weighed twenty-five grains, and sold at first hand for £50. Another one, found at the confluence of the Almond and the Tay in 1865, weighed thirty grains.

While most of these pearls are sold to jewelers in Edinburgh, Aberdeen, Inverness, Perth, and other towns, many of the finest specimens have gone into the possession of prominent Scotch and English families, who have a fancy for collecting them. Queen Victoria possessed a fine collection of Scotch pearls, choice specimens of many years' search, obtained almost exclusively from the Aberdeenshire waters which murmur round her beautiful Highland home. In 1907, a Scotch pearl was sold in Perth for the sum of £80; this was of a good luster with a bluish tint, it was spherical, measured seven sixteenths of an inch in diameter, and weighed twenty-one grains.

The falling-off in the yield of pearls in some streams is credited to a certain extent to the building of bridges and the consequent abandonment of fords. This is based on the theory that injury to the mollusk has something to do with the production of pearls, and that they are to be found more plentiful about fords and places where cattle drink. The theory is beautifully stated by the lamented Hugh Miller: "I found occasion to conclude that the Unio of our river-fords secretes pearls so much more frequently than the Unionidae and Anadonta of our still pools and lakes, not from any specific peculiarity in the constitution of the creature, but from the effects of the habitat which it chooses. It receives in the fords and shallows of a rapid river many a rough blow from the sticks and pebbles carried down in time of flood, and occasionally from the feet of men and animals that cross the stream during droughts, and the blows induce the morbid secretions, of which pearls are the result. There seems to exist no inherent cause why *Anadon cygnea*, with its beautiful silvery nacre—as bright often, and always more delicate, than that of *Unio margaritiferus*—should not be equally productive of pearls; but secure from violence in its still pools and lakes, and unexposed to the circumstances that provoke abnormal secretions, it does not produce a single pearl for every hundred that are ripened into value and beauty by the exposed, current-tossed Unionidae of our rapid mountain rivers. Would that hardship and suffering bore always in a creature of a greatly higher family similar results, and that the hard buffets dealt him by fortune in the rough stream of life could be transmitted, by some blessed internal pre-disposition of his nature, into pearls of great price."[1]

The small blue mussel (*Mytilus edulis*) of the British seas yields opaque pearls of a deep blue color, but most of them are more or less white in some part. Sometimes a shell is found in which a blue pearl will be adhering to the blue lip of the shell while a dull white one adheres to the white portion of the shell. These pearls are commonly flattened on one side, doubtless where they have been adjacent to the shell. None of them is of more than very slight value.

Probably the principal fishery for the salt-water mussel pearls is that in the estuary of the Conway in Wales. These are mostly quite small and well answer the designation of seed-pearls, although a few are of fair size. In color most of them range from dirty white to the dusky or brownish tint noted by Tacitus eighteen centuries ago, but a few are of a pure silvery tint. In some seasons London dealers have agents at Conway for purchasing these pearls. The price is usually from eight to thirty shillings per ounce.

THE CONTINENT OF EUROPE

Après l'esprit de discernement, ce qu'il y a au monde de plus rare, ce sont les diamants et les perles.

LA BRUYÈRE, *Les caractères.*

PEARLS occur in species of mussels found in the streams and lakes of Europe, in some of which the fisheries have been of considerable local interest. It appears that these resources were exploited by the Romans, then by the Goths and the Lombards, and later the natives continued to draw forth the treasures which lay hidden about their

[1] Hugh Miller, "My Schools and Schoolmasters," 1852, p. 201.

homes. These pearls have attracted attention up to the present time; and while they do not compare with those of the seas, either in quality or in aggregate value, yet they are prized on account of their intrinsic worth as well as because they are a product of the fatherland. In the densely populated valleys, the rivers are so polluted by refuse and sewage that the mollusks have been greatly depleted; but in the streams of clear, cool water, draining the mountain regions of France, Germany, Austria, and also in the rivers of Norway, Sweden, Russia, etc., the fisheries are not unimportant.

The most celebrated of the pearl fisheries in France are those of the Vologne, a small river in the extreme eastern part of the country, in the department of Vosges. Its sources are in Lake Longmere in the Vosges mountains on the Alsace frontier, and it flows into the Moselle at Jarmenil, between Remiremont and Épinal. While the pearl-mussel occurs to some extent in nearly the whole length of this river, and, indeed, is to be met with in the wild brooks and forest streams of nearly all the mountainous parts of France, it is most abundant in the vicinity of Bruyères, where the Vologne receives the waters of the Neuré. These resources were described in 1845 by Ernest Puton,[1] and in 1869 by D. A. Godron;[2] to whom—and especially to Godron—we are indebted for much of our information.

The fisheries of the Vologne have been celebrated for nearly four centuries. Writing in 1530, Volcyr stated: "In the river Vologne between Arche and Bruyères, near the ancient castle of Perle, beautiful pearls are found. In the opinion of jewelers and artists they closely resemble the oriental."[3] A few years later Francis Reues wrote: "There is near the Vosges mountains in Lorraine a river fertile in pearls, yet they are not very brilliant. The strange thing is that the quality which they lack by nature is supplied by the aid of pigeons, which swallow them and restore them purer than before."[4] In a publication of 1609, this little river is represented in the frontispiece by the figure of a nymph bearing many pearls, while beneath is the emblem: *Vologna margaritifera suas margaritas ostentat.*[5]

In his paper above noted, Godron recites several orders issued from 1616 to 1619 by the Duke of Lorraine, who then had jurisdiction over the present department of Vosges, showing that a high value was

[1] Puton, "Mollusques terrestres et fluviales des Vosges: Le Département des Vosges, statistique, historique, et administrative, par Henri Lepaye et Ch. Charton," Nancy, 1845, 8vo, 2 vols., Vol. I.

[2] Godron, "Les perles de la Vologne, et le Château-sur-Perle." "Mémoires de l'Académie de Stanislas, 1869," Nancy, 1870, pp. 10-30.

[3] Volcyr, "Cronicque abrégée par petits vers huytains des Empereurs, Roys, et Ducz d'Austraisie," etc., Paris, 1530.

[4] Reues, "De Gemmis aliquot," etc., Tiguri, 1566, p. 47.

[5] Claude de la Ruelle, "Les pourtraicts des ceremonies, . . . et pompe funèbres faitez au corps de feu Charles III, Duc de Lorraine," etc. Nancy, 1609.

attached to these pearls and that the resources were well looked after. Writing in 1699, Dr. Martin Lister alluded to the many pearls taken from the rivers about Lorraine and Sedan. A Paris merchant showed him a fresh-water pearl of 23 grains, valued at £400, and assured him that he had seen some weighing 60 grains each.[1]

In 1779 Durival gave an extensive account[2] of the Vologne fishery. He records that for sixty years pearls had been abundant, but at the time he wrote they were very scarce.

Puton states that, in 1806, when taking the baths at Plombières in the Vosges, Empress Josephine formed a great liking for the Vologne pearls, and at her request some of the mussels were sent to stock the ponds at Malmaison. It does not appear that any favorable result followed this transplanting.

Owing to the extensive fisheries, the mussels became so scarce that in 1826, when the Duchesse d' Angoulême was visiting in the Vosges, it was impossible to secure enough pearls to form a bracelet for her. This scarcity has continued up to the present time; and yet in the aggregate many pearls have been secured, so that there are few prominent families in the neighborhood who do not possess some of them. They are especially prized as bridal presents to Vosges maidens.

While the Vologne pearls are of good form and of much beauty, they do not equal oriental pearls in luster. The color is commonly milky white, but some of them have a pink, yellow, red, or greenish tint. In size they rarely exceed 4 grains. The Nancy museum of natural history possesses one which weighs 5¼ grains and measures 6½ mm. in diameter.

In western France, according to Bonnemere,[3] the pearl-mussel is widely diffused, and in the aggregate many pearls are secured therefrom. They are somewhat numerous in the river Ille near its union with the Vilaine at Rennes; though small, these are commonly of good color and luster. In the department of Morbihan and that of Finistère, many pearls have been secured, especially in the Steir, the Odet, and in the Stang-Alla near Quimper. Small pearls, frequently of some value, are found in the Menech near the town of Lesneven, a few miles northeast of Brest, the great naval port of France.

The *Unio sinuatus* (*pictorum*), the *mulette* of the artists, which has a shorter and smaller shell than the pearl-mussel, has also yielded many small pearls of good quality, as well as shells for manufacturing

[1] Lister, "Journey to Paris in the year 1698," London, p. 143.
[2] Durival, "Description de la Lorraine et du Barrois," Nancy, 1779, Vol. I, p. 280.
[3] Bonnemere, "Les perles fines de l'Ouest de la France," "Revue des sciences naturelles de l'Ouest," 1899, Vol. III, p. 97–99.

purposes. This species has been regularly exploited in the Adour, in the Charente, in the Gironde and its tributaries—the Garonne and the Dordogne and their affluents, and in some other streams in western France.

There is a pearl fishery in the Charente River near the western coast of France, and likewise in the Seugne, a small tributary entering it from the south. The mussel is known locally under the name of *palourde*. In an account of this fishery,[1] Daniel Bellet states that in the Seugne, where the water is shallow and clear, the mussel is secured by entering the pointed end of a wooden staff or stick between the valves of the open shell as the mollusk lies feeding on the bottom; as the shell is immediately closed tightly upon the intruding stick, it is easily removed from the water.

In the deeper waters of the Charente, the fishery is prosecuted on a larger scale. Until recently, the *palourdes* were caught by means of a dredge towed by a small boat, which was raised from time to time and the catch removed. Ten or fifteen years ago the scaphander or diving apparatus was introduced, requiring seven men for its operation, and by its use large catches have been made. The mussels are taken to the bank and there boiled for a time to cause the shells to open, so that the contents may be easily removed.

The shells are examined one by one to find any pearls that may adhere thereto, and then the flesh of the mollusk is crushed between the fingers to locate pearls contained in the mass; this is done largely by children, working under competent supervision. Many pearls of fairly good size and luster are obtained. The flesh of this mollusk is edible and well-liked in southwestern France; and the shells are also of value in the manufacture of buttons and similar objects.

In Germany the pearl fisheries are most important in streams of the southern districts, in Bavaria, Saxony, and Silesia. The pearl-mussel in these waters is not so abundant as formerly; yet, owing to the care which has been given to these resources, it is probably as numerous here as in any other part of the continent. The mussel rarely occurs singly, generally in small beds or banks contiguous to each other, and in some favorable regions these are extensive.

The pearl fisheries of Bavaria have been prominent since the sixteenth century. They exist principally in the districts of Upper Franconia (*Oberfranken*) and Upper Palatinate (*Oberpfalz*), the several tributaries of the Danube between Ratisbon and Passau, and in those tributaries of the Main and the Saale which rise in the Bavarian mountains, such as the Oelsnitz, the Lamnitz, Schwesnitz, Grünebach, Vils, and the Perlbach; also in the district of Lower Bavaria, where in

[1] "La Nature," 1899, pp. 347, 348.

nine districts alone there are one hundred pearl-bearing streams and lakes, of which the most important are the Regen, the Isar, and the Ilz.[1]

Early in the sixteenth century, the river Ilz had the reputation of yielding the choicest pearls in Lower Bavaria. The right to them was reserved to the bishop of Passau, and a decree was made in 1579 that persons convicted of poaching on these reserves should be hanged.[2] Since that time there have been few decades in which the gems have not been found in the woodland brooks and mountain streams that flow through the ravines and past quaint, interesting castles of the wonderful Bavarian highlands. Most of the prominent families in this beautiful region have collections of native pearls, and there is still some trade in them in picturesque Passau. at the junction of the Danube, the Ilz and the Inn.

Tavernier wrote about 1670: "As for the pearls of Scotland, and those which are found in the rivers of Bavaria, although necklaces are made of them which are worth up to 1000 *écus* (£225) and beyond, they cannot enter into comparison with those of the East and West Indies." [3]

The official returns for the Bavarian fisheries, dating from the latter part of the sixteenth century, were examined by Von Hessling in 1858. He noted many gaps in the statements of the yearly returns, partly on account of the loss of the records and partly because the pearls were delivered directly into the hands of the princes. The results of the first fisheries are recorded in the district of Hals for the years 1581–99, in Viechtach for 1581–83 and 1590–93, and in Weissenstadt and Zwiesel for 1583. The range of the fisheries was enlarged through the discovery of new areas during the first half of the seventeenth century; but this was offset by the bad seasons and by disturbed conditions during the Thirty Years' War. From 1650 to 1783 the pearls in the forest lands of the Palatinate were exploited regularly and uninterruptedly, with the exception of the district of Wetterfeld and that of Neunburg vor dem Wald, where they were prosecuted for a few years only. From 1783 to 1814, they were almost entirely neglected, and the take was confined to a few streams in Upper Palatinate and in the Bavarian forests. In the former episcopal principality of Passau, where, according to general accounts, the waters were rich in pearls, the records were scanty previous to 1786; this was probably owing to the fact that the head gamekeeper was obliged to transmit the catch of pearls directly to the prince-bishop. The records for the

[1] Von Hessling, "Ueber die Erzeugung künstlicher Perlen," "Gelehrte Anzeigen der Münchener Akademie," 1856, Vol. II, p. 159.

[2] Weinmann, "Bresslauer Naturgeschichten," 1725.

[3] Tavernier, "Travels in India," 1889, Vol. II, p. 113.

fisheries in the districts of Rehau and Kulmbach began with the year 1733.

From these fragmentary returns—making no estimate for the years for which there were no figures available—Von Hessling found that from 1600 to 1857 there were taken 15,326 pearls of the first class, which were clear white in color and of good luster; 27,662 pearls of the second class, which were somewhat deficient in luster, and 251,778 pearls of the third or poorest class, or "*Sandperlen,*" which, though of poor quality, had sufficient whiteness and luster to be used as ornaments. Had the records been complete, these figures would probably have been at least fifty per cent. greater, or a total of about 445,000 pearls in the 257 years. In the last forty-three years of this period, for which the records are fairly complete, the annual average was 208 pearls of the first, 395 of the second, and 3091 of the third class, a total each year of 3694 pearls of all grades. This was divided among the districts as follows:

ANNUAL AVERAGE

District	First class	Second class	Third class	Total
Upper Franconia	13	34	52	99
Upper Palatinate	38	77	207	322
Lower Bavaria	157	284	2832	3273
Total	208	395	3091	3694

Probably the most interesting of the pearl fisheries in Germany are those prosecuted in the extreme southwestern part of the kingdom of Saxony, in the picturesque region known as Vogtland. This is not on account of their extent, for the output rarely exceeds $2000 in value in any season; but because for nearly three hundred years they have been conducted with the utmost care and regard for the preservation of the resources. Indeed, a record exists of practically every pearl obtained for nearly two centuries.

The waters in which the Saxon Vogtland fisheries are prosecuted are the Elster River, from the health resort of that name to a short distance below Elsterberg; its tributaries, the Mülhaüser, Freiberger, and Marieneyer brooks; the Hartmannsgrüner and the Triebel brooks, the Trieb, the Meschelsgrüner, the Teil, and Loch brooks, and twenty-five or more small ponds.

For most of the data relative to these fisheries, we are indebted to J. G. Jahn's "Die Perlenfischerei im Voigtlande," Oelsnitz, 1854; to Hinrich Nitsche's "Süsswasserperlen, Internationale Fischerei-Ausstellung zu Berlin," 1880, and to O. Wohlberedt's "Nachtrag zur Molluskenfauna des Königreiches Sachsen," "Nachrichtsblatt der deutschen Malakozoologischen Gesellschaft," Frankfurt-am-Main, 1899, pp. 97–104.

In the year 1621, the electoral prince, Johann Georg I, reserved the pearl fishery of the Vogtland in Saxony as a royal privilege, and appointed Moritz Schmerler as superintendent and fisherman. From that time until the present, this fishery has remained a royal prerogative; and, remarkable to state, except at the close of the seventeenth century when the father-in-law of a Schmerler enjoyed the privilege, all the superintendents of the fishery—twenty-four persons in number —have been direct descendants of the second pearler, Abraham Schmerler, who, in 1643, succeeded his brother Moritz. The present superintendent Julius Schmerler has been in charge since 1889.

This fishery is conducted in accordance with regulations of the chief inspector of forests for the district of Auerbach. The present regulations date from June 15, 1827. In compliance therewith an inspection is made of the waters each spring to remove all obstructions and debris that would injure the resources; and, if necessary, entire beds of mussels are removed from one locality to another which appears more favorable. No mussels are opened at that time, for the real search for pearls does not begin until the season is far advanced and the fishermen can wade up to the waist in the water without discomfort.

Dr. Nitsche states that the whole pearling district is not searched over every year, but is divided into 313 sections, each one constituting a day's work for three fishermen, and rarely are more than twenty or thirty of these fished in any one year. Thus each section or district is permitted to rest and recuperate for ten or fifteen years before it is again invaded. Every mussel is opened carefully by hand, with the aid of a peculiarly constructed iron instrument. By inserting the edge of this between the nibs of the shell and turning it at right angles, the valves are opened sufficiently to determine whether a pearl is contained therein. If none is observed, the instrument is released and the mussel returned uninjured to the water; but if a pearl is found within, the shell is forced open and the find removed. In case small pearls are observed which give promise of growing larger in time, they are not removed, but the year is marked upon the shell with the opening implement and the mussel returned to the water. It often happens that good pearls are later removed from shells marked in this manner.

Complete records exist of the yield of this fishery during each year since 1719, when the Vogtland passed to the electorate of Saxony. The following is a summary of these records arranged in series of twenty years each.

DOWAGER CZARINA OF RUSSIA   GRAND DUCHESS VLADIMIR   GRAND DUCHESS MARIE PAVLOVNA

MITER OF PATRIARCH NIKON

Presented by the Czar Alexis Mikhailovitch and the Czarina Marie Illiinichna. Decorated largely
with European fresh-water pearls. Now in the treasury of the Patriarchs, Moscow.

Years	Clear pearls No.	Half clear pearls No.	Sand pearls No.	Damaged pearls No.	Total No.	Average per year No.
1720–1739 ...	1,809	727	1,201	552	4,289	214
1740–1759	1,412	578	484	281	2,755	138
1760–1779	1,042	272	427	219	1,960	98
1780–1799	1,261	243	357	179	2,040	102
1800–1819	1,603	261	325	203	2,392	120
1820–1839	1,659	340	326	326	2,651	133
1840–1859	1,884	610	387	505	3,386	169
1860–1879	1,618	682	450	514	3,264	163
1880–1899	471	394	86	373	1,324	66
1900–1905	79	161	22	86	348	58
Total in 186 years	12,838	4,268	4,065	3,238	24,409	
Average per year	69	23	22	17	131	

In recent years the development of manufacturing industries in Saxony and the resultant pollution of the water has greatly reduced the abundance of the mollusks and consequently the output has been much restricted. The average annual yield in the twenty years ending in 1879 was 163 pearls; in the twenty years ending in 1899 it was 66 pearls, and in the six years ending in 1905 the annual average was 58 pearls. Owing to high water, there was no fishing in 1888; and with a view to permitting the resources to recuperate, the fishery was suspended from 1896 to 1899, inclusive. Omitting these five years, the average yield during each season in the two decades ending 1899 was 88 pearls.

At the end of each season, the pearls secured are turned over to the director of forestry for the district of Auerbach; by him they were formerly sent to the royal cabinet of natural history, or to the royal collection at Dresden, but since 1830 they have been sent to the royal minister of finance, by whom they are sold each year. The total proceeds from these sales now amount to about 55,000 marks.

In former times, according to Dr. Nitsche, it was customary to use these pearls in making royal ornaments. This was the origin of the famous Elster necklace, consisting of 177 pearls, now in the art collection in the Grüne Gewölbe in the palace at Dresden. Another assortment in that collection consists of nine choice, well-matched pearls, weighing 140 grains. For a necklace of Saxon pearls, the property of a duchess of Sachsen-Zeitz, the sum of 40,000 thalers ($28,400) is said to have been refused.

In Prussian Silesia the pearl-mussel is found in the upper tributaries of the Oder, especially in Bober River from Löwenberg to the sources among the foot-hills of the beautiful Riesengebirge, in the Lusatian Neisse to Görlitz, the Queiss above Marklissa, and in the Juppel as far as Weidenau. The Queiss has been famous for its pearls since the sixteenth century, and even yet specimens of great beauty are obtained therefrom. As long ago as 1690, Ledel complained of the diminution of the number of mollusks owing to their wilful destruction by children; and in 1729 the government issued a rescript in Upper Lusatia (*Oberlausitz*) recommending the care of the young mollusks.[1]

Pearls are also found in the White Main a short distance from its source, in the head waters of the Saale, and in numerous other mountain-draining streams of middle Germany. Indeed, references could be made to the discovery of pearls in nearly every stream of Germany at some time during the last three or four centuries.

The records of pearl fisheries in the province of Hanover were traced by Von Hessling as far back as the sixteenth century, when they were prosecuted in the Aller, Ovia or Om, Lua or Low, and in the Seva in the district of Lüneburg. During the reign of Christian Ludwig (1641–65) and in that of George William (1666–1705), pearl fishing was carried on by the state, and old records of the former district of Bodenteich note the customs and practices of that period and of earlier times, and the implements employed. In 1706, for instance, 265 clear and 292 imperfect pearls were taken by three official fishermen from the Gerdauerbach. Gradually, however, owing to indifferent management, the brooks yielded less and less; the government seems to have entirely abandoned supervision of them, so that, according to Taube's "Communication,"[2] slight results were obtained in 1766; indeed, only a few pearls could be shown as curiosities.[3]

Regarding the condition of the Hanoverian pearl-brooks, especially of those in the vicinity of Uelzen, Möbius wrote: "Uelzen lies at the confluence of eleven small rivulets, three of which, the Wipperau, the Gerdau and the Barnbeck, contain pearl-mussels. Fishing has been pursued here for centuries, and there exists an old regulation of the sixteenth century in regard to the pearl fisheries in the Ilmenau. Even at the present day, hundreds of pearls are found here which command a good price when they are bright and of good form. These either have a silvery sheen or they are of a reddish color. The season for fishing is during the months of July and August. The pearls are usually found in deformed shells. Their shape varies greatly; most of them are flat on one side. Naturally those which are spherical are the best, but the pear shapes are highly prized." Möbius

---

[1] Von Hessling, "Die Perlenmuscheln," Leipzig, 1859, p. 179.
[2] "Beiträge zur Naturkunde des Herzogthums Celle," Halle, 1766, Pt. I, p. 70.
[3] Von Hessling, "Die Perlenmuscheln," p. 180.

frequently failed to find one pearl in a hundred shells, but at other times he came across six or eight in this quantity. Most of the mussels are found in the deepest places, especially near the banks of the streams. One end of the shell usually projects out of the sand. The fisherman is represented as feeling about the bottom with his feet, and when he finds a shell, he seizes it between his toes, picks it out, and then places it in the basket suspended from his neck.[1]

In Baden and in Hesse are small pearl fisheries. In 1760, Elector Maximilian III sent to Mannheim, then in the Palatinate, eight hundred living pearl-mussels from the Bavarian forests, and again in 1769, he sent four hundred mussels from Deggendorf on the Danube, so that they might be established in the Palatinate. The mussels were placed in the Steinbach not far from Heidelberg, where they thrived so well that fishing was instituted in 1783. Soon, however, most of the mussels became buried in the sand, and the remainder were transplanted into a quieter portion of the Steinbach, between Kreutzsteinach and Schönau, about five miles northeast of Heidelberg. Here they seem to have been forgotten, and were left undisturbed until, about 1820, a fine pearl valued at two louis d'or was found near Schönau. This discovery soon led to such reckless exploitation that the government reserved the fishery as a state monopoly. The mussels were examined and sorted, and a portion of the brook was specially prepared for their reception. However, the cost of supervision was greater than the proceeds of the fishery, and the business was rented to private parties for a very small amount. This was paid as late as 1840 by the Natural History Society of Mannheim, the annual rate then being ten florins.

An effort was made nearly two hundred years ago to develop the pearl fisheries in Hesse. In 1717, Landgrave Prince William requested his cousin, Duke Moritz of Saxony, to send a pearl fisherman "to examine some streams in his territory where mussels have been found and to determine whether they are fitted for pearl fishing and whether fisheries can be established."[2] In the following year, a member of the famous Schmerler family from the Saxon fisheries was sent to Cassel, but with what result is unknown.

When the pearling excitement developed at Schönau about 1820, Landrath Welker, of Hirschhorn on the Neckar, requested the grand duke of Hesse to place him in charge of the fishery, and when the proposition was declined, he formed a small company for pearl culture. In 1828 his company had 558 mussels, 88 of which showed pearl for-

---

[1] Möbius, "Die echten Perlen," Hamburg, 1858, p. 47.
[2] Jahn, "Voigtländische Perlenfischerei,"
p. 165; Von Hessling, "Die Perlenmuscheln," p. 182.

mations; in 1833, out of 651, 98 contained such objects, and in 1851, 117 mussels were found with pearl formations out of 867 examined.[1] Owing to the policy of the company in selling the pearls only among the members thereof, the profits were altogether insufficient to cover the expenses, and gradually the fishery dwindled down until it was prosecuted only as a pastime.

Pearls are found in the province of Schleswig-Holstein, which formerly belonged to Denmark, but since 1866 has been a part of the kingdom of Prussia. Möbius relates that the Bavarian soldiers in 1864 collected large quantities of pearls from the streams of this province and sold many of them to jewelers in Hamburg.[2] Most of them were of good form and luster; milky white was the prevailing tint, but some were pink and others were rose-tinted.

In Austria, pearl fisheries are most important in the province of Bohemia, where they are prosecuted in the headwaters of the Moldau from Krumau, a few miles above Budweis, to below Turenberg, and to a much less extent in its tributary, the Wottawa, on the northeastern slopes of the Böhmer Wald or Bohemian Forest mountains. From very early times the right of fishery belonged to those domains and estates through which the streams flow, as for example, the cloister of Hohenfurth, the domain of Rosenberg, of Krumau, etc. The Schwarzenberg family formerly drew a considerable revenue therefrom. Over a hundred years ago the fishery was actively prosecuted by Count Adolph Schwarzenberg, who exhibited at the Bohemian Exposition, held in Prague in 1791, an interesting collection of shells, apparatus employed in the fishery, and many beautiful pearls obtained from his domains. The fisheries of the Wottawa were noted in 1560 by the Swiss naturalist Konrad von Gesner,[3] and again in 1582 by the district treasurer, Wolf Huber von Purgstall. In 1679, Balbinus referred to the excellent qualities of the pearls, estimating the value of many of them at twenty, thirty, and even one hundred golden florins each. He described the methods by which they were taken, and also complained of the destruction of the reefs by depredations of poachers.[4]

The Wottawa or Otawa River has long had linked with its name the epithet "the gold- and pearl-bearing brook." Formerly, along its shores gold washing was more or less carried on, as well as the freshwater pearl-mussel industry. At the present time, every third or fourth year, these mussels are gathered, by means of small, fine-woven nets, from the bed of the river, and a goodly number of pearls are collected.

[1] "Von Hessling, "Die Perlenmuscheln," p. 182.
[2] "Die echten Perlen," p. 48.
[3] Gesner, "De aquatilibus," Tiguri, 1560.
[4] Bohuslai Balbini, "Miscellanea historica regni Bohemiæ," Prague, 1679, Vol. I, p. 73.

The reefs in the Moldau from Hohenfurth to Krumau were almost entirely ruined in 1620 by the troops who were cantoned there when the Bohemian Protestants were overthrown near the beginning of the Thirty Years' War, and they never regained the reputation they formerly enjoyed. According to the Vienna "Handels- und Börsenzeitung," the output of the pearls fifty years ago in the upper Moldau, in the Wottawa, and in the Chrudimka—a tributary of the Elbe— reached in some years the sum of one million florins in value, and as much as eighty and sometimes even one hundred and twenty florins were paid for an individual specimen.[1] These pearls closely resemble those from Passau in Bavaria, and some approach the oriental gems in luster.

In the archduchy of Austria, pearls occur in several of the tributaries of the "beautiful blue Danube." They are especially important in streams within the former district of Schärding, such as the Ludhammerbach, the Ranzenbergerbach, the Glatzbachenbach, the Brambach, the Schwarzbergerbach, the Mosenbach, and the Hollenbach; those in the former district of Waizkirchen, including the Pirningerbach, the Kesselbach, and many of their tributary brooks, and the Michel, the Taglinsbach, the Fixelbach, and the Haarbach, in the domain of Marbach.[2] Fishing in the Pirningerbach and the Kesselbach was prosperous about 1765, and Empress Maria Theresa received a beautiful necklace and bracelets of the pearls therefrom. In the district of Marbach, the fishing was prosecuted as long ago as 1685 for the account of the archbishop of Passau.

In Hungary from time immemorial, the native pearls have been popular with the Magyar women, and very many yet exist in the old Hungarian jewelry worn with the national costume. A century ago there was scarcely a family of local prominence which did not possess a necklace of pearls, although these were frequently not of choice quality or of considerable size. With a falling off in the output of the native streams there has been a great increase in the quantity of choice oriental pearls purchased by the wealthy families, and some of the most costly necklaces in Europe are now owned here.

In the kingdom of Denmark no pearl fisheries are now prosecuted, but three centuries ago the gems were taken in the Kolding Fjord in the province of Veile, Jutland. The great Holberg, who ranks first in Danish literature, wrote that the governor of the castle at Kolding employed as a pearl fisherman a Greenlander who had come to Denmark in 1605 or 1606, and who "had given the governor to understand that in his native land he was accustomed to fish for pearls."

[1] "Allg. Zeitung," Nov. 1, 1858, No. 305.   [2] Von Hessling, "Die Perlenmuscheln," Leipzig, p. 178.

Being required to work continuously, both winter and summer, he fell ill and died, and as no one else wished to pursue the occupation, the fishery ceased.[1]

In many of the Norwegian brooks, pearl fishing has been carried on for two or three centuries, and often with satisfactory results. It appears from ordinances dated November 10, 1691, May 14, 1707, and May 28, 1718, that the fisheries were under special supervision as a royal prerogative of the queen of Denmark.[2] Jahn notes that in 1719 and in 1722, Saxon pearl fishermen were sent for. In 1734 Charles VI of Denmark requested the elector of Saxony to send one of the pearl fishermen of Vogtland to examine the brooks of Norway in reference to the pearl resources, and to determine the practicability of establishing fisheries there. In response to this request, C. H. Schmerler was sent to Copenhagen and thence to Christiania, where he began an investigation of the Norwegian waters, the governor himself attending at the beginning of the work. So great was the estimation of its importance, that Schmerler was soon afterward received in audience by the king and queen of united Denmark and Norway at Frederiksborg palace near Copenhagen, and was awarded a gift of one hundred ducats and a life-pension.[3]

In 1751, according to Pontoppidan, Bishop of Bergen, the Norwegian pearl fisheries were placed under the jurisdiction of the diocese of Christiansand. Among the principal pearling regions at that time were the Gon, Närim and Quasim rivers in the Stavanger district or amt; the Undol, Rosseland and other brooks in the Lister and Mandal province; and several streams in the district of Nadenäs.[4]

The returns from the Norwegian fisheries gradually decreased. After 1768 the rights were leased, and the revenue therefrom was paid into the royal treasury. Owing to small returns, this source of revenue received less and less attention, and about a century ago it was altogether neglected, although from time to time choice finds were made. Due to unusually low water in 1841, a number of valuable pearls were found near Jedderen in the province of Christiansand, some selling as high as $300 each; several of these were shown at the London Industrial Exhibition by the diocese of Christiania.

The pearl fisheries of Sweden were noted, nearly four centuries ago, by Olaus Magnus, Archbishop of Upsala.[5] The gems were

[1] Holberg, "Danmarks Riges Historie," Reicharot edition, 1743, Vol. II, p. 632.

[2] Thaaruys, "Versuch einer Statistik der dänischen Monarchie," Copenhagen, 1795, Pt. I, p. 416.

[3] Jahn, "Voigtländische Perlenfischerei,"

p. 175; and Von Hessling, "Die Perlenmuscheln," p. 189.

[4] Pontoppidan, "Versuch einer natürlichen Historie von Norwegen," Copenhagen, 1754, Vol. II, p. 309.

[5] Olaus Magnus, "Historia de gentibus septentrionalibus," Antwerp, 1562, c. 6, p. 192.

PANAGIA OR ORNAMENT WORN ON THE BREAST OF A BISHOP IN RUSSIA

RUSSIAN BOYARD LADIES OF THE SEVENTEENTH CENTURY, SHOWING
CAPS AND OTHER ORNAMENTS OF PEARLS

sought for by expert fishermen in the interior districts, and were brought in large quantities to the coasts for sale, the women and girls of all classes, rich and poor, using them extensively in personal decoration.

The celebrated Linnæus left a detailed account of the method by which mussels were caught in Sweden nearly two centuries ago. He wrote: "In the summer season, if the water is shallow, the fishermen wade in the stream and gather the mussels with their hands. Should the water be deeper, they dive for the mussels and place such as they find in a vessel made of birch bark, which they carry with them. Sunny days are selected, because then they can see deeper into the water. But, should this not suffice, they traverse the river on rafts which are painted white beneath so that the bed of the stream may be illumined by the reflected light. The men lie prone on the rafts and look down into the depths so that they may immediately seize with wooden tongs the mussels which they discover. Or else, hanging by their hands to the rafts, they seize them in the water with their toes. If the water is too deep even for this, they dive and feel around on the bottom with their hands until it becomes necessary to rise again to the surface in order to breathe. However, out of a hundred mussels, scarcely one contains a good pearl; but sometimes as many as twenty pearls of the size of a grain of sand are found in one shell. Many of the larger pearls are reddish or dark, but occasionally a beautiful white pearl is hidden under such a covering; although, naturally, it is rare that this is altogether perfect. It has been noted that mussels seven years old contain pearls; and in each of two mussels eighteen years old, a pearl was found attached to the shell."[1]

The list of streams in Sweden from which pearls were taken, as noted by Olaf Malmer, J. Fischerstein, and Gissler[2] a century and a half ago, seems to cover nearly all the rivers and brooks which flow from the mountains of this beautiful country.

In Russia the love for the pearl has been almost as great as in Persia and India. During the Middle Ages, pearls were worn upon the clothes of nearly all well-to-do Russians. The great head-dresses of the women were ornamented with them; and they were used in decorating the stoles, vestments, crosses, and the priceless relics in the churches.

The pearl-mussel is found in very many of the Russian streams. It occurs throughout Archangel, in most of the rivers which flow into the White Sea, into Lake Ladoga, Lake Onega, and the Baltic Sea; and likewise in the Volga watershed. Von Hessling states that east

[1] Linnæus, "Lach. Lapponica," Vol. II, pp. 104-107.
[2] See "Abhandlungen der Schwedischen Akademie," 1742, Vol. IV, p. 240; 1759, Vol. XXI, p. 136, and 1762, Vol. XXIV, p. 64.

of the Volga its southern boundary extends to Lat. 56°, while on the west it extends further southward, so that in the region of the Dnieper it reaches Lat. 51°. The extreme southern limit is near the mouth of the Don, about 47° north latitude.[1]

In northern Russia pearls are secured in the provinces of Livonia, Esthonia, and Olonetz, and in the grand duchy of Finland, where they have been sought after for three centuries or more. Most of them are bluish gray in color and they attain a maximum weight of about twelve grains. Although not equaling the oriental gems, these pearls are of good quality and are highly esteemed, not only by the peasants but by the nobility and by the royal family of Russia. For reference to most of the historical data relative to the fishery in Livonia, we are indebted to an account written by H. Kawall.[2]

So long ago as 1612, Dionysius Fabricius compared the pearls of Livonia with those of India. Said he: "Nor should I omit to mention that there are rivers in Livonia wherein large pearls are produced in shells; and I myself have seen some as large as the oriental, especially when they are well grown. But because the peasants of this region are too ignorant to determine with certainty when they mature, they are unable to collect them properly, and therefore the pearls have become rarer."[3]

According to Mylius,[4] in the seventeenth century, when Livonia belonged to Sweden, the pearl resources received attention from the government. Charles IX of Sweden decreed October 22, 1694, that the pearls therefrom should not be exported but should be sold to officers of the crown at a definite price. In 1700, an inspector of the fishery in Livonia, whose name was Krey, reported that the peasants collected pearls secretly from the small rivers and brooks, and forwarded them to Moscow for sale. As the peasants objected to selling them to the king's commissioners at the prices fixed, the fishery soon dwindled in extent. However, on the annexation of Livonia to Russia in 1712, and the removal of these restrictions, it revived and became of local importance during the last years of the reign of Peter the Great.

In 1742 the Livonian fishery was reorganized at the suggestion of a Swede named Hedenberg. Furnished by the government with funds and an escort, he began an exploration of the pearl-bearing waters, commencing with Lake Kolk, where he secured many pearls of value, some of which were presented to Empress Elizabeth.[5]

[1] "Die Perlenmuscheln," Leipzig, 1859, p. 194.
[2] Kawall, "La pêche des perles en Livonie," "Annales de la Société Malacologique de Belgique," 1872, Vol. VII, pp. 38-46.
[3] Dionysius Fabricius, "Scriptor rerum Livonicarum," 1612, Vol. II, p. 440.
[4] G. F. Mylius, "Memorabilium Saxoniae subterraneae," Leipzig, 1709-1718, Vol. II, p. 20.
[5] Charles Zeze, "Considerations sur les lièvres blancs en Livonie," 1749, p. 52.

The fishery then came into great favor. To the nobility of Livonia, in whose domains the brooks were situated, the crown accorded sixty rubles for each half ounce of choice pearls secured, and for every half ounce of the second class, thirty rubles; but the nobles were obliged to renounce their rights to the fisheries and to permit the lakes and brooks to be guarded by imperial soldiers. Owing to the very great destruction of mussels which yielded no pearls, a reward was offered to any one who would discover a method of determining from external characteristics those individual shells which contain gems of value.

In 1746, when the Empress Elizabeth passed the summer in Livonia, large quantities of pearls from the neighboring brooks were presented to her. But, owing to the cost of supervision, the expenditures soon exceeded the revenues and the government abandoned the guard and dismissed the fishermen. Little by little the search decreased, and by 1774 relatively few pearls were found.[1]

According to Hupel, the Schwarzbach River, near Werro, was celebrated for its pearls, which were noted for their size and beauty; one of the tributaries of this river is named Perlenbach (Pearl Brook). The Ammat and Tirse streams, and forty other brooks and lakes also yielded them. Pearls of slight value were likewise produced in the Palze and the Rause, near Palzmar; the Paddez, a tributary of the Evest which empties into the Düna, and the Voidau and the Petribach, each of which flows into the Schwarzbach. Near the Tirse was a very old road house, patronized by the peasants, which from time immemorial had borne the name Pehrlu-kroghs (Pearl Tavern).

Formerly some of the brooks of Esthonia on the Gulf of Finland, and principally those near Kolk and the adjacent lakes, furnished beautiful pearls. From these waters came the beautiful necklace which is yet an heirloom in the Kolk family. The choicest of these weighed from five to ten grains, and the color was grayish blue. The Emperor Alexander I is said to have received a present of pearls collected in the vicinity of Tammerfors, in the government of Tavastehus, in the grand duchy of Finland. The development of manufacturing in that region, however, has destroyed most of the mussels.

Von Hessling notes that in the province of Olonetz, pearls are found in the Poventshanka, in the Ostjor, and in the Kums, where they are secured by the neighboring peasants who sometimes make valuable finds.[2] When the brooks dry up, the mussels are easily secured; old inhabitants note that on one occasion of this kind many superb pearls

[1] A. H. Hupel, "Nouvelles topographiques de Livonie et d'Esthionie," 1774, Vol. I, p. 134.

[2] "Die Perlenmuscheln," Leipzig, p. 196.

were found in the Poventshanka, and a necklace of them was presented to the Empress Catherine Alexievna. These pearls rarely leave the province in which they are collected, as the inhabitants are fond of using them for personal decoration. Young girls attend to the fishing, and workmen pierce them for about two copecks each. Choice ones sell for thirty to one hundred rubles apiece.

In the government of Archangel pearls have been collected for centuries from the streams flowing into the White Sea and the Arctic Ocean. An extended account of the fisheries of this region was given by Von Middendorff.[1] He states that the *Unio margaritifera* inhabits all the rivers in which the descent is not too rapid, and especially in the Tjura, the Tuloma, the Kovda, Kereda, the Kanda, etc. The fisheries have been conducted exclusively by the shore Laplanders; but they have been neglected in recent years owing to the small returns. Von Hessling notes that the pearls are dull in color; in the opinion of the fishermen this is caused by the mysterious influence of the copper money which they carry with them. The Tuloma was formerly a productive river; its pearls were sold in Kola, whence they were carried to Archangel, 335 miles distant, where they were pierced by expert workmen. The Tjura also yielded many pearls; but since a Laplander was drowned while fishing for them, a legend has spread that the spirit of the river guards the pearls, and the natives hesitate about seeking them.

Probably the occurrence of so many in the home streams had much to do with developing in Russia that great love for the pearl which has made it the national ornament, all classes finding pleasure in its possession. While the superb gems treasured by the nobility are mostly from oriental seas, a considerable percentage of those worn by the peasantry are from the native waters. An interesting account of this fondness among a certain class of Russian women—the Jewesses of Little Russia—was given sixty years ago by the German traveler Kohl.

In Alexandria, a small city in the government of Kherson in South Russia, a Jew kept a café, and his charming daughter served us with coffee. We paid her compliments on her beautiful eyes and teeth. But she seemed to be much less vain of these natural ornaments than of the acquired ones in the magnificent glittering pearl-cap which she wore upon her head. For all the women through South and Little Russia even as far as Galicia wear a certain stiff, baggy cap which is very disfiguring, and is covered all over with a great number of pearls, upon a foundation of black velvet. It is called a "mushka." This cap, with very unimportant modifications, has almost always the same form; the

[1] Baer and Helmersen, "Beiträge zur Kenntniss des russischen Reiches," St. Petersburg, 1845, Vol. XI, pp. 143, 144.

only difference is that, in the case of the wealthy, the pearls are larger, and sometimes a number of small pearls and precious stones are suspended here and there, set in the same way as the ear-rings of our ladies. It is common for them to wear half their fortune on their heads in this way. For these caps generally cost from five hundred to one thousand roubles, and many are worth five or six thousand and even more; they wear them every day, holidays as well as ordinary days, and strut around the kitchens and cellars with their "mushka." They spend their last penny in order to secure such a pearl-cap, and even when they are clad in rags their head is covered with pearls. In order to furnish the requisite material for this wide-spread fashion, the commerce in pearls of Odessa, Taganrog and some other places in southern Russia is not unimportant. There may live in the region where the pearl-caps of which I speak are worn at least 2,000,000 Jewesses. Let us estimate that among them there are but 300,000 adults, and that only half of these, 150,000, wear pearl-caps (only the most indigent and the most aristocratic do not wear the "mushka"); let us then estimate the average value of such a cap at only five hundred roubles—these are the lowest minima and fall far short of the real figures—and we have a total capital of 76,000,000 roubles, which the Jewesses of this region wear upon their heads. Naturally the annual diminution of this capital is small, since these pearls are transmitted from the mothers to their daughters and granddaughters. Still, if we estimate that they last for a century, the necessary yearly contribution amounts to nearly one million. It is, however, probable that a much larger capital is employed in the commerce of pearls. They are, for the most part, oriental and come by way of Turkey and Odessa or else by way of Armenia and Tiflis. We inquired of our beautiful Jewess whether she was not in perpetual dread on account of her pearl-cap, and how she protected it from thieves. She answered that she wore it on her head all day and at night placed it in a casket which rested under her pillow. So that the whole short life of these Jewesses of the steppes revolves around their pearl-cap as the earth does around the sun.[1]

Several species of marine mollusks on the coasts of Europe yield pearly formations, but none of much ornamental or commercial value. Probably the most interesting of these are from the Pinna on the Mediterranean coasts, and especially on the coast of Sardinia and the shores of the Adriatic. An interesting collection of these Pinna pearls was furnished to the writers by Alexandro Castellani of Rome.

[1] Kohl, "Reisen in Südrussland," 2nd edition, Leipzig, 1846, Vol. I, p. 15.

IX

ISLANDS OF THE PACIFIC

SOUTH SEA ISLANDS, AUSTRALIAN COASTS,
MALAY ARCHIPELAGO

PEARL FISHERIES OF THE SOUTH SEA ISLANDS

Sea-girt isles,
That, like to rich and various gems, inlay
The unadorned bosom of the deep.
MILTON.

GATHERING pearl shells and pearls is the principal industry of the semi-amphibious natives of the hundreds of palm-
crowned and foam-girdled islands of the southern Pacific,
commonly known as the South Sea Islands. Among these
the most prominent for pearl fishing are the Tuamotu Islands or Low
Archipelago, the Society Islands, the Marquesas, the Fiji Islands, Penrhyn or Tongareva, and New Caledonia. These are under the protection of the French government, except Fiji and Penrhyn, which
belong to Great Britain.

Almost ever since the South Sea Islands have been known to civilization they have contributed pearls; and the fishery has been one of
the principal industries, not only for the natives, but also for the not
inconsiderable number of sailors who, preferring the lotus on shore
to the salt pork and monotony of ship life, have yielded to the insular
attractions and formed domestic ties. The industry has been especially
extensive during the last seventy years, when there has been a profitable market for the shells. Most of the natives—men, women, and
children—follow it for a living. Domestic duties rest very lightly
upon the women, and many of these, and even young girls, find employment in diving, in which at moderate depths these dusky mermaids
are nearly, if not quite as expert as the men and boys.

Tahiti, the largest of the eleven Society Islands, is the center of the
pearling industry of French Oceanica. It is situated in about Lat.
17° S. and Long. 150° W., and has an area of approximately 410 square
miles and a population of 11,000, nearly one half of whom live in Papeiti, the principal town. This is one of the most agreeable of the
"Summer Isles of Eden," Nature furnishing food in abundance, and
climate and social customs requiring little in the way of dress and
habitation. Notwithstanding its importance as the headquarters of

the pearling industry, few pearl-oysters are caught at Tahiti, most of them coming from the archipelagoes of Tuamotu, Gambier, and occasionally Tubai.

The Tuamotu Archipelago is the scene of the principal pearl fisheries of the South Seas; and from the local importance of this industry the group is sometimes called the Pearl Islands. These coral-formed islands are strung out for a distance of 900 miles in a northwest and southeast direction, and extend from Lat. 14° to 23° S. and from Long. 136° to 149° W. They number about seventy-eight, many of them made up of small atolls only a few feet above the surface of the ocean, and with an aggregate area of about 360 square miles. The total population is approximately 6000, with many visitors from Tahiti and other neighboring islands during the pearling season. The principal products are pearl shell and pearls, copra, and cocoanut oil; and nearly one half of the islands yield nothing but shell and pearls. The chief port is Fakarava on an island of the same name, and the trade is almost entirely with Tahiti.

As the Tuamotus are of coral formation, they produce little vegetable growth, and the people seem often on the brink of starvation, forming a striking contrast with those of the neighboring Society Islands. Drawing their subsistence entirely from the sea, except for the native cocoanuts and breadfruit, these people have, at times, been in great straits for food, and it was doubtless severe hunger that drove them to the acts of cannibalism with which they have been charged. And the sea which supplies them with food has also visited them with great destruction. As recently as January, 1903, a great storm swept over this group, drowning over 500 of the inhabitants, and destroying a very considerable portion of the pearling fleet and other property.

The pearl-oyster reefs of the Tuamotu Archipelago are very extensive, only eight or ten of the islands failing to contribute to the supply. They occur in the protected lagoons of the atolls, where the bottom is well covered with coral growth, with numerous elevations and depressions of various sizes; and it is about the bases and in the recesses of these coral growths that the best shells are usually found. Most of them are of the black-edged variety of *Margaritifera margaritifera,* which here attains a great size, reaching a diameter of twelve inches in extreme cases.

While pearl-oysters are found about nearly all of the Tuamotu Islands, the reefs are richest at Hikueru or Melville Island. When that lagoon is open it is the scene of the greatest operations, and it is credited with nearly one half of the total product of the archipelago. At the opening of the season, this is the resort of fishermen from all over the group, even from a distance of five hundred miles, and thou-

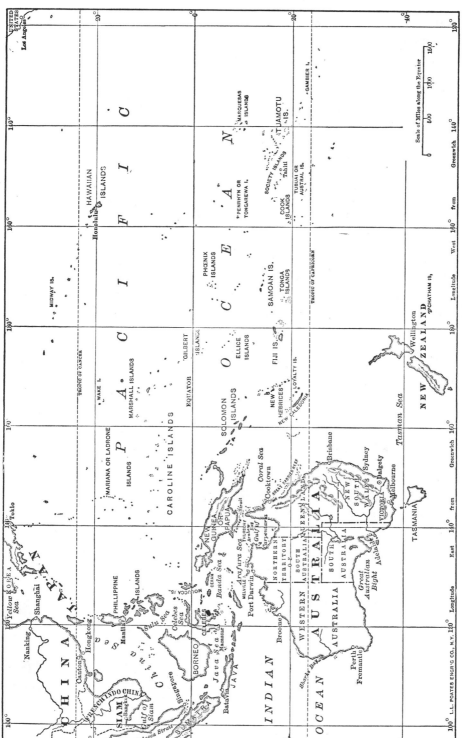

THE PEARLING REGIONS IN OCEANIA AND MALAYSIA

sands of natives camp in temporary leaf-thatched huts among the cocoanut-palms on the beach, those from the different islands congregating in isolated settlements. As many as five thousand persons are sometimes brought together in this way.

The volcanic-formed Gambier Islands, with high peaks reaching, in one instance, an altitude of over 1200 feet, present a striking contrast to the Tuamotu atolls. This group consists of five large and several small islands, surrounded by a coral reef of an irregular triangular figure. The 1100 inhabitants of the Gambier Islands derive a large percentage of their support from the pearl fishery. The patches of pearl-oysters are located between the islands and the barrier reefs. They are numerous about the island of Mangareva, which is well surrounded by them on the north, east, and southeast. Oysters from the reef of Tearae, which extends from the eastern point of Mangareva to the small island of Aukena, a distance of two miles, are especially rich in pearls. On this reef, where the water is from one to four fathoms in depth, the mollusks are small, rarely exceeding five or six inches at maturity, but the shell is very thick and coral covered; these yield many pearls. In greater depths, the oysters attain a larger size, but they yield few pearls.

The first white man to attempt the exploitation of the pearl resources of the Tuamotus appears to have been Mörenhout. In a voyage to the Oceanic Islands in 1827, he learned of the great wealth of pearl shell, and applied to Queen Pomaré at Tahiti for permission to employ the natives in the fishery. With an eye to business, she required a fee of $5000 for herself before granting the desired authority.[1] Considering this excessive, Mörenhout attempted to deal with the natives without permission of the dusky queen, but under these adverse conditions he found the trade unsatisfactory and soon abandoned it.

In 1830, and the years immediately succeeding, desultory pearling voyages were made from Valparaiso, Chile, and these were followed by expeditions from America and elsewhere. An interesting account of the trade at that time is contained in Lucatt's "Rovings in the Pacific from 1837 to 1849," published in London in 1851.

The Mormon influx in 1846 resulted in a further development of the pearl fishery; and Grouard, the local leader of that denomination, is credited with making a fortune in the business.

From the beginning of the industry up to 1880, when control of the islands passed to the French government, it is estimated that about 15,000 tons of pearl-oysters were secured. The extent of the fishery during the few years preceding 1880 made such drains upon the pro-

[1] "Voyage aux Iles du Grand Océan," Paris, 1838; also "Le Correspondant," March 10, 1906.

ductiveness of the reefs that many of them gave signs of exhaustion. With a view to adopting methods for conserving the industry, so essential to the welfare of the natives, the French Ministry of Marine and Colonies in 1883 inaugurated an investigation of its condition, and of the possibilities for improvement. This was made under the immediate direction of G. Bouchon-Brandely, whose interesting report[1] contains much data on this subject.

As a result of these investigations and recommendations, a restricted season for fishing was adopted, and only a portion of the reefs was thrown open each year, a decree of the governor, published in the "Journal Officiel" of the colony, determining the islands in which the fishery might be prosecuted. This interdiction, known locally as *rahui,* is for the purpose of permitting the oysters to develop, and thus prevent the exhaustion of the reefs.

By decree of January 24, 1885, a restriction was made against taking shells measuring less than 17 centimeters in diameter on the interior nacre, or weighing less than 200 grams per valve. But this was repealed in 1890, and since then there has been no restriction on the size of the oysters that may be fished.

The pearl fishery and the isolated leper station are the principal claims which attract the attention of the outside world to the island of Penrhyn or Tongareva, one of the Manahiki group, in Lat. 9° S., and Long. 158° W. This desolate atoll island consists of a ring of land a few hundred yards in width, inclosing a lagoon nine miles long and five miles wide, and it produces little else than pearls and pearl shell. The white gravelly shore yields little vegetation except cocoanuts, which share with fish in furnishing sustenance to the semi-amphibious natives.

At Penrhyn the pearl fishery is carried on in the clear, limpid waters of the atoll where the oysters are undisturbed by storms. The shells belong mostly to the golden-edged variety, and are of good quality, the value in London ranging from £100 to £250 per ton. Relatively few pearls are found, amounting in aggregate value to only about one fourth of the value of the shells. These are the principal objects of the fishery; the finding of pearls is incidental, but careful search is always made for them, and some choice specimens have been secured.

On the coast of New Caledonia, pearling is of recent origin, dating as an industrial enterprise from 1897, although previous to that time some shells and pearls had been secured by native beach-combers. This island is 220 miles in length and 30 in width, situated 850 miles southeast of Australia, and about the same distance from New Zealand.

[1] "La Pêche et la Culture des Huîtres Perlières à Tahiti; Pêcheries de l'Archipel Tuamotu," Paris, 1885.

It is a French colony, and has been used by that government as a penal settlement since 1864.

In 1897, rich beds of pearl-oysters were discovered off the west coast of this island. They are most numerous between the shore and the barrier reefs on the west coast from Pouembout River to Gomen Bay, and especially about the small island of Konienne at the mouth of the Pouembout River. They are also abundant among the Loyalty Islands off the eastern coast of New Caledonia, and especially at the island of Lifu.[1] The shell is similar to that from Torres Straits, and the yield of pearls is very large. Several concessions have been ob-tained to exploit these beds, one of them covering 130 miles in length. The industry is carried on by means of scaphanders, in a manner similar to that of Torres Straits. Virtually all of the catch is sent to France.

The natives of the South Sea Islands, and particularly of Penrhyn and the Tuamotu group, are doubtless the most expert divers in the world. This can be readily appreciated by those who have read of Hua Manu in C. W. Stoddard's thrilling narrative, or have heard the story of the brown woman who swam for forty hours in a storm with a helpless husband on her back. Accustomed to the water from in-fancy, these human otters swim all day long as readily as they would walk, go miles from shore without a boat in search of fish which they take by means of baited hook and line, and boldly attack a shark single-handed. Seemingly fabulous stories are told of their descending, un-aided, 150 feet or more beneath the surface, and remaining at lesser depths for nearly three minutes, far surpassing any modern records of the divers of India.

The water in the South Seas is wonderfully clear, enabling the fishermen to detect small objects at considerable depths, and especially so when using the water-telescope, similar to that employed in the Red Sea fisheries. By immersing this to a depth of several inches and cut-ting off the light from the upper end as he gazes through it down into the waters, the fisherman can readily inspect the bottom at a depth of fifteen fathoms, and thus locate the shells before he descends.

The diving is quite unlike that in Ceylon and Arabia. The men do not descend on stones, but swim to the bottom. The diver is stripped to his *paréu* or breech-clout, his right hand is protected by a cotton mitten or by only a wrapping of cotton cloth, and in his left hand he carries a pearl shell to assist in directing his movements and in detaching the oysters at the bottom. In preparing for a deep descent, he sits for several minutes in characteristic attitude with hands hanging over knees, and repeatedly inflates his lungs to the fullest capacity, exhaling

[1] Seurat, "L'huître perlière," Paris, 1900, p. 133.

the air slowly through his mouth. After five or six minutes of "taking the wind," the diver inhales a good breath, drops over the gunwale into the water to give him a start, and descends feet foremost. At a distance of twelve or fifteen feet below the surface, gracefully as an otter or a seal, he bends forward and turns head downward and, with limbs showing dimly in frog-like motion, he swims vertically the remaining distance to the bottom. There he assumes a horizontal position and swims slowly just above the ground, searching critically for suitable oysters, in this way traversing a distance possibly of fifty feet or more. When he has secured an oyster, or his breath is approaching exhaustion, he springs from the ground in an erect position and rapidly swims upward, the buoyancy of his body hastening his ascent so that he pops head and shoulders above the surface, and falls back with laboring pulse and panting breath. In case the dive has been unusually extended, a few drops of blood may trickle from the nose and mouth. His find—consisting frequently of nothing and rarely of more than one oyster—is carried in a cocoanut fiber sack suspended from the neck, or is held in the left hand, or may be hugged beneath the left arm.

Ordinarily in actual fishing operations, the fishermen do not descend to greater depths than fifteen fathoms, and remain from sixty to ninety seconds. Writing in 1851, a trader who had spent several years in collecting pearls and pearl shells among the Tuamotus stated: "I timed several by the watch, and the longest period I knew any of them to keep beneath the water was a minute and a quarter, and there were only two who accomplished this feat. Rather less than a minute was the usual duration. It is unusual for them to attempt deep diving; and let the shells be ever so abundant, they will come up and swear there are none."[1]

However, in mutual contests or in special exhibitions, reports of twenty, twenty-three, and even twenty-five fathoms are numerous, and they have repeatedly been timed two and a half to three minutes. Bouchon-Brandely speaks of a woman at Anaa, one of the Tuamotus, who would go down twenty-five fathoms and remain three minutes under water.[2] This seems very unusual, but there are numerous reports of two and a half minutes at about seventeen or eighteen fathoms. In October, 1899, at Hikueru Island, another of the Tuamotu group, a young native made an exhibition dive for the officers of the United States Fish Commission steamship *Albatross*. He reached bottom at a depth of 102 feet under the boat's keel, and remained submerged two minutes and forty seconds. The water was so transparent

---

[1] Lucatt, "Rovings in the Pacific from 1837 to 1849," London, 1851, Vol. I, p. 245.

[2] "Bulletin United States Fish Commission," Vol. V, p. 293.

that he was clearly seen from the surface. After he touched bottom at that great depth, he calmly picked over the coral and shells to select a piece to bring up.[1] The diver was ready to go down again only a few minutes after he came up.

In his work on French Oceanica, Chartier states: "There are three women well known in the archipelago [of Tuamotu] who have no equals elsewhere; they explore the depth at twenty-five fathoms and remain not less than three minutes before reappearing at the surface."[2] However, these unusual depths and extensions of time are dangerous, and care must be taken or serious results follow. Most of the catch is obtained in about ten fathoms of water.

At the request of the writer, Mr. Julius D. Dreher, American Consul at Tahiti, made inquiries among the South Sea Islands in regard to the record of the best divers, and wrote as follows:

Mr. J. L. Young, who has lived in these islands for thirty years, informs me that he has never seen a diver remain under water longer than 80 seconds, and that at a depth of twelve to fifteen fathoms. At one time he tested a man who claimed to be able to stay under for three minutes, yet this man could hold his breath on land less than 80 seconds by the watch.

Elder Joseph F. Burton, who has spent many years as a missionary in these islands, states that once in Hikueru, of the Tuamotu group, he went out in a boat with the divers to time them. The best record made was 107 seconds, but he was informed that there were better divers on the island than those he tested. He thinks the water was ten to twelve fathoms in depth. A native of Takaroa, named Metuaro, told Mr. Burton that he could stay under water three minutes or longer. When these divers come up they take a breath and immediately put their head under water to prevent headache.

Mr. J. Lamb Doty, formerly Consul and now Vice-Consul at Tahiti, who has spent eighteen years here, is willing to be quoted as affirming that he once timed a diver who remained under water 2 minutes 35 seconds.

Mr. Henry B. Merwin, a leading trader with the Tuamotu Islands, is willing to be quoted as saying that he saw a diver remain under water 4 minutes 45 seconds by the watch. This is generally regarded, so far as my inquiries go, as improbable; but most persons interviewed believe that men do remain under water 2½ to 3 minutes. A native of Takaroa, named Tai, assured me in the presence of others that there were twenty men in that island who could remain under water 2½ to 3 minutes at a depth of twenty fathoms. He claimed to be able to stay 3 minutes at that depth.

Diving-suits, or scaphanders, have been used at most of the South Sea Islands, but in a very irregular manner. In 1890 the use of scaphanders was restricted in the Tuamotu group, and by decree of De-

[1] Alexander, "Report United States Fish Commission," Vol. XXVII, p. 764.

[2] "Tahiti et les Colonies Françaises de la Polynésie," Paris, 1887, p. 173.

Pearl-divers of the Tuamotu Archipelago ; men, women and children dive in these waters

Settlement of pearl fishermen at Hiqueru, Tuamotu Archipelago

Pearling boats at Hiqueru, Tuamotu Archipelago

Australian pearl-diver (armored) coming up from the depths

cember 28, 1892, it was interdicted altogether with a view to preserving the industry to the natives, as it represents their principal means of livelihood. The suit commonly employed at Penrhyn consists of a helmet and a jumper, neither boots nor trousers being worn. Owing to the absence of weights on the feet, it rarely but nevertheless sometimes happens that a diver turns upside down, and the unwieldy helmet keeps him head downward while the air rushes out under the bottom cord of the jumper and he is suffocated. Also, when a good patch of shells has been located, the temptation to remain down too long is great, and paralysis often results. On the whole, these diving-suits have proven very dangerous to the light, graceful swimmers of these southern seas, to whom they are about as much of an impediment as was Saul's armor to the shepherd lad who slew the giant with the simple pebble from a sling.

And there are dangers also in nude diving, even to those who have spent a lifetime about the water. Sharks and sting-rays and devil-fish there are in abundance, and many of them know the taste of diver's flesh; on the other hand many a daring South Sea Islander could tell of a fierce combat more thrilling than even those pictured by Victor Hugo. One of the chief advantages of the diving-suit is that in case a shark comes along, the diver can bide his time until the fish is ready to leave, or he can frighten it away by ejecting air bubbles from the sleeve of his suit or by other demonstrations; whereas a nude diver is obliged to seek the air without delay, and in the retreat is seized by the fish who, human like, has his appetite increased by the visible retreat of the object of his desire.

Not Schiller nor Edgar Allan Poe ever conjured up a picture more ghastly than that of a Penrhyn diver caught like a rat in a trap by some huge, man-eating shark or fierce kara mauua, crouching in a cleft of the overhanging coral, under the dark green gloom of a hundred feet of water, with bursting lungs and cracking eyeballs, while the threatening bulk of his terrible enemy looms dark and steady, full in the road to life and air. A minute or more has been spent in the downward journey; another minute has passed in the agonized wait under the rock. . . . Has he been seen? . . . Will the creature move away now, while there is still time to return? The diver knows to a second how much time has passed; the third minute is on its way; but one goes up quicker than one comes down, and there is still hope. . . . Two minutes and a half; it is barely possible now, but—the sentinel of death glides forward; his cruel eyes, phosphorescent in the gloom, look right into the cleft where the wretched creature is crouching, with almost twenty seconds of life still left, but now not a shred of hope. A few more beats of the laboring pulse, a gasp from the tortured lungs, a sudden rush of silvery air bubbles, and the brown limbs collapse down out of the cleft like wreaths of seaweed. The shark has his own. (Beatrice Grimshaw in the "Graphic.")

At the end of the day's work, the catch is opened by means of a large knife, and carefully searched for the much prized pearls. Usually the fisherman finds none; occasionally he discovers a small round one or a large baroque, and at long intervals—possibly once in two or three years—his search is rewarded with a fine pearl for which he may receive $50 or $60, and there is always the chance that the very next oyster will disclose a gem which will make him independent for the remainder of his life; and if no pearls whatever are found, there are the shells, the sale of which furnishes sufficient to purchase tobacco, knives, fish-hooks, the gaudy cotton cloths, the flour and other simple articles of food, and especially rum, that fatal gift of civilization which has been the curse of so many primitive peoples.

Some of the individual pearls secured have been remarkably large, weighing 100 grains and over. Returning visitors from Tahiti, with views magnified doubtless in proportion to the distance of the objects of their description, credited Queen Pomaré with the possession of some sufficiently large to be used for billiard-balls. Sixty years ago superb pearls could be obtained from the natives for a few gallons of rum or a small number of pieces of cheap calico, and several shrewd traders made great profits in the business. But as trade at the islands was open to vessels of all nationalities, the competition increased, with the result that the natives gradually learned the high estimation in which pearls are held, and in recent years it has not been unusual for one of medium grade to sell higher in Oceanica than it would in Europe.

It is difficult to form a reliable estimate of the value of the pearling industry of the South Sea Islands. The Tuamotu group, with 4000 fishermen, yields, in an average season, about 450 tons of mother-of-pearl, worth about £65,000 in London, where most of it is marketed. The yield at the remaining French islands is less than that of the Tuamotus. Probably the total yield of mother-of-pearl in all the South Sea Islands is not far from 900 tons, worth about $700,000.

No statistics whatever are available regarding the yield of pearls, and the estimates sent from the islands are small compared with those made by London and Paris firms who import the pearls. A large number of persons living in Papeiti and many traders visiting the islands depend very largely on pearl-dealing for a livelihood. From the yield of pearl shell and estimates made by dealers, we are inclined to put the value of the pearls secured in an average season from all the South Sea Islands at about $125,000, only a small portion of which goes to the fishermen themselves, the greater part representing profits of the traders.

PEARL FISHERIES OF AUSTRALIA

Ocean's gem, the purest
Of nature's works! What days of weary journeyings,
What sleepless nights, what toils on land and sea,
Are borne by men to gain thee!

UNKNOWN.

As regards area of distribution the most extensive pearl-oyster grounds of the world are situated on the northern and western coasts of Australia. These are located within the jurisdictions of Queensland, Western Australia, and South Australia; and extend in irregular patches from near Cooktown on the northeast almost to Fremantle at the southwest, a distance of nearly 3000 miles. Those in Queensland are commonly known as the Torres Straits fisheries, as they are especially important there; but they extend a considerable distance beyond each end of the strait, and pearling expeditions are made from the limits of the Great Barrier coral reef northward to the vicinity of New Guinea.[1] Those of Western Australia are commonly spoken of as the Northwest fisheries.

The fisheries of Queensland and of Western Australia are approximately equal in extent, as regards number of vessels, boats, and men employed, and the quantity and value of the catch, with the advantage slightly in favor of the Northwest fishery in the last four or five years. In 1905, according to the official figures, the Queensland fishery gave employment to 348 vessels and 2850 men, and yielded shell and pearls worth £135,000, which was the smallest output since 1890. The Western Australia fishery, exclusive of Sharks Bay, employed 365 vessels in 1905, and about the same number of men as in Queensland, and yielded £196,000 worth of shell and pearls. The fishery of South Australia employed about 60 vessels and 375 men, and yielded about £25,000 worth of shell and pearls. This makes for the whole of Australia, except Sharks Bay hereinafter noted, a total of 773 vessels, 6075 men, and an output worth £356,000. It should be understood that the South Australia fishery is not prosecuted on the southern coast of the continent, but on the northern coast, in what is known as the Northern Territory of South Australia.

Three species of pearl-oysters are found in Australian waters. The largest species, *Margaritifera maxima,* which is by far the most important and widely distributed, occurs to a greater or less extent throughout the whole of this region. This yields the standard mother-

[1] "Report on Pearl Fisheries of North Queensland," Brisbane, 1890.

of-pearl of commerce. Although the pearls which it yields are among the largest and finest in the world, this mollusk is sought more particularly for the shell, the value of which from season to season averages three or four times as much as that of the pearls. Ordinarily this shell is uniformly white over the entire inner surface, and is commonly known locally as "silver lip"; but some "golden-edged" shell occurs on the muddy grounds in narrow passages between the islands on the northwest coast.

While this species is gregarious, it is not located in densely covered beds, but is scattered in patches over the reefs. Some of these are miles in length and contain scores of tons, but usually they are very much smaller. The oysters occur principally on rocky bottom, and also on clay and sand when well covered with seaweeds, but are rarely found on muddy ground. They are most numerous in the channels where the current is strong. The small oysters are generally loosely attached by the byssus to rock, gravel or other shells; while the mature ones lie loosely on the bottom or slightly turned in the sand.

The second species of the Australian pearl-oysters, *Margaritifera margaritifera*, is smaller, rarely exceeding eight inches in diameter and a weight of two pounds. The distinguishing characteristic is the black edge bordering the inner surface of the shell, whence it acquired the local designation "black lip." This variety is not rare in Queensland, and in Western Australia its range extends as far as Champion Bay in Lat. 29° S. However, the catch is small compared with that of the *Margaritifera maxima,* amounting to only two or three per cent. in Queensland. In 1905, the export of "silver lip" and "golden-edged" from Thursday Island was 527 tons, and of "black lip" only 11 tons; in 1904, these figures were 778 and 7 respectively. In Western Australia the percentage of yield is much larger than this.

The third species, *Margaritifera carcharium,* is confined almost entirely to the limits of Sharks Bay, on the extreme western coast of Australia. At maturity it is the smallest of the three, averaging three or four inches in diameter, and about equals in size the Lingah pearl-oyster of the Persian Gulf. The percentage of pearls therefrom is relatively greater than from the larger variety; but, owing to its small size and lack of thickness, the shell is of little commercial value. The value of the output in recent years has approximated two or three thousand pounds sterling, which is very much less than formerly, the value of the shell having greatly decreased since the introduction of the Mississippi shell in button manufacture.

The pearl fishery on the coast of Australia originated about 1861. It appears that an American sailor named Tays was the pioneer in the business; and on his death by drowning, the business was conducted by his partner named Seubert.[1] This was on the northwest coast, and the output reached the market by way of Singapore. At first the oysters were so abundant in shallow water that they could be picked up at low tide, and beach-combing was profitable, especially when carried on with cheap native labor. As the beach-beds became exhausted, the natives were encouraged to wade out to greater depths, and soon they became accustomed to "bob under" for those oysters visible from the surface. The Australian blacks were thus taught to dive, and in 1867 diving from boats in two or three fathoms was attempted with such success that in the following year the practice was generally adopted, the depth in which they worked gradually extending to six or eight fathoms. In diving from a boat, the men imitated "bobbing under" which they had practised in shoaler water; they slipped off the gunwale feet foremost, and when six or eight feet below the surface, turned and swam downward.

Owing to the close labor relations existing between the natives and the sheep-raisers of northwestern Australia, the latter were brought into the business, and for a number of years pearling and sheep-raising were closely associated. The blacks were employed in various duties in connection with raising and shearing sheep, and it was important to find some occupation for them when ranch-work was slack, not only for their own subsistence but for the protection of the herdsmen and their property. Fortunately, this opportunity was furnished by the pearl fishery, for which these men were well qualified.

The profits of the business soon attracted many outside capitalists, and it became difficult to procure divers. Not only did the pearlers—and particularly new-comers—resort to impressing the blacks into service, but skilled fishermen were brought over from the Malay Archipelago, and in some cases the methods used in securing them were by no means regular.

In 1871 the Northwest pearl fishery gave employment to 12 vessels of 15 to 50 tons each, and yielded about 180 tons of mother-of-pearl. During the same year, in Torres Straits, where the industry had extended about 1868, there were 10 vessels—mostly from the port of Sydney—and the catch of mother-of-pearl approximated 200 tons, valued at £60,000 in London.[2] Each vessel was commonly manned by two or three white men and from ten to fifty divers, who worked from dinghys, in gangs of six or eight each with an overseer in charge.

As the fishery increased rapidly in extent, the problem of securing nude divers became a serious one, and "nigger hunting" became rather common, the Australian black man representing the cheapest form of

---

[1] Garran, "Australasia Illustrated," Sydney, 1892, Vol. II, p. 886.

[2] Gill, "Life in the Southern Isles," London, 1876, p. 294.

labor, working for his food, tobacco, and the simplest articles of clothing. There was no complaint that the men thus impressed were
treated with inhumanity; on the contrary they were well fed and cared
for; yet, with a view to protecting them and preventing even a suspi-
cion of wrong-doing, the Australian government enacted regulations
restricting pearling contracts with the natives. Nearly every year
these regulations became more stringent, affecting the hours for diving, and limiting the work to depths of six and a half fathoms, so
that the employment of Australian aborigines in the fishery became
extremely troublesome and annoying.

The government of the Netherlands also placed severe restrictions
on the employment of natives of the Dutch Indies, requiring security
of £20 per head for the repatriation of each man; and the local chiefs
or rajahs also expected a rake-off before permitting their men to ship.
These Malays—from the islands of Solor, Allor, Adonare, etc.,—also
expected much better pay and better provisions than the Australian
blacks.

The following interesting account by Henry Taunton gives a
graphic description of the fishery as carried on at that time:

The work was far from easy. It was exhausting and perilous for the
divers, and full of privation, exposure, and danger for the white men. Only
the hope of a prosperous season reconciled one to the life. When shells were
plentiful and the weather fine, the work was exciting and interesting enough;
but during rough weather, when one had to be constantly straining at the oar
to keep the dinghy from drifting too rapidly, or when hour after hour might
pass without the men bringing up a single shell, the discouragement was great.
The rays of the vertical sun beating down on one's shoulders at such times
seemed as if it would never reach the western horizon, which was the signal
for returning on board.

As may well be imagined, when three or four white men had to control and
compel some thirty or forty natives to carry on work which they detested, a
very strict discipline had to be maintained. It was the rule that no talking was
allowed amongst the divers when in the dinghy, nor were they even permitted
to address the white man, unless, maybe, to answer a question as to the nature
of the bottom, whether *nanoo* (sand) or *bannin* (shelly bottom), etc., or
unless some urgent necessity arose. Sometimes, indeed, I have pushed off
from the vessel's side of a morning and have not heard a word spoken until
we returned on board at night, unless chance might take me within hail of
some other dinghy, when felicitations or condolences would be exchanged, as
good or bad luck might happen. At times, when the "patch" was small, the
dinghys of the whole fleet might be congregated on a very small area, in which
case the scene was animated enough. On all sides you could see divers slipping
into the water and others just coming to the surface, puffing, blowing, and
coughing to clear their eyes, ears, and mouth from the salt water—some with,

others without shells. Others would be swimming to regain their dinghy or squatting in their places for the few minutes' rest permitted, and, if the wind were at all fresh, shivering with cold; for although the weather might be extremely hot, the constant plunging in and out for many hours at a time tended to reduce the bodily temperature considerably. The white men would be seen standing up in each dinghy. They were lightly clad, with shirt sleeves and trousers rolled up, in all varieties and colours of costume, from the regulation shirt, trousers, and felt hat, with leather belt sustaining sheath-knife and pouch, to the more comfortable pyjama suit, or even the Malay sarong. Some would be straining hard at the end of the scull-oar, forcing the boat against wind and tide in the endeavor to keep it as long as possible on the "patch," which was marked by the discoverer's buoy, which also might be observed nodding on the surface, and canted over by the swiftly rushing tide. Others, their men all being below, just kept the dinghy's head to wind until, by judicious use of the oar and well-calculated drifting, all the divers reappear on the surface within a short distance from their own boat. This is the secret of saving the divers from wasting their powers and time uselessly. . . . As may be supposed, where the tide sweeps the divers along the bottom at the rate of three or four or even six miles an hour, they have to be very smart in seeking and grabbing any shell within reach. I have never tested them with a time-keeper; but by counting seconds on many occasions, from the moment a diver's head sank below until it again came above the surface, I estimated the average time under water was fifty-seven seconds. Part of this is of course expended in swimming to the bottom, where they can remain only a very few seconds, as time must be allowed for reaching the surface before letting go their breath. Practice in ever-varying depths enables them to gauge this limit of time to a nicety. But sometimes they cut things too fine, and then a catastrophe was inevitable, unless much watchfulness was exercised by the white man, who has to keep his eyes turned in all directions once his men are down. So long as a diver can hold his breath the pressure forces him to the surface at a speed which seldom requires accelerating by strokes with the hands or feet; but the moment he lets go his breath—if under water—his upward course is arrested and his body commences to sink. Now, when the white man sees this, either he must plunge in to the rescue himself, or direct such divers who may be on the top to do the needful.

On a calm day, when one can see far into the blue clear depths below, I have often seen one of my men shooting rapidly upwards until within perhaps a foot or two from the surface, when a sudden gush of bubbles from the man's mouth would tell its own tale. Instantly he would begin to sink gently downwards, and only quick action could save this diver who had miscalculated his time. However, as it was not infrequent for divers to go down and never come up at all, one may conclude that, where the time to be allowed is comprised in so few seconds, even the most experienced make fatal errors.[1]

The difficulties in securing labor at length resulted in experiments with the scaphander or diving dress, and gradually its adoption by

[1] "Australind," London, 1900, pp. 233-239.

most of the pearling fleet. The labor problem and the exhaustion of the oysters in medium depths developed more quickly in Torres Straits than on the northwest coast, and diving outfits were introduced there about 1879, while this was delayed about five years longer on the northwest coast. The outfit did not immediately supplant nude diving in either locality. In 1883, only 80 of the 206 Queensland vessels were supplied with scaphanders, the others continuing to use nude diving, and even yet nearly one third of the vessels depend on that form of fishery. Of the 353 vessels fishing in 1904, 108 depended on nude divers and 245 were supplied with armored equipment.

In 1881 the Queensland government took cognizance of the rapidly developing industry, and enacted a license system and other regulations. For every boat under two tons an annual license fee of £1 (in 1886 this was reduced to ten shillings) was enacted, and for every vessel of ten tons or under, the sum of £3, with an additional amount for vessels in excess of that measurement; but not exceeding £20 in any case.[1] In 1886 it was required by the Queensland government that every person employed "as a diver, and using a diving apparatus," must be licensed annually, for which a fee of £1 is exacted.[2] And in 1891 it was required that "every diving dress and air-pump and all air-tubes and gear used in the fishery in connection with diving must be submitted to an inspector for examination once at least in every period of six months."[3] The license system was adopted in Western Australia in 1886, a fee of £1 per annum being exacted for each vessel engaged in the fishery.[4] In 1891, South Australia adopted the license system, requiring that every boat of two tons or under should pay ten shillings, and that each boat over that measurement should pay twenty shillings.

With a view to protecting the reefs, the government of Queensland in 1891 enacted a law forbidding the sale or removal—except for cultivation purposes—of any pearl shell "of the kind scientifically known as *Meleagrina margaritifera,* and of either of the varieties commonly known as 'golden-edged' and 'silver lip,' of which the nacre or mother-of-pearl measures less than six inches from the butt or hinge to the opposite edge or lip, but this does not apply to the variety commonly called 'dwarf-shell.' "[5] Owing to the difficulty in enforcing this regulation, the size restriction was reduced in 1897 to five inches from the hinge to the opposite lip, or six and one half inches exteriorly, shells of this size weighing approximately one pound. It is claimed that many oysters less than five inches in length are raised, opened for

[1] 45 Victoriæ, No. 2.
[2] 50 Victoriæ, No. 2.
[3] 55 Victoriæ, No. 29.
[4] 50 Victoriæ, No. 7.
[5] 55 Victoriæ, No. 29.

pearls, and then cast back into the water.[1] In 1899 the governor of South Australia interdicted the capture in the waters of that territory of any shell of *"Meleagrina margaritifera* measuring less than four inches from the butt or hinge to the opposite edge or lip." Competent evidence exists that a good-sized pearl has been found in an oyster measuring one inch in diameter.

The fishermen of Western Australia rendezvous at Broome, about one thousand miles by water north of Perth, the nearest railway station. With only a thousand or so inhabitants, under normal conditions, this is a scene of great activity, and bears a reputation of being no Sunday-school when the fishermen are in, with tons of shell and many a pickle bottle more or less full of pearls. Cossack and Onslow are also important stations.

In 1905, 340 luggers and 25 schooners were employed in the pearl fisheries of Western Australia, exclusive of Sharks Bay. Of this number about 85 per cent. hailed from Broome. The schooners ranged in size from 13 to 133 tons, and the luggers were mostly about 12 tons, with a minimum of 3 and a maximum of 14 tons. The total number of fishermen approximated 2900, a medley of races, Japanese, Malays, Chinese, Arabs, native aboriginals and South Sea Islanders working together more or less harmoniously. The yield consisted of 1394 tons of mother-of-pearl, with a declared value of £146,225, and about £50,000 worth of pearls, a total of £196,255 for the year, which was an increase of £32,286 over 1904.[2]

The headquarters for the fishery of the Northern Territory of South Australia are at Port Darwin. In 1905 this fishery employed forty-nine sail vessels and two canoes manned by Europeans, and two proas and twelve canoes manned by Malays. The crews, numbering about 375, consisted mainly of Malays, Japanese and Filipinos. In 1905, 42 per cent. were Malays, 24 per cent. were Japanese, and 20 per cent. were Filipinos. Owing to the low price of pearl shell, the fishery was not prosecuted actively, and many of the Asiatics left for the pearling reefs at the Aru Islands. The total value of pearl shells reported among the exports for that year was £18,526; during the preceding year it was £28,391. No record is available for the value of the pearls.

The Queensland pearling fleet has its rendezvous at Port Kennedy, Thursday Island, which was originally maintained by the British, the Queensland, and the New South Wales governments as a harbor of refuge for mariners. Politically this port is important as the strategic key to the northeast of Australia, but its prosperity is almost wholly

[1] "Departmental Commission on Pearl Shell and Bêche-de-Mer Fisheries," Brisbane, 1897.   [2] "Report on the Fishing Industry for the Year 1905," Perth, 1906, pp. 4-7.

dependent on the pearl-oyster fishery. The population approximates 1600, consisting largely of Japanese, Malays, Cingalese, Pacific islanders, and Australian aborigines, with specimens from nearly every Asiatic and European nationality, and some from America and Africa. The Japanese predominate, their influx dating from 1891; and at present the industry is largely dependent on these Scotchmen of the Orient for its most skilful workmen. The heterogeneous nationalities, and the abundance of sand-flies, mosquitos, etc., make this island rather less desirable as a place of residence than it is interesting from a political and ethnological point of view.

The Queensland fishery in 1905 employed 348 vessels, and yielded 543 tons of shell, according to the government returns. In 1904, 353 vessels were engaged, and the catch was 798 tons of shell.

During the last fifteen years there has been a very steady decrease in the average catch of pearl-oysters per boat in the Australian fishery. The average catch in the Queensland fleet in 1890 approximated 7 tons per boat; from 1898 to 1903 it was about 3 tons annually; in 1904 it was only 2¼ tons, and in 1905 a trifle more than 1½ tons. The yearly increasing number of boats would naturally lower the average, but the decrease is generally ascribed to the denudation of the reefs, due to close working for thirty-five years without giving them a chance to recuperate.

The small yield in Queensland in 1904 and 1905 was due largely to the extended rough weather and the accompanying thick or muddy water, which presented an obstacle to the prosecution of the work. Mr. Hugh Milman, the government resident at Thursday Island, states that each year the beds in the more sheltered spots have been extensively fished, rendering it necessary for the fleet to go farther afield in places where the depth of water is greater, and where the vessels are more exposed to the full force of the southeast winds which prevail for about seven months of the year, and which were unusually severe in 1905.[1] The general denudation of the beds is not the principal cause of the decreased take. An additional cause for the falling-off in 1905 was the deflection of a large percentage of the fleet to new fields of operation, 110 vessels leaving for the Aru Islands in the Arafura Sea, when the season was about half finished.

For vessels using diving apparatus, the season continues throughout the year, but it is frequently interrupted by storms, which may cause the boats to lie in harbor for ten days, or even two weeks at a time. The nude divers suspend work from December to March, and also during the season of gales.

[1] During the month of June, 1908, a severe storm destroyed a pearling fleet, with a loss of 40 vessels and 270 lives.

Each vessel is manned by a diver, his attendant, and a crew of four men, who in pairs take alternate shifts at the manual pump for supplying air to the diver. The entire force of men take part in managing the vessel and in caring for the catch. The vessel is provided with full equipment and supplies of food, water, etc., to last two or three weeks, depending on the distance of the fishing-grounds from the shore station, or the frequency of trips made by a supply vessel.

Except a number of owners and their representatives, there are now very few white persons engaged in pearling in Australian waters. Even the persons in charge of the vessels are largely natives of the Pacific Islands. Owing to the hardships encountered and the small remuneration, it is difficult to secure white labor; and aliens from Japan, the Philippines, Java, Singapore, India and New Guinea, are employed.

The divers are of many nationalities, principally Japanese and Malays, and the former are said to be the most efficient. Previous to 1890, they were mostly whites, and were paid at the rate of £40 per ton of shells; but increased competition and the influx of cheaper labor caused a considerable decrease in the rate of compensation, driving most of the white men out of the employment. At present the Japanese almost monopolize the business. Of the 367 divers licensed at Thursday Island in 1905, 291 were Japanese, 32 were Filipinos, 21 were from Rotuma Island, 16 were Malays, and 7 were of other nationalities; this shows how completely the white man has been driven out of this skilled branch of labor.

The oysters are so scattered that considerable walking is necessary to find them. They usually lie with the shells partly open, and in grasping them the fisherman must be careful not to insert a finger within the open shell, or a very bad pinch will result. The progress of the vessel must be adapted to that of the diver, and when a good clump of oysters is found it may even be desirable to anchor. If the current and wind are just right, the vessel may repeatedly drift over a bed, the diver ascending and remaining on board while the vessel is retracing its course to the windward side of the reef. On new grounds, the nature of the bottom is determined by casting the lead properly tipped with soap or tallow, and the prospects for oysters thus determined without descending.

During good weather and in eight or ten fathoms of water, a diver can work almost continually, and need not return to the surface for two hours or more; but as the depth increases, the length of time he may remain at the bottom in safety decreases almost in geometric ratio, and he comes to the surface frequently for a "blow" with helmet removed. Evidence secured by a departmental commission of the

Queensland government in 1897, showed that in good weather at a depth of eight or ten fathoms, a diver works from sunrise to sunset, coming to the surface only a few times. In a depth of over fifteen fathoms the attendant usually has instructions not to let him remain longer than fifteen minutes at a time; yet a diver's eagerness in working where good shell is plentiful sometimes impels him to order the attendant to disregard this rule. The very great pressure of the water —amounting to thirty-nine pounds or more to the square inch—is liable to cause paralysis, and death occasionally results. In working at a depth of twenty to twenty-five fathoms, a diver is rarely under water longer than half an hour altogether during the day. The greatest depth from which shell is brought appears from the same evidence to be "30 fathoms and a little over"; but at that depth—where the pressure is seventy-eight pounds to the square inch—the fisherman remains down only a few minutes at a stretch, and should be exceedingly careful. The work is injurious, and even under the best conditions the diver not infrequently becomes semi-paralyzed and disqualified in a few years. Notwithstanding that the work is performed by men in vigorous health, nearly every year there are from ten to twenty-five deaths in the Queensland fleet alone;[1] three fourths of these are due to paralysis, and most of the remaining result from suffocation, owing largely to inexperience in use of gear. From five to ten years is the usual length of a man's diving career, although in the fleet may be found men who have been diving for twenty-five years or more.

On the vessels manned by Japanese, commonly several members of the crew are competent divers and take a turn at the work, although only one license is secured. Such a vessel carries only one head-piece, but two otherwise complete suits, the helmet fitting either, so that as soon as one exhausted diver comes up to rest, a successor is ready to have the helmet screwed to his body-dress and descend without delay, thus saving about half an hour in the changing.

The nude divers in the Australian pearl fisheries are mostly Malays and Australian aborigines. They work from dinghys operated from a vessel, each dinghy carrying six or eight divers, usually with a white man as overseer. The man in charge sculls against the tide to keep the boat stationary over the ground, and all the fishermen of a particular dinghy descend together for greater safety from sharks, and to cover the ground systematically. On rising, each diver swims to the boat, throws his catch over the gunwale, and climbs in to rest for a few minutes. Sometimes two or possibly even three oysters may be

[1] "Report of Departmental Commission on Pearl Shell and Bêche-de-Mer Fisheries," Brisbane, 1897.

brought up at a single descent, but a diver is doing well if he brings up one oyster in ten descents. The average daily catch of each man is probably two or three oysters, but a fisherman has been known to bring up fifty in one day. On some vessels, those who fall behind in the catch are punished by extra duty aboard ship.

The pearling industry has had a marked effect on the industrial and social condition of the natives of the Australian coast and the adjacent islands. Many of these natives now have boats of their own, and others seek employment on other vessels. Law and order and decent respect for property have arisen, with schools and churches. The result is all the more remarkable when it is considered that scarcely more than a generation has passed since labor among the men was unknown, the women doing all the work necessary to meet their scanty requirements.

As now carried on in Australia, pearling is a hard life, the men working for two thirds of the season in a dead calm and oppressive heat, while in the remaining months they are rolling day and night. The members of the crew are not allowed ashore without a written permission from the captain of the boat, and men and luggage are searched on leaving the vessel. In addition to these objections, life on board is not unusually made intensely disagreeable by the myriads of inch-long cockroaches, which are attracted by and multiply rapidly on the shreds of muscle left on the pearl shell stored in the hold. Storms are frequent on the coast. In February, 1899, three schooners and eighty smaller vessels were wrecked, and eleven white and four hundred colored men were drowned.

At the end of each day's fishing, the oysters are cleaned of submarine growths. Sometimes this is by no means an easy task, as many of the shells are so covered with weeds, coral, and sponge as to bear little resemblance to oysters. After they have been scrubbed and the edges have been chipped, they are washed and stored on deck. Early the following morning they are opened and examined for pearls. This opening is done carefully to avoid injury to any pearl that may be within. The hinge of the shell is placed on the deck and a broad knife forced down so as to sever the adductor muscle, causing the shells to spring open and permitting the removal of the soft parts. The flesh is carefully examined, both by sight and by feeling, to locate all pearls, which are picked out by hand and placed in a suitable receptacle. Within the adductor muscle are found seed-pearls and small baroques; the large pearls are found embedded in the mantle, where their presence may be detected as soon as the shell is opened, the pearly gleam contrasting with the light blue of the mantle. Sometimes, though rarely, large pearls are found loose within the shell, whence they roll out when the shell is opened. Valuable pearls are occasionally removed from blisters on the surface of the shell, or from within the body of the nacre itself. Even when empty, these blisters are valuable, and are especially adapted for brooches and other ornaments requiring a broad and relatively flat surface.

After the flesh has been carefully examined throughout, it is discarded, as it is not considered suitable for food, and the shell is dried for half a day or so to make the hinge brittle in order that it may be broken without injury to the mother-of-pearl. After the shell has been roughly cleaned, it is placed in the hold, if the vessel is operating from a shore station, as is commonly the case in Torres Straits. Since long exposure to the sun affects the quality of the mother-of-pearl, it is important that it be kept under cover. On returning to the station, it is thoroughly cleaned, assorted, dried, the dark edges clipped off, and the cleaned shell is packed in shipping cases, each containing from 250 to 325 pounds. On the west coast, where the vessels at times operate 200 or 300 miles from port, the shell is cleaned, assorted and crated on the vessels; whence it may be delivered direct to the steamers. The Northwest shell is somewhat smaller than the mature shell of Torres Straits, averaging about 1100 to the ton, whereas that of Thursday Island runs about 725 to the ton.

It is very difficult to prevent the theft of pearls by the fishermen as they are liable to treat them as perquisites if not carefully watched. Indeed, on the Torres Straits vessels it has come about that pearls do not constitute a recognized source of income to the proprietors. There the fishery is now conducted almost exclusively for the shells, as the wage-earners secrete probably as many valuable pearls as they turn over to the rightful owners. The hot sun causes many of the oysters to open, and deft fingers quickly pick out such pearls as may be visible. An oyster may be induced to open its shell by being held near the galley fire on the lugger, and the insertion of a piece of cork holds it open while a pearl is shaken out or hooked out by means of a piece of wire. Then the cork is removed and the oyster closes again with no evidence of robbery. The proprietors of boats who themselves open the oysters almost invariably secure larger yields of fine pearls than those who depend on paid employees, who rarely have the luck to find choice pearls, judging from what they turn in. The government of Queensland has endeavored to put a stop to pearl stealing, and by enactment[1] of 1891, it restricted all selling or buying of pearls within the fishing region except through regularly licensed dealers, whose transactions are open to examination.

But the fishermen seem to have little difficulty in evading the laws,

[1] 55 Victoriæ, No. 29.

and throughout the fleet the men have become so adept that they regard the pearls as their contraband perquisites. And the ease with which these may be secreted is surpassed only by the facility with which they may be sold, notwithstanding legislation to the contrary. Indeed, some employers make no claim to the pearls found, thus enabling them to secure fishermen at lower rates of wages.

As previously noted, the pearls constitute only an incidental catch in the fisheries on the Australian coast, but in the aggregate the yield is very large. The yield in the northwest Australian fishery in 1906 is estimated at £50,000, local valuation; in the Queensland fishery £33,000; in that of South Australia £5000, a total of £88,000 or $440,000.[1] Relatively few seed-pearls are obtained, and some of the pearls are of great size. Some beautiful specimens have been found, but usually they have less luster and are more irregular in form than the Persian or the Indian output.

Among the remedies suggested for improving the condition of the Australian pearl reefs may be mentioned the establishment of six inches as the minimum size of the shell that may be taken (five inches is now permitted in Queensland, and there is no restriction in Western Australia), the closure of certain areas for stated periods from time to time, and a limit on the number of vessels employed. The government resident at Thursday Island, Mr. Hugh Milman, who has had long acquaintance with the industry, strongly recommends the adoption of a system of artificial culture; and in the meantime, to foster the industry, "licenses should be granted to a reduced number of boats and certain sheltered areas should be closed altogether for a few years to give the beds time to recover. This latter procedure, however, the pearlers themselves are not in favor of, as they are of the opinion that the weather conditions against which they have to contend are sufficient protection to prevent the denudation of the principal grounds."

A few years ago certain areas in Torres Straits were proclaimed closed for a period against the removal of pearl shell; but, owing to the want of effective patrol, the shell was poached to a very large extent, and consequently the good that should have resulted from the experiment was not apparent. Owing to the impracticability of continuous patrol, and the want of proper legislation to bring the offenders to book, it was decided to remove the restrictions.

The Sharks Bay fishery, to which we have previously referred,[2] is prosecuted by means of small sail-boats using light dredges, except in the case of the very shallow or "pick-up banks," where the oysters are commonly removed by hand. Some years ago this fishery was of

---

[1] To this should be added the output of Sharks Bay, amounting to £2000 in 1906, making a total of $450,000.　　[2] See pp. 70 and 200.

much local importance; but the developing scarcity of the oysters, and the present low value of this grade of shell in Europe, due to the competition with Mississippi shell, have resulted in a great reduction. In 1905, the industry gave employment to 17 small boats and 42 men, of whom 18 were Europeans, 13 Asiatics, and 11 aboriginal natives. The yield of pearls, according to official report of the government of Western Australia, approximated £2000 in value, and of pearl shell there was 88 tons, with a declared value of £607. In 1896 the government of Western Australia surveyed the Sharks Bay reefs, and opened them to preëmption in small areas for cultivating this species of pearl-oyster. At present they are mostly held under exclusive licenses for a period of fourteen years. The business is under an elaborate system of regulations; but as appears from the above figures the results have not been important.

Pearls are more numerous in this pearl-oyster than in the two other Australian species. In removing them from the flesh, a modification of the Ceylon process is adopted. The mollusks are opened by means of a knife, and the contents of the shells are placed in vats or tubs— known locally as "poogie tubs"; and, exposed to the hot sun, are allowed to putrefy. Sea-water is added, and the putrid mass stirred; after several days the water and the thoroughly disintegrated flesh tissues are decanted, leaving the pearls at the bottom. The odor from a number of these "poogie tubs" is said to almost rival that of the "washing toddies" at Marichchikadde.

The Sharks Bay pearls are commonly yellowish or straw colored, and sometimes have a beautiful golden tinge. Although obtained from small shells, they are sometimes of considerable size—twenty grains or more in weight, and fine specimens sell for several hundred dollars each. China and India furnish better markets for them than Europe or America.

PEARL FISHERIES OF THE MALAY ARCHIPELAGO

My thoughts arise and fade in solitude;
The verse that would invest them melts away
Like moonlight in the heaven of spreading day.
How beautiful they were, how firm they stood,
Flecking the starry sky like woven pearl.
                              SHELLEY, *My Thoughts.*

FOR nearly four hundred years, pearls and pearl shells have been the most beautiful objects which have reached the outside world from the many islands of the Malay Archipelago. On his visit to this part of

Opening pearl-oysters and searching for pearls, off the coast of Australia

Grading, weighing, and packing mother-of-pearl, off the coast of Australia

Moro boats, used among the pearl islands of the Malay Archipelago

Raft used for pearl fishing in the Malay Archipelago

the world in 1520, Pigopitta, a companion of Magalhães, reported pearls among the prized possessions of the natives. The fisheries have never been of great importance, although the reefs are widely scattered throughout the archipelago, and the possibilities seem favorable for very great development. Thomas de Comyn stated a century ago, that pearl fisheries had been undertaken "from time to time about Mindanao, Zebu, and some of the smaller islands, but with little success and less regularity, not because of a scarcity of fine pearls, but on account of a lack of skill of the divers and their well-established dread of sharks."[1]

Giacinto Gemmi,[2] writing of Philippine pearls, repeats a strange tale from the "Storia de Mindanao" by the Jesuit father, Combes, to the effect that in a certain spot, under many fathoms of water, there was a pearl of inestimable value, as large as an egg; but, although the king's ministers had made every effort to have it secured, they had always been unsuccessful.

During the last thirty years, pearls and pearl shells have been secured from most of the inshore waters of Malaysia, but the output has not been so regular or so extensive as the conditions seem to warrant. Our observation leads to the conviction that this is not due so much to lack of skill on the part of the divers, or to their dread of sharks, mentioned by Comyn; but to the fact that foreign capital, attracted to this part of the world, has found more security and profit in developing plantations, and the natives have not had sufficient enterprise to systematize and develop the fishery resources.

Throughout Malaysia, including the Philippine Islands, the pearl is known as *mutya, mootara,* or a similar name, closely resembling the Sanskrit *mukta* or the Cingalese *mootoo,* indicating the source of the influence originating the fishery and trade.

The most widely-known pearl fisheries of Malaysia are in the Sulu Archipelago, a group of islands comprising about 1000 square miles in area, and containing a population of 100,000. The beautiful yellow pearls shared with the many acts of piracy in attracting attention to this group previous to 1878, when the islands were brought under the influence of Spanish rule; and since the Spanish-American War, pearl fishing has been the leading industry, though it has received less attention from outside sources, perhaps, than has the existence of slavery and harems as part of the social system.

Writing in 1820, John Crawfurd stated that the annual export of pearls from Sulu Islands to China approximated 25,000 Spanish dollars in value, and the mother-of-pearl similarly exported was worth

[1] Comyn, "State of the Philippine Islands," London, 1820, pp. 38, 39.   [2] "Storia Naturale delle Gemme," Naples, 1730. Vol. I, p. 461.

70,000 dollars. "Considering the turbulent and piratical habits of the natives of the Sulu group, it is certain that a greater share of skill and industry than can at present be applied to the fisheries, would greatly enhance the value and amount of their produce." [1]

In the Sulu Archipelago, the pearl-oyster reefs exist from Sibŭtu Pass to Basilan Strait, and roughly cover an estimated area of 15,000 square miles; that is, in the most favorable localities throughout this area, pearl-oysters occur to a greater or less extent. The fisheries are prosecuted by Malays and Chinese, and are largely centered at Sulu.

Pearl-oysters occur about many other islands. They exist at Maimbun and Parong; and also off the island of Tapul and its neighbor Lagos, both southwest of Maimbun. In the channels among these islands, on the rocky gravelly bottom where there is a good current, oysters are commonly found. They also occur off Laminusa, northeast of Tawi-Tawi, at Cuyo Island, and in the waters about Malampaya and Bacuit.

The large mother-of-pearl oyster (*Margaritifera maxima*) known locally as *concha de nacr,* is by far the most abundant. When full-grown in this region it is ordinarily between ten and thirteen inches in diameter. The young oyster attaches itself to the bottom by means of the green byssus; but after attaining a weight of one pound, it is too heavy to be easily moved by the tide, and the ligature gradually disappears. The Australian "black lip" (*Margaritifera margaritifera*), known here as *concha de nagra,* is also found. In these waters it attains a diameter of about eight inches, but most specimens are considerably smaller.

There is another pearly shell in the Philippines, a spiral gasteropod known locally as *caracoles,* which is ordinarily five or six inches in diameter, and has a beautiful pearly surface. This yields very few pearls; it is sought for pearl-button manufacture, selling for about the same as the *concha de nagra.*

Streeter states that it is declared by the natives of the Sulu Archipelago that pearls of a yellowish hue have been found in the pearly nautilus (*Nautilus pompilius*), one of the group of cephalopodous mollusks. As, however, there is a superstition that they bring ill luck, the natives say that they throw them away, believing that any one who should fight while wearing one of these pearls in a ring, would certainly be killed. If we consider the habits and organism of this remarkable animal, and the splendid nacreous coating of its shell, the assertion that pearls are found in it seems quite natural. Indeed, the occurrence of pearls in the pearly nautilus is generally recognized.

[1] Crawfurd, "History of the Indian Archipelago," Edinburgh, 1820, Vol. III, p. 445.

For many years the successive sultans of Sulu exercised authority over the fisheries and—in addition to exacting certain percentages and presents from the fishermen—claimed as their perquisites all pearls exceeding a designated weight. The fisheries were prosecuted by nude divers, of whom there were a large number. A Chinese company had been particularly fortunate in its relations with the Sulus, and had an extensive equipment in the fishery, consisting of a number of small vessels, each carrying a crew of seven men, who used diving-suits. In addition to these, some of the native Moros owned boats from which diving-suits were employed.

Following the Spanish-American War and the transfer of the Philippine Islands to America, several vessels proceeded to engage in the fisheries without previously consulting the representatives of the Sultan of Sulu. This called forth from that official an appeal to the American authorities for protection in his claims. He gave an account of the pearl fishery in this interesting document, which we quote at length—through the courtesy of the American Bureau of Insular Affairs—because of the light it throws, not only on the industry, but also on the characteristics of these people with whom the American government is now dealing.

STATEMENT MADE BY THE SULTAN OF SULU RELATIVE TO THE PEARL FISHERIES

(Forwarded by the Governor of Moro Province.)

[Translation.]   (SEAL of the SULTAN.)

No date.

I beg to inform my father, the civil governor, Major Scott, as you want to know about the mother-of-pearl shell, why it is the right of all Sulu people, above all my own right, this is the reason:

The forefathers of the Sulu people used to take the mother-of-pearl shell from the downs because the mother-of-pearl shell belonged to the downs, and they took them to eat the oyster with other food; of the shell they made plates and saucers to put the food on, and the pearls they used to make a hole through and put them on a string as necklaces for their children. This was at a time when no other nation had come to Sulu to buy the mother-of-pearl shell.

Later, a big boat, called the *Sampang*, wandered from China to Sulu; there were on board many people, all Chinese; it was loaded with merchandise. The people came ashore and saw the mother-of-pearl shell which the Sulu people were carrying. The captain of the boat said: "Have you many more of these things?" and the people answered, "Plenty; this is what we take from the downs to eat with other food." The Captain said, "Gather me plenty. I will buy them from you. The people went and gathered them and bartered them for plates and saucers. When all the shells from the downs were finished they looked into the deep, and that is how they found the pearling grounds, and the people noted them, and remembered them. This is what they agreed upon; whoever finds pearling grounds they belong to him from generation to generation. That is what they agreed upon. That is the reason why the Sulu people have the right, and that they came to make the dredge (*badja*) to get the mother-of-pearl shell from the deep, because they can not see them.

Later Salips came from Mecca of the Arab nation; they came to Sulu to convert the people into Mohammedans, as they had no religion. And when the Sulu people, including the islanders, adopted the faith, then they agreed to have a sultan and they elected Saripul Hassim to be sultan. Saripul Hassim said: "I don't want you to make me your sultan if I do not know what the rights of the sultan are, and who I have to govern over, because this is not my country, this is your country."

And this is how everybody agreed to accept him as sultan over Sulu and all the islands; this is how he became Sultan and governed over all, and this is how Saripul Hassim accepted to be the sultan of Sulu, to have full power over land and sea, and the people's rights, where they got their living from on land and sea, were left to them, because they were the means of their getting their livelihood.

But a law was made, if they found valuables in the sea, such as pearls, tortoise shell, ambal or anything extraordinary, they have to show it to the sultan, and if the pearls weigh six chuchuk or over they become the share of the sultan; if they do not have that weight, the people can do with them as they please and sell them. If the sultan wants them, he will buy them according to custom. As to tortoise shell, if they weigh two ketties, they go to the sultan, and as to the ambal, whether it is much or little, it falls to the sultan. Whoever finds it must take it to the sultan. Whoever of his subjects violates this law as agreed upon, the sultan can punish him as he pleases.

They accepted this law as agreed upon, to be carried out by them (sultan and people), and their descendants, and not to be changed; but they asked of the sultan not to let any other nation take a share in this industry; it is enough for them; and the sultan agreed to this because they did not know how to earn their living otherwise. This is what the sultan and his subjects agreed to because the Sulu had no other treasures on land beyond the cultivation; the treasures came from the sea only, therefore other people are forbidden because this is the property of all my subjects, and especially my own.

Recently, in my time and in the Spanish time, there came to me Captain Tiana; he wanted to dive for pearl shells. I said "I cannot give you my consent at once because since our forefathers (sultan and people) we have an agreement, I will confer with my people." I sent for the chiefs and the dattos and I told them about it, that Captain Tiana came to me and asked to dive for pearl shells. They said it cannot be done, because there is an agreement between our forefathers that other nations cannot join in this industry of the Sulu seas, because there is no other means of earning a living for your subjects.

I informed Captain Tiana of it. He said: "Allow me to dive for pearl shells, I will give toll to you as sultan and I will also give toll to the owners of the pearl grounds according to what we agree upon."

So I informed all the owners of the pearling ground, and they said, "If he is really in earnest to give toll to us owners of the ground according to what we agree upon, if we don't agree, we will not allow him to fish." Thereupon Captain Tiana and I went to the Spanish governor to bear witness. The governor said: "All right; anything you agree upon; I cannot change the law of the Moro people, and I will not interfere."

That is how I allowed Captain Tiana to fish, and I gave him a letter of the truth according to agreement. Therefore if any person of other nation wants to fish for mother-of-pearl shell, he will have to do as Captain Tiana did, and ask me for a letter of truth, and if he has no letter and does not pay toll to the owners of the ground, and especially to me, he cannot dive, and if he violates this and if anything befalls him, I am not responsible and do not want to be held. responsible, because the mother-of-pearl shells are like the property in our boxes given to us by God. They do not go away from the places where they are put, they are not like fish that go about. Therefore, we forbid it. It is our heritage from our forefathers.

(Signed) HADJI MOHAMAD JAMAUL KIRAM,
Sultan of Sulu.

[SEAL OF THE SULTAN.]

Following these representations, the legislative council of the Moro province, by authority of the Philippine Commission, interdicted all fishing for pearl-oysters within three marine leagues of any land within the territorial limits of the Moro province, without license first obtained from the treasurer of the district within which the vessel carries on the major part of its operations.[1] No license was to be issued to any vessel not owned in the Philippine Islands or in the United States, and not wholly owned by citizens of the United States, by natives of the Philippine Islands, or by persons who have acquired the political rights of natives,[2] except that foreign vessels which for one year immediately preceding had actually engaged in pearl fishing might secure license to continue therein for a period of five years thereafter.

Licenses were of two kinds, according to the nature of the fishery. To engage in fishing with the aid of diving-suits, the fee was five hundred pesos annually, for each of the greatest number of divers beneath the surface of the water at any one time. For fishing with-

[1] Act No. 51, June 7, 1904.
[2] A letter from the Bureau of Insular Affairs, dated November 20, 1906, states: "It is proposed by the officials of the Moro province to amend the regulations so that, under certain restrictions, vessels of foreign build may engage in pearl fisheries."

out submarine armor, the fee was five pesos annually, for each of the greatest number of nude divers to be employed by the vessel during any voyage, and the same sum for each of the greatest number of dredges or rakes to be employed beneath the surface at any one time; but this did not apply to vessels under 15 tons, owned and operated wholly by native Moros, until January 1, 1906.

It was also made unlawful to catch or to have in one's possession within the Moro province "any pearl shell or any bivalvular or lateral plate, or any pearl shell of less than 4½ inches in diameter, measured with a flat, rigid measuring rod along the line of the ligament which joins one binocular or lateral plate to the other at the hinge, unless the lateral plate of such shell be more than 7 inches in diameter measured with a flat, rigid measuring rod from the outer edge of the horny lips to the center of the hinge, the rod being so placed as to form a right angle with the line of the hinge."[1]

According to a report furnished by the Mining Bureau at Manila, there were seven vessels fishing with diving-suits in the Sulu Archipelago in 1905, each representing an investment of about 6000 pesos. In 1906 there were ten vessels engaged in this industry, and the collection on licenses for that fiscal year amounted to 3375 pesos. These vessels are mostly small Moro craft which cannot venture upon distant cruises in the archipelago for prospecting purposes, and their operations are confined for the most part to the immediate vicinity of Jolo. Each vessel carries one diver, a tender, a cook, and four sailors. In addition to food supplies, the sailors and the cook each receive twelve to fifteen pesos per month, the tender thirty to forty pesos per month, and the diver the same amount and in addition thereto a bonus of twenty cents for each shell secured. Near Jolo the vessels work throughout the year, but farther north very little fishing is done from December to April, when monsoons prevail. The man in charge of each vessel is obliged by law to keep an accurate record of the number and weight of shells found, and his figures are checked up by a customs official at either Jolo or Zamboanga, the ports of discharge.

To enable them to secure pearl-oysters at depths of from twenty to forty fathoms, the Sulus have long made use of a dredge (*badja*) peculiarly constructed of native materials, and admirably adapted to the purpose. This consists of five or more long wooden teeth slightly curved and spreading outward, with an expanse at the ends of twenty inches or more. The dredge is properly balanced by two stones, and a bridle rope is so attached to it that, when thrown overboard and towed behind a canoe drifting with the current or the wind, the im-

[1] Act No. 43, amended June 7, 1904.

plement rests on the curve of the teeth, which are in almost a horizontal position. As the teeth enter the gaping shell of an oyster lying on the bottom, the animal instantly closes tightly on the intruder and effects its own capture. The principle is similar to that of the "crowfoot" dredge of the Mississippi River, although the design of the implement is radically different. A second rope is attached so as to raise and lower the implement and to detach it from corals, rocks, and other objects against which it may catch in its course on the bottom. This dredge is designed for very deep areas, where the bottom is relatively smooth.

The Moros employ yet another method of fishing, using a *mag-tung-tung* or three-pronged catcher, which is let down by a rattan rope and by means of which individual shells sighted from the surface are obtained. When the water is perfectly clear this implement can be operated where the depth is fifteen or eighteen fathoms, but its use is impractical where the water is clouded or there is even a slight ripple on the surface.

However, the bulk of the catch is made by the nude divers, of which there are hundreds at Maimbun, Tapul, Lugus and elsewhere. In their small boats these Moro fishermen visit the reefs, where the boats are anchored. Provided only with a short, heavy knife, with which to release the shells from the bottom or, perchance, as a weapon of defense against sharks and other fish, they enter the water feet first, but soon turn and descend head downward, precisely as on the Australian coast, swimming toward the bottom with bold strokes. The Sulu pearl-divers—and especially those at Parang, Patian and Sicubun—are among the most expert in the world. They easily penetrate to twelve fathoms and, if necessary, to eighteen or twenty fathoms. But they are not very industrious, and seldom descend more than twelve or fifteen times a day, preferring rather to go with their wants half satisfied than to satiate them by more active exertions.

Many descents may be necessary to locate and obtain a single oyster, but when this is secured the shell alone may ordinarily be traded for sufficient to supply the fisherman's needs for several days, and there is always the chance of a pearl. After a short day of labor, the fishermen return, and the oysters which they have secured are opened and examined for pearls. After the flesh has been carefully searched it is placed in the sun to dry and, later, to be used for food, and the shells are carefully cleaned and placed under cover until they may be bartered or sold.

The Sulu shell is characterized by a peculiar yellowish tint around the rim, by means of which it is readily distinguished. Its size and beautiful iridescence make it very attractive, and for choice individual specimens high prices are received. It is the largest of the mother-of-pearl shells, single half-shells of "bold" size average one and one half pounds in weight, while some attain a weight of six pounds. The body of the shell furnishes the most beautiful of all mother-of-pearl, yet the necessity for discarding the yellow rim, or, rather, for using it separate from the rest, makes it unpopular with manufacturers. The annual product is estimated at 200 tons, valued in London and New York at $200,000, and of pearls about $30,000 worth.

The Sulu pearls are frequently large and of choice quality, but they are far more inclined to a yellowish tint than those from Australian waters, 1300 miles southward. The sultans accumulated the finest collection of them, and some of these found their way into the markets from time to time as the condition of the exchequer ran low or royal emergency required, as in 1882, for instance, when it was necessary to defray the expense of Sultan Buderoodin's pilgrimage to Mecca. During the last six or seven years, much has been heard of the present sultan's collection, which he largely inherited, and some fairly good specimens have been presented to prominent Americans.

Pearl-oysters are among the important resources of the inshore waters of the Dutch East Indies, including the surrounding seas of Sumatra, Java, Borneo, Celebes, the Aru Islands, the Moluccas or Spice Islands, and Papua or New Guinea. For very many years the natives have gathered pearl shell and pearls from these waters, and especially on the coast of the Aru Islands, at Gilolo or Halmahera, and the islands thereabout, on the east coast of Celebes, and about the Sunda group. The collections were made in the shallow waters by beach-combing and by nude diving, and were bartered with the Chinese and Arab traders sailing from Singapore, Macassar, and other ports. Occasionally a pearling vessel from Singapore or from Torres Straits would try its luck in these waters; but, except for the work of the natives, the reefs were practically untouched previous to 1883.

As the Australian fleet increased in size and the oysters became scarce in Torres Straits and on the northwest coast, some of the vessels occasionally visited the Aru Islands, the coast of Papua, etc. These met with considerable success and the number of trips increased, especially in 1893, when oysters were unusually scarce in Australia.

The following year, 1894, the government restricted the fishery to inhabitants of the Netherlands and of Netherlands India, or to companies established in those countries and operating under the Dutch flag. Owing to the activity of Dutch capital in coffee, tobacco and other plantation enterprises, the pearl resources received very little attention from them. The success of the Australian fishery encouraged the formation in 1896 of an Amsterdam company to exploit the Aru grounds; but apparently without financial success, for it liquidated in 1898.

In the meantime, residents of these islands paid more and more attention to the pearl fishery; also Europeans, Chinamen and Arabs arranged with the native chiefs for fishing in their territorial waters, paying therefor a fixed sum in cash or a percentage of the catch, which was permitted on approval by the governor general of Dutch India. The fleet continued to increase from year to year, and in 1905 there was a very large influx of vessels from the Australian fisheries, 110 luggers and 7 tenders coming from Thursday Island alone.

The species are the same as occur on the northern coast of Australia, the "silver-edge" or "golden lip" (*Margaritifera maxima*) occurring in greatest abundance, and the "black lip" (*M. margaritifera*) to a less extent.

The shells are the principal object of the search, and the pearls found incidentally form an additional source of revenue. These shells divide with those of Australia the reputation of being the most valuable in the world. They are commonly known in the trade by the name of the port from which they are originally shipped, as Manila, Macassar, Banda, Ceram, Penang, Mergui, etc. Before the exploitation of the Australian grounds, they sold at very high prices, and $2000 or more per ton was sometimes realized for those of the best quality. Singapore is the headquarters for supplies for the industry in all this region, and it is from that port that the shells and pearls are mostly distributed.

The pearls obtained in Netherlands India are of choice quality and of relatively large size, a considerable percentage of them weighing over eight grains, and fairly good pearls of fifty grains or more are occasionally reported. Colored pearls are rarely met with, nearly all of them being clear white, like the beautiful Macassar shell.

At Pados Bay, island of Borneo, one hundred or more persons find employment fishing the Placuna oysters, selling the shells for about $2 per picul (139 pounds to the picul), the dried meats at $4 to $6 a picul, and the seed-pearls (*seleesip*) at about $2 per mayam. Many of these pearls are sold in the village of Batu Batu. When a fisherman buys his few necessaries at the Chinese shops, he pulls out his little package of seed-pearls and pays in that currency, the Chinaman making a good profit by the transaction.

Pearling village, with youthful fishermen.   Sulu Islands

Japanese diver in Dutch East Indies, come up to "blow" for a few minutes

X

AMERICAN PEARLS

VENEZUELA, PANAMA, MEXICO, AMERICAN FRESH WATERS,
MISCELLANEOUS

X

PEARL FISHERIES OF VENEZUELA

When I discovered the Indies, I said that they composed the richest country in the world. I spake of gold and pearls and precious stones, and the traffic that might be carried on in them.

Extract from Columbus's Fourth Letter.

THE Caribbean Sea furnishes one of the most interesting chapters in the history of the pearl fisheries. In no region of the world have these resources caused more rapid exploitation or affected the inhabitants to a greater extent than on the shores of Venezuela.

Before the discovery of America, the natives of this region collected pearls from the mollusks which they opened for food in times of necessity, and also sought them for ornamental purposes. And although they had large collections which they used for personal ornamentation and for decorating their temples, it does not appear that they prized them extravagantly, readily bartering them for small returns.

In Columbus's account of his third and fourth voyages to America, he repeatedly refers to pearls. On the third voyage, in 1498, after passing the mouth of the Orinoco River, he entered the Gulf of Paria, where the natives "came to the ship in their canoes in countless numbers, many of them wearing pieces of gold on their breasts, and some with bracelets of pearls on their arms; seeing this I was much delighted and made many inquiries with the view of learning where they found them. They replied that they were to be procured in their own neighborhood and also at a spot to the northward of that place. I would have remained here, but the provisions of corn, and wine, and meats, which I had brought out with so much care for the people whom I had left behind, were nearly wasted, so that all my anxiety was to get them into a place of safety, and not to stop for anything. I wished, however, to get some of the pearls that I had seen, and with that view sent the boats on shore. I inquired there also where the pearls were obtained. And they likewise directed me to the westward and also to the north behind the country they occupied. I did not put this information to the test, on account of the provisions and the weakness of my eyes and because the ship was not calculated for such an undertaking."

In his letter to one of the queen's attendants, written in 1500, Columbus says, in justification of his conduct toward his miserable detractors: "I believed that the voyage to Paria would in some degree pacify them because of the pearls and the discovery of gold in the island of Española. I left orders for the people to fish for pearls, and called them together and made an agreement that I should return for them, and I was given to understand that the supply would be abundant."

And again in the same letter, after speaking of a quantity of gold which mysteriously disappeared when Governor Bobadilla sent him and his brothers loaded with chains to Spain, he says: "I have been yet more concerned respecting the affair of the pearls, that I have not brought them to their Majesties. . . . Already the road is opened to gold and pearls, and it may surely be hoped that precious stones, spices, and a thousand other things will also be found."

A more detailed account of Columbus's pearling adventures, and of the subsequent discoveries and explorations on the Caribbean coast is given by Francisco Lopez de Gomara in his "Historia general de las Indias," published in 1554, of which the following is a literal translation slightly abridged:

Since there are pearls on more than four hundred leagues of this coast between Cape Vela and the Gulf of Paria, before we proceed farther it is proper to say who discovered them. In the third voyage made by Christopher Columbus to the Indies, in 1498, having reached the island of Cubagua, which he called "Isle of Pearls," he sent a boat with certain sailors to seize a boat of fishermen, to learn what people they were and for what they were fishing. The sailors reached the shore where the Indians had landed and were watching. A sailor broke a dish of Malaga ware and went to trade with them and to look at their catch, because he saw a woman with a string of rough pearls (*aljofar*) on her neck. He made an exchange of the plate for some strings of rough pearls, white and large, with which the sailors returned highly delighted to the ships. To assure himself better, Columbus ordered others to go with buttons, needles, scissors, and fragments of the same Valencian earthenware, since they seemed to prize it. These sailors went and brought back more than six marcs (forty-eight ounces) of rough pearls, large and small, with many good pearls among them. Said Columbus then to the Spaniards: "We are in the richest country of the world. Let us give thanks to the Lord." They wondered at seeing all those rough pearls so large, for they had never seen so many, and could not contain their delight. They understood that the

Indians did not care much for the small ones, either because they had plenty of large ones, or because they did not know how to pierce them.

Columbus left the island and approached the land, where many people had collected along the shore, to see if they also had pearls. The shore was covered with men, women, and children, who came to look at the ships, a strange thing for them. Many Indians presently visited the ships, went on board and stood amazed at the dress, swords, and beards of the Spaniards, and the cannon, tackle, and arms of the ship. Our people crossed themselves, and were delighted to see that all those Indians wore pearls on their necks and

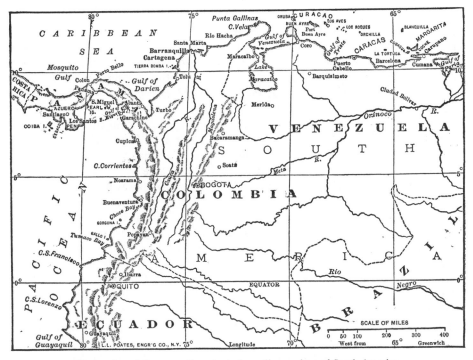

Venezuela and Panama; the principal pearling regions of South America

wrists. Columbus asked by signs where they fished them, and they pointed to the coast and island.

Columbus then sent to the shore two boats with many Spaniards, for greater certainty of those new riches, and because they importuned him. The chief took them to a place where there was a circular building that resembled a temple, where presently much bread and fruits of different kinds were brought. At the end of the feast he gave them pearls for sweetmeats, and took them afterward to the palace to see the women and the arrangement of the house. Of the numerous women there, not one was without rings of gold and necklaces of pearls. The Spaniards returned to the ships, wondering at such pearls and gold, and requested Columbus to leave them there. But he did not wish to do so, saying they were too few to settle. He hoisted sail and ran along the coast as far as Cape Vela, and from there came to Santo Domingo, with the intention of returning to Cubagua after regulating the affairs of the government. He suppressed the joy he felt at having found such treasures, and did not write to the king regarding the discovery of pearls, or at all events did not write it until it was already known in Castile. This was largely the cause for the anger of the king, and the order to bring Columbus a prisoner to Spain. They say that he did not so much intend to conceal this discovery from the king, who has many eyes, as that he thought by a new agreement to get this rich island for himself.

Of the sailors who went with Christopher Columbus when he found the pearls, the greater number were from Palos. As soon as these came to Spain, they told about the country of pearls, displayed many, and carried them to Seville to sell, whence they went to the court and into the palace. Excited by this report, some persons there hurriedly prepared a ship and made Pedro Alonso Niño its captain. He had from the Catholic king license to go in search of pearls and land, provided he should not go within fifty leagues of any discovered by Columbus.

Niño embarked in August, 1499, with thirty-three companions, some of whom had been with Columbus. He sailed as far as Paria, visited the coast of Cumana, Maracapan, Port Plechado, and Curiana, which lies united to Venezuela. There he landed, and a chief, who came to the coast with fifty Indians, conducted him amicably to a large town to take water, refreshments, and the barter he was in search of. He bartered for and secured fifteen ounces of pearls in exchange for pins, rings of horn and tin, glass beads, small bells, and similar trifles. The Spaniards stayed in the town twenty days, trading for pearls. The natives gave a pigeon for a needle, a turtle-dove for one glass bead, a pheasant for two, and a turkey for four. For that price they also gave rabbits and quarters of deer. The Indians asked to be shown the use of needles, since they went naked and could not sew, and were told to extract the thorns with them, for they went barefooted: Niño brought to Galicia ninety-six pounds of rough pearls, among which were many fine, round, lustrous ones of five and six carats, and some of more. But they were not well pierced, which was a great fault. On the route a quarrel arose over the division, and certain sailors accused Niño before the governor in Galicia, saying that he had stolen many pearls and cheated the king in his fifth, and traded in Cumana and other places where Columbus had been. The governor seized Niño, but did not keep him in prison very long, where he consumed pearls enough.[1]

This expedition of Pedro Alonso Niño was the first financially profitable voyage to America. After his return, the Cubagua pearl fishery became the object of numerous speculations, and many other Spaniards fitted out voyages, most of them sailing from Hispaniola or Haiti, nine hundred miles distant. Owing to the ill treatment of the Indians and excessive cruelties toward them, much difficulty was experienced

[1] "Historia general de las Indias," by Francisco Lopez de Gomara, 12mo, 1554, pp. 104-106 b.

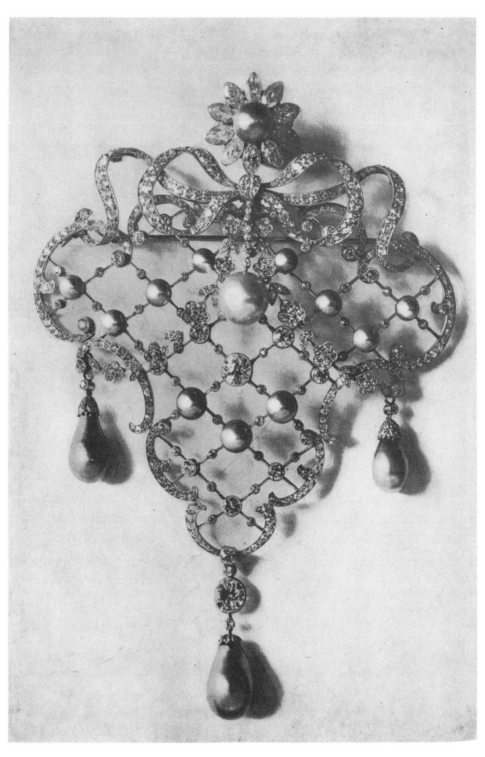

GRAY PEARLS FROM LOWER CALIFORNIA, AND DIAMONDS

Pan-American Exposition, 1901

CLARA EUGENIA, DAUGHTER OF PHILIP II

Painting by Gonzales, in the Galeria del Prado, Madrid
Most of these pearls were doubtless from the early American fisheries

in securing divers. This was relieved in 1508 by transporting large numbers of Indians from the Lucayan or Bahama Islands and impressing them into the service. These were so expert in the work that individuals sold for upward of 150 ducats each.[1] With their aid the fishery prospered so greatly that in 1515 a settlement, called New Cadiz, was established on Cubagua Island by the governor of Hispaniola, Diego Columbus, son of the discoverer. This small island was dry and desolate, without water or wood, which were brought from the mainland twenty miles distant, or from Margarita Island about three miles to the northward.

An interesting description of the manner of securing the pearls by these early adventurers was given by Gonzalo Fernandez de Oviedo y Valdes (1478–1557) in his "Historia natural y general de las Indias," written less than thirty years after the discovery of the mainland of America. A translation of this book was published in 1555 by Richard Eden in his "Decades of the New World"; from which we extract the following account, the retention of Eden's quaint phraseology seeming permissible owing to this being one of the very earliest books on America.

Of the maner of fyshynge for perles

The Indians exercise this kynde of fyschynge for the moste parte in the coastes of the North in *Cubagua* and *Cumana*. And manye of theym which dwell in the houses of certeyne particular lordes in the Ilandes of *San Dominico* and *Sancti Iohannis,* resort to the Ilande of *Cubagua* for this purpose. Theyr custome is to go fyve, syxe, or seven, or more in one of theyr *Canoas* or barkes erly in the mornynge to sume place in the sea there about where it appeareth unto them that there shulde bee great plentie of those shell fyshes (which sume caule muscles and sume oysters) wherein perles are engendered. And there they plonge them selves under the water, even unto the bottome, savynge one that remaynethe in the Canoa or boate which he keepeth styll in one place as neare as he can, lookynge for theyr returne owte of the water. And when one of them hath byn a good whyle under the water, he ryseth up and commeth swymmynge to the boate, enterynge into the same, and leavynge there all the oysters whiche he hath taken and brought with hym. For in these, are the perles founde. And when he hathe there rested hym selfe a whyle, and eaten parte of the oysters, he returneth ageyne to the water, where he remaynethe as longe as he can endure, and then ryseth ageyne, and swimmeth to the boate with his pray, where he resteth hym as before, and thus continueth course by course, as doo all the other in lyke maner, being all moste experte swymmers and dyvers. And when the nyght draweth neare, they

---

[1] Herrera, "Historia general de los hechos de los Castellanos en las islas y Tierra Firme del Mar Oceano," Dec. iii, Book VII, ch. 3.

returne to the Ilande to theyr houses, and presente all the oysters to the master or stewarde of the house of theyr lorde who hath charge of the sayde Indians. And when he hath gyven them sumwhat to eate, he layeth up the oysters in safe custodie untyll he have a great quantitie thereof. Then hee causeth the same fyssher men to open them. And they fynde in every of them pearles other great or smaul, two or three or foure, and sumtymes five and syxe, and many smaule graines accordyng to the lyberalitie of nature. They save the pearles bothe smaule and great whiche they have founde: And eyther eate the oysters if they wyl, or caste them away, havynge so great quantitie thereof that they in maner abhorre them. Those oysters are of hard fleshe, and not so pleasant in eatyng as are owres of Spayne. This Ilande of *Cubagua* where this manner of fysching is exercised, is in the Northe coaste, and is no bygger then the Iland of Zelande. Oftentymes the sea encreaseth greatly, and muche more then the fyshers for pearles wold, bycause where as the place is very depe, a man can not naturally rest at the bottome by reason of the aboundaunce of aery substannce whiche is in hym, as I have oftentymes proved. For althoughe he may by vyolence and force descende to the bottome, yet are his feete lyfted up ageyne so that he can continue no tyme there. And therefore where the sea is verye deepe, these Indian fyshers use to tye two great stoones aboute them with a corde, on every side one, by the weyght whereof they descend to the bottome and remayne there untyl them lysteth to ryse ageine: At which tyme they unlose the stones, and ryse uppe at their pleasure. But this their aptenesse and agilitie in swimming, is not the thynge that causeth men moste to marvaile: But rather to consyder how many of them can stande in the bottome of the water for the space of one hole houre and summe more or lesse, accordynge as one is more apte hereunto then an other. An other thynge there is whiche seemeth to me very straunge. And this is, that where as I have oftentymes demaunded of summe of these lordes of the Indians, if the place where they accustomed to fysche for pearles beynge but lyttle and narrowe wyll not in shorte tyme bee utterly without oysters if they consume them so faste, they al answered me, that although they be consumed in one parte, yet if they go a fyschynge in an other parte or on another coaste of the Ilande, or at an other contrary wynd, and continue fysshing there also untyll the oysters be lykewyse consumed, and then returne ageyne to the fyrste place, or any other place where they fysshed before and emptied the same in lyke maner, they find them ageine as ful of oysters as though they had never bin fysshed. Wherby we may judge that these oysters eyther remove from one place to an other as do other fysshes, or elles that they are engendered and encrease in certeyne ordinaire places. This Iland of *Cumana* and *Cubagua* where they fyshe for these perles, is in the twelfe degree of the part of the said coaste which inclineth toward the North.

The cupidity of the proprietors of the fishery led to most cruel treatment of the divers and, if the accounts of the time are to be relied upon, a large percentage of them died under the harsh régime. About

1515 the unfortunate natives obtained an earnest and influential advocate in Bartolomé de las Casas, who, in 1516, prevailed upon the youthful Charles V to decree that the fishery should be prosecuted only in summer, that the divers should not be required to work more than four hours a day where the depth exceeded six fathoms, that they should receive good nourishment and half a quart of wine daily, should have hammocks or beds in which to sleep, and should be provided with clothes to put on as soon as they left the water.[1] And by later ordinances it was stipulated that death should be inflicted on any one forcing a free Indian to dive for pearls.

In 1528 the resources of Coche Island were exploited with so much success that within six months "1500 marcs (12,000 ounces) of pearls" were secured. Pearl banks were successively found at Porlamar, Maracapana, Curiano, and at various places on the coast from the Gulf of Paria to the Gulf of Coro, a distance of over five hundred miles, which became designated the "Pearl Coast." For a number of years previous to 1530, the output exceeded in value 800,000 piastres annually, approximating one half the produce of the American mines at that time.[2] It was largely these pearls that enriched the cargoes of many of those famous caravels that crossed the Atlantic to Spain. Indeed, for several decades, America was best known in continental Europe as the land whence the pearls came.

An interesting account of an early effort to use dredges in the Cubagua pearl fishery was given by Girolamo Benzoni, who had lived in America from 1542 to 1555, and was familiar with the conditions. He states:

At the time the pearl fishery flourished on this island there came here one Louis de Lampugnan with an imperial license authorizing him to fish such quantities of pearls as he pleased within all the limits and bounds of Cubagua. This man set out from Spain with four caravels loaded with all the necessary provisions and munitions for such an enterprise, which some Spanish merchants furnished him. He had made a kind of rake, the fashion of which was such that in whatever part of the sea it was used, not an oyster would escape. At the same time he would have raked and drawn out all that bore pearls if he had not been disappointed. But the Spaniards in Cubagua all banded against him in the execution of his privilege. They said the emperor was too liberal with other people's goods, and if he wished to give he might give his own as he wished. As for themselves they had conquered and kept that country with great labor and at the peril of their lives, and there were far better reasons why they should enjoy it than a stranger. Poor Lampugnan, seeing that his patents did not avail him the value of a straw, and at the same time not daring to return to Spain, partly through fear of being ridiculed and

[1] Herrera, "Descripcion de las Indias Occidentales," Dec. iv, Book VI, ch. 12.

[2] Humboldt's "Personal Narrative," Vol. II, p. 273.

partly on account of the money he owed, was ruined. In fact, the business and its anxieties drove him crazy and he was exposed to the mockery of all the world as a lunatic. In the end, after dragging out five years in this miserable condition, he died in this isle of Cubagua."[1]

The average size of these pearls derived from the Venezuelan fisheries was small, specimens rarely exceeding twenty grains. In 1577, Urbain Chauveton wrote: "The pearls of Cubagua are mostly 2, 3, 4, and 5 carats. But the quantity of them is so great that the fifth part which is paid to the king of Spain yields every year the value of more than 15,000 ducats; this besides the frauds committed and the pearls which stick to the fingers of those who manage the business, and who pilfer the most beautiful in great numbers, sending them here and there for sale. They place themselves in great danger if the facts become known, but they do it all the same."[2]

The enormous demands made by the Spaniards soon had its effect on the resources, for Chauveton adds: "It is apparent they decrease and not so many are found as in the beginning. The reason for this is that the Spaniards are so eager to gather large quantities of them quickly that they are not content to use their divers to search for them in the depths of the sea, but they have conceived and invented I know not how many machines of rakes and drags to scrape up everything. In fact they have at times collected them all so that another could not be found, and have had to abandon their fishing for a considerable time to give the oysters a chance to lay their eggs and grow their pearls."[3]

The decrease noted by Chauveton was probably not very serious, for the Spanish historian, José de Acosta, reports that in 1581 he saw "the note of what came from the Indies for the king; there were 18 marcs of pearles, besides 3 caskets; and for private persons there were 1265 marcs, and besides them, 7 caskets not pierced, which heretofore we would have esteemed and helde for a lie."[4] Also the records show that in 1597 Spain received from the Venezuelan fisheries "350 pounds' weight of pearls." It is to be regretted that the Spaniards so frequently reported the yield of pearls by pounds' weight, for—owing to the great variation in quality—this is about as unsatisfactory as to report the wealth of an individual by the pounds' weight of his title-deeds or of his stock certificates. The value of "350 pounds of pearls" might have been anywhere from twenty thousand dollars to as many millions. Assuming that all were two grains each in weight and of

---

[1] Translated from "Historia del Mondo Nuovo," Geneva, 1578.
[2] Translation of Chauveton's Notes to Benzoni's "Historia del Mondo Nuovo," Geneva, 1578, fol. 170.
[3] Ibid., fol. 168.
[4] "Natural and Moral History of the Indies," Hakluyt Society, London, 1880, p. 228.

good quality, the total value would approximate $600,000 according to the valuation of that period; and on a basis of eight grains each, it would be $9,600,000, or sixteen times as much. But as original parcels of pearls from the fisheries, these figures should be divided by three.

Following 1597, the productiveness of the Cubagua beds rapidly decreased. By acts of cruelty and oppression the Spaniards had converted the surviving Indians into deadly foes, ready to take advantage of any opportunity to avenge themselves on their oppressors, and thus terrifying the settlers into abandoning the enterprise. Early in the seventeenth century the development of mining resources in Mexico, Peru, etc., attracted the adventurous Spaniards. A considerable decrease in the value of pearls, brought about by the skilful manufacture of imitations at Venice, and elsewhere in southern Europe, also affected the prosperity of the fisheries. As a result of these combined influences, the output in Venezuela was greatly reduced, and it ceased long before the close of the following century. Thus ended an enterprise which, for a number of years, represented the greatest single industry of the European people on the American continent.

According to General Manuel Laudecta Rosales, the Venezuela archives contain no reference to any renewal of the fishery until early in the nineteenth century. At the time of Humboldt's visit in 1799, the fishery was entirely neglected around the islands of Margarita, Cubagua, and Coche, and the only evidence of pearls was a few very insignificant ones picked up about Cumana and sold among the natives at a piaster per dozen.[1]

After the overthrow of Spanish authority on this coast, Messrs. Rundell, Bridge and Rundell, a firm of well-known goldsmiths of London, obtained, in 1823, from the government of Colombia, a ten-year monopoly of the fishery at several places on the coast of the new republic, in consideration of one fifth of the pearls secured.[2] After the independence of Venezuela in 1829, the taxes imposed were so heavy that the industry languished, and about 1833 it was practically abandoned.

Owing to the improved physical condition of the reefs, the fishery developed largely in 1845; and for several years an average of 1600 ounces of pearls were secured, an ounce of good quality selling for 150 to 500 bolivars (one bolivar = 19½ cents), and the inferior quality at 80 to 100 bolivars.[3] At that time there was a tax of sixteen bolivars per boat monthly. In 1853 this was increased to forty-eight bolivars per boat, and the use of dredges (*arrastras*) was interdicted,

[1] Humboldt, "Personal Narrative of Travels to the Equinoctial Regions of the New Continent, 1799–1804."

[2] Findlay, "Directory for the Navigation of the Pacific Ocean," London, 1851, Vol. I, p. 217.

[3] Rosales, "Gran Recopilación de Venezuela," Caracas, 1889.

soon reducing the fishery to a very low stage. Subjected to frequent changes in regulations, and burdened by heavy taxes, the industry remained in poor condition until about 1895. Since then the enhanced value of pearls, and the increased industrial activity on the coast, has resulted in a very large development of the fishery.

In recent years the government of Venezuela has granted concessions to individuals and to companies for the exploitation of defined areas for a limited period, exacting 10 per cent. royalty on the proceeds of the enterprise. In granting these concessions, the government usually reserves the right to examine the books, and to intervene when necessary in any phase of the enterprise. For protecting its revenue, the government requires that shipments of the pearls must be signed by its agent, and bills of sale must be countersigned by the Venezuelan consul in the place where the sales are consummated.

The Venezuelan pearl fishery now gives employment to about 350 boats, manned by five or six men each, sailing from the ports of Juan Griego, Cumana, and Carupano. These are sail craft, measuring from two to fifteen tons each, and are licensed by the Venezuelan authorities at a charge of 15 bolivars ($2.92) each. Most of the boats use dredges, but some of them resort to nude diving, after the manner of the sixteenth century. Attempts have been made to use the scaphander, or diving armor, but without success, owing largely to the difficulty in obtaining experienced workmen, and also to local prejudice against this form of fishery. It is claimed that in using the scaphander, all oysters are removed from the reefs, whereas the *arrastra* or dredge spreads the oysters and thereby enlarges the reefs. This is the principal and, except those at Sharks Bay and the Sulu Islands, the only important pearl fishery in which the oysters are secured by means of dredges. These are made of iron and are similar to those implements used in the scallop fisheries of New York and Rhode Island. They are dragged over the beds, and when filled are lifted and their contents emptied into the boat, the fishermen culling out the desirable oysters from the mass and throwing the refuse material overboard.

The pearl-oyster (*Margaritifera radiata*) secured on the coast of Venezuela is closely related to the Ceylon species. It averages slightly larger in size, and there is a much greater range in coloration. The pearls are of good quality. In color they range from white to bronze, and occasionally a so-called black one is found. The total output is valued locally at about 1,750,000 francs ($350,000) per year. Most of them are sold in Paris.

Owing to their small size and lack of thickness, the shells of the Venezuela pearl-oyster are of little or no value in the mother-of-pearl trade. Thousands of tons of them, the accumulations of scores of fisheries, lie in heaps and ridges along the coast, as though in years long past vast armies of oysters, engaged in deadly combat, had left their innumerable myriads of slain comrades to bleach on the shores.

THE PANAMA PEARL FISHERIES

The bordring Ilands, seated here in ken,
Whose Shores are sprinkled with rich Orient Pearle,
More bright of hew than were the Margarets
That Caesar found in wealthy Albion.
                    ROBERT GREENE, *Orlando Furioso* (1594).

FROM the point of view of the Spaniards of his day, the greatest result of Balboa's immortal journey in 1513 across the Isthmus of Panama to the broad waters of the Pacific, was the discovery of the pearl resources of the Gulf of St. Michael, now known as the Gulf of Panama. Probably the best description of this is given by Lopez de Gomara in his "Historia general de las Indias," published in 1554, from which we translate the following account.

After Balboa had reached the Pacific in 1513, he proceeded a snort distance along the coast until he met with an Indian chief by the name of Tomaco. Being questioned about the gold and pearls which some of his people wore, Tomaco sent for some gold and 240 large pearls and a great number of small ones—a rich present, which filled the Spaniards with pleasure. Seeing the Spaniards so delighted, Tomaco ordered some of his men to go and fish for pearls. These went and in a few days obtained 64 ounces, which also he gave them. The Spaniards were surprised to see such pearls, and that their owners did not value them; they not only gave them away, but their paddles were decorated therewith, for the principal income and wealth of these chiefs was the pearl fishery. Tomaco told Balboa that these riches were nothing in comparison with those of Tararequi, which had pearls larger than a man's eye, taken from oysters the size of sombreros. The Spaniards wished to go there at once, but fearing another tempest, left it for their return. They dismissed Tomaco and rested in the country of Chiape, who, at the request of Balboa, sent thirty of his men to fish. These did it in the presence of seven Spaniards, who looked on and saw them take six loads of small shells. As it was not the season for that fishery, they did not go into very deep water where the shells were. Not only did they not fish in September and the following months, but they did not even travel by water, on account of the stormy weather which then prevails in that sea. The pearls which they extracted from those shells were like peas, but very fine and white. Of those received from Tomaco, some were black, others green, blue, and yellow.

On the return of Balboa's expedition to Darien in 1514, the sight of the pearls and the wonderful reports made by the men, caused his successor, Pedrarias, to fit out another expedition, an account of which we likewise translate from Gomara.

By command of Pedrarias, Gaspar de Morales went in the year 1515 to the Gulf of St. Michael, with 550 Spaniards, in quest of the island of Tara-requi, which was said by Balboa's men to be so abundant in pearls and so near the coast. The chief of that island sallied forth with many people to prevent his entrance, and clamored and fought three times with our people on equal terms, but the fourth time he was defeated. He then made friends, carried the chief of the Spaniards to his house, which was a large and good one, gave him food to eat, and a basket of pearls which weighed 110 marcs [880 ounces]. The chief received for them some looking-glasses, stringed beads, bells, scissors, axes, and small wares of barter, which he valued more than he had the pearls. He promised to give as tribute to the emperor, in whose guardianship he placed himself, 100 marcs of pearls every year. With these the Spaniards returned to the Gulf of St. Michael and from thence to Darien.

Tararequi is within five degrees of the equator. It possessed a great fish-ery for pearls, which are the largest and best of the new world. Many of the pearls which the cacique gave were like filberts, others like nutmegs, and there was one of 26 and another of 31 carats, pear-shaped, very lustrous, and most perfect, which Peter of the Port, a shop-keeper, bought of Gaspar de Morales for 12,000 castilians. The purchaser could not sleep that night for thinking on the fact that he had given so much money for one stone, and so he sold it the very next day to Pedrarias de Avila, for his wife Donna Isabel de Bovadilla, at the same price, and afterwards the Bovadilla sold it to Donna Isabella the Empress.

Pedrarias, who delighted in such fishery, requested the cacique to make his men fish for pearls in the presence of the Spaniards. The fishermen were great swimmers and divers, and seemed to have spent all their lives in that employment. They went in small boats when the sea was calm, and not in any other manner. They cast a stone for an anchor from each canoe, tied by strong, flexible withes like boughs of the hazel. They plunged to search for oysters each with a sack or bag at the neck, and returned loaded with them. They entered four, six, and even ten fathoms of water, for the shell is larger the deeper they go, and if at times the larger ones come in shallow water it is through storms, or because they go from one place to another in search for food, and having found their pasture they stay there until they have finished it. They perceive those who search for them, and stick so close to the rocks or ground, or one to another, that much strength is needed to detach them, and many times the fishermen cannot raise them and leave them, thinking they are stones. In this fishery many persons are drowned, either by remaining too long at the bottom, or because they become entwined or entangled in the cord, or such carnivorous fish as the shark devour them. This is the manner of fishing pearls in all the Indies, and many fishermen die from the dangers aforesaid, and from the excessive and constant labor, the little food, and the maltreatment they have. The emperor was led to enact a law among those whom Blasco Nunez Vela brought, which imposed the penalty of death upon him who should forcibly compel any free Indian to fish for pearls. He thought more of the lives of the men than of his interest in pearls, though they were of great value. The law was worthy of such a prince and of perpetual memory.[1]

Gonzalo de Oviedo referred to the pearl resources of Panama in his "Historia natural de las Indias," Toledo, 1526, mentioned in the chapter on pearl fisheries of Venezuela. After describing the resources of Cubagua and Cumana on the Venezuelan coast, he states, according to Eden's quaint translation:

Lykewise pearles are founde and gathered in the South sea cauled *mare del sur.* And the pearles of this sea [the Caribbean coast] are verye bygge. Yet not so bigge as they of the Ilande of pearles cauled *de las perlas,* or *Margaritea,* whiche the Indians caule *Terarequi,* lying in the gulfe of saincte Michael, where greater pearles are founde and of greater price then in any other coaste of the Northe sea, in *Cumana,* or any other porte. I speake this as a trewe testimonie of syght, havyng byn longe in that South sea, and makynge curious inquisition to bee certenly informed of all that perteyneth to the fysshynge of perles. From this Ilande of Tararequi, there was brought a pearle of the fasshyon of a peare, wayinge xxxi carattes, which Petrus Arias had amonge a thousande and soo many poundes weight of other pearles which hee had when capitayne Gaspar Morales (before Petrus Arias) passed to the saide Ilande in the yeare 1515, which pearle was of great prise. From the saide Ilande also, came a great and verye rounde pearle, whiche I brought owte of the sea. This was as bygge as a smaule pellet of a stone bowe, and of the weight of xxvi carattes. I boughte it in the citie of Panama in the sea of Sur: and paide for it syxe hundredth and fyftie tymes the weyght therof of good gold,[2] and had it thre yeares in my custodie: and after my returne into Spaine, soulde it to the erle of Nansao, Marquisse of Zenete, great chamberleyne to youre maiestie, who gave it to the Marquesse his wyfe, the ladye Mentia of Mendozza. I thyncke verely that this pearle was the greatest, fayrest, and roundest that hath byn seene in those partes. For youre maiestie owght to understande that in the coaste of the sea of Sur, there are founde a hundredth great pearles rounde after the fasshyon of peare, to one that is perfectly rounde and greate. This Iland of Terarequi which the Christians caule the Ilande of pearles, and other caule it the Ilande of floures, is founde in the

---

[1] Gomara, "Historia general de las Indias," 1554, pp. 268, 269 b.

[2] 111½ ounces of gold; present value about $2300.

eyght degree on the southe syde of the firme lande in the provynce of golden
Castyle or Beragua. (Arber, "The First Three English Books on America,"
Birmingham, 1885.)

In addition to the gems noted by Oviedo, these waters furnished
many other beautiful pearls in the sixteenth century, and added largely
to the collections of the Spanish court and of the cathedrals of Seville,
Toledo, etc. The Italian traveler, Gemelli-Careri, who visited the
Panama fisheries in 1697, reported that they yielded pearls equal to
those of Ceylon. He mentioned one weighing 60 grains, for which
the owner—a Jesuit priest—refused 70,000 pesos.[1]

In 1735, the Spanish admiral, Antonio de Ulloa visited the Panama
pearl fisheries and wrote an extended description of them.[2] According to his account the pearls were then found in such plenty that there
were few slaveholders in the vicinity who did not employ at least a
portion of their Negroes in the fishery. These were selected for their
dexterity in diving, and were sent to the islands in gangs of from eight
to twenty men each, under the command of an overseer. They lived
in temporary huts on the shore, and visited the pearl reefs in small
boats. Anchoring in eight or ten fathoms of water, the Negroes would
dive in succession to the bottom, returning with as many oysters as
possible. It was laborious work, attended with danger owing to the
numerous sharks.

Every one of these Negro divers is obliged daily to deliver to his master
a fixed number of pearls; so that when they have got the requisite number of
oysters in their bag, they begin to open them, and deliver the pearls to the
officer, till they have made up the number due to their master; and if the pearl
be but formed, it is sufficient, without any regard to its being small or faulty.
The remainder, however large or beautiful, are the Negro's own property,
nor has the master the least claim to them, the slaves being allowed to sell
them to whom they please, though the master generally purchases them at a
very small price. . . . Some of these pearls, though indeed but few, are sent
to Europe, the greater part being carried to Lima, where the demand for them
is very great, being not only universally worn there by all persons of rank,
but also sent from thence to the inland portions of Peru.[3]

During the hundred years following, the pearl reefs of Panama were
not very productive, and relatively little attention was paid to them.
The development of a market for the shells in the mother-of-pearl
trade, about 1840, enhanced the profits of the few natives engaged in

[1] Gemelli-Careri, "Giro del Mondo," Venezia, 1719, p. 240.
[2] Ulloa, "Relación histórica del viage á la
América meridional," Madrid, 1748.

[3] "Ulloa's Voyage to South America,"
translated by J. Adams, London, 1758.

pearling in a desultory manner, and led to an increase in the number of fishermen. During some years when industrial and market conditions were favorable, large quantities of shells were exported. In 1855, for instance, 650 tons of these shells were shipped to England alone, and in 1859 the reported quantity was 957 tons. Those from the Island of San José, one of the Pearl Archipelago, were said to be the largest and choicest in the bay. Many of them were used in decorating the twin towers of the stately old cathedral at Panama.

Since then the industry has fluctuated greatly, depending on the market for the shell. Many outsiders have experimented in the fishery, but most of these attempts have resulted in financial loss, through mismanagement, storms, sickness, or other causes. A story is told locally of a party of thirty men, principally from Scotland, who arrived at Panama equipped with a diving-bell and such necessary machinery as air-pumps, windlasses, etc. Much was expected of their operations, but soon yellow fever broke out among them, and within six weeks two thirds of the members of the party had died. The remaining members, becoming disheartened, and in fear of the dread disease, lost no time in leaving the country. The diving-bell and machinery remained for several years as a curiosity at Panama, for no one returned to claim them, nor has the use of similar apparatus been attempted since then.

The scattered pearl reefs extend from the east side of the Bay of Panama nearly to the Costa Rica boundary. However, this gives an exaggerated idea of their area, as much of this territory yields no pearl-oysters whatever. The principal reefs and the headquarters of the fishery are at Archipelago de las Perlas or Pearl Islands, which are from thirty to sixty miles southeast of the Pacific terminus of the projected Panama Canal. This archipelago contains sixteen small islands, on which are about twice that number of small settlements of Negro and Indian descendants, with a total population of perhaps one thousand. About half of these live on Isla del Rey, the largest island, about fifteen miles long and half that in width. The chief village, San Miguel, is the center of the pearling industry, and consists mostly of palm-thatched huts and a handsome stone church, more costly than all the remaining buildings of the town combined. While the soil is fertile and some vegetables are raised, the inhabitants depend almost wholly on the fisheries.

In 1901, the Republic of Colombia invited bids for the right to operate the pearl and coral fisheries for a term of fifteen years, but nothing seems to have come of it, and the establishment of the Panama Republic in 1903 terminated the authority of Colombia in these resources.

The Panama fisheries differ widely in their character from those of Venezuela. The mollusk is much larger, averaging about six inches in diameter when fully grown, thus furnishing a valuable quality of mother-of-pearl. The shell constitutes the principal object of the fishery; the pearls themselves are of incidental importance, but are always looked for and anxiously expected.

The season extends from May to November, with a rest during the remaining five months of the year. The fishery is open to natives and to foreigners alike. While the leading fishermen employ diving-suits, which were introduced here about 1890, nude diving is yet practised to a considerable extent, the men descending in eight or ten, and some even in twelve fathoms of water. There is no restriction whatever on the nude fishermen, but for each machine diver an annual license fee of $125 United States currency is exacted.

Owing to the low market price for Panama shell during recent years, the fishery has not been vigorously prosecuted, and it has even dwindled to low proportions. A letter from one of the leading pearling companies in Panama states that the machine divers number about twenty, while there are about four hundred nude fishermen; and another firm likewise prominent, estimates these fishermen at twenty and three hundred respectively.

Yet a third pearling company writes that there are fifteen machine divers and two hundred head divers; and adds that the small demand for this quality of mother-of-pearl has made the condition of the industry about as bad as it could be; many who have capital invested are getting out of the business, and unless the market improves, the industry may be abandoned. Probably with the introduction of new capital and methods in the infant republic, the pearl resources may receive greater attention and a large development ensue.

The Panama pearls are of good quality and frequently of large size. In color they range from white to green and lead-gray, and frequently greenish black. Valuable pearls are not common, but occasionally the fisherman is amply rewarded. A letter from the American consul at Panama states that in 1899 a native boy, fifteen years old, fishing in shallow water, as much for sport as for profit, found a pearl which he sold to a local speculator for 4000 silver dollars ($1760); this speculator delivered the same pearl to a dealer in Panama for 10,000 silver dollars ($4400), and an offer of 30,000 francs was refused for it later in Paris. A pearl worth $2400 was reported as found within half a mile of the steamship anchorage at Panama. A pearl from a giant oyster resembling Tridacna, was an absolute egg-shape, pure cocoanut white, and weighed 169 grains; it was 21 mm. at the longest and 16.5 mm. at the narrowest part. The surface showed very distinctly a wavy structure, occasionally with a tiny, brighter central point; the surface under the glass resembling a honeycomb network. At the smallest point there was a radiated center with quite a brilliant field. It was worth only $100.

Not always, however, does the poor, ignorant fisherman receive the full value of his find; and many a story is told of some thoughtless improvident native, who, for less than a mess of pottage, "like the base Indian, threw a pearl away, richer than half his tribe."

Most of the Panama pearls are sold in Paris, relatively few of them coming to America direct. This is not because of any greater estimation of them in Paris or higher prices obtained; but the trade relation has been long continued and the credits are well established. From Paris many of these pearls reach the American market.

THE PEARL FISHERIES OF MEXICO

> Then, too, the pearl from out its shell,
>   Unsightly in the sunless sea,
> (As 't were a spirit, forced to dwell
>   In form unlovely) was set free,
> And round the neck of woman threw
> A light it lent and borrowed too.
>   THOMAS MOORE, *The Loves of the Angels.*

PEARL-BEARING oysters are found at various places on the Pacific coast of Mexico, and especially along the coast of Lower California, where extensive fisheries are prosecuted. The pearls are noted for the great variety of colors which they display. A large percentage are black, others are white, brown, peacock green, etc. Generally they are small and of irregular form, yet sometimes very large ones are secured, weighing 100, 200, and even 300 grains.

European knowledge of the pearl resources of Mexico dates from the conquest of that country by Hernando Cortés about 1522. The diary of his lieutenant, Fortuno Ximines, tells of finding native chiefs living in primitive huts along the sea-shore, with quantities of beautiful pearls lying carelessly around. From a tribe near the present site of Hermosillo, in the State of Sonora, Cortés secured great quantities of the gems. It appeared that the fishery had been in existence for centuries. The location of the pearl reefs was prominently noted on Cortés' map of this coast, made in 1535, a copy of which was procured by the Rev. Edward E. Hale when in Spain in 1883.

Following Cortés' explorations of the Pacific coast of Mexico (1533–1538), a number of expeditions were fitted out for securing pearls by trading with the natives, by forcing them to fish, and by even more questionable means. Several of these expeditions found record in history either by reason of their unusual success or through the extreme cruelty with which they were conducted. The contact of the Spaniards with the Indians resulted in very bitter feelings on the part of the latter, so that it became risky for small traders to venture among them. From time to time, successful expeditions were made, especially the one of 200 men sent in 1596 by the viceroy of Mexico to "the rich Isles of California," mentioned by Teixeira.[1] Antonio de Castillo, a Spanish colonist, with headquarters south of Mazatlan, was one of the most successful of the early adventurers, and Iturbide Ortega and José Carborel were also among the fortunate ones of that period.[2] Ortega marketed his pearls in the city of Mexico, and the reported sale of one for 4500 dollars had considerable effect in stimulating the industry.

The advent of the Jesuits to western Mexico in 1642, developed amicable relations with the Indians; and although the missionaries were agriculturists rather than fishermen, the restoration of harmony resulted in a more favorable prosecution of the fisheries. The colonists of Sinaloa and Nueva Galicia, who had formerly, in small vessels and with great danger, made occasional visits to the pearl beds, built larger vessels and made more frequent visits without apprehension. The skilful Yaqui and Mayo Indians were employed or impressed as divers, just as natives of the Bahamas had served in the fisheries of Venezuela. Great profits resulted from the operations. Venegas wrote that "it was certain that the fifth of every vessel was yearly farmed for 12,000 dollars."[3]

So profitable was the fishery that the Spanish soldiers and sailors stationed in the Gulf of Cortes—as the Gulf of California was then called—were frequently charged with devoting more attention to pearling than to their official duties. In order to put a stop to this evil, in 1704, Father Silva-Tierra, who was in authority in that part of the country, ordered that no soldier or sailor should engage in the fishery. With a view to removing the demoralizing influences of promiscuous adventurers among the Indians, the industry was later restricted to persons specially authorized.

Probably the most successful of the early pearlers was Manuel

[1] Hakluyt's "Voyages," Glasgow, 1904, Vol. IX, pp. 318, 319.
[2] Clavigero, "Storia della California," Venezia, 1789, Vol. I, p. 161.
[3] Venegas, "Noticia de las Californias," Madrid, 1757, p. 454.

Osio, who is credited with having marketed "127 pounds' weight of pearls in 1743," and "275 pounds' weight" in 1744.[1] He operated in the vicinity of Mulege and northward, employing the Yaqui Indians;

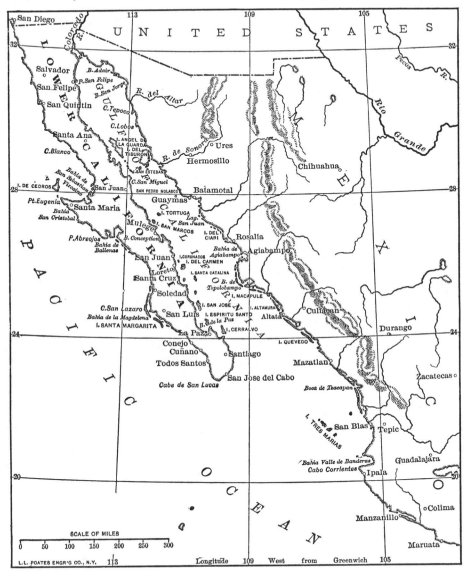

Gulf of California and the pearling territory of western Mexico

and through his pearling interests is said to have become the richest man in Lower California.

The revenue from the royal fifth, somewhat later, was reported by

---

[1] Clavigero, "Historia de la Baja antigua California." Esteva, "Boletin de la Sociedad Mexicana de Geographia y Estadística," Vol. X, pp. 673-697.

Alvarado[1] at 12,000 dollars per year; but this was disputed by Jacob Baegert, a Jesuit priest. Baegert spent seventeen years in Mexico and, returning to Europe on the expulsion of his order from that country in 1767, published a report in 1772, containing rather an unfavorable view of the fishery. He stated that each summer eight, ten, or twelve poor Spaniards from Sonora, Sinaloa, and elsewhere on the mainland, crossed the gulf in small boats to the California shore for the purpose of obtaining pearls. They carried supplies of Indian corn and dried beef, and also a number of Indians who served as divers, the Spaniards themselves showing little inclination to engage in the work when native fishermen could be employed so cheaply. Provided with a sack for receiving the oysters which they removed from the bottom, the fishermen dived head first into the sea, and when they could no longer hold their breath they ascended with the gathered treasure. The oysters were counted before opening; and, when the law was complied with, every fifth one was put aside for the king's revenue. Most of the oysters yielded no pearls; some contained black pearls, others white ones, the latter usually small and ill-shaped. If, after six or eight weeks of hard labor and deducting all expenses, a Spaniard gained a hundred American pesos, he thought he had made a little fortune, but this he could not do every season. "God knows," said Baegert, "whether a fifth of the pearls secured in the California sea yields to the Catholic king an average of 150 or 200 pesos in a year, even without frauds in the transaction. I heard of only two persons—with whom also I was personally acquainted—who had accumulated some wealth, after spending 20 or more years in the business. The others remained poor notwithstanding their pearl fishing."[2]

Father Baegert's statement of the returns seems to be substantiated by the reports of the royal fifth a few years later. For the period from 1792 to 1796 this was placed at "2 lbs. 2 ozs." by some writers; and according to others, from 1788 to 1797 it amounted to only "3 lbs. 9 ozs.," which is the quantity assigned by some accounts to 1797 alone.[3] These returns apparently indicate that a great decrease had occurred since the days of Osio; but it seems very doubtful whether, under the conditions existing in Mexico at that time, the royal treasury received its due share of the proceeds.

Shortly following the independence of Mexico in 1821, and after a period of little activity, several attempts were made to exploit the pearl resources. The great prosperity in England, ensuing upon the

[1] Pedro Alvarado, "Historia California," Vol. I, p. 10.
[2] Baegert, "Nachrichten von der Amerikanischen Halbinsel Californien," Mannheim, 1772.
[3] Arch. Cal. Prov. St. Pap. xvi. Ben. Mil. xvi, xvii, xviii.

termination of the Napoleonic Wars, resulted in much speculation and the promotion of stock subscriptions in many visionary schemes. Among these was "The General Pearl and Coral-Fishing Association of London," which in 1825 equipped and sent out to Mexico, by way of Cape Horn, two vessels prepared to exploit the pearl resources by the use of diving-bells similar to those formerly employed in submarine construction. This expedition was under the direction of Lieutenant R. W. H. Hardy, whose report thereon presents an interesting exhibit of the condition of the pearl fishery at that time.

Hardy found the fishery at a very low ebb, owing, largely, to the scarcity of oysters and the uncertainty of depending on the native divers. He adds with peculiar naïveté: "I had almost forgotten to mention a very curious circumstance with respect to the pearl-oyster, namely that on the coast of Sonora there are none at all, except at Guaymas." He states also that to the northward of 28° 30' not the trace of a shell could be discovered on either side of the gulf.

The center of the industry was then at Loreto, a village of 250 inhabitants; but another small station existed at La Paz. At Loreto six or eight vessels of twenty-five tons each were employed, each having three or four sailors and fifteen or twenty Yaqui Indians who served as divers. Head-diving was in vogue, the work proceeding from 11 A.M. to 2 P.M., and the depth ranging from three to twelve fathoms. The annual catch of pearls was "4 or 5 pounds' weight, worth from $8000 to $10,000."[1] After the government's claim of one fifth had been set apart, the owner and captain of the vessel received one half and the divers the other half.

It was found impossible to use diving-bells when the sea was at all rough, and even during calm weather they were impracticable on account of the unevenness of the ground and the strong undercurrents. An effort was made to employ native divers, but owing to the disorganized state of affairs only four could be secured. In the Gulf of Mulege a large number of oysters were collected, but when these were opened "six very small pearls" were all that could be found. After spending about three years on the coast, Hardy returned to England, and the company abandoned the enterprise.

In the early history of the Mexican pearl fishery, the shells were of no market value; but about 1830 a French trader named Combier made experimental shipments to France, securing cheap freight rates by using the waste shells largely as ballast for the vessels.[2] The best quality sold for about 600 francs per ton, and the market was found

[1] Hardy, "Travels in Mexico," London, 1829, pp. 231-238.

[2] Diguet, "Bulletin de la Société Centrale d'Aquiculture," Paris, 1895, Vol. VII, pp. 1-18.

sufficient for regular shipments. The value gradually increased, and in 1854 it approximated 2000 francs per ton in France, placing the industry upon a very remunerative basis. This resulted in much activity in the fishery, and an increase in the number of boats and divers.

In 1855, the fishery gave employment to 368 divers, and yielded $23,800 worth of pearls, and 350 tons of shells worth $13,500.[1] It was estimated by Lassepas that from 1580 to 1857, inclusive, 95,000 tons of oysters were removed from the Gulf of California, yielding 2770 pounds of pearls, worth $5,540,000.[2]

For protection of the reefs, the Mexican government in 1857 divided the Gulf of California into four pearling districts, and provided that only one of them should be worked each year, and then only in areas leased for the season to the highest bidders, thereby permitting the reefs successively to remain undisturbed for three years.

The yield of pearls in 1868 approximated $55,000, and that of shells $10,600 in value; while in 1869 these items were given as $62,000 and $25,000, respectively.[3] The local prices ranged from $15 per ounce for seed-pearls to $1500 for a choice gem.

At that period the fishery was carried on from shore camps or from large vessels, each carrying twenty to fifty divers, who were mostly Yaqui Indians from the eastern shore of the gulf. The camp or vessel was located in the vicinity of the reefs or beds, and the fishing was prosecuted from small boats, each carrying three or four nude divers. Fastened to the waist or suspended from the neck was a net for the reception of oysters, and each diver carried a short spud or stick with which to detach them from the bottom, and to some extent for use as a weapon of defense against sharks and similar enemies. The diving progressed mostly in the morning, when the sea was unruffled by the breeze which usually begins shortly after noon. The season lasted from May to late in September, when the water became too cold for further operations.

The divers were paid a definite share of the catch, and kept in debt-bondage by means of advances and supplies. Little clothing was necessary, and the provisions consisted principally of corn, beans, and sun-dried beef. Luxuries were added in the form of tobacco, and of mescal distilled from the maguey plant, indulgence in these constituting the chief remuneration for the season's labor. The finding of an unusually choice pearl brought to the lucky fisherman a gratuity of a few dollars, and shore leave for several days in which to spend it. Dress-

[1] Esteva, "Memoria sobre la Pesca de la Perla," "Boletin de la Sociedad Mexicana de Geographia," Vol. X, pp. 681-688.

[2] Lassepas, "Historia de la Baja California," Mexico, 1859, p. 65.

[3] Pujol, "Estudio Biológico sobre la ostra Avicula margaritiferus," "Boletin de la Sociedad de Geographia," Epoc. 2, Vol. III, p. 139 et seq.

ing in his best calico garments, he hastened to the nearest town to indulge in release from restraint, in drunkenness and debauchery—the highest dreams of happiness of a Yaqui Indian—thoughts of which served to bring him to the fishery each year from his home across the gulf.

From the Spanish conquest until 1874, the Mexican pearl fishery was conducted exclusively by nude divers. The experiments with the diving-bell in 1825 had been without favorable result, and also an attempt by an American in 1854 to use a diving-suit with air-pump, etc., this failure being credited to imperfection of apparatus. In 1874, through the influence of European pearl merchants, two schooners, each of about 200 tons' measurement, one from Australia and the other from England, visited the Mexican grounds, with a dozen boats fully equipped with scaphanders or diving armor, including helmets, rubber suits, pumps, etc. Owing to their working in deeper water than the nude divers were able to exploit, their success was remarkable, and they secured upward of a hundred thousand dollars' worth of pearls and shells during the first season.

The hitherto somnolent inhabitants of Lower California were amazed at seeing their resources thus easily removed, and were awakened to the opportunities afforded them to acquire the wealth which nature had scattered at their very doors. With this object-lesson before them, companies were formed for raising sufficient capital for the business, and the leading operators equipped their men with scaphanders, to the great annoyance of the would-be independent fishermen, who had not sufficient means to purchase the costly equipment. Many of these continued to employ nude divers, but after 1880 this method of fishery was subordinate to the use of diving apparatus. The change was accompanied by many accidents, and rarely did a month pass without the loss of a man, due in most cases to faulty apparatus or to inexperienced management.

In 1884 President Gonzalez inaugurated the policy of granting exclusive concessions to the pearl reefs. On February 28 of that year, five concessions were granted to as many persons, giving them and their associates and assigns the exclusive right to all shell fisheries in their respective zones of large area, for a period of sixteen years, in consideration of a royalty and export duty, amounting altogether to about $10 per ton of shells exported in the first three years, and $15 per ton for the remaining thirteen years of the term. Immediately these five grants were consolidated, forming the Lower California Pearl Fishing Company ("Compañia Perlífera de la Baja California"), incorporated under the laws of California with an invested capital of $100,000.

Other concessions were given covering the ocean shore of Lower California, the eastern side of the gulf within the States of Sonora and Sinaloa, and the ocean shore of Mexico southward from Sinaloa. In addition to these, certain territorial rights of fishing are claimed through grants dating back very early in the history of the country. So eagerly have these concessions been sought in recent years, that there is now little pearling ground on the coast which is not under corporate or private claim. And, owing to speculation in these concessions and in the formation of companies to develop them, it is somewhat difficult to obtain wholly reliable data relative to the condition and extent of the industry.

Two species of pearl-bearing mollusks occur on the Mexican coast. The principal one is the *M. margaritifera mazatlanica*, known locally as the *concha de perla fina*. This species is closely related to the "black lip shell" of the Australian coast. It is considerably larger than the Venezuelan oyster, averaging four or five inches in diameter and attaining an extreme diameter of seven or possibly eight inches. It occurs to some extent all along the Pacific coast of Mexico, in detached beds intercalated in places. The principal reefs, which have been exploited for nearly four centuries, are in the shallow waters of the Gulf of California and especially within the 300 miles between Cape San Lucas and Mulege Bay. The fisheries have centered about the islands of Cerralvo, Espiritu Santo, Carmen, and San José, and in the bays of Mulege, Ventana, and San Lorenzo. The depth of water on the reefs ranges from two to twenty-five fathoms, with an average of probably six or eight fathoms. The species is generally isolated, and firmly attached by the byssus to the bottom rocks or the stone corals, from which it may remove in case of necessity, though it probably does not do so frequently.

The second species is known locally under the name *concha nacar,* and has been named *Margaritifera (Avicula) vinesi* (Rochebonne).[1] It occurs only in the northern part of the gulf near the mouth of the Colorado River. Formerly it was abundant in that region, occurring in large areas, but it has become much reduced and is now little sought after. It is claimed that this species is far more productive of pearls than the *M. margaritifera,* and that it yielded the large quantities obtained by Osio in the eighteenth century. Although iridescent, the shell is so thin and convex that it is without commercial value.

The headquarters of the Mexican pearl fishery are at La Paz, the capital of Lower California, 240 miles northwest of Mazatlan and 150 miles north of Cape San Lucas. This "Mantle of Peace"—the literal translation of La Paz—contains about 5000 inhabitants, nearly all of

[1] Diguet, "Bulletin de la Société Centrale d'Aquiculture," 1895, Vol. VII.

THE ADAMS GOLD VASE

Ornamented with American gems and fresh-water pearls, rock crystal, gold quartz and agatized wood
Top of vase and side view

Now in the Metropolitan Museum of Art

Negro pearling camp on bank of an Arkansas river

Group of Arkansas pearl fishermen; photographed shortly after the woman in the center
of the group had found a pearl for which she received $800

whom are more or less dependent on the pearl fishery. It presents an
attractive picture, with the cocoanut-palms extending down almost to
the water's edge, and the high mountains forming a background. The
low, stone houses, the tile roofs, the plaza with tropical trees, and the
beautiful flower beds under perennial sunny skies, give it a quaint appearance. The most conspicuous objects from the harbor are the large
old warehouses, with thick walls and iron-barred windows, for the
storage of the pearls and the shells. During the season, from April to
November, the arrival and departure of the pearling vessels presents a
scene of great animation.

The present methods of the fishery on the Mexican coast are quite
different from those of thirty years ago when nude diving was the only
method in vogue. Instead of the haphazard work, largely in shallow
water, the industry is conducted systematically, and the limit of depth
is increased, much of the diving being in depths of ten to fifteen
fathoms. The fishermen operate either from a large vessel making a
cruise two or three months in length, or from a camp on the shore near
the reefs. A vessel visits them frequently to furnish supplies and to
transport the catch to La Paz. The fishing boats are undecked craft,
each equipped with an air-pump and a crew of six men: a diver, a cabo
de vida or life-line man, who is usually the captain, two bomberos at
the air-pump, and two rowers.

The greatest depth at which armored diving is attempted in Mexico
rarely exceeds twenty fathoms; twenty-five fathoms is fully as deep as
it is practicable to go, and it is not advisable to remain at that depth
more than a very few minutes. At fifteen fathoms a diver may remain
half an hour or more, and at six or eight fathoms he may work uninterruptedly for several hours. When the water is very cold, the diver
comes up frequently to restore his numbed circulation by vigorous
rubbing. The occupation is especially conducive to rheumatism, and
paralysis is more or less general, due, not only to the compressed atmosphere, but to the abrupt changes of temperature. The work is very
debilitating, with particular effect on the nerves, and partial deafness
is common. It is important that the diver be careful about overeating
before descending, as heavy foods, and meats especially, make respiration difficult; therefore, breakfast consists of little more than bread
and coffee. The risks and dangers from sharks, devil-fish, etc., have
greatly diminished since the introduction of scaphanders; for a stout
diver in his waterproof dress, with leads on the breast, shoulders, and
shoes, and on his head a massive helmet containing great gaping windows for eyes, is enough to cause even a hungry shark to hesitate and
to seek a more digestible meal.

There are yet many nude divers in Mexico, who operate in shallow waters, their cheap labor making them successful competitors of the armored divers. In arranging with these, the pearling company commonly grub-stakes a crew, pays a stipulated sum per hundredweight for the shells, and bargains for the pearls. If the fishermen are not satisfied with the price offered for these, they are at liberty to sell to other buyers under certain restrictions.

Nude diving is confined to the warm months, beginning about the middle of May and continuing until October. Owing to the cloudy or muddy condition of the water in the gulf, the nude diver can not inspect the bottom from the surface and select the best oysters before descending, nor can he work satisfactorily at depths greater than seven or eight fathoms. While the work is hard, it is more remunerative than the average branch of labor in this region.

Each day the boats deliver their catch of oysters at the fishing-camps or on board the receiving vessels. After they have been freed from marine growths and refuse, the mollusks are opened and searched for pearls. This operation is performed by trusted employees, usually elderly men who have become physically disqualified for diving, and who, seated together at a low table, work under the watchful eyes of overseers. A knife is introduced between the valves of the oyster, the adductor muscle is severed, and the valves are separated by breaking the hinge. The animal is removed from the shell and carefully examined with the eyes and the fingers, and then squeezed in the hands to locate any pearl which may be concealed in the organs or tissues. The debris is passed to other persons, who submit it to further examinations. A man may work all day long and find only a few seed-pearls, but occasionally there is the excitement of discovering a beautiful gem.

In some localities the flesh of the pearl-oyster is a source of profit through its sale to Chinamen, who dry and otherwise prepare it for sale among their countrymen in Mexico and America, as well as in the Orient. Frequently the large adductor muscle is dried for food, making excellent soup-stock, and, indeed, it is quite palatable when stewed.

It is difficult to approximate the output of the Mexican pearl fisheries, other than the pearl shell, because the dealers place a merely nominal value on the pearls in their invoices when sending them to Europe, an invoice of $500 sometimes representing gems valued in Paris at several thousand dollars. Furthermore, it is difficult to obtain satisfactory information from the pearling companies, owing, presumably, to the fear of developing greater competition. According to the estimates at La Paz, the local value of the pearl-yield now approximates $250,000 annually, and the value of the same over the counters in Europe and America probably exceeds one million dollars.

Some remarkably large pearls have been secured in the Mexican fisheries, especially considering the small size of the oysters. In 1871 a pearl of 96 grains, pear-shaped and without a flaw, sold at La Paz for 3000 pesos. In March, 1907, a beautiful pinkish white one, found near the lower end of the peninsula, sold for 28,000 pesos or $14,000. One of the best years for choice finds was 1881, when the scaphanders were first employed to their greatest efficiency. A black pearl was then secured which weighed 112 grains, and which brought 40,000 francs in Paris. In 1882 two, weighing 124 and 180 grains respectively, sold for 11,000 pesos. In the following year a light brown pearl, flecked with dark brown, and weighing 260 grains, sold for 7500 pesos. These are the prices which the La Paz merchants received for these pearls, and not the much greater amounts for which they were finally sold by the jewelers.

One of the finest pearls was found in 1884 near Mulege. This weighed 372 grains. The Indian fisherman is said to have sold it for $90; the purchaser declined an offer of 1000 pesos, and also a second offer of 5000, and soon sold it to a La Paz dealer for 10,000 pesos. Its value in Paris was estimated at 85,000 francs. Probably the most famous of all pearls obtained from these grounds was "the 400-grain pearl" found near Loreto, and "which is now among the royal jewels of Spain." It is said that this was offered by the lucky fisherman to the Mission of Loreto, and by the Director of Missions in Lower California was presented to the Queen of Spain.[1]

As in every other fishery, one hears in Mexico of fishermen who have grasped a prize only to lose it through inexperience or improvidence. The account given above of the sale of the 372-grain pearl found near Mulege furnishes an instance of this. It is related in La Paz that in 1883 an Indian sold for ten pesos a gem weighing 128 grains, for which the purchaser received 27,500 francs in Paris. On another occasion a Mexican sold two pearls, easily worth $4000, for $16 worth of groceries.

In the eighteenth century, the Notre Dame de Loreto possessed a remarkable collection of Mexican pearls, which had been presented from time to time by the fishermen. During the régime of the Jesuits, it was customary to devote the proceeds of the last day of the fishery to the decoration of the altar of that mission. After the expulsion of this religious order in 1767, the mission was pillaged and the collection dissipated. From the old aristocracy of Mexico, family heirlooms of many choice pearls were placed on the European market during the civil wars in Mexico to contribute to the support of the contending armies. One lady in Sonora is said to have disposed of her collection for

[1] Lassepas, "Historia de la Colonización de la Baja California," Mexico, 1859.

550,000 francs. A fine collection of these pearls, accumulated from 1760 to 1850, and showing them in a great variety of colors, shapes, and sizes, was in Chihuahua until recently.

AMERICAN FRESH WATERS

And my pearls are pure as thy own fair neck,
With whose radiant light they vie.
WHITTIER, *The Vaudois Teacher.*

THE most recently developed pearl fisheries are within the limits of the United States, in the rivers and fresh-water lakes, and especially those in the Mississippi Valley. As an important industrial enterprise, these fisheries are less than two decades old, yet they are very productive, yielding annually above half a million dollars' worth of pearls, many of which compare favorably in quality with those from oriental seas.

The prehistoric mounds in the Mississippi Valley present evidence of the estimation in which pearls were held by a race of men who passed away ages before America was first visited by Europeans. In some of these mounds, erected by a long-forgotten race, pearls have been found not only in hundreds and in thousands, but by gallons and even by bushels. Some of these equal three quarters of an inch in diameter, and in quantity exceed the richest individual collections of the present day. Damaged and partly decomposed by heat and through centuries of burial, they have lost their beauty, and are of value only to the archæologist and to indicate the quantity of pearly treasures possessed by these early people.

Owing to the great wealth of pearls which had been uncovered on the Spanish Main, at Panama, and in the Gulf of California, Eldorado explorers, in the sixteenth century, were particularly eager in searching for them within the present limits of the United States; in the reports of their wanderings, much space is given to these gems, and these reports aided largely in inducing and encouraging other expeditions. Some of these accounts read like the marvelous stories of Sindbad the Sailor, quantities of pearls—hundreds of pounds in some instances—being secured by the exchange of trinkets and by more questionable means. It would be easy to bring together numerous accounts of apparently reliable authorities to show that in the sixteenth century pearls were obtained here in far greater quantities than were ever known in any other part of the world; but this conclusion seems not wholly correct.

The unfortunate wanderings of Hernando de Soto from 1539 to

1542 gave rise to most of the reports of rich pearl finds within the limits of this country. Of this voyage there are three principal accounts. The first was by Luis Hernandez de Biedma, who had accompanied De Soto as factor for Charles V of Spain. His brief report was presented to the king in 1544, although it was not published until 1841, nearly three centuries later, when it appeared in a French translation.[1] The second, and in our opinion the most reliable account,[2] published at Evora in 1577, was by an unnamed Portuguese (in English editions, commonly spoken of as the Gentleman of Elvas), who was a member of the expedition. The third account,[3] by far the longest and most widely known, but which was not written until 1591, was by Garcilasso de la Vega, who represented that his information was from a Spanish cavalier who had accompanied De Soto.

The only reference made to pearls in Biedma's report seems to be his allusion to the large quantity secured at the village of Cofaciqui, on the east bank of the Savannah River. He states: "When we arrived there, the queen . . . presented the governor with a necklace of pearls of five or six rows, procured for us canoes to pass the river, and assigned the half of the village for our quarters. After having been in our company three or four days, she escaped into the forest; the governor caused search to be made after her, but without success; he then gave orders to break open a temple erected in this village, wherein the chiefs of the country were interred. We took out of it a vast quantity of pearls, which might amount to six or seven arrobas,[4] but they were spoiled by having been underground."[5]

The Portuguese narrative alludes to the pearls at Cofaciqui, stating that the queen "took from her own neck a great cordon of pearls, and cast it about the neck of the governor. . . . And the lady, perceiving that the Christians esteemed the pearls, advised the governor to search certain graves in the town, where he would find many; and that if he would send to the abandoned towns, he might load all his horses. He sought the graves of that town and there found fourteen rows of pearls, and little babies and birds made of them."[6] This account makes no further mention of pearls, except to state that at the battle of Mavilla this great collection was burned, and that when the Queen of Cofaciqui escaped from the Portuguese she carried with her a little chest full of unbored pearls, which some of the

[1] "Recueil des pièces sur la Floride," Paris, 1841.
[2] "Relaçam verdadeira dos Trabalhos q ho gouernador dõ Fernãdo de Souto e certos fidalgos portugueses passarom no descobrimẽto da provincia da Frolida. Agora nouamente feita per hũ fidalgo Deluas."
[3] "La Florida del Ynca," Lisbon, 1605.
[4] One arroba = twenty-five pounds' weight.
[5] "Discovery of Florida," Hakluyt Society, 1851, Vol. IX, p. 181.
[6] Ibid., p. 50.

Spaniards thought were of great value;[1] and further, that on one or two other occasions a few pearls were received from the Indians as presents.

The account of De Soto's wanderings, given by Garcilasso, the Peruvian historian, contains many references to pearls, which read more like romance than reality. With his knowledge of the jewels, temples, etc., in Mexico and Peru, and recognizing some similarities in the manners of the people of those countries and the ones with whom De Soto came in contact, Garcilasso was easily led to statements which, though possibly true in the one case, seem fictitious in the other.

He gives the story of the Queen of Cofaciqui, with some additional particulars. The string of pearls which she presented to the governor made three circuits of her neck and descended to her waist. In his account, the graves in Cofaciqui became a temple containing, among other riches, more than a thousand measures of pearls, of which they took only two. Near Cofaciqui was the temple of Talomeco, over a hundred steps long by forty broad, with the walls high in proportion. Upon the roof of the temple were shells of different sizes, placed with the inside out, to give more brilliancy, and with the intervals "filled with many strings of pearls of divers sizes, in the form of festoons, from one shell to the other, and extending from the top of the roof to the bottom." Within the temple, festoons of pearls hung from the ceiling and from all other parts of the building. In the middle were three rows of chests of graded sizes, arranged in pyramids of five or six chests each, according to their sizes. "All these chests were filled with pearls, in such a manner that the largest contain the largest pearls, and thus, in succession, to the smallest, which were full of seed-pearls only. The quantity of pearls was such, that the Spaniards avowed, that even if there had been more than nine hundred men and three hundred horses, they all together could not have carried off at one time all the pearls of this temple. We ought not to be too much astonished at this, if we consider that the Indians of the province conveyed into these chests, during many ages, all the pearls which they found, without retaining a single one of them."[2] In the armory attached to this temple were long pikes, maces, clubs, and other weapons mounted with links and tassels of pearls.

Garcilasso has an interesting story of an incident said to have occurred a few days after leaving Cofaciqui, when the troops were passing through the wilderness.

[1] "Discovery of Florida," Hakluyt Society, 1851, Vol. IX, p. 60.
[2] Bernard Shipp, "The History of Hernando de Soto and Florida from 1512 to 1568," Philadelphia, 1881, 8vo, p. 364.

Juan Terron, one of the stoutest soldiers of the army, toward noon, drew from his saddle-bags about six pounds of pearls, and pressed a cavalier, one of his friends, to take them. The cavalier thanked him and told him that he ought to keep them, or rather, since the report was current that the general would send to Havana, send them there to buy horses and go no longer afoot. Offended at this answer, Terron replied that "these pearls then shall not go any farther," and thereupon scattered them here and there upon the grass and through the bushes. They were surprised at this folly, for the pearls were as large as hazel-nuts, and of very fine water, and because they were not pierced they were worth more than six thousand ducats. They collected about thirty of these pearls, which were so beautiful that it made them regret the loss of the others, and say, in raillery, these words, which passed into a proverb with them, "There are no pearls for Juan Terron." [1]

At the capital of Iciaha, De Soto received from the cacique or chief, a string, five feet in length, of beautiful and well matched pearls as large as filberts. Upon De Soto's expressing a desire to learn how the gems were extracted from the shells, the chief immediately ordered four boats to fish all night and return in the morning.

In the meantime they burnt a great deal of wood upon the shore, in order to make there a great bed of live coals, that at the return of the boats they might put thereon the shells, which would open with the heat. They found, at the opening of the first shells, ten or twelve pearls of the size of a pea, which they took to the cacique, and to the general who was present, and who found them very beautiful, except that the fire had deprived them of a part of their lustre. When the general had seen what he wished, he returned to dine; and immediately after, a soldier entered, who instantly said to him that, in eating oysters which the Indians had caught, his teeth had encountered a very beautiful pearl of a very lively color, and that he begged him to receive it to send to the governess of Cuba. Soto politely refused this pearl, and assured the soldier that he was as obliged to him as if he had accepted it; and that some day he would try to acknowledge his kindness, and the honor which he did his wife; and that he should preserve it to purchase horses at Havana. The Spaniards valued it at four hundred ducats; and as they had not made use of fire to extract it, it had not lost any of its lustre. [2]

Notwithstanding the strong indorsement given to Garcilasso's narrative by Theodore Irving and some other writers, his tendency to exaggerate depreciates greatly the historical value of his account, and it seems wholly unreliable as an authority relative to early resources in America. We may reasonably doubt whether De Soto's expedition came in contact with more pearls than those mentioned by Biedma and the Portuguese writer.

[1] Bernard Shipp, "The History of Hernando de Soto and Florida from 1512 to 1568," Philadelphia, 1881, 8vo, p. 369.

[2] Ibid., p. 372.

The account of the first voyage along the coast of the United States, that of the Italian, Juan Verrazano, in 1524, contains no reference to pearls, although he penetrated into the interior a score or two of miles, and was frequently in contact with the natives, who lived largely by fishing, and who prized many ornaments of different colored stones, copper rings, etc.

The first expedition which went far into the interior was the ill-fated one under command of Pánfilo de Narvaez in 1528. A thrilling account[1] of this journey was written by Cabeza de Vaca, who was one of the four survivors, after eight years' wandering through America to Mexico. Cabeza had been controller and royal treasurer of the expedition, and in that position it was his particular duty to acquaint himself with all the pearls, gold, and similar riches found by the party. Notwithstanding his tradings with the Indians and their efforts to gain his friendship by means of presents, his account makes no mention of pearls, except to refer to a statement made by some Indians that on the coast of the South Sea there were pearls and great riches.

Hernando D'Escalante Fontaneda, who was shipwrecked on the Florida coast about 1550, and was detained there a prisoner for seventeen years, wrote:

"Between Abolachi [Appalachicola] and Olagale is a river which the Indians call Guasaca-Esqui, which means Reed River. It is on the sea-coast, and at the mouth of this river the pearls are found in oysters and other shells; from thence they are carried into all the provinces and villages of Florida."[2]

The European narrators also reported great stores of pearls along the Atlantic seaboard. Among the first of these may be mentioned David Ingram, who is represented as traveling by land from the Gulf of Mexico to the vicinity of Cape Breton in the years 1568 and 1569. As it appeared in the first edition of Hakluyt's Voyages, this relation states:

"There is in some of those Countreys great abundance of Pearle, for in every Cottage he founde Pearle, in some howse a quarte, in some a pottel [half a gallon], in some a pecke, more or less, where he did see some as great as an Acorn; and Richard Browne, one of his Companyons, found one of these great Pearls in one of their Canoes, or Boates, wch Pearls he gave to Mouns Campaine, whoe toke them aboarde his shippe."[3]

Estimation of Ingram's wonderful relation is decreased by Purchas's comment:

[1] "Relation of Alvar Nunez Cabeça de Vaca," translated by Buckingham Smith, New York, 1871.

[2] Bernard Shipp, "History of Hernando de Soto and Florida," Philadelphia, 1881, p. 586.

[3] Hakluyt, "The Principall Navigations, Voiages, and Discoveries of the English Nation," London, 1589.

As for David Ingram's perambulations to the north parts, Master Hakluyt, in his first edition printed the same; but it seemeth some incredibilities of his reports caused him to leave him out in the next impression; the reward of lying being, not to be believed in truths.[1]

Even the members of Raleigh's Roanoke Colony of 1585 reported pearls. Hariot stated:

Sometimes in feeding on Muscles we found some Pearle: but it was our happe to meet with ragges, or of a pide colour: not having yet discovered those places where we heard of better and more plenty. One of our company, a man of skill in such matters, had gathered from among the Savage people about five thousand: of which number he chooses so many as made a faire chaine, which for their likenesse and uniformity in roundenesse, orientnesse, and piednesse of many excellent colours, with equality in greatnesse, were very faire and rare: and had therefore been presented to her Majesty, had we not by casualty, and through extremity of a storme lost them, with many things els in coming away from the countrey.[2]

So far as we can learn, there is no evidence to show that, during the sixteenth or the seventeenth century, any pearls of value were received in Europe from within the present limits of the United States, as was the case with the resources of Venezuela, Panama and Mexico. Many of the accounts quoted above seem wholly fictitious, some of them possibly drawn up for the purpose of promoting exploring expeditions. It is also probable that knowledge of the enormous collections at Venezuela and Panama misled some of the narrators into recognizing as pearls the spherical pieces of shell or even the cylindrical wampum which the Indians made in large quantities and used as money.

However, it is unquestionable that pearls of value were in the possession of some of the wealthier tribes. Biedma's account of the 150 pounds or more of damaged pearls in the graves at Cofaciqui seems wholly reliable, and likewise many other statements; and it is an interesting problem to determine the source from which the Indians obtained them.

Most of the narratives refer to the pearls as coming from the coast of the South Sea or Gulf of Mexico. The evidence of Fontaneda, who had spent seventeen years in the country, throws some light on this. He states that pearls were obtained at the mouth of Reed River near Appalachicola, whence they were distributed throughout Florida. This seems to indicate that on the west coast of Florida there might have been extensive reefs of pearl-bearing mollusks, which have since become extinct, although existing shell-heaps do not confirm this.

---

[1] "Purchas's Pilgrims," London, 1625, Vol. IV, p. 179.　　[2] Hakluyt's "Voyages," Glasgow, Vol. VIII, p. 357.

While it is possible and even probable that many of these pearls in the possession of the Indians came from the Gulf of Mexico or even from the Caribbean Sea, it seems much more likely that they came largely from the Unios of the inland lakes and rivers.

The voyages of Narvaez, Ayllon, De Soto, Ribault, etc., had been so unfortunate that for a century little exploration was made in the territory of the southern part of the United States. When this territory was again invaded, little was seen in the way of pearls.

Iberville, who established the French settlement near the mouth of the Mississippi in 1699, was specially directed to look for them. His instructions state: "Although the pearls presented to his Majesty are not fine either in water or shape, they must nevertheless be carefully sought, as better may be found, and his Majesty desires M. d'Iberville to bring all he can; ascertain where the fishery is carried on, and see it in operation."[1] Pearls were found in the territory of the Pascagoulas, but they were not worth the trouble of securing them. It appears that from these the Pearl River in Mississippi derived its name.

The only reference to pearls in the seventy-one volumes of Travels and Explorations of the Jesuit Missionaries in New France, from 1610 to 1791, is a note by Father Gravier stating that he saw no choice pearls: "It is true the chief's wife has some small pearls; but they are neither round nor well pierced, with the exception of seven or eight, which are as large as small peas, and have been bought for more than they are worth."[2]

Daniel Coxe's description, in 1722, of pearl resources in America, is of special interest because of the extended experience of his father as a trader in the country. He states:

Pearls are found to be in great abundance in this country; the Indians put some value upon them, but not so much as on the colored beads we bring them. On the whole coast of this province, for two hundred leagues, there are many vast beds of oysters which breed pearls, as has been found in divers places. But, which is very remarkable, far from the sea, in fresh water rivers and lakes, there is a sort of shell fish between a mussel and a pearl oyster, wherein are found abundance of pearls, and many of an unusual magnitude. The Indians, when they take the oysters, broil them over the fire till they are fit to eat, keeping the large pearls they find in them, which by the heat are tarnished and lose their native lustre; but, when we have taught them the right method, doubtless it would be a very profitable trade. There are two places we already know within land, in each of which there is a great pearl fishery. One about one hundred and twenty leagues up the River Meschacebe [Mississippi], on the west side, in a lake made by the river of the Naches, about forty miles

[1] P. F. X. de Charlevoix, "History of New France," New York, 1900, p. 129.
[2] Voyage of Father Gravier in 1700 From the Country of Illinois to the Mouth of the Mississippi," Cleveland, 1900, p. 141.

BROOCH, RENAISSANCE STYLE, SET WITH BAROQUE PEARLS, FROM AMERICAN STREAMS

Pan-American Exposition, 1901

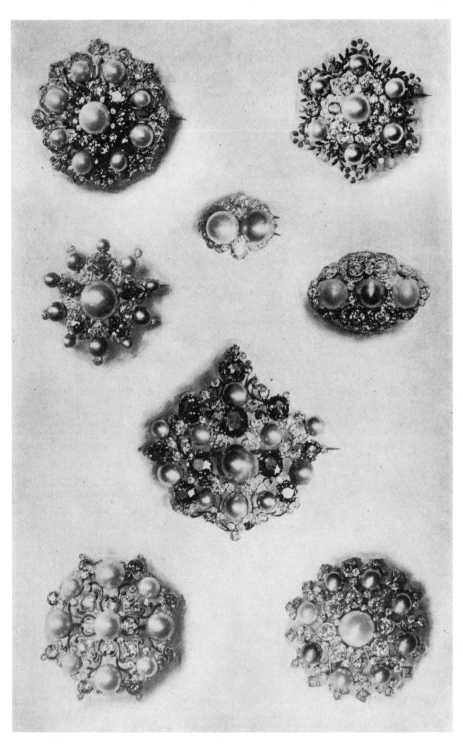

BROOCHES AND RINGS OF FRESH-WATER PEARLS FROM WISCONSIN AND TENNESSEE

Paris Exposition, 1900

from its mouth, where they are found in great plenty and many very large. The other on the River Chiaha, which runs into the Coza or Cussaw River (as our English call it), and which comes from the northeast, and, after a course of some hundred miles, disembogues into the Gulf of Florida, about one hundred miles to the east of the Meschacebe.[1]

It is interesting to note that the first place mentioned by Coxe as the location of a great pearl fishery is not far from one of the most productive pearling regions of the last fifteen years, *viz.*, the eastern part of Arkansas. The second place noted by him appears to be identical with the Iciaha, where, nearly two centuries before, the Indians exhibited the methods of their fishing to De Soto and his companions.

Excepting Coxe's notice, for 250 years following 1600, little was heard of the occurrence of pearls within this country. This does not indicate necessarily that the gems were absent from the waters; but, not using the Unios for food as did the aborigines, the residents had little occasion to open them and in this way learn of their contents. And even where pearls were occasionally found in mollusks opened for fish-bait, the people were in few instances informed as to their market value, and did not attempt to sell them, although the most attractive ones may have been treasured as ornaments or as keepsakes. This was paralleled in the diamond fields of South Africa, where gems worth thousands of dollars were used as playthings by the farmers' children. A jewel, like a prophet, is frequently without honor in its own country until the residents of that country learn of the great esteem in which it is held elsewhere.

And yet, in some localities a few pearls were collected from time to time. The Moravians—familiar with the pearls of their native streams in Europe—gathered many from the Lehigh River near Bethlehem, Pennsylvania, over a century ago;[2] and from Rhode Island and elsewhere a few were obtained.

The first awakening to a realization of the value of fresh-water pearls in America occurred fifty years ago, when several beautiful gems were marketed from the northern part of New Jersey. The story of this find has been frequently told. A shoemaker named David Howell, who lived on the outskirts of Paterson, occasionally relieved the monotony of his trade by a fishing excursion to some neighboring stream, where he would usually collect a "mess" of mussels. Returning from one of these visits to Notch Brook in the spring of

---

[1] Coxe, "A Description of the English Province of Carolana, by the Spaniards call'd Florida, and by the French La Louisiane, as also of the Great and Famous River Meschacebe or Missisipi," London, 1722, pp. 82, 83.

[2] "Allgemeine Handelszeitung," Leipzig, April, 1789, p. 218.

1857, the mussels were fried with the usual abundance of grease and heat. After this preparation, one of them was found to contain a large, round pearl weighing "nearly 400 grains," which possibly might have proven the finest of modern times, had not its luster and beauty been destroyed by the heat and grease.[1] Had the pearl been discovered in time, its value might have exceeded $25,000, thus making poor Howell's fried mussels one of the most expensive of suppers.

Hoping to duplicate his wonderful find, Howell collected and searched other mussels, and his example was followed by several of his neighbors. Within a few days a magnificent pink pearl was found by a Paterson carpenter named Jacob Quackenbush. This weighed ninety-three grains, and was bought by the late Charles L. Tiffany for Messrs. Tiffany & Co., New York City, for $1500. Mr. Tiffany later described with much interest the feelings he experienced after making the purchase. Said he: "Here this man finds a pearl within seventeen miles of our place of business! What if thousands should be found, and many perhaps finer than this one! However, we risked buying the pearl, and as no one in New York seemed interested in it, we sent it to our Paris house for sale, and a French gem dealer offered for it a very large advance on the original price, paying 12,500 francs." From this dealer it passed into the possession of the young and beautiful Empress Eugénie, from whom and from its great luster it derived the name "Queen Pearl." Its present market value would doubtless amount to $10,000 or more.

When news of the very large price received for Quackenbush's find became public, great excitement developed in the vicinity of Notch Brook. Persons came from all directions to search in the shallow streams for valuable pearls. Farmers of the neighborhood tried their luck, and also mechanics and other residents of the adjacent villages and towns, and even some from Newark, Jersey City, and New York. An old resident, who was an eye-witness, describes the scene as one of great animation, the crowds of people and the horses and wagons along the shore giving "an appearance of camp-meeting time." At least one schoolmaster in the vicinity is said to have closed his school to give his pupils an opportunity to engage in the hunt.

With trousers rolled up, the people waded into the shallow water and sought for the mussels in the mud and sand on the bottom. Many pearls were secured, but none approached in size or value the two above noted.[2] During 1857, the New York City market received about $15,000 worth of pearls from these waters, and in addition many were

[1] "Frank Leslie's Magazine," New York, May 23, 1857, Vol. III, pp. 384-386.
[2] "Gems and Precious Stones of North America," by George F. Kunz, New York, 1889- 1892, pp. 211-257. "The Fresh-Water Pearls of the United States," Washington, 1898, 50 pages and plates.

sold locally or retained as souvenirs of the hunt. At the low price of pearls existing then, this figure would mean possibly ten times as much at present, or $150,000.

The active search soon depleted the resources of the little stream, so that in the following year the reported value of the yield was only a few thousand dollars. The decrease continued until in a few years practically every mussel was removed, and at present scarcely a single Unio is to be found in these waters.

The interest in pearling extended far from the place of the original find; and in Pennsylvania, Ohio, and even as far away as Texas, search was made in the streams. In the Colorado and its tributaries, about 20,000 were found in a short while. Most of these were small and unattractive, but a considerable number were reported "as large as pepper-corns" and a few "the size of a small rifle ball," the number decreasing with the increase in size. A correspondent in the "Neue Zeit" wrote:

Sometimes they are round, sometimes cylindrical, elliptical, hemispherical, or of an altogether irregular shape. The finest have a milk-white, silvery sheen; many, however, are reddish yellow, bluish brown, or quite black; the last naturally have no value whatever. As to their value, there is considerable uncertainty, and it can easily be understood that those who have a great number of them in their possession greatly overestimate them. So far they are found principally in the Llano and the San Saba.[1]

After the resources in northern New Jersey were depleted and the excitement had died out, little was heard of pearling in this country until 1878, when many were found in Little Miami River in southwestern Ohio. The fishing was carried on at low water, and principally by boys, who would wade out in the water and feel for the mollusks with their feet, and then bob under and pick them up with their hands. The senior author spent a day in this fishery with a party of six boys with some success. During 1878 about $25,000 worth of pearls were collected in the vicinity of Waynesville on that stream. Mr. Israel H. Harris, a banker of Waynesville, then began collecting these pearls; and by purchasing during several years nearly every interesting specimen found in the vicinity, he made his collection one of the largest and best known in the country. When sold in 1888, it contained several thousand pearls, mostly of small size, averaging in weight little more than one grain each. A large portion of this collection was exhibited in the American section of the Paris Exposition of 1889, and was awarded a gold medal. Included in this exhibit was a series of ornaments in which the gems were arranged according to color, so that in one the pearls were green, in another purplish brown, in another pink,

[1] "Neue Zeit," in Ausland, 1858, No. 8, p. 192.

in another waxy white, and in one a cream white. It also contained a button-shaped pearl weighing thirty-eight grains and several pink ones almost translucent. A pink pearl of eight grains was admired by all who saw it; by reflected light this had the color and translucency of a drop of molten silver. Many of the pink pearls found in the Little Miami and its tributaries were of the most beautiful rose-petal pink; pearls of this peculiar color have never been found in any other waters.

From Ohio the industry gradually extended westward and southward, and new fields were developed, pearls to the value of about $10,000 annually coming on the market from such widely separated States as Vermont, Kentucky, Tennessee, Florida, Texas, Washington, etc. However, little general interest was taken in fresh-water pearls, and few choice ones were found until the magnificent resources of the upper Mississippi Valley were discovered. Owing to the ease with which the mollusks may be collected by wading, it was in the relatively shallow tributaries that the fishery first developed, rather than in the deep channels of the main stream and of the large affluents.

The first region in the Mississippi Valley to attract attention was southwestern Wisconsin. Early in the summer of 1889, many beautiful pearls were found in Pecatonica River, a tributary of Rock River, which in turn empties into the Mississippi. Within three months, $10,000 worth of gems were sent from this region to New York City alone, including one worth $500, which was a very considerable sum for a fresh-water pearl at that time. The interest quickly spread to neighboring waters, and within a short time pearls were found also in Sugar River, in Apple River, in Rock River, in Wisconsin River, and in the Mississippi in the vicinity of Prairie du Chien. The fact that little experience and no capital was required for the business drew large numbers of persons to the newly-found Klondike; and the finds were so numerous and of such high quality that about $300,000 worth of pearls were collected before the end of 1891, greatly exceeding all records for fresh waters.

The Wisconsin pearls are remarkable for their beauty, luster, and diversified coloring, and some lovely shades of pink, purple, and especially metallic green have been found. Several of them have weighed in excess of fifty grains each, and some individual values ran well into four figures. One shipment made from Sugar River to London in September, 1890, contained ninety-three pearls, weighing from four to twenty-eight grains each, for which £11,700 was received in payment. In the limits of one county in the following year, pearls to the value of nearly $100,000 were secured.

Shortly following the outbreak of pearling in Wisconsin came the development of interest in certain parts of Tennessee. For many years pearls had been secured from the Cumberland and Tennessee rivers and their tributaries, especially Caney Fork, Duck, Calf Killer, and Elk rivers, the headquarters of the fishery and the local markets being Carthage, Smithville, Columbia, and Arlington. The search had been conducted in a moderate way by pleasure parties in the summer, and by farmers after the crops had been laid aside.

In 1901 pearling excitement developed in the mountain regions of eastern Tennessee, especially in Clinch River. These newly-discovered resources proved so valuable that the local interest became very great. Vivid and picturesque accounts published in the local papers reported hundreds of persons as camping at various points along the streams, some in tents and some in rough shanties, and others going from shoal to shoal in newly-built house-boats. They were described as easy-going, pleasure-loving people, the men, women, and children working hard all day, subsisting largely on fish caught in the same stream, and dancing at night to the music of a banjo around the camp-fires. The center of the new industry was Clinton, the county seat of Anderson County, whither the successful hunters betook themselves each Saturday, the preferred time for selling the catch.

The next outbreak of pearling excitement was in Arkansas, in the region referred to by Daniel Coxe two centuries ago as the location of great pearl resources.[1] Although in recent times little had been heard of pearls in Arkansas previous to 1895, they were not unknown in that State. For years they had been picked up by the fishermen, and used as lucky stones or given to the children for playthings. Some had come into the possession of persons acquainted with their value. About 1875, a few pearls were collected by a party of men engaged in cutting cedar poles on White River; in 1888, a brilliant pear-shaped pink pearl of twenty-seven grains was secured from the same river, and sold to a prominent resident. Little had been said about these finds, and in general the people of Arkansas had slight idea of the occurrence or the value of pearls in those waters.

In 1895, a surveying party on White River found pearls in the Unios of that stream, and collected them to the value of about $5000. News of this discovery attracted attention to the resource, and other persons sought for the gems in the White River and its tributaries, in the St. Francis and the Arkansas rivers. The unusually low water in 1896 facilitated the fishery, and resulted in the discovery of many large and valuable gems. The interest developed rapidly, and within twelve months nearly every stream of water in Arkansas yielded pearls, with the finds most extensive and valuable in White River and

[1] See p. 258.

its tributary the Black River, which has proved to be the richest pearling region in America. The industry centered at Black Rock, more than a thousand persons fishing within twenty miles of that place. It is estimated that within three years following the development of this fishery, this State yielded pearls to the value of more than $500,000.

When the Arkansas fishery was at its height, it was reported that ten thousand persons were employed therein. The fishermen were from nearly every class and condition in the State. Women were not absent; even children participated in the industry, and some proved more fortunate than the older hunters. It was not uncommon to see several hundred persons congregated at one bar or in one stretch of the river, all intent on making a fortune, and all occupied in fishing or in opening the shells. So complete was the absorption of the people in this pursuit, and so many of the farm-hands were occupied in the eager search for anticipated fortunes, that the local papers reported much apprehension and difficulty in harvesting the cotton and other crops.

Within the main channel of the Mississippi, the relative scarcity of pearls in the Unios, and the greater preparation required for collecting the mollusks in the deep waters, retarded the fishery until the establishment of button manufacture afforded a market for the shells, this originating in 1891. The industry developed rapidly, and for several years has consumed about 35,000 tons of shells annually, obtained principally in the Mississippi between Quincy and La Crosse, and to a much less extent in other streams in this valley. This is more than twice the total product of mother-of-pearl shell in all parts of the world. However, the value per ton is very much less than that of the best grade of mother-of-pearl; that from Australia, for instance, commonly selling for $1200 per ton, whereas the Mississippi shell usually sells for less than $20, although the very choicest may bring upward of $50 per ton.

The gathering of shells for manufacture has extended to many of the large tributaries of the Mississippi, especially the Arkansas, the White, the St. Francis, the Ohio and the Illinois rivers, and this industry has added largely to the pearl yield in these waters.

In the last three years, the scenes of greatest activity have been the Wabash River and its tributaries, where shell-collecting developed in 1903, and the Illinois River, where the industry was of little importance previous to 1906. On the Wabash, camps were established at almost every town, from the mouth up to St. Francis, Illinois, and about one thousand persons found employment. Some of the most beautiful American gems have come from this river. They are usually silvery white in color and of the sweetest luster. A single pearl weighing only ten grains has been sold at the river for $1000; but it is frequently the case that a fine gem will sell for more at the place where found than in the great markets. During the spring of 1907, three pearls were found in the Wabash near Vincennes, which weighed forty-one, fifty-one, and fifty-three grains respectively. One of these was white, one faint pink, and the third was yellow. The finest pearls have been reported from the vicinity of Mount Carmel near the lower end of the river. Very large quantities of baroques or slugs are found in the Wabash and the Illinois; 30,000 ounces were reported from those rivers and their tributaries in 1907, for which the fishermen received a total of $50,000. A large symmetrical pearl found during 1907 weighed a trifle under 150 grains, and a slug was found which weighed fully one ounce, or 606 grains.

The pearl-hunting excitement has been felt even on the Atlantic seaboard, as a result of the publication of the discoveries in the Mississippi Valley. In Maine many pearls have been reported, especially in the vicinity of Moosehead Lake. In 1901 over one hundred were found in that vicinity; most of them were of little value, but more than a dozen were worth $10 or $15 each. Three found by Kineo guides were sold for an aggregate of $300. The choicest one reported in that year weighed twelve and one half grains and sold for $150; had it been perfect in form and luster its value would have been several times that amount. Most of these pearls were found by Moosehead guides, who found purchasers among the visiting fishermen and hunters.

Since 1901 many farm-boys as well as guides have devoted much attention to the business, some of them deriving as large a revenue therefrom as from the use of the rifle. Good finds have been made, during the last year or two especially. In 1906, one choice pearl sold for $700, and many have sold for $10 to $75 each. The search has proven so alluring that returning visitors have complained that some of the guides care to do little more than search every rill, brook, and creek they come across looking for the mollusks. Just at present the principal attention seems to be directed to the streams in the western part of Maine, where the river-beds are more sandy and the shell-fish more abundant than in the northern and eastern part of the State.

In Massachusetts pearls have been collected from many of the ponds and brooks. In Nonesuch Pond in Weston, the *Unio complanata* has yielded many small ones of attractive appearance, but not of sufficient size or luster to sell for more than $10 each. Ponds in the town of Greenwich and also in Pelham in Hampshire County are among the best in Massachusetts for pearls. The Sudbury River above Concord also yields many. Relatively few of the Unios contain pearls, and the gem-bearing individuals seem to be grouped in special localities. Outside of these places, thousands of mussels may be opened without revealing a single gem. A collection of small Massachusetts pearls was brought together a few years ago by Mr. Sherman F. Denton of Wellesley Farms, who has devoted much time to exploring the inland waters of Massachusetts.

Connecticut also has had a slight touch of the pearl fever. In 1897, Mr. C. S. Carwell of Ledyard, explored the headwaters of Mystic River, and in a few weeks collected a number of pearls, one of which he is reported as having sold for $500, and two others were estimated at $400 each. And from the other end of the State, along the Shepaug River, is reported a similar account of the success of Mr. Arlo Kinney of Steep Rock. Attracted by these reports, crowds of seekers have proceeded in the usual reckless manner to make wholesale destruction of the mollusks. The finds have been especially large and valuable in the lakes and streams of Litchfield County, particularly in Bantam Lake.

In New York State, pearls have been found in the swift shallow streams in the Adirondack region, and in several of those entering the St. Lawrence, particularly the Grass River in St. Lawrence County. Pearls were first reported from this region in 1894. In 1896 the Grass yielded one pearl weighing fifty-eight grains, worth $600 locally; and in 1897 one weighing sixty-eight grains was found, the fisherman selling it for $800. A resident of Russell township devoted most of his time to pearling in Grass River during 1896 and 1897, from which he is said to have realized $2000. In this region the mussels are found by wading in the shallow water and scanning the bottom through a water-telescope. Most of the pearls are of slight value, but many individuals are reported as worth from $30 to $60 each.

Pearl River in Rockland County, New York, has furnished a number of brown pearls. These are commonly small, weighing from one eighth to one half grain each, although some weigh seven or eight grains each. Most of these are not lustrous, but occasionally a bright brown or a bright copper-brown specimen of from one to four grains is met with. At the Paris Exposition, in 1900, were exhibited one hundred of these pearls, with an aggregate weight of 281 grains; these now form part of the Morgan-Tiffany Collection, in the American Museum of Natural History.

Even in the rich coal regions of Pennsylvania pearls are found. Possibly the most productive section in that State has been the headwaters of the Schuylkill River in the vicinity of Tamaqua, Quakake, and Mahony City. Of the tributaries of the Schuylkill, those contributing largely to the yield have been Lewiston, Nipert, Still, Locust, and Hecla. These rise in the mountains and are rivulets of fair size by the time they reach their common outlet.

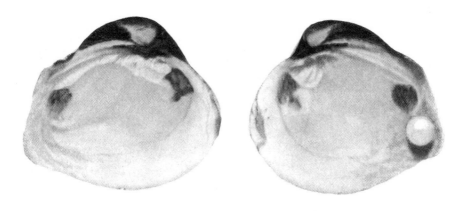

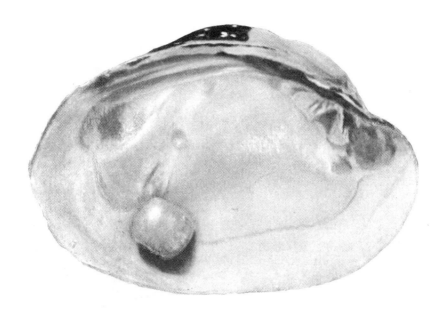

PEARL-BEARING UNIOS

From the Mississippi Valley

The upper pictures show the two valves of the same shell, and the pearl is detachable

Pearling scene on White River, Arkansas

The fishermen are using scissor-tongs from flat-bottom skiffs

Pearling camp on upper Mississippi River

Crowfoot drags are shown on the flat-bottom skiffs at the river-bank

The original pearl finds in the Schuylkill date from half a century ago, when they were secured by farmers who used the mussel shells in removing hair from the hides of slaughtered pigs. During the Mississippi pearling excitement in 1897, several persons from New York, who were summering in Schuylkill County, searched the small streams for pearl-bearing mussels with such success that within a short while many farmers became enthusiastic hunters during their spare time. Half a dozen or more men did very well, their catch amounting to thousands of dollars' worth. Mr. Frank M. Ebert, of Quakake, has put most of his spare time in the business in the last ten years, and has secured many good pearls. It is estimated that the total catch in Schuylkill County alone approximates $20,000 at local values. So actively has the search been conducted 'that at present few adult mussels of the pearl-bearing species remain, and a day's work may result in finding less than a dozen.

The best price reported as received by a local fisherman was $200 for a twenty-grain pearl in the year 1904. Many individual specimens have been sold at prices ranging from $100 to $175. It is claimed that a pearl sold by a fisherman in Schuylkill for fifty cents was later marketed in Philadelphia for $125, and with slight mounting was ultimately sold for $1600. The most attractive weigh from ten to twenty grains each; larger ones have been found, weighing up to thirty-eight grains, but as a rule the luster is not so good as that possessed by pearls of medium size. The common colors are dark blue, pink, lavender, and white. A few are black and some are brown. The brown pearls are seldom of value, owing to deficiency in luster.

In Maryland pearls have been collected from the brooks near the head of Chesapeake Bay, and especially in Kent and Cecil counties. These are of almost every conceivable color, ranging from a clear white to a dainty pink, and to very dark colors, especially bronze and copper. Most of them are too small for commercial value, and only a few reach sufficient size to command more than $5 or $10 each, but single specimens have sold as high as $50.

Georgia has yielded some pearls, chiefly in the vicinity of Rome, at the junction of the Etowah and Oostanaula rivers. This is believed to be the site of the Indian town Cofaciqui, where, in his memorable expedition of 1540–1541, De Soto found the natives in possession of so many pearls. The general news of finds in the Mississippi Valley stirred up local interest in this region in 1897, and when the streams were low and clear in the autumn many persons engaged in hunting the mussels. An ex-sheriff of Rome is reported as having secured about fifty pearls, lustrous but irregular. A few miles above Rome, a farmer made a trial on Johns Creek, a tributary of the Oostanaula;

and from a basketful of Unios he reports finding several marketable pearls for which he received $180 from a Baltimore jeweler. Others followed, and many fine specimens were secured. Unios are especially abundant in the Flint, Ocmulgee, and Oconee rivers, and it seems probable that many pearls might be found in these streams.

Florida has not yet been actively exploited, but it may prove a productive region ere long. The reports of De Soto's expedition make special reference to the size and beauty of the pearls found at a point where he crossed the Ocklocknee River about thirty miles above its mouth, near the present site of Langston, Wakulla County. And there seems little doubt that pearls may be found in the Ocklocknee and also in its affluent, the Sopchoppy River. The banks of these streams are full of shells, and pearls of choice color have been sent from there.

It is unnecessary to refer in detail to the origin of pearling in each of the States. The general interest in this industry from 1889 to the present time has resulted in the examination of most of the rivers and creeks, and in few has the search been entirely unrewarded, although the finds have been relatively much greater in some waters than in others. As a rule, pearl-bearing Unios are most numerous in clear, swift streams, with sandy or gravelly bottoms and which flow through calcareous rocks. With pearlers as with miners, there is a stampede to the places where a good find is reported, since the rivers are free for all; consequently, there is much variation from year to year in the amount of attention which the individual streams and localities receive.

While many of the pearlers operating in the Mississippi River are professional fishermen or rivermen, most of those in the smaller streams have had no previous experience in similar work. Frequently whole families come twenty or thirty miles, and even greater distances, and camp on the river bank. In many instances farm-hands are there who have abandoned their crops, mechanics who have left steady jobs, railway men who have taken a lay-off, teachers, merchants, all eager and expecting to find a fortune. In some localities, pearl fishing has been used as an attraction in big picnic advertisements, and has drawn larger crowds than a public orator.

The mollusks are removed from the river bottoms in various ways and by many forms of apparatus. In the shallow streams the fishermen simply wade out in the water and pick up the shells by hand. If not readily visible from the surface, the shells may be located with the bare feet or by the use of a water-telescope. Where the water is too deep for wading, the fishermen work from small boats, and use garden rakes or other convenient inplements.

Where pearling has developed into more of an industry, special forms of rakes and drags are employed. A shoulder rake, with a handle twelve to twenty feet in length, is used extensively under the ice in frozen rivers, and in lakes and other places where the water is still and from eight to fifteen feet in depth. This is simply an overgrown or enlarged garden rake, armed with twelve or fifteen iron teeth about five inches in length. A wire scoop or basket is attached to receive the catch as it is pulled from the bottom by the teeth, and when this scoop is well filled it is lifted and the contents dumped on the ice or into the skiff. This method is laborious, and is employed only where the water is shallow and the mollusks are abundant. Scissor tongs—similar to those used by oystermen on the Atlantic coast—are also employed in some localities, especially in Arkansas, where it is estimated that 1700 pairs were manufactured and sold in 1899 and 1900, at about $7 each.

In the large streams of the Mississippi Valley, with their slow and steady currents, and where the Unios are taken largely for their shells to be used in button manufacture, the most popular form of apparatus since 1896 has been the crowfoot drag. This ingenious contrivance consists of a crossbar of hollow iron tubing or common gas-pipe, six or eight feet long, to which are attached, at intervals of five or six inches, stout twine or chain snoods or stagings, each about eighteen inches in length. To each of these are attached three or four prongs or "hooks," about six inches apart. These "hooks" are four-pronged, and are made of two pieces of stout wire bent at right angles to each other. According to the depth of the water, from twenty-five to seventy-five feet of three quarter inch rope is attached to the drag for the purpose of towing it behind the boat, which is permitted to drift down the stream with the current. This contrivance costs about $3, and each fisherman generally has at least two of them, as well as a wide flat-bottom boat costing $5 or $10.

Sometimes, when the current is light, the fisherman prepares a "mule" to assist the boat in towing the resisting drag. This "mule" consists of a wooden frame, hinged in V shape, and is fastened several feet in advance of the boat with the V end pointed down the stream. It sinks low in the water, and the current pressing against the angle carries it along, and thus tows the skiff and the resisting drag at a uniform rate of speed. When there is not sufficient current even for this contrivance, as in the wide reaches and in the lakes, oars, sails, and even power engines may be used for propelling the boat.

As the crowfoot drag is slowly drawn along the bottom, it comes in contact with the mollusks feeding with open shells. When a hook or other part of the drag enters an open shell, the mollusk immediately closes firmly upon the intruding object and clings thereto long enough to be drawn up into the boat.   In this way, where the Unios are thick, nearly every hook becomes freighted, and some may have two or three shells clinging thereto.   It is easy to collect fifty mollusks in passing over a length of two hundred feet.   Two drags are carried by each fisherman, and the second one is put overboard as soon as the first one is ready to be raised.   This is suspended with the bar across two upright forks on either side of the boat with the prongs swinging freely, and the mussels are removed therefrom.   When this operation is completed, the drag is put overboard and the other one is ready for lifting. This apparatus is very effective, and as much as a ton of shells has been taken by one man in twelve hours, but the average is very much less, probably not over four or five hundred pounds.   Objection is made to this manner of fishing, since many mollusks not brought to the surface are so injured that they die.

A cruder implement of similar type has long been employed on many logging streams.   The weighted branch of a tree is dragged on the bottom behind a raft of logs, and the mussels attach themselves to the twigs in the same manner as on the crowfoot hooks.

During the pearling excitement in Arkansas, a considerable portion of the choice pearls were found, not in the mussels, but lying loosely in the mud of the shores, indicating that under some circumstances, as agitation by freshets or floods, the loose pearls are shaken out from the Unios.   In some instances, indeed, the pearls were found upon or in the soil at some distance from streams or lakes.   It is reported that in October, 1897, Mr. J. W. McIntosh, of the northern part of Lonoke County, while digging post-holes in the old bed of Cypress Bayou, found a number of pearls, some "as large as a 44-caliber Winchester ball," lying within the shells at a depth of a foot and a half below the surface.   This peculiar occurrence is partly explained by the wide extension of the waters in flood times over the low region, and by the shifting of streams and the isolation of cut-offs.

Stray pearls have been found in many other odd places, as in the viscera of chickens and ducks, in the stomachs of fish, and even within a pig's mouth.   It is not an uncommon scene in the pearling region to see men raking over the muck in hog-pens along the river banks, hoping there to find a stray pearl lost from the mussels with which the animals had been fed by persons who had indeed "cast pearls before swine."   It is related that a Negro near Marley, Illinois, in this way secured a pearl weighing 118 grains, for which he received $2000 from a St. Louis buyer, and which was ultimately sold to a New York dealer for $5000.

During the height of the Arkansas pearling excitement in 1897, the speculative spirit was so rife that many persons—unwilling to engage in the labor of fishing—purchased unopened mussels from the fishermen in the venture for aleatory profits. The price for these ranged from twenty-five cents to $2 per hundred, and fluctuated rapidly, according to the immediate results, increasing several hundred per cent. in a few minutes under the influence of a valuable find. One fisherman sold mussels to the value of $28 in one.day, and thought he had made an excellent bargain until over $1000 worth of pearls were revealed when the shells were opened.

While some pearlers work in southern streams throughout the year, generally the season is coincident with warm weather, when the water is low and the work may be conducted with comfort. In the vicinity of Muscatine and Rock Island about twelve years ago, large quantities of Unios were taken during the winter when the river was frozen over, the men working with long rakes from the surface of the ice.

When only a few mollusks are taken, they are readily opened with a knife to permit a search for the pearls. But where there are many, as in the Mississippi River, the opening is facilitated by heating. After a sufficient catch has been obtained, they are subjected to the action of steam in a box, or they are heated in an ordinary kettle; a few minutes of steaming or cooking are sufficient to cause the shells to spring open. The fleshy parts are removed and thoroughly searched, the interior surfaces of the shells are likewise examined for attached pearls, and the liquid at the bottom of the vessel is strained so that nothing of value may escape.

This cooking is a convenient method of opening the shells, but unquestionably it injures the quality of many pearls. In some instances when the shells open, the pearls fall out and descend to the heated iron bottom, where they are quickly injured. The surface of one exposed too long to the heat shows numerous minute cracks, which increase in number and size when subjected to changes of temperature. Some choice gems have in this manner been rendered almost valueless. If a jacket boiler, or one with a double bottom, were used, there would be less danger of injuring the pearls; or a similar result could be accomplished by placing a wire screen a few inches above the bottom.

Several fishermen have endeavored to devise mechanical methods for removing the pearls and thus avoid the painstaking search among the flesh tissues now necessary; but these contrivances have not proved satisfactory, and have not been employed except experimentally.

In the Mississippi and its tributaries, where the fishery is very extensive, after the pearls have been secured, the shells are sold to button manufacturers and to exporters at prices ranging from $4 to $40 per ton, according to species, quality, and market conditions. This provides a fairly remunerative income to the fishermen even if no pearls

whatever are found.   But in the small tributaries and where the mollusks are less numerous, the shells are of little value owing to the
expense of bringing them together and conveying them to market.

Not every mollusk contains a pearl, and the village belle, intent on
her evening toilet, need not buy a bushel of clams with the pleasant
anticipation of finding a sufficient number of gems for a necklace.
Small and irregular pearls are not at all uncommon, but choice ones
are decidedly scarce, and each one represents the destruction of tens
of thousands of mollusks.   Quantities of irregular and imperfect
nodules known as slugs are collected, which sell for only a few dollars
per ounce.   In some sections of the Mississippi, the slugs are so very
numerous that their aggregate value exceeds that of the choice pearls.

In the Mississippi, the percentage of pearls found in a definite quantity of mollusks is less than in the tributary streams, yet the much
greater quantity of shells collected raises the total yield to a very considerable amount.   Pearling is subordinate and incidental to gathering
the shells for manufacture.   In that length of the river from St. Paul
to St. Louis, a fair average yield to the fishermen is about fourteen
dollars' worth of pearls and slugs to each ton of shells.   Of course,
this is not the individual experience, for a single Unio may contain a
gem worth $5000, and on the other hand several tons of shells may
yield only a few cents' worth of baroques.   The market for the shells
places the Mississippi fishing upon an industrial basis, and guarantees
a substantial income to every fisherman even when no pearls whatever
are found.

Unios from the upper part of the Mississippi yield a much greater
percentage than those from below Davenport.   In 1904, for instance,
from the 4331 tons of shells taken in Wisconsin the fishermen secured
pearls which they sold for $91,345, an average of $21 per ton; from
the 822 tons in Minnesota the average was $16 per ton; in Iowa the
average was $12 for each of the 7846 tons; in Illinois, $5 per ton for
the 2364 tons, and in Missouri less than $1 worth of pearls was secured
by the fishermen for each ton of shells which they took in the year
named.   A large number of choice pearls weighing over thirty grains
each were found in the vicinity of Prairie du Chien and McGregor.
Within a river length of one hundred miles in that region, the fishermen in 1904 gathered pearls which ultimately sold for $300,000.   It
is therefore apparent that the returns vary greatly in the different
regions; nevertheless, even in the less productive localities fine pearls
are sometimes found, which contribute to make the industry a profitable one.

Success in pearling is like that in mining.   In the White River in
Arkansas, for instance, one man found $4200 worth in one month.

Another discovered a $50 pearl in the first shell he opened. A Negro found an $85 pearl the first day he worked, while another fisherman worked seven months and secured less than $10 worth. It is a question of finding or not finding; the finding brings riches sometimes, and though the failures reduce the average profits as low as in other local ventures, the big prizes affect the mind, and the average is lost to sight. Taking the country as a whole, it is probable that the total find has been sufficient to pay the average fisherman little if any more than $1 for each day's work.

The fresh-water pearls range in size from that of the smallest seed to that of a pearl weighing several hundred grains. There is relatively only a small quantity of seed-pearls, especially when compared with the output in the fisheries of Ceylon and Persia. Possibly this is due largely to a scarcity of the parasites which seem to perform so important a function in the regions noted. A further reason may be found in the manner in which the mollusks are opened and searched. Were the Ceylon method of opening employed here—which, however, is not at all practicable—it seems probable that the quantity of seed-pearls found in this country would be greatly increased.

The pearls from the tributaries of the Mississippi are noted for their great range of coloration. From a dead white, the color is gradually enhanced to faint shades of pink, yellow, or salmon tints, then to a more decided form of these. From the light shades, the range extends to purple and to bright copper red, closely resembling a drop of molten copper. Some are very light green; others rose, steel blue, or russet brown, while purplish and very dark brown are not uncommon. White pearls are probably the most numerous; but pink, bronze, and lavender are by no means rare.

A large percentage of the Mississippi River pearls are very irregular in form, many of them resembling dogs' teeth, birds' wings, the heads or bodies of different animals, etc.

As a rule the fresh-water pearls do not rank so high in value as those from oriental seas, since ordinarily they are not so lustrous. However, some of them have sold at very high figures. A round pearl weighing 103 grains, found in Black River, Arkansas, in 1904, was eventually sold for $25,000; and one of 68 grains, found, in 1907, on the Wisconsin side of the Mississippi River, was recently marketed at $15,000.

One of the largest American pearl necklaces, brought together in 1904, consisted of thirty-eight pearls weighing 1710 grains in the aggregate, an average of 45 grains for each pearl. The central gem weighed 98½ grains and those on the left of it respectively 85¾, 79⅝, 65¼, 59⅝, 49⅜, 46¼, 45⅛, 43¾, 41½, 40½, 40⅝, 35⅛,

31⅝, 30, 25⅛, 22¼, 20¼, and 19 grains. The pearls on the right were graduated as follows: 85½, 76⅛, 64⅞, 59½, 47¼, 46, 45⅛, 44½, 42½, 41¾, 38, 37⅞, 36, 35, 34⅝, 29½, 25¼, 21, and 20⅜ grains. This necklace was exhibited at the St. Louis World's Fair. It was sold to a London merchant, who in turn sold it to a Parisian dealer, and it was finally purchased by a Spanish nobleman at a price said to be about 500,000 francs.

Another necklace shown at the St. Louis World's Fair, was of American fresh-water pearls from the rivers of Arkansas. The total weight of these pearls, sixteen in number, was 861⅝ grains, an average of 61½ grains for each pearl. Of these one drop pearl weighed 77 grains, and two others each 65½ grains. A round pearl of 70 grains completed the adornment of the pendant. The circlet consisted of ten round pearls alternating with precious stones. The central pearl weighed 98½ grains and on each side were two of 61 grains, then two of 56 grains, two of 54⅞, and two of 48 grains, one of 45⅜ grains being at the back of the necklace.

In the early days of pearl hunting in American streams, the fishermen had little idea of their value, and sold choice gems for insignificant sums. In 1887, a fisherman on Rock River, Illinois, found a 40-grain pearl which he carried in his pocket for several months. Showing it one day in Davenport, he was offered $20 for it. He quickly accepted the offer, and on his return home told his friends about "the sucker who gave $20 for the shell slug." At present this "shell slug" is worth more than one hundred times that amount. Numerous instances of a similar nature occurred until the average fisherman lost all confidence in his judgment as to the values, and extravagant ideas prevailed regarding even almost worthless nacreous concretions. Thus, when a choice pearl is found, an exorbitant price is set upon it and the seller feels for the market value by repeated dickerings with several buyers. And unless one is an expert, he is quite likely to pay two or three times as much for a pearl at the river bank as in a metropolitan market. Some of the fishermen collect everything in the shape of nacreous concretions, and very often pearl buyers in New York and elsewhere receive packages which are not worth the postage; in many other packages nine tenths of the lot is worthless; and the practical joker and the swindler have solicited bids on bright marbles, rounded pieces of pearl shell, and even sugar-coated pills.

While many pearls of fine luster and beautiful and regular form have been derived from these fisheries, it occasionally happens, in the case of pearls consigned to the city pearl dealer, that cracks, breaks or marks, which might detract from their value, are closed or removed, either by means of water or oil, the pearls having been kept in

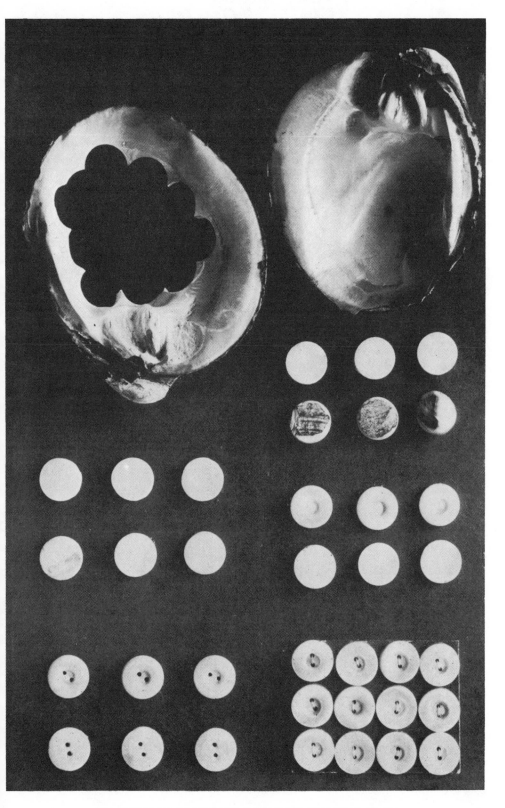

THE EVOLUTION OF BUTTONS, MADE FROM MISSISSIPPI SHELLS

NECKLACE OF FRESH-WATER PEARLS

Paris Exposition, 1900

one or the other until a few moments before they were shown to the merchant. Pearls worth hundreds of dollars have sometimes shown breaks, and in one instance a pearl valued at $7000 showed these cracks even a very short time after the sale.

In many of the pearling regions of the Mississippi Valley, inquiry of almost any fisherman will result in his bringing forth from an inside pocket a small box padded with raw cotton and containing an assortment of pearls and slugs. Most of the slugs he will sell at prices ranging from fifty cents to $5 per ounce, for several of the small pearls he will likely ask from $2 to $20 each, and one or two of the largest he may value at $50 or more. At very rare intervals, a choice pearl will be found, for which he may expect anywhere from $200 to $5000.

While the highest prices are not received by the fishermen, there are many who have been so fortunate as to obtain $1000 or more for a single pearl, and several have received double that amount. Probably the highest figure obtained by the original finder was $3800, notwithstanding exaggerated stories of enormous five-figure prices. Recently the press credited a lad sixteen years of age with securing $20,000 for a pearl he had found.

A particularly striking yarn relative to a so-called "Queen Mary" pearl went the rounds of the press some time ago. According to the newspaper report, this pearl was found by the wife of a fisherman who was a cripple or something equally pathetic, and, fortunately, when the family resources were at the lowest. With tears of joy, the fisherman embraced his wife and told her it was her very own and she should wear it. However, by means of a check for $17,500, he was induced to part with it, but only on condition that it be named Queen Mary in honor of the hard-working wife. The report continues that the original buyer sold it for $25,000, and at last accounts it was held by a Chicago dealer who had "refused $40,000 and probably would not accept $50,000 for it." The facts seem to be that this pearl, which was found near Prairie du Chien in 1901 and weighed 103 grains, was originally sold for $250, and the local buyer sold it in Chicago for $550, where for many months it was offered at $1000.

All sorts of stories of valuable finds are told in the pearling regions: stories of mortgages that have been released, of homes bought, of college educations secured from the proceeds of a single gem; but these tales are offset by the untold stories of the undermining of fine, strong character in awaiting the turn of fortune which never comes. The public is quickly apprised of the valuable finds, but it does not hear of the time and labor lost by the hundreds who are unsuccessful. Pearling excitement has many of the features of a mining craze. While a few are benefited, hundreds are made poorer, and in many instances reduced to absolute want. Persons have given up their established business to devote their time to pearling, staking all on the aleatory profits, and have squandered days and months in the hope that one great, immense, all-rewarding find will be made. The monotony of continued disappointment is occasionally brightened by the news that some one—possibly a near neighbor—has made a lucky find, and then the work is continued with renewed enthusiasm. A spirit akin to that which dominates the gambler takes possession of the fisherman, and the days go on and the seasons go by while the gem that is to bring the fortune still eludes him. In many localities the pursuit yields far less profit than pleasure, and many a man who spends a summer in pearling is in a fair way to spend the winter at the expense of some one else.

The pearls are collected for the trade by a score or more of buyers, who visit the fisheries at intervals and purchase of the individual fishermen by personal dickering and bargaining. The buyers endeavor to keep informed of all choice pearls discovered, and when an especially valuable find is reported each one endeavors to have the first chance to secure it. The principal local centers of the pearling industry and marketing are Prairie du Chien, Wisconsin; McGregor, Clinton, and Muscatine, Iowa; Newport, Black Rock, and Bald Knob, Arkansas; Clinton, Carthage, and Smithville, Tennessee; St. Francisville, Illinois; and Vincennes and Leavenworth, Indiana.

However, a large number of the pearls from American rivers are consigned by the finders to well-known gem dealers, the owners depending for fair treatment on the integrity and high standing of these experts. An interesting story is told of the pearl and the accompanying shell in which it was found, which was sent to a New York dealer by a poor woman. The price she received pleased her immensely; and in writing her appreciation, she added that she was especially gratified at receiving so good a price because it enabled her to send her boy to school. The dealer sent another check as a gift, and a few days before the next Thanksgiving Day a thirty-five-pound turkey was received by the four-score-year-old jeweler as an evidence of the mother's gratitude.

The outbreak of pearl hunting in various parts of the country is frequently chronicled by the newspapers. These despatches are much alike, usually telling how some fisherman discovered a beautiful pearl which he sold to some responsible jeweler for an amount varying from $100 to $2000. The despatches generally state further that the effect of the find has been remarkable; the whole region is seized with the fever, and into the rivers and creeks swarm the hunters of both sexes, of all ages, and from all classes of the community. Factory-men leave their mills, farmers their crops, and merchants their stores, and with the members of their families join in searching for the gems. The mussels are secured by whatever means is most convenient. If valuable finds continue, thousands and thousands of mollusks are destroyed in the search, and when the efforts begin to prove futile the excitement subsides almost as quickly as it began. In very many localities the industry has run the whole gamut of the feverish excitement of its beginning, the humor and romance of its existence, and the pathos of its ending.

If disturbed labor conditions at the height of the excitement were the only disagreeable attendant, these pearling furors could be viewed more favorably. But, unfortunately, in many localities, especially in shoal waters of restricted area, the fishery has been prosecuted so vigorously that it appears probable the resources will be very materially impoverished if not ruined in a few years, unless prompt and decisive protective measures are adopted. In some waters the crowds engaged in the search have removed practically every mussel without regard not only to protecting the immature mussels, but even to the necessity for preserving breeding mollusks. Many ponds and small river basins have been so denuded that not for many years, if ever, can they recover their former wealth of pearl-bearers.

This state of affairs has not come about without opposition on the part of those interested in the industry and the general welfare of the localities. Intelligent and well-directed efforts have been made to provide a system of regulations for protecting the mussels so that the maximum yield of pearls may be secured. But this is a very difficult problem to deal with. It involves not only the methods of fishery, but the question of sewage disposal by the cities and the large factories, through which great quantities of mussels have been destroyed.

Undoubtedly it will be difficult to devise regulations that will be satisfactory alike to the fishermen, the button manufacturers and the farmers. The great desideratum in the pearl fisheries—of the seas as well as in the fresh-water streams—is a restriction of the gathering to such mollusks and to such seasons and periods of years as produce the largest results with the least injury to the permanency of the resources.

It is generally agreed that the young or immature mollusks should be protected; but it is not easy to determine what is an immature Unio, as some species never grow large. Likewise, the beds should not be disturbed when the mollusks are loaded with young, but it is difficult to select particular months which would be better for close season than any others. The propositions which seem to be most actively advocated impose restrictions on the number and size of the mussels to be taken, a cessation of fishing from January 1 to May 31, closing certain areas when partly depleted, and prohibiting the use of especially injurious forms of apparatus. But whatever is done should be done without delay, before the pearl hunters and the button manufacturers kill the goose which for some years has been laying the golden eggs.[1]

MISCELLANEOUS PEARL FISHERIES OF AMERICA

The deep's wealth, coral, and pearl, and sand
Like spangling gold, and purple shells engraven
With mystic legends by no mortal hand.
SHELLEY, *The Revolt of Islam.*

THE beautiful pearls of the conch (*Strombus gigas*) are sought for in the West Indies and on the neighboring continental coasts. They are found most abundantly about the Bahamas, a group of more than four hundred islands off the Florida coast, where many of the fishermen devote a considerable portion of their time to collecting them. It is from this industry that the beach-combers of this group of islands, as well as those of the Florida reefs, have received the designation "Conchs."

Near the shores, where they formerly abounded, a few conchs are yet picked up by wading fishermen. In waters of medium depth they are secured either by diving or by means of a long pole with a hook at the end. In great depths, the mollusks are located by means of a water-glass similar to the type employed in the Red Sea or among the South Sea Islands.

The animal is readily removed from the shell after crushing the tip end of the spire where the large muscle is attached. The flesh forms an important article of food to the fishermen and to the residents of the outlying islands. It is said that a "Conch" can make a visit to Nassau of a week or ten days, and subsist almost entirely on this dried meat, with which he fills his pockets on starting. A large demand exists for the beautiful shells for ornamenting flower-beds, garden-walks, etc. Many of them are burned into lime for building purposes. Formerly several hundred thousand shells were exported annually to England for use in porcelain manufacture.

The pearls are generally found embedded in the flesh of the mollusk; quite often they are in a sac or cyst with an external opening, from which they are sometimes dislodged by the muscular movement of the

[1] Illinois State has passed a bill to regulate pearl fishing. See Addendum on p. 513.

animal. The yield is small, a thousand shells in many cases yielding only a very small number of seed-pearls or perhaps none at all. Most of them are oval, commonly somewhat elongated. The usual size is about one grain in weight, but some of them weigh over twenty, and a very few exceed fifty grains each. These pearls are generally of a deep pink color, shading toward whitish pink at each end. While this is the usual color, yellow, white, red, and even brown conch pearls are occasionally obtained; these are not so highly prized as the pink ones. Conch pearls present a peculiar wavy appearance and a sheen somewhat like watered silk, a result of the reflections produced by the fibrous stellated structure. While many are beautifully lustrous, they are commonly deficient in orient, and the color is somewhat evanescent.

Most of the Bahama conch fishermen sell their catch of pearls at Nassau. According to the late Mr. Frederick E. Stearns, there are in Nassau four dealers who have an arrangement with Paris and London houses, to whom they can ship pearls in any number and draw against them with a bill of lading. In addition to these, there are a dozen dealers in Nassau who buy what pearls they can secure and offer them for sale.

The value of conch pearls is as variable as their form, color, and size, and they are sold by the fishermen at prices ranging from twenty-five cents to one dollar or more each. Those weighing from three to ten grains, and of good color and luster, but not quite regular in form, sell for about $10 per grain; those of exceptional perfection in color and form, and of about the same weights, sell for from $15 to $30 per grain. In other and exceptional cases, where the size is very large, the form perfect, and the color and luster choice, the value is enhanced to several hundred and even several thousand dollars each. A perfect conch pearl is among the most rare and most valuable of gems. An unusually choice one has sold in New York City for more than $5000. The yield fluctuates considerably, but perhaps averages about $85,000 in value annually. One of the finest conch pearls ever found is shown on the plate with the conch shell.

There are two important materials that have occasionally been sold and mistaken for the conch pearl. First, the pale Italian, Japanese, or West Indian coral, with a color very closely approaching that of the pearl. By means of a lens it can readily be seen that the coral is in layers, and does not possess the concentric structure of the pearl, or the peculiar interwoven structure, with its characteristic sheen, so frequent in conch pearls. Secondly, the pink conch shell in which the pearl itself is found; this is frequently cut to imitate the pearl and sold as such in the West Indies and elsewhere. This can also be detected by the fact that the layers are almost horizontal and the structure is not concentric or interwoven, as it is in the conch pearl, while the luster is more like that of the shell than that of the pearly nacre.

Streeter relates that many years ago an ingenious American turned out some bits of conch shell into the shape of pearls and placed them in the conch shells. A slight secretion formed over them, but it was not the true pearly secretion, and the layer was very thin, so that the deception was easily detected.

Not the least interesting of the American pearl fisheries is that which has the abalones (Haliotis) for its object. These occur in many inshore tropical and semi-tropical waters, and particularly in the marginal waters of the Pacific. They attach themselves to the rocks by means of their large muscular disk-shaped foot, which acts like a sucker or an exhaust-cup.

On the California coast the abalones are gathered in large quantities for the pearls, for the shells, and especially for the flesh, which is dried and used for food. The principal fishing grounds are at Point Lobos in Monterey County, and along the shores of Catalina and Santa Rosa islands in Santa Barbara County, with smaller quantities from Half-moon Bay and from the rocks along the shores of Mendocino County. At low tide the fishermen wade out in shallow water, and, by means of a knife, separate the mollusk from its resting-place. Unless this is done quickly and before the mollusk has time to prepare itself for the attack, it closes down on the rock by means of its sucker-like foot, from which it cannot be removed without breaking the shell. A story is told at Santa Barbara of a Chinese fisherman having been drowned off one of the outer islands by having his hand caught underneath the shell of an abalone.

A few years ago, Japanese fishermen introduced the use of diving-suits in taking these mollusks in three fathoms of water; but in March, 1907, the California legislature interdicted this form of fishery. That legislature also interdicted the capture of black abalones measuring "less than twelve inches around the outer edge of the shell, or any other abalone, the shell of which shall measure less than fifteen inches around the outer edge."

The animal is removed from the shell by thrusting a thin blade cf soft steel between the flesh and the shell, and thus loosening the great muscle. The flesh is salted and boiled, and then strung on long rods to dry in the open air. When properly cured, the pieces are almost as hard and stiff as sole leather. Most of it is packed in sacks and exported to China, but large quantities are sold on the Pacific coast at from five to ten cents per pound. The catch is much less than it was forty years ago.

Many pearly masses are obtained from the abalones, and a few of

SHELL OF PEARL-BEARING ABALONE

From the coast of California

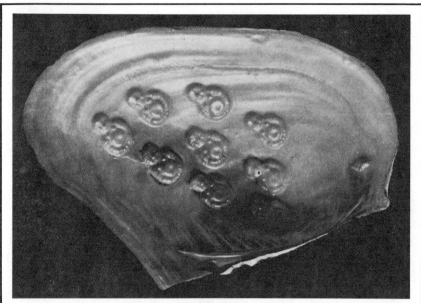

Shell of *Dipsas plicatus*, with attached metal figures of Buddha coated with nacre

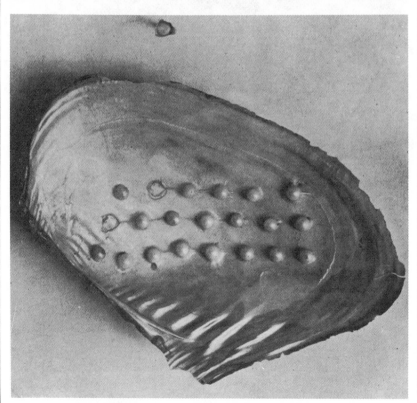

Shell of *Dipsas plicatus*, with attached porcelain beads coated with nacre

these are of considerable beauty. Some are very large, measuring two inches in length and half an inch or more in width; but they are rarely of good form, and their value is commonly far less than that of choice Oriental pearls. Owing to their irregularity in form, they are scarcely suitable for necklaces. One of the best necklaces of these pearls ever brought together sold a few years ago for $2000; but individual specimens have exceeded $1000 in market value. While abalone pearls are not on the market in any great quantities, one resident of Santa Barbara has a collection of more than a thousand specimens, ranging in value from several hundred dollars to less than one dollar each. Most of the objects sold in curio and jewelry stores on the Pacific coast as abalone pearls are simply irregular knots or protuberances cut from the surface of the shell. The California fishermen are credited with having received $3000 for the abalone pearls in 1904; but it is safe to say that this represents only a small fraction of their final sale value.

In the river mussels of Canada, and especially in those from the Province of Quebec, and the Ungava Region, pearls are occasionally found. These are usually white and of good luster. They are not the object of systematic search, but in the aggregate many are secured by Indians and Eskimos, and some by the trappers and fishermen who operate from Quebec and Montreal. A number, weighing from one to sixty-five grains each, were shown at the Colonial Exhibition in London in 1886, and received favorable notice. Recently, two beautifully matched pink pearls, weighing about fourteen grains each, were obtained from one mussel. A single pearl found in Canada has sold for $1000, but as a general rule they are of relatively little value. The Hudson Bay traders are represented as having secured a fair share of these pearls.

During the last few years, many pearls have been found in the streams of Prince Edward Island and of New Brunswick Province, and also in those of Nova Scotia. Most of them are well formed, but their color is generally inferior and their luster deficient. Many of them are buff or brown in color, some are bright and fairly good, a few are rose-tinted, and others are slate-colored and even almost black. Toronto jewelers report that many Canadian pearls are in the possession of farmers and others in the lower provinces, held by them for higher prices than the jewelers are willing to pay. The Nova Scotia pearls are from a bivalve which has been identified as *Alasmodon margaritifera*. They are especially abundant in Annapolis and King counties.

Even in the streams of northern Labrador and of the Caniapuscaw watershed, pearls are obtained by the natives, and by the hunters and fishermen who resort to that desolate country. These closely resemble the pearls of Scotland in color, size, and luster. A story is told of a fisherman who by chance found in one shell two well-matched pearls, which he later sold for $150; so pleased was he with his success that he spent a fortnight in diligent search, but secured only half a dozen small ones, worth perhaps $3 for the lot. Most of these pearls are silvery white, but beautiful pink ones are not rare. An unusually choice 20-grain pearl from this region sold in 1905 for $1000.

On the coast of Ecuador, pearl fisheries of minor importance have been prosecuted from time to time. Dr. H. M. Saville, of the American Museum of Natural History, states that in his explorations in that country he frequently came across evidence of pearls and the information that fisheries had existed on the coast centuries ago.

An interesting letter from that world-wide traveler and interesting writer, William E. Curtis, states that formerly there was a pearl fishery on the coast of Ecuador at the little town known as Manta, in the Province of Manabi; but it had to be abandoned on account of a particularly voracious species of fish called *el manti,* which abounds in that locality and gives the place its name. Pearls are said to be even more abundant at Manta than in Panama Bay. It is reported that this is the place where the Incas obtained those splendid gems which the Spaniards found in the palaces and temples of Peru.

In the waters of Costa Rica, pearl-oysters are found, and at times the fishery has been of considerable local importance. Owing to fear of injury to the reefs, the use of diving machinery was interdicted there a few years ago; but in 1906 its employment was authorized under certain restrictions. Licenses good for six months were authorized for a maximum of thirty machines, which may work at a minimum depth of thirty-seven feet.

On the coast of Colombia, South America, scattered reefs of pearl-oysters occur. A lease of the pearl fisheries and those for corals and sponges was granted July 2, 1906, but it is unknown what results have followed. This lease lasts five years, beginning August 1, 1906.

There is almost an absolute paucity of information in regard to the occurrence of fresh-water pearls in other parts of South America. The only data we have obtained are from Prof. Eugene Hussak of the Mining School of Sao Paulo, Brazil, who writes us that some pearls have been obtained from one of the Bahia rivers. Possibly, when the resources of the interior of that continent are better known, many pearls may be found.

XI

PEARL-CULTURE AND PEARL-FARMING

XI

PEARL-CULTURE AND PEARL-FARMING

> Some asked how pearls did grow, and where.
> Then spoke I to my girl,
> To part her lips, and show them there
> The quarelets of pearl.
> HERRICK, *The Quarrie of Pearls.*

THE great profit that would accrue from an increased output of pearls has long directed attention to the problem of bringing this about by artificial means.

In his life of Apollonius of Tyana, Philostratus, a Greek writer of the third century, repeats a story afloat at the time, which credited the Arabs of the Red Sea with possessing some method of growing pearls artificially. The story as it reached Greece was that they first poured oil upon the sea for the purpose of calming the waves, and then dived down and caused the oysters to open their shells. Having effected this, they pricked the flesh with a sharp instrument and received the liquor which flowed from the wounds into suitable molds, and this liquor there hardened into the shape, color, and consistence of the natural gems.[1]

While the description given by Philostratus is charged with many improbable details, and could scarcely develop belief, even in the most credulous, as to the exact method of procedure, it seems that the story may not have been wholly without foundation, and that attempts were made at that remote date to stimulate the growth of pearls.

In more modern times, the possibility of aiding or starting pearly formations in mollusks seems first to have been conceived by the Chinese about the fourteenth century. In 1736 there appeared in that storehouse of Oriental information, "Lettres édifiantes et curieuses écrites des missions étrangères,"[2] a communication from F. X. de Entrecolles, dated Pekin, 4th November, 1734, which set forth that there were people in China who busied themselves with growing pearls,

---

[1] Philostratus, "Vita Apollonii," *Lib.* III, c. 57, edit. Olearii, p. 139. Also see Konrad von Gessner, "Historiæ natura," *Lib.* IV, p. 634. [2] Vol. XXII, pp. 425-437.

and the product was not only vastly superior to the imitations manufactured in Europe, but were scarcely to be distinguished from the genuine. From Father Entrecolles's very detailed quotation of his unnamed Chinese authority, we condense this account. In a basin one half full of fresh water, place the largest mussels obtainable, set this basin in a secluded place where the dew may fall thereon, but where no female approaches, and neither the barking of dogs nor the crowing of chickens is to be heard. Pulverize some seed-pearls (*Yo tchu*), such as are commonly used in medicine, moisten this powder with juice expressed from leaves of a species of holly (*Che ta-kong lao*), and then roll the moistened powder into perfectly round pellets the size of a pea. These are permitted to dry under a moderate sunlight, and then are carefully inserted within the open shells of the mollusks. Each day for one hundred days the mussels are nourished with equal parts of powdered ginseng, china root, *peki,* which is a root more glutinous than isinglass, and of *pecho,* another medicinal root, all combined with honey and molded in the form of rice grains.

Although extremely detailed in some particulars, the Chinese account omits much to be desired as to the method in which the shells were opened to receive the pellets and the nourishment, and as to the importance of seclusion from females and loud noises. Admitting that it is "inaccurate and misleading," this letter seems to indicate very clearly that the Chinese had some method of assisting nature in growing pearls in river mussels.

The first person in Europe whose suggestion of the possibility of pearl-culture attracted general attention was Linnæus, the Swedish naturalist (1707–1778). In a letter to Von Haller, the Swiss anatomist, dated 13th September, 1748, he wrote: "At length I have ascertained the manner in which pearls originate and grow in shells; and in the course of five or six years I am able to produce, in any mother-of-pearl shell the size of one's hand, a pearl as large as the seed of the common vetch."[1] There was much secrecy about Linnæus's discovery, and even yet there is uncertainty as to the details of the method.

The Linnean Society of London apparently possesses some of the very pearls grown by Linnæus, as well as several manuscripts which throw much light on this subject. It appears from the latter that, under date of 6th February, 1761, Linnæus wrote that he "possessed the art" of impregnating mussels for pearl-production, and offered for a suitable reward from the state to publish the "secret" for the public use and benefit. A select committee of the state council of Sweden was appointed to confer with him, and on 27th July, 1761, the

[1] Pulteney, "General View of the Writings of Linnæus," London, 1805.

naturalist appeared and verbally explained his discovery. After various meetings, the select committee approved the "art" and recommended a compensation of 12,000 dalars (about $4800). It does not appear that the award was paid, and the following year the secret was purchased by Peter Bagge, a Gothenberg merchant, for the sum of 6000 dalars. On 7th September, 1762, King Adolph Frederick issued a grant to this merchant "to practice the art without interference or competition."[1]

Peter Bagge was unable to exercise the rights which he had acquired, nor was he able to dispose of them to advantage. On his death the memorandum of the secret became lost, and it was not found until about 1821, when it was discovered by a grandson, J. P. Bagge. Under the date of 27th February, 1822, the King of Sweden confirmed to this grandson the privileges which his ancestor had purchased in 1762. Fruitless efforts were again made to dispose profitably of the rights either to individuals or to the Swedish government.

The details of Linnæus's "secret" have never been published authoritatively. In his "History of Inventions," Beckmann states that before the naturalist thought of the profits that might accrue from his discovery, he intimated the process in the sixth edition of his "Systema naturæ," wherein he states: "Margarita testæ excrescentia latere interiore, dum exterius latus perforatur."[2] "I once told him," says Beckmann, "that I had discovered his secret in his own writings; he seemed to be displeased, made no inquiry as to the passage, and changed the discourse."[3]

In the second volume of his edition of "Linnæus's Correspondence,"[4] Sir J. E. Smith remarks: "Specimens of pearls so produced by art in the *Mya margaritifera* are in the Linnean cabinet. The shell appears to have been pierced by flexible wires, the ends of which perhaps remain therein." Referring to this remark, J. P. Bagge comments: "This is the nearest I have seen any one come to truth, but still it will be remarked by reading the 'secret' that more information is required to enable persons to practice the art."

After a thorough examination of the manuscripts and other material, Professor Herdman concludes that the essential points of Linnæus's process are to make a very small hole in the shell and insert a round pellet of limestone fixed at the end of a fine silver wire, the hole being near the end of the shell so as to interfere only slightly with the mollusk, and the nucleus being kept free from the interior of the

---

[1] "Proceedings of the Linnean Society of London," October, 1905, p. 26.

[2] Pearl: an excrescence on the inside of a shell when the outside has been perforated.

[3] Beckmann, "History of Inventions," London, 1846, Vol. I, p. 263.

[4] London, 1821, p. 48.

shell so that the resulting pearl may not become adherent to it by a deposit of nacre.[1]

Shortly after Linnæus communicated with the Swedish government and before his death, it was learned in Europe that the art of producing "culture pearls" by a somewhat similar process had been practised by the Chinese for centuries.[2] They used several forms of matrices or nuclei, but principally spheres of nacre and bits of flat metal or molded lead, which were not infrequently in conventional outline of Buddha. In the spring or early summer, these were introduced under the mantle of the living mollusk after the shell had been carefully opened a fraction of an inch, and the animal was then returned to the pond or lake. The mollusk did its work in a leisurely way, like some people who have little to do, and many months elapsed before it was ready for opening and the removal of the pearly objects.

The most satisfactory description we have seen of this process appears to be that communicated nearly a century later to the London Society of Arts by Dr. D. T. Macgowan,[3] through H. B. M. plenipotentiary in China, from which this account is abridged and modified.

The industry is prosecuted in two villages near the city of Titsin, in the northern part of the province of Che-kiang, a silk-producing region. In May or June large specimens of the fresh-water mussels, *Dipsas plicatus,* are brought in baskets from Lake Tai-hu, about thirty miles distant. For recuperation from the journey, they are immersed in fresh water for a few days in bamboo cages, and are then ready to receive the matrices.

These nuclei are of various forms and materials, the most common being spherical beads of nacre, pellets of mud moistened with juice of camphor seeds, and especially thin leaden images, generally of Buddha in the usual sitting posture. In introducing these objects, the shell is gently opened with a spatula of bamboo or of pearl shell, and the mantle of the mollusk is carefully separated from one surface of the shell with a metal probe. The foreign bodies are then successively introduced at the point of a bifurcated bamboo stick, and placed, commonly in two parallel rows, upon the inner surface of the shell; a sufficient number having been placed on one valve, the operation is repeated on the other. As soon as released, the animal closes its shell, thus keeping the matrices in place. The mussels are then deposited one by one in canals or streams, or in ponds connected therewith, five or six inches apart, and where the depth is from two to five feet under water.

[1] "Proceedings of the Linnean Society of London," October, 1905, p. 29.

[2] See Grill, Abhandlungen der königlichen Schwedischen Akademie der Wissenschaften auf das Jahr 1772," Leipzig, Vol. XXXIV, pp. 88-90.

[3] "Journal of the Society of Arts," Vol. II, pp. 72-75.

If taken up within a few days and examined, the nuclei will be found attached to the shell by a membranous secretion; later this appears to be impregnated with calcareous matter, and finally layers of nacre are deposited around each nucleus, the process being analagous to the formation of calculary concretions in animals of higher development. A ridge generally extends from one pearly tumor to another, connecting them all together. Each month several tubs of night soil are thrown into the reservoir for the nourishment of the animals. Great care is taken to keep goat excretia from the water, as it is highly detrimental to the mussels, preventing the secretion of good nacre or even killing them if the quantity be sufficient. Persons inexperienced in the management lose ten or fifteen per cent. by deaths; others lose virtually none in a whole season.

In November, the mussels are removed from the water and opened, and the pearly masses are detached by means of a knife. If the matrix be of nacre, this is not removed; but the earthen and the metallic matrices are cut away, melted resin or white sealing-wax poured into the cavity, and the orifice covered with a piece of shell. These pearly formations have some of the luster and beauty of true pearls, and are furnished at a rate so cheap as to be procurable by almost any one. Most of them are purchased by jewelers, who set them in various personal ornaments, and especially in decorations for the hair. Those formed in the image of Buddha are used largely for amulets as well as for ornaments. They are about half an inch long, and while in the shell have a bluish tint, which disappears with removal of the matrix. Quantities of them are sold as talismans to pilgrims at the Buddhist shrines about Pooto and Hang-chau.

In some shells the culture pearls are permitted to remain by the Chinese growers, for sale as curios or souvenirs; specimens of these have found their way into many public and private collections of Europe and America. These shells are generally about seven inches long and four or five inches broad, and contain a double or triple row of pearls or images, as many as twenty-five of the former and sixteen of the latter to each valve. That the animal should survive the introduction of so many irritating bodies, and in such a brief period secrete a covering of nacre over them all, is certainly a striking physiological fact. Indeed, some naturalists have expressed strong doubts as to its possibility, supposing the forms were made to adhere to the shell by some composition; but the examination of living specimens in different stages of growth, with both valves studded with them, has fully demonstrated its truth.

It is represented that in the northern part of the Che-kiang province about five thousand families are employed in this work in connection with rice-growing and silk-culture. To some of them it is the chief source of income, single families realizing as much as 300 silver dollars annually therefrom. In the village of Chung-kwan-o, the headquarters for culture pearls in China, a temple has been erected to the memory of the originator of this industry, Yu Shun Yang, who lived late in the thirteenth century, and was an ancestor of many persons now employed thereby.

The method in vogue in China for so many centuries has been the starting-point for similar attempts in various other countries. During the New Jersey pearling excitement in 1857, there were found several spherical pieces of nacre which had been introduced into Unios apparently for experimental pearl-culture; and in the collection of shells bequeathed to the United States National Museum by the late Dr. Isaac Lea, is a hemispherical piece of candle grease partly coated with pinkish nacre. Kelaart applied the Chinese method to the Ceylon pearl-oysters with much success in 1858. At the Berlin Fisheries Exhibition, in 1880, appeared the results of experiments in growing culture pearls in the river mussels in Saxony. Small foreign bodies had been introduced in the mantle, and others had been inserted between the mantle and the shell. These nuclei consisted of shell beads, unsightly pearls from other mussels, etc.; but unfortunately the shape of these was such that the mantle could not fit closely around them, consequently the result was so irregular as to be of no value except to show that German Unios as well as those of China could be made to cover foreign objects with pearly material.

Professor Herdman notes that, between 1751 and 1754, an inspector named Frederick Hedenberg received an annual salary "to inoculate the pearl mussels of Lulea (in the northern part of Sweden) with 'pearl-seeds' which he manufactured, and then to replant the mussels. Certain pearls were produced by the inspector, which it is recorded were sold for some 300 silver dollars."[1]

As noted by Broussonnet, in Finland artificial pearls were produced by inserting a round piece of nacre between the inner face of the shell and the mantle. The owner of the pearl fisheries at Vilshofen has succeeded in producing pearly figures by introducing into the mollusk flat figures of pewter, most of them representing fish in form.

In 1884, Bouchon-Brandely made experiments in pearl production at Tahiti. Gimlet holes about half an inch in diameter were drilled through different places in the shells of pearl-oysters, and through each of these holes a pellet of nacre or of glass was inserted and held by brass wire passing through a stopper of cork or burao wood, by means of which each opening was hermetically closed, so that the

[1] "Proceedings of the Linnean Society of London," October, 1905, p. 28.

Artificial rearing-ponds for the development of pearl-oysters on the Island of Espiritū Santo, Gulf of California

Trays containing small pearl-oysters prepared for placing at the bottom of artificial rearing-ponds at Espiritū Santo Island, Gulf of California

pellet was the only foreign substance protruding on the inside of the shell.[1] The oysters were returned to the sea without further injury, and after the lapse of a month the pellets were found covered with thin layers of nacre.

Experiments in growing pearls in the abalone or Haliotis were made in 1897 by Louis Bouton, an account of which was given at the meeting of the Paris Académie des Sciences in 1898.[2] The tenacity of life in this mollusk makes it especially desirable for experiments of this nature. Through small holes bored into the shell, pellets of mother-of-pearl were inserted and placed within the mantle, the small holes being afterward closed up. Other nacreous pellets were introduced directly into the bronchial cavity. The objects were soon covered with thin, pearly layers, resulting in a few months in spheres of much beauty, resembling somewhat the pearls naturally produced by this mollusk. In six months, according to M. Bouton, the layers became of sufficient thickness to be attractive. Within limitations, the size of the pearl produced is in proportion to the length of time it is allowed to remain within the mollusk. The results of the experiments seem to encourage further efforts in this line, and possibly in course of time there may be a profitable business in growing pearls in abalones on the Pacific coast of the United States. Indeed, the experiments in transplanting and cultivating the pearl-oyster in Australia leads one to fancy that the culture of that species in the warm coastal waters of America is by no means an impossibility.

Many other experiments along similar lines have been made more recently. An interesting feature of attempts made by Mr. Vane Simmonds of Cedar Rapids, Iowa, in 1896–1898, is that in order to avoid straining the adductor muscles by forcibly opening the shell while the mollusk resisted the intrusion, each selected Unio was exposed in the open air and sunshine until the valves opened; then a wooden wedge was carefully inserted in the opening, and the mollusk immediately immersed in water to revive it or to sustain life. After a few moments of immersion, the operator carefully raised the mantle from the shell, inserted the pellet of wax or other small article to be covered with nacre, drew the mantle to its normal position, removed the wedge, and returned the mollusk to a selected place in the stream at sufficient depth to avoid danger of freezing in winter.

Probably it would be more satisfactory to stupefy the mollusks by means of some chemical in order to insert the pellets. Marine mollusks have been successfully stupefied by slowly adding magnesium sulphate crystals to the sea water until the animals no longer respond

[1] "La Pêche et la Culture des Huîtres Perlières à Tahiti," Paris, 1885.

[2] "Comptes Rendus de l'Académie des Sciences," Vol. CXXVII, pp. 828-830.

to contact. If treatment is not too prolonged, they may be returned to normal sea water with good prospects of recovery. To stupefy fresh-water mollusks, either chloral hydrate or chlorosone may be employed, although the latter is expensive to use in great quantity. Dr. Charles B. Davenport, of the Carnegie Institution, suggests that it might be well to experiment with pouring ether or chloroform over them.

In Japan the production of these pearly formations in *Margarit-ifera martensi,* which is closely related to the Ceylon oyster, has developed into some prominence since 1890, and the results have been well advertised. The industry is located in Ago Bay, near the celebrated temple of Ise in the province of Shima, and gives employment to about one hundred persons. It is stated that the proprietor, Kokichi Mikimoto, has leased about one thousand acres of sea bottom, on which are a million oysters of this species, which yield from 30,000 to 50,000 culture pearls annually.

As described by Dr. K. Mitsukuri, the shoal portions of this area are used for breeding the oysters and raising them to maturity, and in the deeper parts—covered by several fathoms of water—the oysters are specially treated for producing the culture pearls. In the former, the spat is collected on small stones, weighing six or eight pounds each, placed during May or June. The following November these stones, with the attached spat or young, are removed, for protection from cold, to depths greater than five or six feet, where they remain for about three years. At the end of that period, the growing oysters are taken from the water, the shells opened slightly, and rounded bits of pearl shell or nacre are introduced under the mantle without injury to the mollusks. About 300,000 are thus treated annually, and placed in the deeper water at the rate of about one to each square foot of bottom area. After the lapse of about four years more, the oysters are removed from the water and opened, when a large percentage of the pellets are found covered on the upper or exposed surface with nacre of good luster.

Most of these culture pearls are button-shaped and weigh two or three grains each. Although somewhat attractive and superior to the culture pearls of China and other fresh waters, they by no means compare favorably with choice pearls. They are rarely, if ever, spherical, and only the upper surface is lustrous; consequently they serve only the purpose of half-pearls. A cross section shows the nacreous growth in a thin concentric layer, forming a fragile hemispherical cap, the concave wall of which is covered with a brownish granular secretion which prevents perfect adhesion. Compared with choice pearls, they are not only deficient in luster, but are fragile, and are beautiful only on the upper surface, and not available for necklaces. Good specimens sell for several dollars each, and some individuals reach $50 or more. Specimens exhibited at the Paris Exposition in 1900 were awarded a silver medal; at the St. Petersburg Exhibition in 1902 they were awarded a gold medal; at the Tokio Exhibition a grand prize, and a medal at the St. Louis Exposition in 1904. The awards were given in the fisheries, and not the gem divisions.

The work of Mikimoto is not the only attempt now being made in Japan to produce pearls. A letter from Dr. T. Nishikawa, of the Tokio Imperial University, states: "It is a great pleasure for me to tell you that I am studying pearl formation and pearl-oyster culture in the university laboratory, and recently I have got my pearl laboratory at Fukura, on the Island of Awaji, where I began the pearl culture work this summer (1907). Fortunately, I found the cause of Japanese pearl formation, *i.e.*, the reason why and how the pearl is produced in the tissue of an oyster. I made practical application of this theory with great prospects for producing the natural and true pearls at will."

Among the most interesting of the pearl-culture enterprises are those of the Compañia Criadora de Concha y Perla, under the direction of Sr. Gaston J. Vives, in the Gulf of California. This company has an extensive station at San Gabriel, near La Paz, where breeding oysters are placed in prepared chests or cages for collecting the spat on trays. After remaining there for several weeks or months, the young mollusks are removed to prepared places (*viveros*) for further growth. Experiments are now made in depositing them between a series of parallel dams alternately touching each shore of a lagoon, thus developing a current of water over the oysters for conveying food to them, and thus hastening their growth.

In efforts to increase the output of pearls, attention has been given to the possibilities for extending the area and production of the reefs, and for stocking new areas and replenishing exhausted ones, thus bringing the pearl-bearing mollusks to maturity in greater abundance.

Although theoretically it does not seem a very difficult undertaking to cultivate the pearl-oysters by methods somewhat similar to the cultivation of edible oysters and clams, in no part of the world has this been successfully done on an extensive scale. While in certain minor cases, the areas of some species of pearl mollusks have been extended indirectly through man's agency—as the range of the Red Sea pearl-oyster into the Mediterranean since the Suez Canal was opened—there is no well-known instance in which new areas have been abundantly populated through direct efforts.

In the chapter on the pearl fisheries of Asia are noted the hitherto unsuccessful efforts made in Ceylon and India to preserve the young and immature oysters on the storm-swept reefs by removing them to less exposed areas. This has received close attention from the Ceylon authorities during the last two years. Other practical measures which are recommended for that region include "cultching," or the deposit of suitable solid material, such as shells or broken stone, to which the young oysters can attach themselves; thinning out over-crowded reefs, and cleaning the beds by means of a dredge, thereby removing starfish and other injurious animals. The attempts made by individuals and associations to extend the range of the reefs on the coast of Australia, among the Tuamotu Islands, in the Gulf of California, and some other localities, are noted in the appropriate chapters. But it may be stated that in most instances lack of adequate police protection has been not the least of the difficulties with which these experiments have had to contend.

Nor has much greater success followed upon efforts to prevent the exhaustion of the reefs and productive grounds through overfishing, except in those instances in which the government exercises a proprietory interest and determines the season, the area to be fished, and the quantity of mollusks to be removed. The most prominent instance of this is in Ceylon, where the fishery has been restricted to such seasons and periods as appeared to insure the maximum yield of pearls. Without restriction upon the fishery, the pearl-oyster in that populous region would doubtless become almost extinct in a few years. Another instance of proprietory interest on the part of the government is in some of the German States, where pearl fishing has been regulated and restricted for centuries. But there the sewage from cities and factories has accomplished almost as effectively, if less rapidly, what unrestricted fishing would have done.

Much attention has been given to the subject of pearl-culture in Bavaria, where the government has granted a small subsidy to encourage this industry, and a model pearl-mussel bank has been established in one of the brooks for the rational culture of the mussels.

On the Australian coast, the only theoretical protection of consequence is the restriction on taking small or immature oysters; but, owing to the great area over which the fisheries are prosecuted there, it has not been possible to enforce the regulations. At some of the Pacific islands and elsewhere, interdictions exist as to use of certain apparatus of capture, but this is intended for the purpose of reserving the industry to dependent natives rather than for protecting the reefs. Several efforts have been made to insure adequate protection for the Unios in our American rivers, but nothing in this direction has yet been accomplished by legislative enactment, except in Illinois.

Reference has already been made to the parasitic stage of Unios.[1] The attachment of the newly-hatched mollusks to the gills or fins of a fish is entirely a matter of chance, and unless this takes place they die within a few days. Under natural conditions the fish thus infected will rarely be found carrying as many of the parasitic Unios as they can without serious injury. If the fish are placed in a tank or a pond containing large numbers of newly-hatched Unios, it is possible to bring about the attachment of hundreds of them for every one that would be found there by chance of nature. A fish six inches in length may thus be made to carry several hundred parasitic Unios, and thus a thousand fish artificially infected may do the work of several hundred thousand in a state of nature. Experiments with small numbers of fish under observation in the laboratory indicate that their infection on a large scale is entirely possible, and the experiment by Messrs. Lefevre and Curtis now in progress at La Crosse, Wisconsin, in which over 25,000 young fish have been infected, gives every indication that such work may be begun even with the scanty knowledge now possessed.

Since it has already been shown that the production of pearls is an abnormal condition, it does not follow that an increase in the quantity of mollusks would necessarily result in a corresponding increase in the yield of pearls. Indeed, it might even be that the artificial conditions bringing about an enhanced prosperity and abundance of the mollusks would result in a corresponding decrease in the product of gems, the improved surroundings impairing if not destroying the conditions to which the pearls owe their origin. This has resulted in directing efforts toward abnormally increasing the abundance of pearls in a definite number of mollusks.

The development of the parasitic theory of pearl formation has naturally invited attention to the possibilities of increasing the yield of pearls by inoculating healthy mollusks with distomid parasites. It does not appear that this has yet advanced beyond the experimental stage, and virtually all that has been accomplished has been set forth in the chapter on the origin of pearls. It seems that there are great possibilities in the artificial production along these lines; and that under skilful management it could be made a profitable industry, especially if carried on concurrently with the systematic cultivation of mother-of-pearl shells.

Although there is scientific basis for the belief that it may be possible in time to bring about pearl growth in this manner, the public should not be too hasty in financing companies soliciting capital for establishing so-called "pearl farms." Every once in a while announcement

[1] See p. 73.

is made in the public press of wonderful success which has been attained by some investigator, who surrounds his discovery with as much mystery as enveloped the Keeley motor, and who is as anxious to sell stock as was the owner of that mythical invention. A prospectus of one of these "pearl syndicates," which is now before us, claims to "increase and hasten pearl production by forcing the oyster, through doctoring the water in which it is immersed and also by irritating the mollusk itself." So far as the writers are aware, aside from the inexpensive but somewhat attractive culture pearls, no commercial success has yet followed the many attempts at artificial production.

This chapter should not close without reference to the so-called "breeding pearls," probably the most curious of all theories of pearl growth, regarding which many inquiries have been made. Throughout the Malay Archipelago there exists a generally accepted belief that if several selected pearls of good size are sealed in a box with a few grains of rice for nourishment they will increase in number as well as in size. If examined at the expiration of one year, small pearls may be found strewn about the bottom of the box, according to the theory; and in some instances the original pearls themselves will be found to have increased in size. If again inclosed for a further period of a year or more, the adherents of the theory say, the seed-pearls will further increase in size, and additional seed-pearls will form. Furthermore, the grains of rice will present the appearance of having been nibbled or as though a rodent had taken a bite in the end of each.

It is claimed that the breeding pearls are obtained from several species of mollusks, mostly from the Margaritifera, but also from the Tridacna (giant clam) and the Placuna (window shell). While cotton is the usual medium in which the pearls and rice are retained, some collectors substitute fresh water and yet others prefer salt water. It seems that rice is considered essential to success.

The earliest account we have seen of this extraordinary belief was given by Dr. Engelbert Kæmpfer,[1] who was connected with the Dutch embassy to Japan from 1690 to 1696, and since that time it has been referred to by many travelers in the Malay Archipelago.

A correspondent in the time-honored "Notes and Queries," 20th September, 1862, writes:

Nearly five years ago, while staying with friends in Pulo Penang (Straits of Malacca), I was shown by the wife of a prominent merchant five small pearls, which had increased and multiplied in her possession. She had set them aside for about 12 months in a small wooden box, packed in soft cotton and with half a dozen grains of common rice. On opening the box at the expiration of that time, she found four additional pearls, about the size of a

[1] Kæmpfer, "History of Japan," London, 1728, Vol. I, pp. 110-112.

small pinhead and of much beauty, which I saw and examined not long after the lady made the discovery. While my story may be received with laughter, I can most solemnly assure you of the truth of my having seen these pearls, and I have not the slightest doubt of the perfect truthfulness of the lady who possessed them. I questioned an eminent Malay merchant of Penang on this subject, and he assured me that one of his daughters had once possessed a similar growth of pearls. [1]

Notwithstanding the apparent absurdity of this pearl-breeding theory, belief in it appears to be not only sincere but wide-spread, as can be attested by any one familiar with affairs in the archipelago. A critical examination into the matter was made in 1877 by Dr. N. B. Dennys, curator of the Raffles Museum at Singapore, the result of which was communicated to the Straits branch of the Royal Asiatic Society, 28th February, 1878.[2] From his numerous quotations of persons who gave the results of their experiences we extract two instances. One gentleman had 120 small pearls in addition to the five breeding ones with which the experiment had started twenty years before, and during the entire period the box had not been molested except that it was opened occasionally for inspection by interested persons. Another experimentor inclosed three breeding pearls with a few grains of rice on 17th July, 1874; on opening the box on 14th July, 1875, nine additional pearls were discovered, and the three original ones appeared larger.

The belief has many curious variations. It is stated that in Borneo and the adjacent islands, many of the fishermen reserve every ninth pearl regardless of its size, and put the collection in a small bottle which is kept corked with a dead man's finger. According to Professor Kimmerly, nearly every burial-place along the Borneo coast has been desecrated in searching for "corks" for these bottles, and almost every hut has its dead-finger bottle, with from ten to fifty "breeding pearls" and twice that number of rice grains.[3] A correspondent at Sandakan, North Borneo, writes that at the time of his death at Hongkong in 1901, Dr. Dennys had in his possession a small box containing "breeding pearls"; but these disappeared after his death, and his brother, the crown solicitor, was unable to find them. This correspondent also states that the Ranee of Sarawak, a British protectorate in western Borneo, has a collection of "breeding pearls" numbering about two hundred, and that this is the only large collection known at present.

[1] "Notes and Queries," 3rd Series, Vol. II, p. 228.
[2] "Journal of the Straits Branch of the Royal Asiatic Society," Singapore, 1878, Vol. I, pp. 31-37.
[3] "Jewelers' Review," May 10, 1892.

As contrasted with abundant and unquestionably sincere testimony that pearls do "breed," it may be stated that absolutely no result has followed one or two native experiments made under supervision. While it must be admitted that negative evidence is always weaker than positive, and twenty failures would be outweighed by one successful experiment, yet the scientific objections to the possibility of pearls "breeding" cannot be overcome. The phenomenon is doubtless one of those curiosities of natural history in which some important factor has been overlooked.

Another curious theory is that peculiar pearls continue to grow after removal from the mollusk in which they originate. Quite recently it was reported from New Durham, North Carolina, that a pearl found there in 1896 had been growing continually since it was found and removed from the water. Unfortunately, it was weighed only when the last observation was made, and its increased size doubtless existed only in the imagination of its possessor.

XII

MYSTICAL AND MEDICINAL PROPERTIES OF PEARLS

MYSTICAL AND MEDICINAL PROPERTIES OF PEARLS

> Divers are the virtues of gems; some give favor in the sight of
> lords; some protect against fire; others make people beloved; others
> give wisdom; some render men invisible; others repel lightning;
> some baffle poisons; some protect and augment treasures, and others
> cause that husbands should love their wives.
>
> Arabic version of Solomon's writings.

WHILE no special gems are mentioned in the tribute which the Arabs credit to Solomon, it seems that pearls must certainly have been included, for in nearly all countries where these gems have been prized and from the earliest period, they have been credited with mystic properties and healing virtues.

In the first chapter of this book, reference was made to the Atharva-veda, dating from at least 2400 years ago, and its allusion to the use of an amulet of pearl shell and of pearls among the Hindus in bestowing long life and prosperity upon young Brahmanical disciples. As this amulet is fastened upon the youth, the following hymn is recited, according to this ancient Veda of the Atharvans:

Born of the wind, the atmosphere, the lightning, and the light, may this pearl shell, born of gold, protect us from straits!

With the shell which was born in the sea, at the head of bright substances, we slay the Rakshas and conquer the Atrins [devouring demons].

With the shell [we conquer] disease and poverty; with the shell, too, the Sadanvas. The shell is our universal remedy; the pearl shall protect us from straits!

Born in the heavens, born in the sea, brought on from the river [Sindhu], this shell, born of gold, is our life-prolonging amulet.

The amulet, born from the sea, a sun, born from Vritra [the cloud], shall on all sides protect us from the missiles of the gods and the Asuras!

Thou art one of the golden substances, thou art born from Soma [the moon]. Thou art sightly on the chariot, thou art brilliant on the quiver. (May it prolong our lives!)

The bone of the gods turned into pearl; that, animated, dwells in the waters. That do I fasten upon thee unto life, luster, strength, longevity, unto a life lasting a hundred autumns. May the amulet of pearl protect thee![1]

The mystical Taoists, in their pursuit of immortality, made much of pearls as an important ingredient in formulæ for perpetuating youth. According to an old Taoist authority, in preparing one of these elixirs, an extra long pearl which has been worn for many years is steeped in some infusion of malt, or a preparation of serpents' gall, honeycomb, and pumice-stone. When the pearl becomes plastic, it is drawn out to the length of two or three feet, cut into suitable lengths, and formed into pills, the taking of which renders food thenceforth unnecessary.[2]

The myth of the dragon and the pearl has been a far-reaching theme of the artists in Japan and China, whether in color, metal, or stone. There has been much written as to how the myth became so fixed in the minds of the Orientals, and Prince Rupprecht of Bavaria, who has made an exhaustive study of the myth of the dragon in all its phases, has very courteously communicated to us the following facts. Personally he had never been able to learn of a true or clear description of the origin of the myth other than the well-recorded legend given by Legge in the "Sacred Books of the East" (Vol. XL, p. 211), in which there is a quotation from Shuangtze, a writer of the fourth century before Christ, who says: "Near the Ho river there was a poor man, who supported his family by weaving rushes. His son, when diving in a deep pool, found a pearl worth a thousand ounces of silver. The father said: 'Bring a stone and beat it in pieces. A pearl of this value must have been in a pool nine khung deep and under the chin of the black dragon. That you were able to get it must have been owing to your having found him asleep. Let him awake, and the consequences will not be small.'" Prince Rupprecht says:

This legend has nothing to do with the illustration to which you refer; it belongs to a cycle of myths concerning a stone in the head of a serpent, or the crown of the king of the serpents or dragons; myths which also exist in Germany since the days of old. I should rather be inclined to think that the commonly accepted pearl between the two dragons is not a pearl at all. At least this pearl is always surrounded by ornaments in the shape of flames or claws, and Professor Hirth discovered on such a representation in woodcut, an explanation of the flames by the sign for Yangsui, a very ancient kind of metallic mirrors, of concave form, that were used to produce the heavenly fire.

[1] Bloomfield, "Hymns of the Atharvaveda," Oxford, 1897, p. 62.

[2] Macgowan, "Journal of the Society of Arts," Vol. II, p. 73.

JAPANESE LEGEND OF THE DRAGON AND THE PEARL, IDEALIZED IN JADE

Heber R. Bishop Collection, Metropolitan Museum of Art

RUSSIAN EIKON OF THE MADONNA

Ornamented with pearls

This explanation is probably erroneous and due to a misunderstanding of the signs for flames. In my opinion, another explanation, that the pearl is not really a pearl but a spider, is nearer to the truth. As an argument in favor of this theory the following sentence may be quoted from an encyclopedia of the eleventh century ("Pieu-tzi-lei," chap. 223): "The pearl of a fish is its eye, the pearl of a tortoise is its foot, the pearl of the spider is its belly." Pearl, as well as spider, are both called in Chinese by the same word but are written in a different way.

I, for my part, believe that the pearl is the belly not indeed of a spider, but of Garuda, the eagle of Vishnu, known in the old Hindu mythology as the foe of the Vagas, beings with human bodies and the tails of serpents. At least, I found on an old Chinese gateway, dating back to the times of the Mongol emperors, a sculpture showing the contest between Garuda and the Vagas. On another sculpture of the late King epoch the Vagas are already changed into dragons, and the wings, the limbs and the head of Garuda have become quite insignificant, while his belly is prominent like a ball.

A beautiful metaphor occurs in ancient Chinese writings, in the Book of the Later Han,[1] for instance, which regards this gem as the hidden soul of the oyster.

There is no end of legends and myths regarding the pearl in oriental literature. One fable credits it with a peculiar magical power: by speaking the right word, a spirit can be called therefrom which makes the owner a possessor of all the happiness of the earth. Browning notes this in two exquisite stanzas, "A Pearl, a Girl," published on the day of his death in 1889, in which he compares this characteristic with a woman's love called forth by the mystic word.

> A simple ring with a single stone,
>   To the vulgar eye no stone of price;
> Whisper the right word, that alone—
>   Forth starts a sprite, like fire from ice,
> And lo, you are lord (says an Eastern scroll)
> Of heaven and earth, lord whole and sole,
>   Through the power in a pearl.
>
> A woman ('t is I this time that say)
>   With little the world counts worthy praise;
> Utter the true word—out and away
>   Escapes her soul: I am wrapt in blaze,
> Creation's lord, of heaven and earth,
> Lord whole and sole—by a minute's birth—
>   Through the love in a girl.

[1] Pfirzmaier, "Kaiserliche Akademie der Wissenschaften," Wien, 1868, Vol. LVII, p. 623.

In the folk-song of Servia is a pretty little poem which testifies to the love they bear to pearls:

A youth unmated prays to God,
To turn him to pearls in the sea,
Where the maidens come to fill their urns;
That so they might gather him into their laps,
And string him on a fine green thread,
And wear him pendant from the neck;
That he might hear what each one said,
And whether his loved one spoke of him.

His prayer was granted and he lay
Turned to pearls in the dark blue sea,
Where the maidens come to fill their urns;
Then quickly they gather him into their laps,
And string him on a green silk thread,
And wear him pendant from the neck;
So he hears what each one says of her own
And what his loved one says of him.[1]

In the days when romance and chivalry held sway in Europe, pearls and other favors were presented by ladies for the brave knights to wear at tournaments. And we are told in the Arthurian legends how Elaine, "the lily maid of Astolat," gave to Sir Lancelot "a red sleeve of scarlet, embroidered with great pearls," for him to wear on his helmet: and "then to her tower she climbed and took the shield, there kept it and so lived in fantasy"; while he fought and won at the tilt, "wearing her scarlet sleeve, tho' carved and cut, and half the pearls away."[2]

The sweet sentiment of purity associated with the pearl ennobles it above all other gems. Rabanus Maurus, archbishop of Mainz, wrote, about 850, that "mystically, the pearl signifies the hope of the Kingdom of Heaven, or charity and the sweetness of celestial life."[3] True, it was not among the twelve gems which adorned the breastplate of the high priest of the Temple, symbolical of the twelve apostles. A Father of the Church—St. Augustine, we believe—explains this by saying that it was reserved for a more sacred office, that of representing Christ himself.

Pearl signifies purity, innocence, humility, and a retiring spirit. All stones of the gray color of the pearl have the significances which are given to this beautiful gem.[4]

---

[1] Translated from Klenn, "Culturgeschichte," Leipzig, 1852, Vol. X, p. 318.
[2] Tennyson, "Idylls of the King."
[3] "Opera omnia," Paris, 1864, Vol. V, p. 473.
[4] W. & G. Audsley, "Handbook of Christian Symbolism," London, 1865, p. 140.

Unlike other gems, the pearl comes to us perfect and beautiful, direct from the hand of nature. Other precious stones receive careful treatment from the lapidary, and owe much to his art. The pearl, however, owes nothing to man. Perhaps this has much to do with the sentiments we cherish for it. It touches us with the same sense of simplicity and sweetness as the mountain daisy or the wild rose. It is absolutely a gift of nature, on which man cannot improve. We turn from the brilliant, dazzling ornament of diamonds or emeralds to a necklace of pearls with a sense of relief, and the eye rests upon it with quiet, satisfied repose and is delighted with its modest splendor, its soft gleam, borrowed from its home in the depths of the sea. It seems truly to typify steady and abiding affection, which needs no accessory or adornment to make it more attractive. And there is a purity and sweetness about it which makes it especially suitable for the maiden.

The idea of pearly purity is inseparably linked with the name Margaret, derived from the Persian *Murwari* (pearl, or child of light) through the Greek μαργαρίτης. This name—beautiful in sound as well as in origin—is popular in all European countries, and likewise are its abbreviations and diminutives: in Italian, Margherita and Rita; in French, Marguerite, Margot, and Groten; in German, Margarethe, Gretchen, and Grethel; and in English, Margaret, Marjorie, Madge, Maggie, Peggy, etc.

The use of the word as a proper name among the early Christians was doubtless suggested by the sweet simplicity and loveliness of the pearl, and by the beautiful symbolical references to this gem in the Scriptures; and the meaning of the name has been strengthened by the pure lives and the good deeds of the many beautiful Margarets in all lands, including the virgin martyr, St. Margaret of Antioch, "the mild maid of God" referred to in the Liturgy, who, before the fifth century, was the embodiment of feminine innocence and faith overcoming evil, and who is often represented wearing a string of pearls; also St. Margaret Ætheling of the eleventh century, who endeared the name in Scotland, was canonized in 1215, and was adopted as the patron saint of Scotland in 1673; and Margaret, "Pearl of Bohemia," so beloved by the Danes.

Especially among the Germans has the name a tender significance; with them it is symbolical of maidenly sweetness and purity associated with richness of womanhood, such as was typified by Goethe in the heroine of his "Faust." This idea may have impelled Wordsworth in the selection of a name for the lovely, girlish character in his "Excursion"; and Tennyson for his "Sweet pale Margaret," and likewise Scott for "Ladye Margaret, the flower of Teviot." With the memory of these lives and characters before her, many a loving mother has crystallized the hope entertained for a baby daughter by enriching her with this beautiful name.

Poets seem never to tire of using the pearl as a symbol of perfection in form, in purity, in luster, and in sweetness. But probably none has made a more lovely comparison than Owen Meredith:

> As pure as a pearl,
> And as perfect: a noble and innocent girl.[1]

The Oriental poets unite with those of the West in their love for this gem, and those gifted writers are lavish in its use. Let us but add the lament of Shabl Abdullah on the death of Nozami:

> Nozami 's gone, our fairest pearl is lost.
> From purest dew, kind Heaven had given her birth,
> And then had fashioned her the pearl supreme.
> She softly shone, but hidden from mankind,
> So God has now restored her to her shell.

Far more crude, but possibly equally pathetic, is that old epitaph from Yorkshire, England:

> In shells and gold pearls are not kept alone,
> A Margaret here lies beneath a stone.

In the seventeenth century, Pierre de Rosnel wrote in a burst of enthusiasm:

The pearl is a jewel so perfect that its excellent beauty demands the love and esteem of the whole universe. Suidas expresses himself in regard to it thus: "The possession of the pearl is one of love's greatest delights; the delight of possessing it suffices to feed love." In a painting, Philostratus, who had the same ideas, has represented cupids with bows enriched with pearls; and the ancients were all agreed to dedicate the pearl to Venus. Now, to my thinking, the reason for their so doing was, that inasmuch as this goddess of love, the fairest of all divinities, is descended from heaven and is formed of the sea, so in like manner the pearl—the loveliest of all gems—is formed in the sea and is the offspring of the dew of heaven. But he that would learn more of the excellence of the pearl, let him inquire of the ladies, who will relate much more in its praise than I can write, and who will doubtless confess that nothing else so well adorns them.[2]

Emblematic as the pearl is of maidenly purity and sweetness, it is deemed especially appropriate as a wedding gift. This use dates

---

[1] "Lucile," Pt. II, c. 6, st. 16.

[2] "Le Mercure Indien, ou le Trésor des Indes," Paris, 1672, p. 160.

from the earliest dawn of Hindu civilization, when the beloved Krishna drew it from the sea to decorate his beautiful daughter on her nuptial day. And among the Hindus not uncommonly the presentation of a virgin pearl and its piercing forms part of the marriage ceremony. In most of the European royal weddings in recent years, pearls have been prominent among the bridal gifts; nor have they been overlooked among the presents to American brides, including one much in the public print about 1906, for whom a necklace of them was selected by a neighboring republic as an appropriate present.

The dedication of the pearl to love and marriage appears to have been recognized by the artistic Greeks. One of the choicest engravings preserved from classic times is a magnificent sardonyx showing the marriage of Cupid and Psyche, in which the lovers are united by what some authorities consider a string of pearls—emblematic of conjugal bonds—by means of which the god Hymen leads them to the nuptial couch.[1] This engraved gem now forms one of the choicest objects in the Boston Museum of Fine Arts, having been purchased at the sale of the Marlborough gems, London, 1898, at a cost of about $10,000.

And yet in Western countries the ill omen of pearls as bridal ornaments has been widely recognized, these determining the tears that will be shed in the married life. As Milton says, referring to the Marchioness of Winchester:

> And those pearls of dew she wears,
> Prove to be presaging tears.

It was told that when the Empress Eugénie of France was finishing her toilet preparatory to her wedding in Notre Dame in 1853, a personal attendant reminded her of the omen, and begged that she refrain from wearing her pearl necklace on that occasion. Eugénie paid no heed to the warning and wore the beautiful jewels just the same; and, as all the world knows, her life has been one long tragedy. Since that necklace was a lengthy one, containing very many pearls, the bride who wears only a few on her wedding day need not dread the adage so much, for, unfortunately, no woman's life is wholly free from grief; and most brides would doubtless risk a few tears rather than refuse to wear a wedding gift of pearls.

It was a very old idea that to dream of pearls betokens tears. A suggestion of this occurs in John Webster's "Duchess of Malfi" (1623), Act III, sc. 5:

---

[1] Many Greek scholars maintain that this is a knotted cord or fillet; but this view is contested by others.

> *Duchess:* I had a very strange dream to-night;
> Methought I wore my coronet of state,
> And on a sudaine all the diamonds
> Were chang'd to pearles.
>
> *Antonio:* My interpretation
> Is, you 'll weepe shortly;
> For to me the pearles
> Doe signifie your teares.

And we quote also from "The Parson's Wedding" (1663), Act II, sc. 5, where Jolly exclaims: "What! in thy dumps, brother? The captain sad! 'T is prophetic. I 'd as lieve have dreamt of pearl, or the loss of my teeth."

Tradition relates that Queen Margaret Tudor, wife of James IV of Scotland, just before the battle of Flodden Field (1513), had many fears as to the disastrous issue of that conflict, owing to having dreamed on three nights in succession that all her jewels were suddenly turned into pearls. This was interpreted as a sign of coming widowhood and sorrow, which was soon verified; and a similar story is told of Marie de' Medici shortly before the murder of Henry IV of France in 1610.

The employment of pearls medicinally dates from an ancient period. This use is mentioned in the oldest existing Sanskrit medical work, the "Charaka-Samhita,"[1] composed early in the Christian era; and likewise in the somewhat more modern "Susruta,"[2] which probably originated before the eighth century.

It is particularly in Oriental countries that therapeutic properties have been credited to pearls. The powder of these gems has been rated very highly there, and is still used to some extent. It was considered beneficial in cases of ague, indigestion, and hemorrhages, and was regarded as possessing stimulative qualities. Medical literature of the Orient contains many accounts of the uses of pearls and of the methods of forming them into pills, ointments, etc.

According to a treatise written by Narahari, a physician of Kashmir, about 1240 A.D., the pearl cures diseases of the eyes, is an antidote to poisons, cures consumption and morbid disturbances, and increases strength and general health.[3]

In China, as well as in other Asiatic countries, a distinction was made in the therapeutic effects of so-called "virgin" pearls and of

---

[1] Edited by Jibananda Vidyasagara, Calcutta, 1877.
[2] Edited by Vidyasagara, 1873.
[3] Garbe, "Die Indischen Mineralien." Narahari's "Raganighantu," Varga XIII, Leipzig, 1882, p. 74.

those pierced or bored for stringing. The Chinese natural history of Li Shi Chin, completed about 1596, states that bored pearls will not serve for medicine, for which unpierced ones should be used. It further adds that the taste is saltish, sweetish, and cold; and that they benefit the liver, clear the eyes, and cure deafness. Dr. T. Nishikawa informs us that at the present time many Mytilus seed-pearls are exported from Japan to China for medicinal purposes.

Quoting principally from Ahmed Teifashi, Whitelow Ainslie wrote in 1825 that Arabian physicians suppose the powder of the pearl to have virtues in weak eyes; and they credit it with efficacy in palpitations, nervous tremors, melancholia, and hemorrhage. Also they have this strange notion, that when applied externally and while in the shell, it cures leprosy.[1]

Statements of the curative properties of pearls come also from Japan at a somewhat recent date. The catalogue of the National Exhibition at Yedo in 1877, Part V, page 78, notes that they soothe the heart, lessen phlegm, are an antidote to poison, and cure fever, smallpox, and blear-eyedness.

The popular modern idea in India as to the therapeutic value was thus expressed by a native prince, Sourindro Mohun Tagore, Mus. Doc., the Maharajah of Tagore, in 1881:

The use of pearls conduces to contentment of mind and to strength of body and soul. The burnt powder of this gem, if taken with water as sherbet, cures vomiting of blood of all kinds. It prevents evil spirits working mischief in the minds of men, takes off bad smell from the mouth, cures lunacy of all descriptions and all mental diseases, jaundice and all diseases of the heart, intestines and stomach. Burnt pearl mixed with water and taken into the nostrils, as a powder, takes away headsickness, cures cataract, lachryma and swelling of the eyes, the painful sensation such as is caused by the entry of sand into them, and ulcers. Used as a dentifrice, it strengthens the gums and cleanses the teeth. Rubbed on the body with other medicines, it cures all skin diseases. It stops bleeding from cuts and ultimately heals them up. Whether taken internally or externally, it is a sure antidote to poison. It drives away all imaginary fears and removes all bodily pain. To prevent its tendency to affect the brain, it should always be used with the burnt powder of basud, and in its absence with that of white mother-of-pearl. The dose of pearl powder should not exceed 2¼ mashas [19.68 grs.].[2]

The Hindus credited specific virtues to pearls of different colors: the yellow brought wealth, the honey shade fostered understanding, the white attracted fame, and the blue, good luck. Defective pearls caused leprosy, loss of fortune, disgrace, insanity, and death, according

[1] Ainslie, "Materia Indica," London, 1826, Vol. I, p. 292.   [2] "Mani-málá," Calcutta, 1881, p. 871.

to the degree of defect. The "Mani-málá," previously quoted, states that "pearls possessed of every valuable quality shield their master from every evil, and suffer nothing harmful to come near him. The house which contains a perfect pearl the ever-restless Lakshmi (goddess of wealth) chooses to make her dwelling for ever and a day."[1]

A similar idea is expressed in an old Hindu treatise on gems by Buddhabhaṭṭa, where we read: "The pearl from the shell ought always to be worn as an amulet by those who desire prosperity.[2]

Pearls still find a place in the pharmacopœia of India. One of the latest standard works, that of R. N. Khory and N. N. Katrak,[3] credits the powder as a stimulant, tonic, and aphrodisiac. It is one of the ingredients in numerous Indian prescriptions used in curing impotence, heart-disease, consumption, etc. According to these authorities, the dose is from one fourth to one half grain of the powdered pearl.

Owing to the high cost of sea pearls, even those of the smallest size, a substitute for medicinal and similar purposes is found in the Placuna pearls of Ceylon, Borneo, etc. These are of such slight luster that only the choicest are of ornamental value, consequently they are sold at relatively small prices. A considerable demand exists for them to be placed in the mouths of deceased Hindus of the middle class, instead of the sea pearls which are used by the wealthy, or the rice which is employed in a similar manner by persons of poorer rank. This custom seems to be analogous to that of the ancient Britons, and also to that of the American Indians, in depositing food and other requisites for a journey in burial graves. The practice is an old one in India and was noted by Marco Polo more than six hundred years ago.

Most of the Placuna pearls are calcined and are used with areca-nuts and betel-pepper leaves in a very popular masticatory, one of the "seven sisters of sleep," which is to the Hindu what opium is to the Chinaman, or tobacco to the American or European. The hard white areca-nut (*Areca Catechu*) is about the size and shape of a hen's egg. Three or four thousand tons of the small, tender nuts are annually shipped from Ceylon to India for this masticatory, which is chewed by a hundred million persons. After boiling in water, pellets of them are placed in a leaf of the betel-pepper (*Piper betle*) with a small quantity of lime made from pearls or shells, according to the desired quality and value of product. It is credited with hardening the gums, sweetening the breath, aiding digestion, and stimulating the nervous system like coffee or tobacco; its most visible effect is tingeing the saliva and blackening the teeth, which is far from attractive, especially in an otherwise beau-

[1] "Maṇi-málá," Calcutta, 1879, p. 315.
[2] Finot, "Les Lapidaires Indiens," Paris, 1896, p. 15.
[3] "Materia medica of India and their Therapeutics," Bombay, 1903, p. 98.

tiful woman.   A more recent use for these Placuna pearls is as an ingredient in a proprietary face powder and enamel, which is marketed in Europe.

It is not alone the Orientals that have found medicinal virtues in pearls.   Even in Europe they have occupied a prominent place in materia medica, especially during the Middle Ages when a knowledge of the occult properties of gems was an important branch of learning.   Indeed, they could scarcely have been overlooked by people who at one time or another swallowed pretty much everything, from dried snake's eyes to the filings of a murderer's irons, in their quest for the unusual and costly with which to relieve and comfort themselves.   During the Middle Ages in Europe, writers who gave attention to pearls, as well as to other gems, treated almost exclusively of their reputed efficacy in magic and in medicine; and most of the accounts from the ninth to the fourteenth century seem wholly without scientific value, and at times reach the climax of extravagance and absurdity in their claims for the wonderful potency of the gem.

Albertus Magnus, the Dominican scholar born in Germany in the twelfth century, wrote that pearls were used in mental diseases, in affections of the heart, in hemorrhages, and dysentery.[1]

The "Lapidario" of Alfonso X of Castile (1221–1284), called "The Wise," the father of the Spanish language, states:

The pearl is most excellent in the medicinal art, for it is of great help in palpitation of the heart, and for those who are sad or timid, and in every sickness which is caused by melancholia, because it purifies the blood, clears it and removes all its impurities.   Therefore, the physicians put them in their medicine and lectuaries, with which they cure these infirmities, and give them to be swallowed. They also make powders of them, which are applied to the eyes; because they clear the sight wonderfully, strengthen the nerves and dry up the moisture which enters the eyes.[2]

Anselmus de Boot, physician to Emperor Rudolph II, and one of the great authorities at the beginning of the seventeenth century, gave the following directions for making *"aqua perlata,* which is most excellent for restoring the strength and almost for resuscitating the dead.   Dissolve the pearls in strong vinegar, or better in lemon juice, or in spirits of vitriol or sulphur, until they become liquified; fresh juice is then added and the first decanted.   Then, to the milky and turbid solution, add enough sugar to sweeten it.   If there be four ounces of this solution, add an ounce each of rose-water, of tincture of strawberries, of borage flowers and of balm and two ounces of cinnamon water.   When you wish to give the medicine, shake the mixture so that the sediment

[1] "Alberti Magni Opera omnia," ed. Augusti Borgnet, Paris, 1890, Vol. V, p. 41.

[2] "Lapidario del Rey D. Alfonso X," Codice original, Madrid, 1881, p. 4.

may be swallowed at the same time.   From one ounce to an ounce and a half may be taken, and nothing more excellent can be had.   In pernicious and pestilential fevers, the ordinary *aqua perlata* cannot be compared to this.   Care must be taken to cover the glass carefully while the pearls are dissolving, lest the essence should escape."[1]

A curious book on the medicinal use of pearls was written in 1637 by Malachias Geiger,[2] in which he especially praises the efficacy of Bavarian pearls.   It was true that their material value was less than that of oriental pearls, but this was compensated by their therapeutic qualities.   He had accomplished many cures of a very serious disease and had used these pearls successfully in cases of epilepsy, insanity, and melancholia.

Quotations might be given from a hundred medieval writers as to the therapeutics of pearls.   The diseases for which they were recommended, as noted by Robert Lovell's "Panmineralogicon, or Summe of all Authors," published at Oxford in 1661, seems to have included a large portion of the entire list known at that period.   This summary states:

Pearls strengthen and confirme the heart; they cherish the spirits and principall parts of the body; being put into collyries, they cleanse weafts of the eyes, and dry up the water thereof, help their filth, and strengthen the nerves by which moisture floweth into them; they are very good against melancholick griefes; they helpe those that are subject to cardiack passions; they defend against pestilent diseases, and are mixed with cordiall remedies; they are good against the lienterie, that is, the flux of the belly, proceeding from the sliperiness of the intestines, insomuch that they cannot retaine the meat, but let it passe undigested; they are good against swoounings; they help the trembling of the heart and giddinesse of the head; they are mixed with the *Manus Christi* against fainting (called *Manus Christi perlata*) in the London *Pharmacopaea*); they are put into antidotes or corroborating powders; they help the flux of bloud; they stop the terms, and cleanse the teeth; they are put into antidotes for the bowels, and increase their vertue, make the bloud more thin, and clarify that which is more thick and feculent; they help feavers.   The *oile of Pearles* or unions helpeth the resolution of the nerves, convulsion, decay of old age, phrensie, keepeth the body sound, and recovereth it when out of order, it rectifieth womens milk, and increaseth it, corrects the vices of the natural parts and seed.   It cureth abseses, eating ulcers, the cancer and hemorrhoides. . . . The best are an excellent cordial, by which the oppressed balsame of life and decayed strength are recreated and strengthened, therefore they resist poyson, the plague, and putrefaction, and exhilarate, and therefore they are used as the last remedie in sick persons.[3]

[1] De Boot, "Gemmarum et Lapidum Historia," Hanover, 1609, *Lib.* II, c. 38, p. 87.

[2] Margaritologia, Monachii, 1637.

[3] Lovell, "Panmineralogicon," Oxford, 1661, pp. 77, 78.

So powerful and mysterious were their alleged virtues, that in some instances it was necessary only that the pearls be worn to make effective their prophylaxis against disease.   This belief was by no means confined to the ignorant and inexperienced, for we are told that even Pope Adrian was never without his amulet made of the extraordinary combination of oriental pearls, a dried toad, etc.[1]   Leonardo, in the fifteenth century, wrote that pearls render true and virtuous all who wear them.[2]   Although we wonder at what we call the superstitions of the Middle Ages, perchance future generations will smile at many of our mistaken follies.

A prominent historical instance of administering pearls medicinally was in the treatment of Charles VI of France (1368–1422), to whom pearl powder mixed with distilled water was given for the cure of insanity.

A far more illustrious patient was Lorenzo de' Medici, "The Magnificent" (1448–1492), the celebrated ruler of Florence.   When this plebeian prince lay dying of a fever at Careggi, just after that famous interview with Savonarola, his friends called in Lazaro da Ticino, a physician of reputation, who administered pulverized pearls.   Politian, who was present, is credited with the statement that when the medicine was administered, to the inquiry as to how it tasted, Lorenzo replied: "As pleasant as anything can be to a dying man."[3]

Even the English philosopher, Francis Bacon (1561–1626), mentioned pearls among medicines for the prolongation of life.   He adds: "Pearls are taken, either in a fine powder or in a kind of paste or solution made by the juice of very sour and fresh lemons.   Sometimes they are given in aromatic confections, sometimes in a fluid form.   Pearls no doubt have some affinity with the shells wherein they grow; perhaps may have nearly the same qualities as the shells of crawfish."[4]

Powdered pearl or mother-of-pearl mixed with lemon juice was used as a wash for the face, and was considered "the best in the world."[5]   The pearl powder and lemon juice were permitted to stand for a day or two and the combination was then filtered before using.   Another method of preparing this was:

Dissolve two or three ounces of fine seed pearl in distilled vinegar, and when it is perfectly dissolved, pour the vinegar into a clean basin; then drop some oil of tartar upon it, and it will cast down the pearl into fine powder; then pour the vinegar clean off softly; put to the pearl clear conduit or spring water; pour that off, and do so often until the taste of the vinegar and tartar

[1] Jones, "Credulities Past and Present," London, 1880, p. 166.

[2] "Speculum lapidum," Venice, 1502, p. 37.

[3] Yriarte, "Florence," Paris, 1881, p. 39.

[4] Bacon, "Historia Vitæ et Mortis," Londini, 1623, p. 100.

[5] Grew, Nehemiah, "Musæum Regalis Societatis," London, 1681, p. 145.

be clean gone; then dry the powder of pearl upon warm embers, and keep it for your use.[1]

Through their composition of carbonate of lime, pearls possibly possess some slight therapeutic value, which, however, can easily be supplied by other materials—as the shell, for instance—and is entirely out of proportion to their market value as ornaments.

Although pearls have lost their therapeutic prestige and no longer have a recognized place in materia medica, their healing qualities are not to be denied, for there are few ills to which women are subject that cannot be bettered or at least endured with greater patience when the sufferer receives a gift of pearls; the truth of which any doubting Thomas may easily verify in his own household to the limit of his purse-strings.

Owing to their beauty and great value, pearls have been deemed particularly appropriate as a sacrifice in enriching a drink for a toast or tribute. Shakspere alludes to this in the words of King Claudius, the pearl being frequently designated *union* in the fifteenth and sixteenth centuries:

> The king shall drink to Hamlet's better breath;
> And in the cup an union shall he throw,
> Richer than that which four successive kings
> In Denmark's crown have worn.[2]

It is stated that a pearl worth £15,000 was reduced to powder and drunk by Sir Thomas Gresham, the English merchant, in the presence of the Spanish ambassador, as a tribute to Queen Elizabeth, by whom he had been knighted.[3]

The most celebrated instance of enriching a drink with a pearl was doubtless Cleopatra's tribute to Antony, Pliny's account of which we give in the words of old Philemon Holland:

This princesse, when *M. Antonius* had strained himselfe to doe her all the pleasure he possibly could, and had feasted her day by day most sumptuously, and spared for no cost: in the hight of her pride and wanton braverie (as being a noble courtezan, and a queene withall) began to debase the expense and provision of Antonie, and made no reckoning of all his costly fare. When he thereat demanded againe how it was possible to goe beyond this magnificence of his, she answered againe, that she would spend upon him at one supper ten million Sestertij. *Antonie* laid a great wager with her about it, and shee bound it againe, and made it good. The morrow after, *Cleopatra*

---

[1] "A Queen's Delight," London, 1671, pp. 75, 76.
[2] "Hamlet," Act V, sc. 2.
[3] W. J. Lawson, "History of Banking," London, 1850, pp. 24, 25.

made *Antonie* a supper which was sumptuous and roiall ynough: howbeit, there was no extraordinarie service seene upon the board: whereat *Antonius* laughed her to scorne, and by way of mockerie required to see a bill with the account of the particulars. She again said, that whatsoever had been served up alreadie was but the overplus above the rate and proportion in question, affirming still that she would yet in that supper make up the full summe that she was seazed at: yea, herselfe alone would eat above that reckoning, and her owne supper should cost 60 million Sestertij: and with that commanded the second service to be brought in. The servitors set before her one only crewet of sharpe vineger, the strength whereof is able to resolve pearles. Now she had at her eares hanging these two most precious pearles, the singular and only jewels of the world, and even Natures wonder. As *Antonie* looked wistly upon her, shee tooke one of them from her eare, steeped it in the vineger, and so soon as it was liquified, dranke it off. And as she was about to doe the like by the other, *L. Plancius* the judge of that wager, laid fast hold upon it with his hand, and pronounced withall, that Antonie had lost the wager.[1]

Elsewhere has been set forth the impracticability of dissolving a pearl in a glass of vinegar without first pulverizing it.[2] It seems probable that if Pliny's interesting story has any foundation, Cleopatra might have swallowed a solid pearl in a glass of wine—certainly a more pleasing draught as well as a more graphic sacrifice; and we should accept its reported value with a grain of salt, for it would scarcely have been safe for the court gossip to belittle the value of this tribute of love.

Pliny, and other Roman writers, mention another instance, that of Clodius "the sonne of Aesope the Tragedian Poet," who took two pearls of great price "in a braverie, and to know what tast pearles had, mortified them in venegre, and drunke them up. And finding them to content his palat wondrous well, because he would not have all the pleasure by himselfe, and know the goodnesse thereof alone, he gave to every guest at his table one pearle apeece to drinke in like manner."[3] The chronicler fails to tell what the guests thought of the flavor of pearls, or whether some would not have preferred them for a more appropriate use.

---

[1] "The Naturall Historie of C. Plinius Secundus," *Lib.* IX, c. 35. This anecdote is mentioned also by Macrobius (*Circa* 400 A.D.) in "Saturnaliorum conviviorum libri septem," *Lib.* II, c. 13.

[2] See p. 55.

[3] "The Naturall Historie of C. Plinius Secundus," *Lib.* IX, c. 35.

VALUES AND COMMERCE OF PEARLS

XIII

VALUES AND COMMERCE OF PEARLS

A pearl,
Whose price hath launch'd above a thousand ships,
And turn'd crown'd kings to merchants."
Troilus and Cressida, Act II, sc. 2.

TO trace the markets of the pearl is to trace the routes of commerce from early times. The first routes from the Far East seem to have been two: one by the Persian Gulf and the Euphrates to Babylonia and Assyria, and thence by caravan through Damascus to Tyre and Sidon; the other by the Red Sea and Suez to Egypt. As regards the former route, Sir George Birdwood furnishes positive evidence that the Phenicians visited India as early as 2200 B.C. It seems highly probable that pearls were introduced by this route at an early period, although it is difficult to find material proof of the fact.

By means of this commerce, the great ancient civilizations of Phenicia, Mesapotamia and the Nile valley doubtless became familiar with the gem treasures of eastern Asia. Then came the opening of the Mediterranean with first "the great Sidon," and later Tyre, as the starting-points of commerce, exploration, and colonial settlement among the islands and on the shores of what, to the Asiatic peoples, was the great western sea. However, as the Greek islands and their colonies developed, the Phenicians were more strictly confined to the coasts of Africa and Spain. Gades, Tartessus, and Carthage were their great colonies and trading-ports, and their adventurous sailors passed on through the Straits of Gibraltar and directed their course northward to the British Isles, where they very probably obtained the pearls of the Scotch rivers.

Meanwhile, the campaigns of Alexander had carried Greek influence and authority over all western Asia, reaching even to India itself, and had led to a widely increased intercourse. Although he died at the age of thirty-two, Alexander the Great did more than any single individual in the world's history to bring the nations of the Eastern and the Western worlds into contact with each other, and it is certainly due to this circumstance that we find much greater evidences of the use of pearls in the western countries after his time. Besides this, the founding of Alexandria provided a mart, in whose bazaars the traders of India, Persia, and Arabia bartered their treasured gems, just as their descendants do in the same place at the present day.

It was not, however, until the establishment of the Roman empire that this commercial intercourse reached its highest development. The Romans, with their marvelous capacity for organization, were the first to build a great system of permanent and well-kept roads to facilitate land travel and land traffic. These great roads, starting from the Forum, reached out in every direction, even to the limits of the empire; and, as a result of increased commercial activity, more gems were engraved, mounted, and set during the five hundred years of Rome's commercial supremacy than during any other early epoch of the world's history.

In Rome, the trade in pearls was so important that there was a corporation of "margaritarii." The *officinæ margaritariorum* were installed in the Forum, in the neighborhood of the *tabernae argentariæ;* some were also on the Via Sacra.[1] However, the name *margaritarius* did not only apply to the jewelers, merchants, and setters of pearls, but also to those who fished for them and to the guardians of the gems and jewels wherein pearls were used.

With the fall of the Western empire, the Dark Ages settled down like a cloud over Europe for five hundred years. Only among the Saracens and at Byzantium did the culture of the old civilization survive, and eventually the light of knowledge and of progress was rekindled from these sources. The Crusades were the chief factors in this new development; they gave a mighty stimulus, by means of which Europe was aroused from her lethargy and once more brought into contact with the Orient. Venice and Genoa now became the great carriers, and from this time, and to this source, may be traced many of the oriental gems in Europe. The Venetian fleet of three hundred merchant ships brought the products of the East and distributed them over Europe, by way of the German cities of Augsburg and Nuremberg, where the great jewelers and silversmiths made world-famed ornaments.

When Constantinople fell into the hands of the Turks, the treasures of the Eastern empire were scattered throughout Europe; but, at the same time, the establishment of the Turkish empire served to close the way to India and the far East for the merchants and travelers of Europe, and, hence, new means of access had to be sought by sea.

[1] See the epitaph of Tutichylus "qui fuit margaritarius," Orelli, 4076.

PECTORAL CROSS OF CONSTANTINE IX, MONOMACHUS (1000–1054 A.D.)

Containing some wood attributed to the true cross

GREAT PEARL NECKLACE OF THE FRENCH CROWN JEWELS

Composed of 362 pearls, weighing 5808 grains.   Actual size.   Worn by the Empress Eugénia.

This, as is well known, was the cause of the voyages of De Gama and Columbus.   The unexpected result of these voyages—the discovery of a new continent—ushered in the wonderful period of Spanish and Portuguese development and their colonization of both the East and the West Indies; and to this epoch belongs the introduction of American pearls to the markets of Europe.   The gradual decline of the power of Spain and Portugal—largely owing to bigotry and to the reckless exploitation of the regions under their control—brings us to the beginning of the present phase of commercial intercourse in which all the nations of the civilized world are engaged in varying proportion, according to their power and aptitude.   Never before have the different regions of the earth been more closely in touch with each other, and we may safely say that nothing is likely to occur which can permanently interrupt the progressive development of the world's commerce.

With the various means of transportation and locomotion that have existed in the past twenty-three or twenty-four centuries, there is no doubt that the commerce of pearls has varied more or less, but there has ever been, in some part of the world, a great potentate, a great collector or dealer who has influenced the finest gems to gravitate his way.   Never has there been a time when some person was not prepared to encourage—and to richly encourage—the sale of fine jewels to him.   The history of the commerce of precious stones is a history of travel and exploration, of hardship, pleasure, reward, and sometimes of serious disappointment.

The lesson we derive from these decorative objects of natural beauty and softness—treasured alike by savage, barbarian, ancient warrior, statesman, king, emperor, peasant, bourgeois, magyar, lady, and queen—always carries with it the moral that the gifts of creation are ever prized by some one in every age or place.

The necessary qualifications affecting the value of a pearl are: first, that it should be perfectly round, pear-shaped, drop-shaped, egg-shaped, or button-shaped, and as even in form as though it were turned on a lathe.   It must have a perfectly clear skin, and a decided color or tint, whether white, pink, creamy, gray, brown or black.   If white, it must not have a cloud or a blur or haze, nor should the skin have the slightest appearance of being opaque or dead.   It must be absolutely free from all cracks, scratches, spots, flaws, indentations, shadowy reflections or blemishes of any kind.   It must possess the peculiar luster or orient characteristic of the gem.   The skin must be unbroken, and not show any evidence of having been polished.

Diamonds and the more valuable precious stones generally are bought and sold by the weight called a carat. This carat, whatever its precise value, is always considered as divisible into four diamond or pearl grains, but the subdivisions of the carat are usually expressed by the vulgar fractions, one fourth, one eighth, one twelfth, one sixteenth, one twenty-fourth, one thirty-second, and one sixty-fourth. The origin of the carat is to be sought in certain small, hard, leguminous seeds, which, when dried, remain constant in weight. The brilliant, glossy, scarlet-and-black seed of *Abrus precatorius* constitutes the Indian rati, about three grains; the *Adenanthera pavonina* seed weighs about four grains. The seed of the locust-tree, *Ceratonia siliqua,* weighs on the average three and one sixth grains, and constitutes, no doubt, the true origin of the carat.

Another[1] of the more notable of these weight-units used for precious stones and precious metals is the *candarin, condorine,* or *cantarai,* also termed by the Chinese *fun* or *fan,* and by the south Indians a *fanam,* and used all over the Indo-Chinese archipelago. This is by origin a large lentil or pea of a pinkish color dotted with black, about double the size of the *gonj,* and possessing the same quality of very slight variability of weight when dried. It is probably a variety of the same botanic genus or species as the *Abrus precatorius.* The value when reduced to absolute standard became a subsidiary part or submultiple of the weight of some local coin, rupee, or pagoda, or a decimal fraction of some local tchen, as in China and Japan.

The following derivation of the word carat is given by Grimm: "Carat. Italian: *carato;* French: *carat;* Spanish and Portuguese: *quilate;* Old Portuguese: *quirate,* from Arabic *qirat,* and this from the Greek, κεράτιον."[2]

The carat is not absolutely of the same value in all countries. Its weight, as used for weighing the diamond, pearl, and other gemstones in different parts of the world, is given in decimals of a gram, by the majority of the authorities, as follows:

	Grams	In Grains Troy
Indian (Madras)	.2073533	3.199948
Austrian (Vienna)	.20613+	3.18107+
German (Frankfort) . . .	.20577+	3.175514
Brazil and Portugal	.20575+	3.175206
France	.2055+	3.171347
England	.205409	3.169943
Spain	.205393	3.169696
Holland	.205044	3.16431+

---

[1] Lowis d'A. Jackson, "Modern Metrology," London, 1881, p. 370.

[2] Grimm, "Deutsches Wörterbuch," Leipzig, 1873, Vol. V, p. 205.

	Pearl Grains in Grams	In Grains Troy
Indian (Madras)	.0518383	.799987
Austrian (Vienna)	.05153+	.79526+
German (Frankfort) . . .	.05144+	.793878
Brazil and Portugal	.05143+	.793801
France	.051375	.792836
England	.051352	.792485
Spain	.051348	.792424
Holland	.051261	.791077

Assuming that the gram corresponds to 15.43235 English grains, an English diamond carat will nearly equal 3.17 grains. It is, however, spoken of as being equal to four grains, the grains meant being "diamond" or "pearl" grains, and not ordinary troy or avoirdupois grains. Thus a diamond or pearl grain is but .7925 of a true grain. In an English troy ounce of 480 grains there are 151½ carats; and so it will be seen that a carat is not indeed quite 3.17 grains, but something like 3.1683168 grains, or less exactly, 3.168 grains. Further, if we accept the equivalent in grains of one gram to be, as stated above, 15.43235, and if there be 151½ carats in a troy ounce of 480 grains, it will follow that an English diamond carat is .205304 of a gram, not .205409, as commonly affirmed. The following exact equivalents, in metric grams and grains troy, of the diamond carat as used in different parts of the world in 1882, are given by Mr. Lowis d'A. Jackson:

DIAMOND CARATS

	Grams	Grains Troy
Turin	.2135	3.29480
Persia	.2095	3.23307
Venice	.2071	3.19603
Austro-Hungary	.2061	3.18060
France (old)	.2059	3.17752
" (later)	.2055	3.17135
" (modern)	.2050	3.16363
Portugal	.2058	3.17597
Frankfort and Hamburg	.2058	3.17597
Germany	.2055	3.17135
East Indies	.2055	3.17135
England and British India	.2053	3.16826
Belgium (Antwerp)	.2053	3.16826
Russia	.2051	3.16517
Holland	.2051	3.16517
Turkey	.2005	3.09418
Spain	.1999	3.08492
Java and Borneo	.1969	3.03862

DIAMOND CARATS—*Continued*

	Grams	Grains Troy
Florence	.1965	3.03245
Arabia	.1944	3.00004
Brazil	.1922	2.96610
Egypt	.1917	2.95838
Bologna	.1886	2.91054
International carat	.2050	3.16363
Proposed new international carat .	.2000	3.08647

Recalculating the above figures into pearl grains we have:

PEARL GRAINS

	Grams	Grains Troy
Turin	.053375	.823700
Persia	.052375	.808267
Venice	.051775	.799007
Austro-Hungary	.051525	.795150
France (old)	.051475	.794380
" (later)	.051375	.792837
" (modern)	.051250	.790907
Portugal	.051450	.793992
Frankfort and Hamburg . . .	.051450	.793992
Germany	.051375	.792837
East Indies	.051375	.792837
England and British India . . .	.051325	.792065
Belgium (Antwerp)	.051325	.792065
Russia	.051275	.791292
Holland	.051275	.791292
Turkey	.050125	.773545
Spain	.049975	.771230
Java and Borneo	.049225	.759655
Florence	.049125	.758112
Arabia	.048600	.750010
Brazil	.048050	.741522
Egypt	.047925	.739595
Bologna	.047150	.727635
International	.051250	.790907
Proposed International	.050000	.771617

With the present system of diamond carats and pearl grains it is necessary to keep two entirely different sets of weights or to resort to troublesome calculations. The stock-book of a jeweler, at the present time, will contain the following fractions, expressing the weight of a single pearl: ½, ¼, ⅛, 1/16, 1/32, 1/64, when the weight could be much better stated as 63/64 of a carat. It requires but a glance to see how much easier this would be. Certain dealers have therefore proposed the use of sets of fractions arranged in a similar way. In this manner a stock-book can be kept much more easily and with greater precision. Others, again, have adopted ·a decimal notation of the fractions of a carat, which is even more simple and feasible, since the common fractions ½, ¼, ⅛, etc. can be expressed as .5, .25, .125, etc., of a carat, this being either a carat of .2053 of a gram or the English carat of .20534 of a gram.

On the other hand, an agreement was arrived at, as the result of a conference between the diamond merchants of London, Paris, and Amsterdam, by which the uniform weight of a diamond carat was fixed at .205 of a gram, making the pearl grain .05125 of a gram. This standard, which was suggested in 1871, by a syndicate of Parisian jewelers, goldsmiths, and others dealing in precious stones, was subsequently (1877) confirmed. But there is still a lack of uniformity in the standard by which diamonds and pearls are bought and sold, and very serious discrepancies exist in the sets of carat weights turned out by different makers, although the international carat is almost universally used.

At the International Congress of Weights and Measures held at the World's Fair at Chicago in 1893, the writer suggested that the carat should consist of 200 milligrams, so that ½ of a carat would be 100 milligrams and ¼ of a grain would be 12.5 milligrams. This would mean 5 carats or 20 grains to a French gram, and 5000 carats or 20,000 pearl grains to a French kilogram. This would depreciate the present diamond carat or pearl grain only about one per cent., and it would do away with the needless series of carats and grains of the many nationalities. It could be simply explained to any private individual in any country, especially as there are only two countries which do not use the metric system.

This carat has been earnestly indorsed, its introduction advocated, and its merits clearly shown, by M. Guilliame, of the French Bureau des Arts et Métiers, whose energetic work has found a reasonable co-operation, in this country as well as in Europe, in introducing what will be a scientific, logical, comprehensive, and possibly the final and international carat; and any ancient, obsolete, or foreign carat can be readily reduced to this carat once the metric value of the former is computed.

The Association of Diamond Merchants of Amsterdam has already, to avoid confusion, fixed the value of the carat (17th October, 1890) at 1 kilogram = 4875 carats, or 1 carat = 3.16561 grains troy = 205.128 mg. One pearl grain = .7914 grains troy = 51.282 mg.; but the association has decided that, in case of litigation, these values shall be determined by appointed bureaus, which would express them in grams and milligrams, a most important and valuable decision, as the gram and the milligram will always be known as weights of constant value.

In view of the difficulty of inducing the abolition of the carat in different countries, the German Federation of Jewelers decided to petition the imperial government for authority to use the carat, in order that it might be legally recognized. Such a proposition not being in accord with the German laws in force on the subject of the metric system, it was proposed to substitute for the carats then in use one carat only, weighing two hundred milligrams. This proposal was very favorably received in trade circles and may be taken into consideration by the International Committee of Weights and Measures. The Commission des Instruments et Travaux, to which this proposition was referred, recommended its adoption to the committee in the following terms:

"The Commission recognizes that it would be very desirable that the unit of weight of precious stones (the carat) which varies in different countries, should be made uniform, and should be reduced to the nearest metric equivalent. The weight of 200 mg., which is very close to the carat most in use (205.5 mg.), would seem to be the best for this purpose. The Commission believes that there can be no objection to this standard of 200 mg. being called 'the metric carat' in order to facilitate the abolition of the old carat."

This proposition, adopted at the meeting of the International Committee on the 13th of April, was communicated to the more important associations. The Chambre Syndicale de la Bijouterie, Joaillerie et Orfèvrerie de Paris, and the Chambre Syndicale des Négociants en Diamants, Perles, Pierres Précieuses et des Lapidaires de Paris assured the committee of their support of this measure.

The following is the text of the resolution which was passed by both the above associations in January, 1906:

"The Council, recognizing the advantages which would result to the international trade in precious stones from the use of a unit based on the metric system, desires that the metric carat of 200 mg. be universally adopted."

The German Federation of Jewelers passed the following resolution in August, 1906:

"The German Federation considers that it is both necessary and advantageous to replace the old carat by the metric carat of 200 mg.; it authorizes its president to approach the imperial government and the International Bureau of Weights and Measures, and the foreign associations in order that the metric carat may be introduced as soon as possible in all countries."

The Chamber of Commerce of Antwerp promised, in a letter dated the 7th of December, 1906, to rescind a decision of 29th of April, 1895, approving the adoption of a carat of 205.3 mg., when the metric carat of 200 mg. should come into universal use in the markets.

The Association of Jewelers and Goldsmiths of Prague formally authorized the German Federation to act in its name, in order that the reform should come about as soon as possible by international agreement, and the Association of Goldsmiths of Copenhagen has declared its willingness to support the reform. The Committee of Weights and Measures in Belgium prepared a law for the adoption of the metric carat in December, 1906.

Mr. Larking, president of the Chamber of Commerce of Melbourne, Australia, has transmitted by letter of September 16, 1907, the following resolution of the Association of Manufacturing Jewelers of the Colony of Victoria:

"It is desirable that the carat weight should be the same in all countries, and our association approves a metric carat of 200 milligrams."

On October 16, 1907, the Association of Societies for the Protection of Commerce in the United Kingdom passed the following resolution:

"The Committee of the Association approves the attempt to urge the adoption in all countries of an international carat of 200 milligrams, and hopes that, in the interest of the unification of weights, it will prove successful."

The fourth General Conference of Weights and Measures, held in Paris in October, 1907, passed this resolution:

"The Conference approves the proposition of the International Committee and declares that it sees no infringement of the integrity of the metric system in the adoption of the appellation 'metric carat' to designate a weight of 200 milligrams for the commerce in diamonds, pearls, and precious stones."[1]

The following resolution was passed by The Birmingham Jewelers' and Silversmiths' Association, January 23, 1908: "That the best thanks of this Committee be conveyed to the Decimal Association for the good work they are doing, and this Committee expresses the hope that all countries will adopt an International Carat of 200 milligrams in weight." Finally, on March 11, 1908, the metric carat of 200 milligrams was adopted in Spain as the official carat for diamonds, pearls, and precious stones.

Pearls have become of so much importance to so many dealers that a special form of weight has been proposed for them. This would have a diamond form and not a square form, and it would be stamped

[1] Guillaume, "Les récents progrès du système métrique," Paris, 1907, pp. 62-66, "La réforme du carat."

"Grain" instead of "Carat." Another set would be stamped in milligrams, the regular milligram weight with the pearl fraction above it, and they could even be made round so as better to designate the pearl.

The great value of pearls has suggested the making of a gage, called the Kunz gage, by means of which round pearls can be very accurately measured. Pearls of a given weight and perfectly spherical form have been weighed and then measured by this gage, and the theoretical diameters as computed from the measurement of a single pearl are in the majority of instances in exact accord with these actual measurements, the occasional variations in the smaller pearls barely exceeding the thousandth part of an inch. These discrepancies may be due to imperceptible divergencies in sphericity or, possibly, to trifling differences in specific gravity.

The following table gives the diameters of round pearls by measurement, from 1/16 to 500 grains, in millimeters and inches:

Weight Grains	Diameter Millimeters	Inches	Weight Grains	Diameter Millimeters	Inches	Weight Grains	Diameter Millimeters	Inches
1/16	1.30	.0512	4½	5.44	.2141	20	9.01	.3547
⅛	1.66	.0653	5	5.65	.2224	25	9.71	.3823
¼	2.09	.0823	5½	5.86	.2283	30	10.31	.4059
½	2.65	.1043	6	6.03	.2374	35	10.86	.4275
¾	2.99	.1187	6½	6.20	.2442	40	11.35	.4468
1	3.32	.1307	7	6.36	.2504	45	11.82	.4653
1¼	3.60	.1417	8	6.64	.2614	50	12.23	.4815
1½	3.80	.1496	9	6.90	.2716	60	13.00	.5118
1¾	3.98	.1567	10	7.15	.2815	70	13.68	.5386
2	4.18	.1645	11	7.38	.2905	80	14.30	.5630
2¼	4.32	.1701	12	7.60	.2992	90	14.89	.5862
2½	4.47	.1759	13	7.81	.3074	100	15.42	.6071
2¾	4.63	.1823	14	8.00	.3149	125	16.60	.6535
3	4.80	.1889	15	8.18	.3220	150	17.63	.6941
3¼	4.88	.1921	16	8.36	.3291	200	19.41	.7641
3½	5.01	.1972	17	8.53	.3358	300	22.22	.8748
3¾	5.17	.2035	18	8.70	.3425	400	24.46	.9630
4	5.23	.2058	19	8.86	.3488	500	26.35	1.0374

The new and finer analytical balances weigh to the tenth part of a milligram, the two thousandth part of a carat, the five hundredth part of a grain; but this is not necessary. If the 200-milligram carat were used, the two hundredth part of a carat could readily be ascertained, and then a short-beam, rapid-weighing balance would answer every purpose and save much time for the dealer who must make many weighings in the course of a day. In an office where thousands of weighings were made in a month, the task was accomplished with such minute accuracy that the margin of error did not exceed one carat during that time.

The *mina,* the sixtieth part of the lesser Alexandrian talent of silver, was divided by the Romans, when they occupied Egypt, into twelve ounces (*unciae*), and, weighing as it did 5460 grains, it became the predecessor of the European pounds of which the troy pound is a type. If we may believe a Syrian authority, Anania of Shiraz, who wrote in the sixth century, the carat or diamond weight was originally formed from one of these ounces by taking the 1/144 part.[1]

We find in Murray[2] that the Greek κεράτιον was originally identical with the Latin siliqua, and was called the *siliqua Graeca.* As a measure of weight and fineness the carat represents the Roman siliqua as 1/24 of the golden solidus of Constantine, which was 1/6 of an ounce, hence the various values into which 1/24 and 1/144 enter, or originally entered. As a measure of weight for diamonds and precious stones, it was originally 1/144 of an ounce or 3 1/3 grains. It is stated in Hakluyt (Voy. II, pp. 1, 225, 1598): "Those pearls are praised according to the caracts which they weigh; every caract is four graines."

There have been at all times men who possessed a delicate touch or a fine sense of feeling, but probably few men are living to-day who would be able to accomplish the feat attributed to Julius Cæsar, namely, that of estimating the weight of a pearl by simply holding it in his hand. There are very few who can tell the weight of a pearl in this way, and while the story may be historically interesting, it is rather dubious.

To attempt to formulate a list of prices, comparative or otherwise, of pearls, is almost an impossibility, as probably no two authors of the past three centuries have ever seen the same lot of pearls, nor have their estimates always been the same as to quality, rarity and value.

As interesting statistics from an historical point of view, there will be presented here a list of the values of pearls dating back some ten centuries. That there always has existed a higher valuation for the larger pearls, which are the rarest, will readily be apparent, but that the correct value of a pearl of one, ten, twenty or fifty grains be definitely given for the years 1602, 1702, 1802, or 1902 is an impossibility. However, we believe this to be the first attempt to present so large a body of carefully selected quotations, and they are given to the reader, whether he be layman or professional, for what they are worth.

In regard to the smaller pearls, as is the case with the smaller diamonds, prices have been dependent upon the changes of fashion; that

---

[1] William Hallock and Herbert T. Wade, "Outlines of the Evolution of Weights and Measures and the Metric System," New York, 1906, p. 25.

[2] "A New English Dictionary," Oxford and New York, 1893, Vol. II, Pt. I, p. 105.

is, whether the prevailing style of jewelry was such that the smaller pearl or diamond was in demand. In other words, if they were used as a decoration forming a border, a flower, a scroll ornament, or a pave requiring many small gems, the demand naturally increased and the prices were higher or lower as the occasion required.

It is not the project of this book to fix the prices of pearls at the present time, for any such attempt would prove misleading, owing to the fact that pearls vary in the estimation of the different dealers, and a figure given here for the highest standard, if applied to an inferior grade, would necessarily mislead the buyer to his positive injury. This much, however, can be said: during the year 1907 pearls from five grains upward have been sold according to their quality, at a base of five, eight, ten, fifteen, or even twenty dollars in very exceptional cases; that is to say, twenty, thirty-two, forty, sixty, or eighty shillings, or twenty-five, forty, fifty, seventy-five or one hundred francs. Nevertheless, it would be impossible, without considerable experience, for a layman to apply these valuations to objects that require much practice in determining their quality and perfection.

With diamonds, rubies, and emeralds there may be a stated price per carat for stones of a certain size, but a gem of unusual perfection or brilliancy, or of exceptionally fine color, will often command a price far beyond that generally quoted. It is the same with the pearl. Sums which may seem exorbitant in comparison with those that are paid for ordinary pearls, are often given for specimens remarkable for their beauty, size, or luster.

Pearls of one hundred grains are even more rare at the present time than are diamonds of one hundred carats. Until the middle of the nineteenth century, the diamonds of the world weighing one hundred carats or over could be counted on the fingers, but since the opening of the African mines in 1870, the number of large diamonds has increased at a much greater ratio than have the pearls of one quarter of their weight. It would thus seem that pearls of great size are worth four times as much as diamonds of equal weight. For instance, a 100-carat diamond of the finest quality would be worth at least from $1000 to $1500 a carat, making a total value of $100,000 to $150,000; and a pearl of 100 grains at a base of $10 would be worth $100,000. But no such high price has ever been paid.

The usual method of estimating the value of pearls is by establishing a base value for those weighing one grain and then multiplying this amount by the square of the number of grains that the pearl weighs. For instance, if the base value of a one-grain pearl should be fixed at $1, a pearl weighing two grains would be worth $4 ($2 \times 2 = 4$), or $2 per grain; one weighing five grains would be worth $25, or $5 per grain, etc. Naturally, these values increase in proportion to the increase in the value of the base. A base of $3 would give a value of $75 for a five-grain pearl, or $15 per grain, while a $10 base would make the value $50 per grain, or $250.

This method of estimating pearls by squaring their weights has been credited by many authors to David Jeffries, who published an interesting treatise on diamonds and pearls in 1750–1753. It has also been credited to Tavernier, the oriental traveler of the middle of the seventeenth century. We have, however, traced this method back to Anselmus de Boot, in his treatise on precious stones, dated 1609. Before this date we have not been able to find any mention of the computation of the value of diamonds and pearls by squaring their weight and multiplying the product by a base of a franc, guilder, crown, dollar, or of many dollars, as would be necessary at present. It is probable, however, that this system is of oriental origin and it may have come to Europe through some of the oriental traders, with the precious stones, as did the use of the carat.

De Boot makes the carat (four grains) his unit of comparison, increasing his base value by one third for pearls weighing eleven carats (forty-four grains) or over. In Pio Naldi's treatise, published in Bologna in 1791, the unit is the grain, the base being the fourth part of the value of four pearls weighing together one carat. Naldi, also, increases his base value making it 1½ lire ($.30) for pearls weighing less than ten grains, and 2½ lire ($.50) for those weighing twenty grains and upward.

A curious method of valuing pearls by their weight is shown in a treatise by Buteo, published in 1554.[1] The writer states that a pearl weighing two carats was valued at 5 gold crowns; one of four carats at 25 crowns; and so on, the price increasing fivefold when the weight was doubled. The intermediate figures were obtained by computing the proportional mean of any two known weights and values. For example: $8 \times 4 = 32$, the square root of which is 5.656. Now, the value of a four-carat pearl is 25 and that of an eight-carat pearl 125 crowns, and $125 \times 25 = 3125$, the square root being 55.9; hence a pearl weighing 5.656 carats was worth 55.9 crowns.

The base value of a necklace can be determined in the following way. Should the center pearl weigh 25 grains, multiply 25 by 25; the result is 625; then, take the next two, three, or four pearls, as many as are of approximately the same weight, add their weights together, multiply the resulting figure by itself and divide the product by the number of pearls in the group. Proceed in exactly the same way with the remainder of the necklace, always grouping the pearls so that there shall not be a considerable difference in weight between the smallest and the largest pearl, and then add together the figures obtained for

---

[1] Buteonis, "Opera Geometrica," Lugduni, 1554, pp. 88-96.

the center pearl and for the various groups and divide the price of the necklace by this total; the quotient will represent the multiple or base.

As may be seen by comparison of the first with the second and third of the accompanying tables, the result arrived at in this way will, if there is any difference in the weight of the pearls in the various groups, vary slightly from that obtained by calculating the weight of each pearl separately, but it represents a satisfactory approximation.

NECKLACE OF 41 GRADUATED PEARLS ON A $10 BASE

1	pearl, weighing	25	grs.	25	$\times 25$	=		625.000
2	pearls, each of	22	"	44	$\times 44$	= 1936	$\div 2 =$	968.000
2	"	"	20	"	40	$\times 40$	= 1600	$\div 2 =$ 800.000
2	"	"	19	"	38	$\times 38$	= 1444	$\div 2 =$ 722.000
2	"	"	18	"	36	$\times 36$	= 1296	$\div 2 =$ 648.000
2	"	"	$17\frac{1}{2}$	"	35	$\times 35$	= 1225	$\div 2 =$ 612.500
2	"	"	17	"	34	$\times 34$	= 1156	$\div 2 =$ 578.000
2	"	"	$16\frac{1}{2}$	"	33	$\times 33$	= 1089	$\div 2 =$ 544.500
2	"	"	16	"	32	$\times 32$	= 1024	$\div 2 =$ 512.000
2	"	"	$15\frac{1}{2}$	"	31	$\times 31$	= 961	$\div 2 =$ 480.500
2	"	"	15	"	30	$\times 30$	= 900	$\div 2 =$ 450.000
2	"	"	$14\frac{1}{2}$	"	29	$\times 29$	= 841	$\div 2 =$ 420.500
2	"	"	14	"	28	$\times 28$	= 784	$\div 2 =$ 392.000
2	"	"	$13\frac{1}{2}$	"	27	$\times 27$	= 729	$\div 2 =$ 364.500
2	"	"	13	"	26	$\times 26$	= 676	$\div 2 =$ 338.000
2	"	"	$12\frac{1}{2}$	"	25	$\times 25$	= 625	$\div 2 =$ 312.500
2	"	"	12	"	24	$\times 24$	= 576	$\div 2 =$ 288.000
2	"	"	$11\frac{1}{2}$	"	23	$\times 23$	= 529	$\div 2 =$ 264.500
2	"	"	11	"	22	$\times 22$	= 484	$\div 2 =$ 242.000
2	"	"	$10\frac{3}{4}$	"	$21\frac{1}{2}$	$\times 21\frac{1}{2}$ =	$462\frac{1}{4}$	$\div 2 =$ 231.125
2	"	"	$10\frac{1}{4}$	"	$20\frac{1}{2}$	$\times 20\frac{1}{2}$ =	$420\frac{1}{4}$	$\div 2 =$ 210.125
41				624				10,003.750

$$\$10 \times 10,003.75 = \$100,037.50$$

THE SAME NECKLACE FIGURED IN GROUPS

1	pearl, weighing . .	25	grs.	$25 \times 25 =$	625.00
2	pearls, total weight	44	"	$44 \times 44 = 1936 \div 2 =$	968.00
4	"	"	"	78	"	$78 \times 78 = 6084 \div 4 =$ 1521.00
4	"	"	"	71	"	$71 \times 71 = 5041 \div 4 =$ 1260.25
6	"	"	"	99	"	$99 \times 99 = 9801 \div 6 =$ 1633.50
6	"	"	"	90	"	$90 \times 90 = 8100 \div 6 =$ 1350.00
6	"	"	"	81	"	$81 \times 81 = 6561 \div 6 =$ 1093.50
6	"	"	"	72	"	$72 \times 72 = 5184 \div 6 =$ 864.00
6	"	"	"	64	"	$64 \times 64 = 4096 \div 6 =$ 682.67
				624	9997.92

$$\$10 \times 9997.92 = \$99,979.20$$

On a $5 base this necklace would be worth $50,018.75 according to the first reckoning, and $49,989.60 according to the second; on a base of $2.50 the figures would be $25,009.37 and $24,994.80 respectively.

THE SAME NECKLACE FIGURED IN OTHER GROUPS

1 pearl, weighing . .	25 grs.	25 × 25 =	625.00
4 pearls, total weight	84 "	84 × 84 = 7056 ÷ 4 =	1764.00
6 " " "	109 "	109 × 109 = 11881 ÷ 6 =	1980.16
6 " " "	99 "	99 × 99 = 9801 ÷ 6 =	1633.50
6 " " "	90 "	90 × 90 = 8100 ÷ 6 =	1350.00
8 " " "	106 "	106 × 106 = 11236 ÷ 8 =	1404.50
10 " " "	111 "	111 × 111 = 12321 ÷ 10 =	1232.10
	624		9989.26

$$\$10 \times 9989.26 = \$99,892.60$$

On a $5 base this would represent a value of $49,946.30 and one of $24,973.15 on a base of $2.50. The different grouping of the pearls accounts for the slight reduction in value.

A system of estimating the value of pearls which has recently been introduced into Germany, is an adaptation of the ordinary method of squaring the number of grains and then multiplying the result by a certain base figure. The pearls are first grouped according to quality and size, and a figure is agreed upon as the multiplicator of each class. In Germany the carat is employed as the weight-unit for pearls as well as for diamonds, and in this new system the total weight of a given number of pearls of the same class is first reduced to grains; the number of grains is then multiplied by four and the quotient is multiplied by the figure agreed upon. The resulting sum, after being divided by the number of pearls, gives the carat value of such pearls. For example, if the base figure agreed upon is 5, and we wish to find the carat worth of 4 pearls of similar size, weighing together $3\frac{14}{64}$ carats, the sum would be as follows:

$$\frac{206 \times 4 \times 4 \times 5}{64 \times 4} = 64.37$$

At this rate per carat, reckoning in marks, the value of the $3\frac{14}{64}$ carats would be 207.20 marks. This result is identical with that obtained by the ordinary method, but the calculation is perhaps a trifle simplified.[1]

[1] See "Edelsteinkunde," Wilhelm Rau, Leipzig, 1907, p. 137.

A curious Hindu treatise on gems has been preserved for us in the Brhatsaṃhitâ of Varâhamihira (505–587 A.D.). It is the earliest work of this kind that we have in Sanskrit, and M. Louis Finot,[1] who has published it, together with several other similar treatises, believes that it was based upon an original composed at a much earlier period. In his introduction M. Finot says: "It would be an error to regard the ratnaçastra [treatise on gems] as a simple manual for the use of jewelers. Without doubt this subject formed one of the principal branches of commercial instruction, . . . but it was also taught to princes and it is for their use that the ratnaçastras we publish seem to have been composed."

This treatise only describes four gems, although a larger number are enumerated. These gems are the diamond, the pearl, the ruby, and the emerald. One of the most interesting portions is that treating of the valuation of pearls. The system described is peculiar, and, unfortunately, there is some difficulty in finding an absolutely correct equivalent for the values expressed.

A price is first placed upon a pearl weighing 4 mâsakas (about 45 grains). This is estimated at 5300 kârsâpanas (about $1600). As the weight diminishes the valuation decreases as follows:

4 mâsakas . .	5300 kârsâpanas	1½ mâsakas .	353	kârsâpanas
3½ " . .	3200 "	1 " .	135	"
3 " . .	2000 "	4 guñjas[2] .	90	"
2½ " . .	1300 "	3 " .	50	"
2 " . .	800 "	2½ " .	35	"

Smaller pearls were grouped together in dharanas (one dharana = about 72 grains). If there were thirteen fine pearls in a dharana, they were valued at 325 rûpakas (about $100); the other values were as follows:

16	pearls	in a	dharana	were	worth	200	rûpakas
20	"	" "	"	"	"	170	"
25	"	" "	"	"	"	130	"
30	"	" "	"	"	"	70	"
40	"	" "	"	"	"	50	"
55–60	"	" "	"	"	"	40	"
80	"	" "	"	"	"	30	"
100	"	" "	"	"	"	25	"
200	"	" "	"	"	"	12	"
300	"	" "	"	"	"	6	"
400	"	" "	"	"	"	5	"
500	"	" "	"	"	"	3	"

[1] "Les Lapidaires Indiens," Paris, 1896.

[2] The guñja was one fifth of a mâsaka and equaled about 2¼ grains.

It would be extremely interesting if we could find at this early date (sixth century A.D.) an indication of the use of the system of computing the value of pearls by the square of their weight as expressed in some weight unit, and it is singular that the three valuations given for the weight in guñjas are graduated in accordance with this system. A pearl weighing 2½ guñjas and valued at 35 kârsâpanas would have a base value of 5.6 kârsâpanas. Estimated at this ratio we would have the following figures:

$$3 \text{ guñjas } \ldots \ldots \quad 50.4 \text{ kârsâpanas}$$
$$4 \text{ " } \quad \ldots \ldots \quad 89.6 \text{ "}$$

Now, the values actually given are 50 and 90 kârsâpanas, respectively, and these figures are easily obtained by rejecting the fraction that is less than one half and counting the fraction that is in excess of one half as a unit. After this, however, the progression becomes irregular. A pearl weighing 1 mâsaka (5 guñjas) is valued at 135 kârsâpanas, while the equivalent according to the system would be 140. However, it is possible that the writer may have changed this figure intentionally so as to add exactly one half to the preceding valuation $(90 + 45 = 135)$. The succeeding values bear no relation to the system and appear to be entirely arbitrary. Still, it can scarcely be due to hazard that the first three figures are practically in exact accord with the system and the fourth in close approximation. As the change seems to come when the weight is expressed in mâsakas instead of guñjas, we are tempted to think that the system may have been used for single pearls weighing less than twelve grains (1 mâsaka = 11¼ grains), while the value of those over that weight was estimated in a different way.

In a much later Hindu treatise, by Buddhabhatta, after certain values have been given for pearls of the best quality, a pearl of this class is described as follows:

White, round, heavy, smooth, luminous, spotless, the pearl gifted with these qualities is called qualified (*gunavat*). If it be yellow, it is worth half this price; if it be not round, a third; if flat or triangular, a sixth.[1]

One of the earliest records we have of a system of prices for pearls is the treatise on precious stones written in the year 1265, by Ahmed ibn Yusuf al Teifashi, who was probably a native jeweler of Egypt. In his time pearls were sold in Bagdad in bunches of ten strings, each string comprising thirty-six pearls. If one of these strings weighed one sixth of a miskal (four carats or sixteen grains), the ten strings

[1] Finot, "Les Lapidaires Indiens," Paris, 1896, p. 22.

were valued at four dinars (about ten dollars). The values increased progressively as follows:[1]

Average weight of each pearl Grains			10 strings of 36 pearls, weight of each string		Value	
			Carats	Grains	Dinars	U. S. money
½	.	. .	4	16	4	$10.00
⅔	.	. .	6	24	5	12.50
1⅓	.	. .	12	48	6	15.00
2	.	. .	18	72	10	25.00
3⅓	.	. .	30	120	15	37.50
4	.	. .	36	144	20	50.00
4⅓	.	. .	42	168	25	62.50
5⅓	.	. .	48	192	35	87.50
6	.	. .	54	216	40	100.00
7⅓	.	. .	66	264	70	175.00
8	.	. .	72	288	80	200.00
9⅓	.	. .	84	336	110	275.00
10	.	. .	90	360	150	375.00
10⅔	.	. .	96	384	200	500.00
12	.	. .	108	432	400	1000.00
12⅔	.	. .	114	456	550	1375.00
13⅓	.	. .	120	480	650	1625.00
14	.	. .	126	504	750	1875.00
14⅔	.	. .	132	528	800	2000.00
16	.	. .	144	576	1000	2500.00
18⅔	.	. .	168	672	1500	3750.00

Al Teifashi then proceeds to describe a pearl of the first quality; it must be "perfectly round in all its parts, colorless and gifted with a fine water. When a pearl possesses these requisites and weighs one miskal [24 carats or 96 grains] it is worth 300 dinars [$750]. If, however, a match is found for this pearl and each one weighs one miskal and has the same form, the two pearls together cost 700 dinars [$1750]." This writer also mentions that in the shops of the Arab jewelers, the pearl which exceeded the weight of a drachma (12 carats or 48 grains) even by one grain, was called *dorra*, while the name *johar* was used for that which did not reach the above weight.

In 1838, Feuchtwanger gave the price of a one-carat pearl as five dollars, and used this amount as the multiplier of the square of the weight; therefore, a four-carat pearl would cost four times four multiplied by five dollars, the value of the first carat; that is to say, a sixteen-grain (four-carat) pearl would have been worth eighty dollars in 1838, according to this computation.

[1] "Fior di Pensieri sulle Pietre Preziose di Ahmed al Teifascite," text and translation by Antonio Raineri, Florence, 1818, pp. 8, 9.

THE SIAMESE PRINCE IN FULL REGALIA

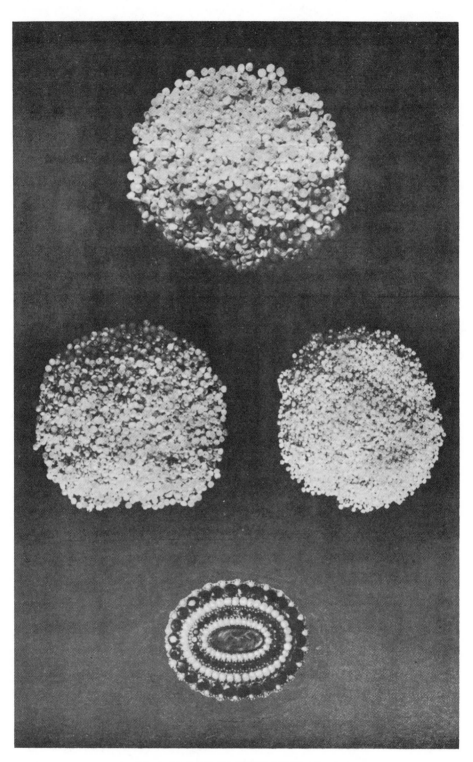

HALF-PEARLS: LOTS OF THREE DIFFERENT SIZES
BROOCH OF HALF-PEARLS AND ONYX. UNITED STATES, 1860

In 1858, Barbot[1] gave the value of pearls under ordinary conditions, but very indefinitely, as follows:

Grains	Carats	Francs per carat	U. S. currency
1	¼	4	$0.80
2	½	10	2.00
3	¾	25	5.00
4	1	50	10.00

Above four grains they sold by the piece, and below, by the ounce. Baroque pearls sold for 300 to 1000 francs per ounce. Seed-pearls, if quite round, were worth about 120 francs per ounce.

Emanuel[2] gave the following table of prices for the pearl, reduced to United States currency:

Grains	1865	1867
3 . . .	$2.88— $3.84	$4.32— $4.80
4 . . .	5.28— 6.72	6.72— 8.40
5 . . .	8.40— 10.80	9.60— 12.00
6 . . .	13.20— 15.60	16.80— 19.20
8 . . .	21.60— 26.40	24.00— 28.80
10 . . .	38.40— 43.20	48.00— 52.80
12 . . .	57.60— 72.00	67.20— 76.80
14 . . .	72.00— 86.40	86.40— 96.00
16 . . .	96.00—144.00	96.00—144.00
18 . . .	144.00—192.00	144.00—192.00
20 . . .	192.00—240.00	192.00—240.00
24 . . .	288.00—345.60	288.00—345.60
30 . . .	384.00—480.00	384.00—480.00

The following values appear in the "Encyclopedia Hispano-Americana," Barcelona, 1894, Vol. XV, p. 180 (Louis Dieulafait):

Grains	Value, 1865 Pesetas	U. S. currency	Value, 1867 Pesetas	U. S. currency
3	17— 18	$3.40— $3.60	21— 23	$4.20— $4.60
4	25— 32	5.00— 6.40	32— 40	6.40— 8.00
5	41— 52	8.20— 10.40	46— 58	9.20— 11.60
6	64— 75	12.80— 15.00	81— 93	16.20— 18.60
8	104— 128	20.80— 25.60	116— 139	23.20— 27.80
10	202— 227	40.40— 45.40	252— 277	50.40— 55.40
12	302— 378	60.40— 75.60	352— 403	70.40— 80.60
14	378— 453	75.60— 90.60	455— 504	91.00—100.80
16	504— 756	100.80—151.20	504— 756	100.80—151.20
18	756—1005	151.20—201.00	756—1005	151.20—201.00
20	1005—1260	201.00—252.00	1005—1260	201.00—252.00
24	1512—1815	302.40—363.00	1512—1815	302.40—363.00
30	2117—2521	423.40—504.20	2117—2521	423.40—504.20

[1] Charles Barbot, "Traité Complète des Pierres Précieuses," Paris, 1858, p. 467.

[2] Emanuel, "Diamonds and Precious Stones," 2nd edition, London, 1867, p. 6.

COMPARATIVE STATEMENT SHOWING THE VALUES OF PEARLS AT STATED TIMES

Weight Grains	1609[1] Thal.	Kreutz.	1672[2] Livres	1675[3] £	s	1751[4] £	s	1774[5] £	s	1791[6] Lire
1	0	13		0	½	0	1	0	⅓	1½
2	0	52	2	0	2	0	4	0	2	6
3	1	47	5	0	6	0	9	0	7½	13½
4	3	0	10	0	12	0	16	0	18	24
5	4	48	18	1	5	1	5	1	10	37½
6	6	52	28	2	10	1	16	2	5	54
7	9	13	38	4	10	2	9	3	1	73½
8	12	0	55	6	0	3	4	4	10	96
9	15	23	75	8	0	4	1	6	0	121½
10	18	52	100	10	0	5	0	8	5	150
11	22	48	130	12	0	6	1	9	15	242
12	27		175	14	0	7	4			288
13	31	48		16	0	8	9	13	15	338
14	36	52	270	18	0	9	16			392
15	42	13		21	10	11	5	21	0	450
16	48		380	25	0	12	16			512
17	54	13		30	0	14	9	27	10	578
18	60	52	500	35	0	16	4			648
19	67	48		37	10	18	1			722
20	75		650	40	0	20	0	37	10	800
22	90	52		50	0	24	4	52	10	1210
24	108			60	0	28	16	82	10	1440
26	126	52				33	16	99	0	1690
28	147					39	14	150	0	1960
32	192					51	4	225	0	2560
36	243					64	16	262	10	3240
40	300					80	0	300	0	4000
45	506	17				101	5			5062½
50	625					125	0			6250
60	900					180	0			9000
70	1225					245	0			12250
80	1600					320	0			16000
90	2025					405	0			20250
100	2500					500	0			25000

[1] Anselmi de Boot, "Gemmarum et Lapidum Historia," Hanoviae, 1609, pp. 88-90.

[2] De Rosnel, "Le Mercure Indien," Paris, 1672, Pt. III, pp. 17, 18.

[3] Rice Vaughan, "A Discourse of Coin and Coinage," London, 1675, p. 241.

[4] David Jeffries, "A Treatise on Diamonds and Pearls," London, 1751, pp. 128-141.

[5] "Encyclopédie de Diderot et d'Alembert," Neuchâtel and Paris, 1774, Vol. XII, p. 385.

[6] Pio Naldi, "Delle Gemme e delle Regole per Valutarle," Bologna, 1791, p. 207.

COMPARATIVE STATEMENT SHOWING THE VALUES OF PEARLS AT STATED TIMES,
REDUCED TO UNITED STATES CURRENCY

Weight Grains	1609	1672	1675	1751	1774	1791
1	$0.20		$0.12	$0.24	$0.09	$0.30
2	0.81	$0.80	0.48	0.96	0.50	1.20
3	1.82	1.90	1.44	2.16	1.87	2.70
4	3.24	3.80	2.88	3.84	4.50	4.80
5	5.06	6.84	6.00	6.00	7.50	7.50
6	7.28	10.64	12.00	8.64	11.25	10.80
7	10.92	14.44	21.60	11.76	15.25	14.70
8	12.96	20.90	28.80	15.36	22.50	19.20
9	16.40	28.50	38.40	19.44	30.00	24.30
10	20.25	38.00	48.00	24.00	41.25	30.00
11	24.50	49.40	57.60	29.04	48.75	48.40
12	29.16	66.50	67.20	34.56		57.60
13	34.22		76.80	40.56	68.75	67.60
14	39.69	102.60	86.40	47.04		78.40
15	45.56		103.20	54.00	105.00	90.00
16	51.84	144.40	120.00	61.44		102.40
17	58.52		144.00	60.36	137.50	115.60
18	65.61	190.00	168.00	77.76		129.60
19	73.10		180.00	86.64		144.40
20	81.00	247.00	192.00	96.00	187.50	160.00
22	98.01		240.00	116.16	262.50	242.00
24	116.64		288.00	138.24	412.50	288.00
26	136.89			162.24	495.00	338.00
28	158.76			188.16	750.00	392.00
32	207.36			245.76	1125.00	512.00
36	262.44			311.04	1312.50	648.00
40	324.00			384.00	1500.00	800.00
45	546.75			486.00		1012.50
50	675.00			600.00		1250.00
60	972.00			864.00		1800.00
70	1323.00			1176.00		2450.00
80	1728.00			1536.00		3200.00
90	2187.00			1944.00		4050.00
100	2700.00			2400.00		5000.00

Giving the pearl values in 1867, Emanuel[1] says: "It would be almost useless to give any value for drop-pearls, as when of large size and fine quality they are of so rare occurrence as to command fancy prices; still, as a slight guide, it may be mentioned that perfect white drop-pearls, of 80 to 100 grains, may be estimated at from £7 to £11 [$35–$55] per grain; those of 50 to 80 grains at from £4 to £7 [$20–$35] per grain, and those of 30 to 50 grains at from £3–£5 [$15–$25] per grain; smaller sizes bring from 20s. to 60s. [$5–$15] per grain."

Emanuel also states that misshapen pieces called "baroque pearls" (*perles baroques*), are sold by the ounce, the price varying from £10 to £200 ($50–$1000) per ounce, depending on quality, color, and size.

PRICES IN NEW YORK CITY IN 1878

Grains	Value per grain	Total value	Grains	Value per grain	Total value
1	$1.00	$1.00	10	$8.25	$82.50
2	1.83	3.66	11	9.62	105.82
3	2.75	8.25	12	10.45	125.40
4	3.60	14.40	13	11.68	151.84
5	4.03	20.15	14	12.55	175.70
6	4.69	28.14	15	14.20	213.00
7	6.32	44.24	20	19.70	394.00
8	6.87	54.96	24	24.75	594.00
9	7.42	66.78			

HALF-PEARLS
I QUALITY. PER HUNDRED

Size No.	Diameter Millimeters	Inches	1873	1876	1878	1885	1908
4	....		$1.10		$0.85	$0.50	$1.55
5	1.20	.047	1.35	$0.70	1.00	.60	1.95
6	1.22	.048	1.80	.90	1.35	.70	2.90
7	1.24	.049	2.25	1.10	1.70	1.12	3.88
8	1.26	.049	2.70	1.35	2.00	1.80	5.27
9	1.28	.050	3.35	1.80	2.50	2.00	6.65
10	1.80	.071	4.50	2.25	3.40	3.00	9.15
11	1.83	.072	5.60	2.70	4.20	4.00	11.36
12	1.86	.073	8.00	3.35	5.90	5.00	13.86
13	1.90	.075	9.00	4.50	6.75	5.75	15.51
14	2.00	.078	11.00	5.60	8.40	6.75	17.50
15	2.10	.082	14.00	8.00	10.00	8.25	20.80
16	2.25	.088	17.00	9.00	12.50	10.50	25.00
17	2.40	.094	19.00	11.00	14.00	12.00	30.50
18	2.60	.102	23.00	14.00	17.00	14.50	37.40
19	2.75	.108	28.00	17.00	21.00	16.25	48.50
20	2.90	.114	33.00	19.00	24.00	18.25	61.00

[1] Emanuel, "Diamonds and Precious Stones," London, 1867, p. 197.

I QUALITY. PER HUNDRED—*Continued*

Size No.	Diameter Millimeters	Inches	1873	1876	1878	1885
22	3.05	.120	42.00	28.00	31.00	33.00
24	3.15	.124	53.00	38.00	39.00	48.00
26	3.30	.130	67.00	45.00	50.00	69.00
28	3.55	.140	101.00	56.00	75.00	98.00
30	3.90	.153	124.00	79.00	92.00	150.00

HALF-PEARLS
II QUALITY. PER HUNDRED

Size No.	1873	1876	1878	1885	1908
4	$0.55		$0.45	$0.30	$0.84
5	.70	$0.35	.50	.35	1.22
6	.90	.45	.70	.50	1.87
7	1.10	.55	.85	.80	3.05
8	1.35	.70	1.00	1.05	4.43
9	1.80	.90	1.35	1.45	5.82
10	2.25	1.10	1.70	1.80	8.32
11	3.35	1.35	2.50	2.60	10.53
12	4.00	1.80	3.00	3.00	12.75
13	4.50	2.25	3.40	3.75	14.41
14	5.60	3.35	4.20	4.25	15.51
15	6.75	4.00	5.00	4.75	18.00
16	9.00	4.50	6.75	5.25	20.80
17	10.00	5.60	7.50	6.00	26.35
18	11.00	6.75	8.40	7.00	31.90
19	14.00	9.00	10.00	7.75	41.60
20	17.00	10.00	12.50	8.75	52.70
22	20.00	14.00	15.00		
24	27.00	19.00	20.00		
26	34.00	23.00	25.00		
28	51.00	28.00	38.00		
30	62.00	40.00	46.00		

HALF-PEARLS
III QUALITY. PER HUNDRED

Size No.	1876	1907	Size No.	1876	1908
4		$0.47	15	2.70	8.93
5	$0.25	.70	16	3.35	11.20
6	.35	1.11	17	4.00	13.90
7	.40	1.94	18	4.50	18.00
8	.45	2.77	19	5.60	22.20
9	.70	3.86	20	6.75	27.75
10	.80	4.99	22	9.00	40.00
11	.90	5.82	24	14.00	75.00
12	1.10	6.65	26	17.00	85.00
13	1.60	7.48	28	19.00	100.00
14	2.25	8.32	30	28.00	200.00

VALUE OF IRREGULAR PEARLS IN 1774[1]

Pearls to the ounce	Value in English money £ s.		Equivalent in U. S. currency	Average for each pearl
500	3	0	$15.00	$0.03
300	6	0	30.00	.10
150	11	2	55.50	.37
100	18	0	90.00	.90
60	33	15	168.75	2.81
30	75	0	375.00	12.50

The following values for the smaller oriental pearls are given in the "Museum Brittanicum" of John and Andrew van Rymsdyck, 1778, p. 9.

No. to the ounce	Rix dollars	Equivalent in U. S. currency	Average for each pearl
200	70	$75.60	$0.378
300	50	54.00	.18
900	10	10.80	.012
2000	3	4.24	.00212
4000	2½	2.70	.006755
8000 } 10,000 }	2	2.16	{ .00027 { .000216

Pio Naldi's treatise of 1791 gives the following rule for estimating the value of small, round pearls, weighing less than one carat or four grains. As the carat value of four such pearls is given as five lire and 576 one-grain pearls were counted as one ounce, these two numbers were used to determine the value of an ounce of small pearls. The product of 576 multiplied by 5 is 2880, and this number was then divided by 2000, 1000, 500, or whatever might be the number of pearls in a given ounce. If there were 2000 pearls, the carat value would be 1.44 lire or $.29; if there were 1000, the carat would be worth 2.88 lire or $.57; if 500, 5.76 lire or $1.15, etc.

The same author[2] gives tables expressing the values of pearls not perfectly spherical in form, which he designates as "perle dolce." These pearls he considers to be worth half the price of good round pearls; that is to say, 2½ lire (about $.50) per carat for four weighing together one carat. Where there are as many as three thousand of these "perle dolce" in an ounce, the 2½ lire base is multiplied by 576, the number of grains given to the ounce; this makes the value of an ounce of one-grain pearls $288. This amount is then divided by 3000, and the quotient, $.096, represents the value of one carat of these small

[1] "L'Encyclopédie ou Dictionnaire Raisonné des Sciences," Neuchâtel and Paris, 1774, Vol. XII, p. 385.   [2] "Delle Gemme," etc., 1791.

pearls. Multiplying this by 144 we obtain, as the value of an ounce of such pearls, $13.82. An ounce consisting of two thousand would be worth $20.73, while if there were but one hundred to the ounce it would be valued at $414.72, or $4.15 for each pearl and $.72 per grain of weight. In this latter case the pearls would average 5¾ grains. Another class of pearls denominated by this author as "scaramazzi," pearls of an irregular form and with protuberances, are estimated in a similar way, but at exactly half of the above values. The baroque pearls were not considered to be worth even half as much as the "scaramazzi."

Scotch pearls (fresh-water) are mentioned by De Boot (1609, p. 88 *sq.*) among the other western pearls—Bohemian, etc. He remarks that they were valued much less than the oriental pearls, but if they were of especially pure color their value was greater, although they lacked the silvery hue characteristic of the eastern pearl. Fine pearls of this sort were valued on a carat base of one fourth of a thaler ($.27), so that a forty-grain pearl was worth $27, and one of eighty grains, $108. The author of the Bologna treatise, "Delle Gemme," 1791, attributes the lack of luster in the Scotch pearls to the presence of a dark mass in the interior which interfered with the passage of light. He estimates Scotch pearls to be worth one half the value of oriental pearls of mediocre quality, provided the former are fairly good.

A Scotch writer of the seventeenth century is more enthusiastic in regard to these pearls; he mentions having paid one hundred rix dollars for an exceptionally fine one, but he does not specify its weight. This is the value given by De Boot for a pearl of this class weighing eighty grains, as we have just mentioned. The Scotch writer asserts that he could never sell a necklace of fine Scotch pearls in Scotland itself, as every one wanted oriental pearls; he continues: "At this very day I can show some of our own Scots Pearls as fine, more hard and transparent than any Oriental. It is true that the Oriental can be easier matched, because they are all of a yellow water, yet foreigners covet Scots Pearls."

In Ceylon[1] and India, pearl-grading and valuing has received close attention, and an elaborate system has been evolved by the pearl merchants. This system has been in use for generations and possibly for centuries. Although apparently very complicated, it is in reality quite simple, if we only remember that the value of inferior pearls is determined by their weight, whereas the value of superior pearls is computed from the square of their weight.

[1] See "Report to the Government of Ceylon on the Pearl Oyster Fisheries of the Gulf of Manaar," by W. A. Herdman, F.R.S., Pt. V, London, 1906, pp. 34–36.

The pearls are first grouped according to the size, of which ten grades are made. This is done by passing them successively through ten brass saucer-like sieves or baskets (*peddi*), each about three and a half inches in diameter and one inch deep. The holes in the bottom of each sieve are of uniform size, but they are graduated in size for the different baskets. The pearls are sifted in the basket with the largest holes, and those which will not pass through are of the first size. The pearls which pass through are then sifted in the second basket, and those retained are of the second size; and so on through the entire series of ten sieves or baskets. Those which pass through the tenth sieve are known as *masi-túl*, or powder pearls; they are of little value owing to their very small size, and are not subject to further classification. Of course, the attached pearls or very irregular baroques—the *oddumuttu*—are not subject to the sifting process, and are valued independently of this.

Sometimes in India, as well as in western countries, false measures are used, and an oriental pearl merchant may have one set of sieves for use in buying and another for selling. The rule for determining the proper size of the holes in the first sieve is that they may pass pearls weighing 20 to the *kalañchŭ*, whence this sieve is commonly known as the "*20 peddi.*" The second sieve is the "*30 peddi,*" since it passes pearls weighing 30 to the *kalañchŭ*. In the proper order the other sieves respectively pass pearls requiring 50, 80, 100, 200, 400, 600, 800, and 1000 to the *kalañchŭ*.

This use of sieves for grading the Ceylon pearls was mentioned by Cleandro Arnobio, a writer of the latter part of the sixteenth century, in his "Tesoro delle Gioie," and he took his description from an older writer, Garzia dell' Horto.

After the sifting, each of the ten graded lots of pearls are placed on pieces of cloth for classification as to quality, shape, and luster. This classification requires much skill and judgment on the part of the valuer. Not only will two persons commonly fail to class a large lot of pearls exactly alike, but one person is not likely to class the same lot twice in precisely the same manner.

From long established custom, recognition is made of twelve classes into which the ten grades or sizes of pearls are divided with respect to shape and luster, the local names of these classes giving a fair indication of their respective characteristics. These names are:

1 *Ani*, "best": perfect in sphericity and luster, the true orient pearl.
2 *Anatári*, "follower": failing slightly in sphericity and luster.
3 *Masanku* or *Masaku*: badly colored pearls, usually gray, symmetrical, and with luster.

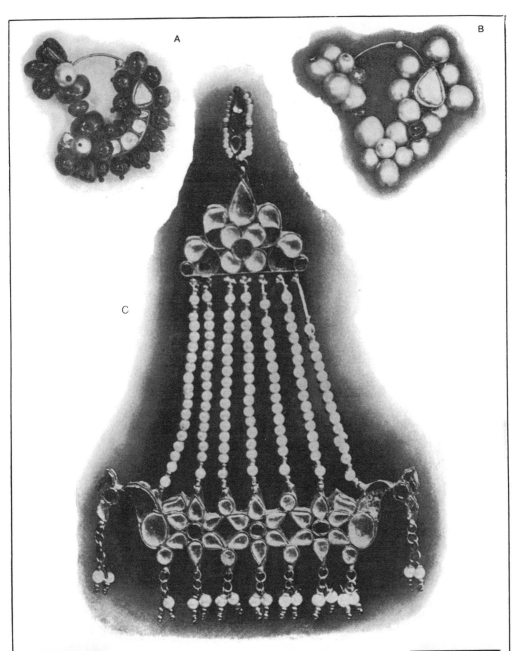

A, B.   Pearl nose rings.   Baroda, India.
C.   East Indian earring of strings of pearls and table diamonds.
Collection of Edmund Russell, Esq.

D, E.   Grape pendants.   Oriental pearls.

NECKLACE CONTAINING 126,000 SEED-PEARLS. LOUIS XVI PERIOD

Property of an American lady

4 *Kaiyéral,* "the clasp of a necklace": a dark-colored treble pearl, not quite round.

5 *Machchakai.*

6 *Vadivu,* "beauty," also "decreasing": that which is strained or sifted; found in the 100, 200, and 400 sieves. These small pearls, regular in shape, and of good luster, are especially favored in the East.

7 *Madanku,* "folded," or "bent": all pearls of *vadivu* size that are imperfect in form or color.

8 *Kŭrŭval,* "short": deformed and double pearls; they may, however, be of excellent luster. *Ani Kŭrŭval:* where two *áni* are fused together, but so formed that if separate they would be perfectly spherical. *Pisal Kŭrŭval:* where several pearls of good luster and color are fused partially and irregularly together. *Pampara Kŭrŭval:* a pearl grooved regularly, like a top.

9 *Kalippu,* "abundance," or "rejected": inferior to *Anatári;* a good pearl, may be lens-shaped or elongated; usually flattened.

10 *Pisal,* "torn": a deformed pearl or cluster of small misshapen pearls; of poor color and of little value.

11 *Kurál:* very misshapen and small.

12 *Túl,* "powder": the seed-pearls, those retained by the 600, 800, and 1000 sieves.

In addition to the above designations, the following are also used:

Samadiam: a pearl of a reddish hue; pear-shaped but of dull color.

Nimelai: a nose-pearl, perfect skinned, and pear- or egg-shaped.

Sirippu: a pearl grooved with irregular wrinkle-like furrows.

Kodai, "brown": like a nut, with no nacreous luster; formed of prismatic shell; may be large, is usually spherical, and includes pearls of various colors. This name is also used for white pearls with black or brown marks. *Van Kodai:* a *kodai* pearl with one side nacreous. *Karunk Kodai:* a black or blue-black slag-like pearl.

Masi-túl, "ink-dust," or "chalk-powder": smaller than the 1000 sieve. Generally used for medicinal purposes, or burnt and eaten with areca-nut and betel by the natives.

Oddu—or *Ottumuttu,* "shell-pearl": an attached pearl or nacreous excrescence on the outside of the shell.

Of the twelve classes named above, the first four are known as the *chevvŭ,* or superior classes; the next three as the *vadivu,* or beautiful classes; and the last five as the *kalañchŭ,* or inferior classes. The *chevvŭ* pearls are found only in the first four sieves or baskets; and for this reason these are known as the *chevvŭ peddi* or "chevvŭ baskets," although they may also retain inferior pearls. A name used to indicate the class of pearls found in the first four sieves is *mel* or *mel-muttu*, "upper" or "superior pearl," while *vadivu* designates those retained by the next three and *túl* those of the last three.

After the pearls have been graded according to size and classified according to quality, they are weighed. The unit of weight is the *manchádi*, the seed of *Abrus precatorius,* a small, red berry of practically uniform weight when ripe. H. W. Gillman of the Ceylon Civil Service reports the weight of the *mañchádi* to be 3.35 grains troy. Fractional parts of a unit are obtained by using a berry called *kundumani*, grains of rice, etc., whose weights have been determined beforehand. A brass weight —the *kalañchŭ*—is also employed; it equals 67 grains or 20 *mañchádi*.

However, choice pearls—those of the superior classes—are not valued in this manner, but at so much per *chevvŭ* of their weight, which is three fourths of the square of the weight in *mañchádi*. Thus, to find the value of an *anatári* pearl in the second sieve, if the weight be found to be three *mañchádi,* three fourths of the square of three, or 6¾, is multiplied by the base value of the *anatári* class.

The actual process of the calculation of value is as follows: owing to the small size of the pearls, many fractions enter into the computations; to preserve uniformity it is customary to increase all fractions so that each may have 320 as a denominator, this being a common multiple of those that ordinarily arise in *chevvŭ* calculations. The weight in *mañchádi* of the pearls is increased to a fractional figure having 320 as a denominator. Three fourths of the square of the numerator of this fraction is divided by the number of pearls, and this quotient is divided twice consecutively by 320, giving the *chevvŭ* of the weight. The market value then follows from the quoted price of the pearls per *chevvŭ* at the time.

In actual practice, these computations are not made; but each merchant provides himself with sets of tables showing the calculations for different weights, analogous to the use of interest tables by bankers, or of tables of logarithms by surveyors. Some of the merchants commit these tables to memory, and at times may be heard reciting them quietly to themselves to refresh the memory.

If a pearl of a particular grade and class is of exceptional merit, the merchant adds somewhat to the money value computed by the above process. This applies especially to double pearls of the *kŭrŭval* class, which sometimes consist of two fine bouton pearls suitable for setting, but not for stringing.

Pearls of one of the inferior or *kalañchŭ* classes are valued by simple weight, at so much per *kalañchŭ,* the market price, of course, differing for pearls of the various classes. The weight having been ascertained, each in its class as before noted, the value is determined by multiplying that weight by the current market price per unit of such pearls, at so many rupees per *kalañchŭ*.

The star pagoda is used in calculating the values. This small gold coin was current in south India in the early part of the last century. In the computations it is considered to be worth three and a half rupees, although its intrinsic value as a gold coin is about six rupees.

It is considered probable that the London syndicate,[1] which has lately leased the Ceylon pearl fisheries for a period of twenty years, will do away with the complicated calculations employed for so many generations, surviving all changes of administration, Portuguese, Dutch, and British. This is only one of the many instances showing the tendency of the British Government to abolish time-honored usages in India, without regard to the wishes of its population; and, unimportant as many of these changes may seem to us, they all serve to foster a spirit of discontent that may lead to serious trouble. This conduct on the part of Great Britain is all the stranger in view of the stubborn opposition of that country to the adoption of the scientific and logical metric system.

In Bombay, the weight of pearls in tanks is made the basis of their valuation; the tank equals 24 ratti or about 72 grains troy. The square of the number of tanks is multiplied by 330 and the quotient divided by the number of pearls; this gives the number of *chevvŭs*, or *chows*, as they are sometimes called, and the market price of the *chevvŭ* for a given class of pearls shows their value. If, for instance, we have 56 pearls of a certain quality, weighing 5 tanks, and the *chevvŭ* of these pearls is worth 14 rupees, the sum would be as follows:

$$\frac{5 \times 5 \times 330 \times 14}{56} = 2062.5 \text{ rupees, or about \$825.}$$

In this case, as in the other system of weighing which we have mentioned, the *chevvŭ* is only a nominal weight; but there is in India a real weight unit which bears this name.[2]

The high esteem in which the pearl was held by the Hindus is well illustrated by the following statement from an old treatise on gems: "A pearl weighing two kalañjas (about 180 grains) should not be worn even by kings. It is for the gods, it is without equal."[3]

An interesting account of a great savant's experience, in the early part of the sixteenth century, regarding the value of pearls, is given

[1] See pp. 124-127.
[2] See "Modern Metrology," Lowis d'A. Jackson, London, 1882, p. 369.
[3] From "Navaratnapariska," in Finot, "Les Lapidaires Indiens," p. 158.

by Guillaume Budé[1] (1467–1540), the celebrated French Hellenist who lived during the reign of Francis I and who is regarded as the founder of the Collège de France. In his work entitled "De Asse," he states that he once inquired of a gem dealer in Paris whether the latter could recall the weight of some remarkable pearl which had passed through his hands. The dealer replied that he had seen one weighing 30 carats (120 grains), whereupon another gem dealer, who was present, remarked that he had in his possession one of 40 carats (160 grains). This pearl was sold a few days later for 3000 gold crowns ($6750). On another occasion Budé was told that a pearl of exquisite beauty weighing 30 carats, had been sold to the Duchesse de Bourbon, daughter of Louis XI of France, for the sum of 4000 gold crowns ($9000).

In regard to the manner of computing the value of pearls Budé writes: "I think the ratio of these prices can be calculated. When I asked a gem dealer what was the value of a pearl of four carats [sixteen grains], according to the formula, he replied: 'I have seen such a pearl sell for thirty gold crowns [$67.50].' Whereupon I asked: 'How much would you estimate one weighing eight carats [thirty-two grains]?' 'At least two hundred gold crowns [$450],' he answered; and as I continued to ply him with questions, gradually increasing the weight, he responded in such a way that I could understand that the increase of the price bore not a numerical, but a proportional relation to the weight; so that the above mentioned eight-carat pearl, having double the weight of a four-carat pearl, was valued at seven times as much. The same was true of a pearl weighing twelve carats, twenty carats, and so on; the price augmenting by a greater and greater increment as the weight increased."

In the "Coronae Gemma Noblissima" of Wilhelmus Eo (1621, pp. 32, 33), an instance is given of the rapid changes that are possible in the worth of a pearl. A large and beautiful pearl was brought to Nuremberg by a merchant who had paid 500 florins for it; he soon found a purchaser among the merchants there, who was willing to pay him 800 florins. This latter merchant in his turn disposed of his gem for 1000 florins, and shortly after it again changed hands twice, the first time at an advance of 200 florins and the second at an advance of 300 florins. All this happened within a few days. The writer tells us that the last purchaser, who paid 1500 florins for the pearl, took it with him to Venice "where the wealthy dames wear a great treasure of beautiful pearls as necklaces upon their bare skin, and he will not have lost anything on his pearl there."

In 1884, Mr. Edwin Streeter was asked by a member of a London

[1] Guillielmi Budaei, "De Asse," Venice, 1522, *Lib.* V, pp. 67, 68.

syndicate to proceed to the East, to value a large quantity of jewels, as a heavy sum of money was about to be advanced to a certain Power, to provide the sinews of war. On his way he was requested to stop at one of the principal towns in Germany to purchase some jewels which had been valued for probate but were not easy of sale in that market. The valuation paper was shown to him, and after examining the ornaments, he agreed to take them at the prices named. Among them was an old gold brooch of Russian manufacture, valued at £4; in the center of this brooch was what appeared to be a piece of hematite, but was in reality a fine, round, black pearl, weighing 77 grains. The color had faded from exposure to the sun. This pearl was brought to London, and the outer layer was taken off, when a perfect black pearl of 67 grains was uncovered. This was sold to a manufacturing jeweler in London for £400; but, having heard that in Paris there was a pearl that would exactly match it, Mr. Streeter bought it back again for £600, and then sold it at a large profit to one of the Paris crown jewelers, who, in his turn, sold the pair to a rich iron merchant for 50,000 francs (£2000 or $10,000). Since then the sum of 100,000 francs (£4000 or $20,000) has been refused for this pair of matchless black pearls. At present values they may be worth double this sum.

At different times the values assigned to the different forms and colors of pearls have varied. For instance, in the French Encyclopédie of 1774 (Vol. XII, p. 385), it is stated that pear-shaped pearls, although they might be equally perfect and of the same weight as round pearls, were valued much less than these. Even in the case of well-matched pairs, their price was a third less than that of round pearls.

As early as the sixteenth century it was not uncommon that jewelers who had in their possession a fine pear-shaped pearl would have a replica of it molded in lead, and then send the casts to the large cities of Europe and the East. If a mate was found for it, the respective owners soon came to terms, for such pearls command a much higher price together than they do separately.

An interesting story is told of no less a collector than the Duke of Brunswick, who was so generous to the city of Geneva. For many years every pear-shaped pearl from every land had been submitted to him for examination. He always claimed the privilege of examining it alone for a moment or two and in every instance he returned it. At last a new pear-shaped pearl of marvelous size and beauty was heard of in a distant country. It was sent to Germany, where the duke was visiting at that time, to a local dealer who acted as agent for the owner. The price demanded for it seemed excessive, but the duke took the pearl, stepped aside for a moment, and said, quick as a flash, "The pearl is mine." The next day he showed it with a mate he had owned for many years and that was a most faultless match. Through all the years of his search he had never informed any one of his intention to match the pearl he already owned.

In 1879, at the time of the death of the father of Sultan Buderuddin of the Sulu Islands, a box of large and fine pearls was among the treasures he left behind him. Many of these disappeared, but some of them came into the hands of Sultan Buderuddin and his mother. The former sold those which he had inherited, in order to defray the expenses of a pilgrimage to Mecca, in 1882. His mother, who exerted a great influence over the conduct of affairs, retained a number of the pearls, and it was always difficult to induce her to part with any of them. When, as very rarely happened, she was persuaded to do so, she invariably got a higher price for them than they would have commanded in London, because she was never anxious to sell, and always said: "Why should I sell my pearls? If the Spaniards come to attack us, I can put them in a handkerchief and go into the hills; but if I had dollars I should need a number of men to carry them." We do not yet know what became of the stolen pearls.

Many times has a dealer put nearly all that he possessed into a fine pearl or necklace, frequently without a reward; often gradually buying more and more, hoping for some great patron to relieve him. When the client appears, there is happiness, but when he does not, there is woe. This instance is well illustrated when Philip IV of Spain asked of the merchant Gogibus: "How have you ventured to put all your fortune into such a small object?" "Because I knew there was a king of Spain to buy it of me," was the quick reply. And Philip rewarded the faith of the jeweler by purchasing the pearl.

Caire and Dufie[1] state:

We need have no fear that either the price or the use of pearls will diminish when we consider the great demand for them both on account of luxury and superstition. There is no Hindu who does not regard it as a matter of religion that he should pierce at least one pearl on the occasion of his marriage. This must be a new pearl which has never been perforated. Whatever may be the mysterious signification, this very ancient usage is, at least, very useful for the commerce of pearls.

In 1898, one of the writers had a long talk with his late chief, who had, at that time, devoted sixty years of his life to the jewelry profession. In the course of the conversation the latter remarked: "It seems to me that pearls are too dear"; to which the writer rejoined: "Have pearls ever gone down in price during your entire connection with the jewelry profession?" The answer was: "No, they have always ad-

[1] Caire and Dufie, "La Science des Pierres Précieuses appliquée aux Arts," Paris, 1833.

vanced." Whereupon the writer said: "I can give you statistics for two hundred years preceding your earliest experience, which prove that pearls constantly advanced in value during that period."

The following are the names given to the different kinds of pearls, according to their origin.

The term "oriental" designates those pearls that are found in the true pearl-oyster, and have a marine or salt-water origin, being found either in the ocean or one of its adjacent tributaries, and belonging to one of the numerous species of the Margaritiferæ.

The term "fresh-water" is given to those pearls that are found in the fresh-water brooks, rivulets, rivers, or fresh-water lakes, and not in salt water, and which belong to the Unionidæ.

The term "conch" is applied to that variety of pearl which is usually pink, or yellow, in color, and that is either found in the univalve shell, known as the common conch (*Strombus gigas*), or in the yellow shell (*Cassis madagascarensis*).

The word "clam pearl" is used to designate those pearls that are found in the common clam of the Atlantic coast, and are either black, dark purple, purple, or mixed with white, more especially if they are boiled.

"Placuna pearl" designates those pearls that are found in the Placuna, or window-glass shell, in the East. They have a micaceous luster, are rarely of much value, and are sold entirely in the Orient, almost exclusively for medicinal purposes.

"Oyster pearl" signifies those concretions that are found in the common edible oyster (Ostrea). They are generally black, purple, or with a mixture of black and white, or purple and white. They are devoid of nacreous luster and possess neither beauty nor value.

"Coque de perle" designates the globuse walls of the nautilus and possibly other shells that have a pearly nacre; they are almost hemispherical and are either round or long, having a pearly effect.

"Abalone": a name applied to those pearls that are found in the univalve "ear-shell" or *awabi,* as it is called in Japan. They are generally green, blue-green, or fawn-yellow, and have an intense red, flame-like iridescence. They are rarely round, generally flat, or irregular, and are occasionally worth several hundreds of dollars each.

"Pinna pearls": those pearls that are found in the Pinna, or wing-shells of the Mediterranean and adjacent seas. These possess no orient, but are more highly crystalline than any other pearls. They are almost translucent and have a peculiar red or yellow color, and are of little value except locally.

"Cocoanut pearl": this name is given to those pearls that are found in the giant oyster or clam of the vicinity of Singapore; they are erroneously called cocoanut pearls because they have the appearance of the meat of the cocoanut. They are often of great size, but have no commercial value.

The following are special designations of the different varieties of pearls according to their forms and appearance:

Paragon: this term was formerly used to designate large and exceptionally perfect or beautiful pearls, usually weighing over one hundred grains.

Round: when the pearl is absolutely spherical, as if turned on a lathe, without any flattening or any indentations on the sides.

Button or Bouton: if the pearl is domed on top and has either a flat or slightly convex back.

Pear-shaped: when the pearl is formed like a pear, terminating in a point, and is either flat at the lower end or rounded.

Drop-shaped: when the pearl is elongated like a pear, but is larger at the lower end than a pear-shaped pearl.

Egg-shaped: when ovate in form, rounded more or less at each end, or formed like an egg.

Cone-shaped: applied to pearls that are elongated and rounded with one flat end, and have the form of a cone.

Top-shaped: a name given to those pearls that are broad, flattened at the top and rounded on the sides, terminating in a point, like a top.

Seed-pearls is a name given to pearls that are round or irregular, and weigh one fourth grain or even less. They are frequently so small that 18,000 are contained in a single ounce, and they are often sent from the East in bunches of about a dozen or so of strings.

Dust-pearls. When seed-pearls are very small they are known as "dust-pearls"; they are really as fine as dust and have very little value; still, their form is in many cases wonderfully perfect.

Petal pearls are those which are somewhat flat, frequently more pointed at one end than at the other, and have the appearance of a petal or leaf.

Hinge pearls are those pearls that are long, generally pointed at either or both ends, and are found near the hinge part of the shell. They are divided into two distinct forms, namely dog-tooth, and wing-shaped.

Wing pearls: those that are elongated or irregular, resembling a wing or part of a wing.

Dog-tooth: applied to pearls with pointed ears, elongated, and which are narrower than the wing pearls.

Slugs: a name used for the very irregular, distorted pearls, frequently made up of masses or groups of small pearls; usually without luster or form, and of little value except for medicinal purposes.

Nuggets: when the pearls are somewhat round, but are indented or slightly irregular.

Haystacks: when the pearls are either round or oval, with the top considerably elevated.

Turtlebacks: when the pearls are a trifle longer than they are wide, with a domed surface not much elevated. This form is quite prevalent among American pearls.

Strawberry pearls: those that are round or elongated and entirely covered with prickly points, somewhat resembling a strawberry or pickle. It is believed that these irregular marks are frequently produced by minute pearls.

"Blister" and "Chicot" are names applied to those pearls that are found embedded within a nacreous coating, often containing mud, water, or imperfect mother-of-pearl. After these "blisters," as they are termed, are broken, and layer after layer has been removed from the contents, very fine pearls have frequently been found.

Peelers: a term applied to pearls having imperfect surfaces or skins that may have some inner layers which are perfect. Pearls having opaque bands or rings are rarely peeled with much success as this opaque layer frequently extends to some depth.

Cylindrical pearls: for pearls that have the form of a cylinder, being elongated and flattened at each end.

Hammer pearls: when pearls are long and somewhat rounded and assume the shape of a hammer or barrel. These are rounded or domed at the side and flattened at the ends.

Baroque (Wart pearls in German): when pearls are not of any perfect form such as round, pear, ovate, or any regular form, they are termed baroque, and this term covers a large class of varieties, such as all that follow (except seed- and half-pearls).

Double, triple, or twin pearls are those that are made up of two or more pearls united together in a single nacreous coating, showing, however, that they are still separate pearls.

Monster pearls: this name was formerly applied to very large, irregular, pearly masses which either resembled some animal or were adapted to form the head, trunk, or other part of an animal: these are also occasionally called "Paragons."

Bird's-eye: a name used for a pearl that has dull spots, giving it the appearance of a bird's eye.

"Ring-a-round" is a term applied to such pearls as are black, brown, pink, or white, and have a circle running around the pearl itself of some distinctive contrasting color, as white on black, pink on brown or black on white.

Embedded pearls are those that are partly or entirely surrounded by mother-of-pearl, having been enveloped and passed outward from the interior of the shell by the mollusk so that in time the pearl would have been lost on the outside of the shell. These embedded pearls are occasionally found in the manufacture of mother-of-pearl articles. When the mother-of-pearl is split, the pearl will fall out from between the layers.

Half-pearls is the name given to such pearls as are round and spherically domed, and are either somewhat flat or almost the shape of one half of a whole pearl of the same diameter. They are usually made by cutting off the best part of a hemispherical bright spot from a large irregular pearl; frequently two to four cuttings are made from the bright spots of a single pearl, each of the cuttings having the appearance of half a pearl.

The so-called Indian pearls have a faint rosy tint with much orient. These are generally pearls from the Ceylonese fisheries that are sold from the Bombay side. The term "Madras white" describes the whiter varieties, there being a preference for these in Madras, while the rosy, yellow, and darker shades are favored in Bombay.

Australian pearls are generally a pure waxy white and lustrous, often with a silver-white sheen, extremely brilliant and beautiful.

Nearly all the Venezuela and Panama pearls have a faint golden-yellow tint, very often extremely lustrous, and are especially desired by the darker skinned people and brunettes.

The preference at various times has varied with different peoples: in China and India, golden-yellow and satin-yellow pearls are preferred; from Panama we have the very white; in Bombay the yellow pearls from the Persian Gulf are highly appreciated.

Yellow pearls from other shells than the pearl-oyster are frequently offered for sale in the East, where they are greatly appreciated, although they find little favor in England. Some of these pearls are attributed to the pearly nautilus (*Nautilus pompilius*). This may be the case with those that have a pearly luster, but those that have the appearance of porcelain, and are as bright as polished china, are certainly not from this shell, but evidently from the large Melo or other shells of that character. Some may come from the large conch (*Cassis madagascarensis*). A yellow pearl, very perfect in form and color, and weighing more than one hundred grains, was shown at the Paris Exposition of 1889 and was valued at 50,000 francs.

Wonderful golden-yellow pearls with a saffron tint are unusually lustrous and beautiful. One of the most remarkable pearls of this character is of a brilliant golden-yellow color which belongs to an American lady, and weighs 30½ grains. These pearls are from Shark's Bay, West Australia, and only a limited number of them are found annually.

Black pearls do not seem to have been regarded with any favor by the ancients, and we find no mention of them by medieval writers. Only fifty years ago a perfectly round, black pearl, weighing 8 grains, was sold for £4 ($20); to-day this pearl would easily bring £100 ($500). Empress Eugénie, the wife of Napoleon III, may be said to have brought them into favor; she owned a splendid necklace of black pearls which was sold at Christie's, after the fall of Napoleon, for the sum of £4000 ($20,000). Some time later, the Marquis of Bath bought, at Christie's, the pearl which formed the clasp of the necklace, paying £1000 ($5000) for it; he destined it for the center of a bracelet.

Greenish-black pearls are perhaps valued higher than any other colored pearls, if they have the proper orient; this is probably partly owing to their rarity. A bluish-black pearl possessing a fine orient commands almost the same price as a pure black pearl. Those which are found in the *Placuna placenta* are often of a dull gray hue, while those produced by the *Pinna squamosa* are generally brown in color.

Baroque pearls were formerly much worn and appreciated in Spain and Poland. Their price varies greatly, according to their size, their beauty, and also to their scarcity in any particular place. The pieces of pearl detached from the shells—often half-pearl and half mother-of-pearl, and called "de fantaisie"—are always very irregular in form, and sometimes offer a certain resemblance to a part of the human or animal form.[1]

How is it that such quantities of jewels are continually brought from the East, and such a wealth of them continues to exist there, when there are now no very extensive mines that maintain a constant supply? The reason is that from time immemorial, precious stones have been the form in which wealth, in those lands, has been hoarded and preserved. Until very recently, in the Orient, interest-bearing securities have been unknown; and hence jewels have been sought and kept as an investment, and sold only when money was needed for special purposes, as in times of war, famine, or other emergency.

Their small bulk made them easy to conceal and to transport, and hence they were well adapted for such use. How long this condition will last, is perhaps dependent only upon the introduction of interest-paying investments, and of the new forms of Western civilization that involve greater expenses and require means of income in excess of the older and simpler conditions.

The wealth of jewels possessed by Oriental monarchs, notables, and dealers, has been the theme of story and tradition, time out of mind. We of the West have been disposed to regard these tales as largely ex-

[1] Charles Barbot, "Traité Complète des Pierres Précieuses," Paris, 1858, pp. 464, 465.

aggerated, and to some extent they may be; yet any one who has witnessed an important social function or state occasion where East Indian rajahs and nabobs are present, knows that the profusion of jewels which they wear is simply astounding to our Western eyes. These objects represent, moreover, the gatherings of generations and centuries; they are heirlooms and ancestral treasures, priceless to their owners as the pride of their houses; handed down from fathers to sons in long succession; and they have also the investment feature already noted, in that whenever necessity arises they can be turned into available funds.

The manner of keeping and of selling such objects is also different from ours. If it be a question of buying gems from an Eastern owner, the best are never shown first, but on the contrary, the most inferior. The purchaser must either be content with these, or else must prove clearly that he is a substantial buyer or evince a knowledge and appreciation that mark him as a judge of such objects. The order in which they are produced is, first the poorest, then successively, poor, medium, fair, good, fine, and at last the rare and wonderful prizes.

In visiting an Oriental dignitary, his jewel-treasures are not all shown at once, as at an American reception or an Indian durbar, or even as a collector or connoisseur among us exhibits his cabinet, arranged for choice display. The method is far different. The visitor may be shown a few objects in the first day or hour; perhaps a few more later in the day; some on the next day or the one following, and so on; and he may remain a guest for weeks, and never see all, or the finest of the jewels belonging to his host. When they are produced, moreover, they are not in iron caskets or in gold or silver jewel-cases, covered or lined with fine leather or with silk or satin. On the contrary, they are often in old ginger jars, shabby boxes, tin cans, and all sorts of unsightly or unpromising receptacles, which, when placed between the owner and his guest, may well cause the latter to wonder. Nor is his surprise lessened as the wrappings are unfolded, one after another, perhaps a dozen old cloths, until the piece of jewelry or the splendid pearl is at last brought to view, after having been hidden from sight in its manifold wrappings for months or perhaps for years.

But this method of keeping such treasures is not in reality so strange as it appears. There are none of the provisions that we have for the responsible safe-guarding of investments or valuable objects, —no fire-proof safes, no banks, no deposit-vaults. Security is best attained by concealment in unattractive and improbable receptacles, and by dividing and distributing the treasured objects. The owner, too, must learn to know his visitor quite well before he exhibits to him all, or the best, that he possesses. Hence the oriental method, though so

Seed-pearls and gold; Chinese ornaments of the nineteenth century

Complete set of seed-pearl jewelry in original case
New York, 1860

PERSIAN PRINCESS AND LADIES IN WAITING

From a Persian illuminated manuscript of the eighteenth century, in the library of Robert Hoe, Esq.

peculiar to us, has been the best adapted to the conditions among those peoples.

As an illustration of the interest taken by Oriental potentates in the collection of jewels, we quote an instance from Marco Polo, who, centuries ago, wrote the following:[1] "Several times every year the King of Maabar sends his proclamation through the realm that if any one who possesses a pearl or stone of great value will bring it to him, he will pay for it twice as much as it cost. Everybody is glad to do this, and thus the King gets all into his own hands, giving every man his price."

Great quantities of pearls, the result of centuries of accumulation, and exceeding in splendor the collections of the present day, must have been garnered up in many cities of the Orient during the period of their prosperity. But these cities have disappeared, wrecked and ruined by fire and sword, and no vestige of their former wealth remains with them. Their treasures have been looted, hoarded, buried, or scattered to the four ends of the Orient, frequently finding their way in former times to Europe, but now more often to America, where fine gems always find a generous buyer.

In Syria, and some of the Oriental countries, until recently, and perhaps at the present time, it has been the custom, when a native wished to embark in the pearl business, for him to allow himself to drift gradually into a state of vagrancy, becoming a veritable tramp for fully a year. Then, with the money that he had himself or that which was supplied by his backer, he would visit the pearl fisheries and shrewdly acquire the gems to the best advantage, returning again as a vagrant; for if it were known at any point along the route that he carried with him sums of money his life would be in jeopardy, and he would probably never reach the fisheries; or, if he did, the chances are that he would never return. This may remind us of Marco Polo's old coat, in which he had concealed some valuable gems, the gift of the Grand Khan. His wife heedlessly gave the coat to a beggar and it was only regained by a clever stratagem.

The product of the pearl fisheries, either that of entire fisheries where they are managed by a company, or the gatherings of merchants, or even the single gems which may be acquired by the smaller merchants, all these usually find their way to the great markets, although occasionally they change hands at once. In the East they are sent either to Bombay, Calcutta, Madras or Colombo; frequently they are intended for a higher market. Many of them remain in the East, for in the East to-day a fine pearl is as much prized as ever, and there are those who love pearls as much as did the King of Maabar in

---

[1] "The Book of Ser Marco Polo, the Venetian." Trans. and ed. by Col. Henry Yule, London, 1871, Vol. II, p. 275.

the time of Marco Polo. However, the world over, there is a feeling that if things are sent to the greatest market there will be an opportunity for disposing of them at the greatest price. Therefore, the larger number of parcels of exceptionally fine pearls are sent to the London market, a few of them going to Paris, the cable, often within a few days after their arrival, informing the sender of the acceptance or rejection of a parcel, or of a new offer which is often accepted. In this market they are acquired by the dealers, who frequently exhibit many times before the lot is purchased.

Pearls from a fishery are in many cases of mixed quality; that is to say, they are of different sizes and varying grades of perfection as regards skin, color, and orient. These parcels are often sold directly on offers to dealers, but generally they are sold by brokers who show the various parcels to the dealers, each of the latter in turn making his offer on that portion of the parcel which is of most value to him. Thus a single dealer may want one pearl, a dozen, or even twenty or more, to complete a great necklace, or else to add to, or improve the necklace, by better graduation or by increasing the evenness of the color. When the broker receives enough offers to give him the desired price for the entire parcel, the sale is consummated, and each one who has made an offer and who has sealed his particular parcel until his offer is accepted or rejected, receives his portion. Pearls do not grow in the form of necklaces, although they are frequently seen in this form only, and to create a large necklace means not only the use of the pearls of one fishery alone, but it often requires a selection from pearls of various sizes, the product of many fisheries.

It is needless to say that even the shrewdest dealers do not always succeed in their purchases of lots which are to be broken up when the proper number of bids are obtained.

When the pearl revival came in 1898 there was a sudden and rapid upward tendency in the prices, because at that time, in England, money could be borrowed upon a very low rate of interest,—as low as 3 per cent.,—and it was a temptation to a number of young men to enter as dealers into the pearl trade. The result was that a number of new stocks were created, not for a regular, but for a speculative demand, and this tended to advance the price spasmodically, rather than gradually, as it would have risen by regular consumption. However, when the foreign market became higher, the demand for pearls was not as great as had been anticipated, and there was a sudden adjustment of prices and a readjustment of the pearl stocks, resulting in the elimination of a certain number of speculative dealers; and, notwithstanding the state of the fisheries, pearls have not advanced so rapidly in the past two years as they did from 1898 to 1905.

More than 90 per cent. of the pearls of commerce, whether they are round, perfect, half- or seed-pearls, are of oriental origin; that is, pearls from the true pearl-oyster. About 8 per cent. are probably from the fresh-water mussels, three fourths of which are from the United States.

American fresh-water pearls have had many prejudices to overcome, often because of the natural indifference in regard to anything that is found at home or is easily obtainable. It has been said that, in comparison with foreign pearls, they had less specific gravity; that they were not so hard, and that their luster was not as good. It is certain, however, that the skin is generally smooth, and although they may not have so peculiar an orient, their brilliancy equals that of any known pearls. Sometimes they are translucent and either pink or of a faintly bluish tint, like molten silver. More frequently their hue is white, rose, pale yellow, or pale copper, deepening to copper red until they resemble the most intense and highly polished copper button.

According to the estimates of the value of European fresh-water pearls given by seventeenth and eighteenth century writers, their worth was considered to be one half that of oriental pearls of approximately the same quality. Few European pearls, we feel sure, were ever found that possessed the wonderful beauty and brilliancy of the pearls found either in the Miami or the Mississippi and its many tributaries.

So great a quantity of the poorer quality of pearls have been found, principally in the Mississippi Valley, that a foreign dealer has bought 30,000 ounces of baroque pearls at $1 an ounce, and of the slightly better grades fully 100,000 dollars' worth were obtained in the year 1906. The exportation was strictly limited to the poorer qualities. When pearls are worth from $1 to $6 a grain and upward, they are rarely sent abroad, as the regular pearls of this quality are much appreciated by Americans, and find a ready sale in the United States. The poor pearls above mentioned were principally sent to New York, either from the local fishermen, or else through the dealers in sweet-water shells, in lots of a fraction of an ounce, or in bags weighing a number of pounds. Thirty thousand ounces would equal 18,180,000 grains.

After all the fine pearls have been selected—buttons, baroques, turtlebacks, haystacks, wings, petals and other pearls that can be used in any way as a jewel on this side of the water—the balance of the material is sold by the ounce, varying in price from $1 to $5. These are shipped to Germany, France, and Austria, where they are again selected for cheaper forms of jewelry than are made in the United States. Of these pearls the baroques and slugs go mainly to

Germany, while the somewhat finer ones are sent to France, where they are used in artistic but inexpensive work, such as flowers and other imitative forms, and in *art nouveau* jewelry. Some, again, are shipped to Algiers, Morocco, and Egypt, for the decoration of saddles, garments, etc., and quantities go to India to be used for medicinal purposes. In this way all the material is utilized and even the poorest is not wasted. No better proof can be required of the wide-spread appreciation of the pearl among all the races of mankind.

So extensive has become the finding of American pearls that great quantities have been gathered together of all varieties. At the time of this writing there are many large single lots of these pearls, slightly irregular, and not of fine quality, but yet of sufficient regularity of size to be termed baroques. At one time such quantities were gotten together that single papers of pearls, weighing one fourth, one half, one, two or three grains each, contained more than 10,000 grains, and quantities of the wing and dog-tooth varieties weighing as much as 20,000 grains were inclosed in a single paper.

So prolific has been the yield of these common American pearls that the markets of Europe and Asia have almost been flooded with them. In 1906, a single shipment of 3500 ounces, troy (equaling over 2,100,000 grains), were sent abroad, at prices varying from $1 to $15 per ounce, according to the quality. This alone would represent a worth of $30,000 at one time.

The turtleback is a form quite prevalent among American pearls, and they are often matched in pairs slightly resembling each other and weighing from 10 to 100 or more grains for each pair. Some of them are lustrous and many are of very good color and regular in form. Although differing but little in shape, they naturally are much less expensive than a finer formed pearl, and many of them have been sold for link buttons, and more especially for earscrews. Although they formerly sold for 50 cents a grain, they are now held at from $1 to $8 per grain.

In regard to the prices of some of the finer American pearls, one of 15 grains, of wonderful brilliancy, luster, and perfection, was sold for more than $2500—$166 a grain, or a base value of over $11 a grain. Two extraordinarily well matched button pearls, weighing a trifle over 30 grains, were held at about $3500, or $115 a grain, a base value of about $8 a grain.

At the time of this writing there are for sale in the United States a pair of button earrings, almost round, not of absolutely perfect color, weighing about 140 grains, the price being $6000; a round, slightly ovate pearl, not of the finest color, weighing 85 grains, held at $3500; and a wonderful pearl with a rich, faintly pink luster, round, but slightly button on each side, weighing about 44 grains, and beautiful as are American pearls, is held at a fanciful valuation of over $6000.

The cupidity of many of the American pearl finders and pearl dealers cannot be exceeded even by that of the foreign pearl finder in any other land, and this is shown by the variety of materials that from time to time are sold to the unsuspecting public, or that are sent to pearl dealers in the large cities. This is surprising and suggests either that the sender believes the pearl dealers are not familiar with these deceptions, or else that he himself has been imposed upon, and is innocent in his commercial deceit. Among the notable examples are, first, spheres made out of the various shells, either from a good part of the material or from hinge-material, or else from the spot where the mussel is attached, these pieces of the shell being rounded and polished; such spheres vary in color from white to pink or yellow, just as the shell itself may have been colored. Second, the pupils of fish-eyes. Third, imitation pearls. Fourth, yellow or brown translucent or transparent masses of hinge-binding material having no greater hardness than horn, and about the same appearance. The most interesting, however, are the absolutely beautiful, smooth spheres of anthracite coal, which admits of a rich polish and has a peculiar luster; these they attempt to pass off as black pearls.

It is interesting to note that in Arkansas a negro sold a very valuable pearl for a few dollars, under the persuasion of a white man, who, it is said, resold the pearl for nearly a hundred times more than what he paid for it. The local authorities investigated the matter; the case was brought to court, and the negro received a large advance on the price that had originally been paid him.

If a list were kept of the thousand and one different methods of wrapping American pearls for shipment to the larger cities, it would show how much ingenuity is displayed in environments that frequently differ very much from each other. A box that has contained the pills that relieved him of fever, ague, and other ills due to swamps and damp climates, serves a secondary purpose for the fortunate finder of a pearl in forming a receptacle in which he can ship it to the greater market. Sometimes they are sewed in leather cut from gloves and shoes, or in strips of cloth, generally of the humbler varieties, such as calico or blue jean; in other cases they are wrapped in tissue-paper and newspaper; and occasionally they are packed in boxes made by hollowing out a bit of wood, a cover being nailed over the opening. In almost every instance they have been treated with a certain degree of care.

The majority of conch pearls which are carried by individuals to New York, London, or Paris, are generally brought in small papers or bits of cloth, each pearl being wrapped separately. Usually, there are a few white ones, a few yellow, a few pale pink, occasionally a few of a very beautiful rich pink, and once in a great while a fine, large pearl appears. Many of these pearls, commonly the inferior ones, are sold in the West Indies directly to the tourists who wish to purchase something in the country through which they are traveling, with the result that better prices are generally obtained than would have been secured if the pearls had been sent to the great markets.

The tariff on pearls at present operative in the United States is so indefinite as to have led to much serious misinterpretation and misunderstanding, as well as to an endless chain of lawsuits, often resulting in serious loss to the dealer or client who imports. As a consequence of the enforced outlay of large sums for unexpected and additional duties, the importer, who was both ready and willing to pay what seemed to him a just duty, often found that, where he had quoted a price to a customer, he was a loser by the transaction; and if, to escape this loss, he endeavors to dispute the payment of the duty, he becomes involved in an expensive and occasionally unsuccessful lawsuit. On the other hand, a private buyer who has paid all that he feels he can afford at the time for a necklace, expecting to pay a duty of 10 per cent. and interpreting the law to mean a duty of 10 per cent., may be called upon to pay a duty of 60 per cent., or have the notoriety of a public lawsuit, because the pearls have been strung, or because it is held that they had recently or at some former time been assembled as a necklace. In other words, if the pearls constituting such a necklace are bought at various times from various people, either here or in Europe, and not as a necklace, the duty is held to be 10 per cent., but if they are sent in one shipment, a duty of 60 per cent. is levied. As it is held that pearls assembled in the form of a necklace have a greater value than before they were so assembled, the purchaser might naturally expect to pay the 10 per cent. duty on this higher value, but instead of this a 60 per cent. duty is demanded on the higher assembled value.

The ambiguity of this clause of the tariff is such that a logical ruling should be made by some superior official such as the Secretary of the Treasury. As the law is now interpreted, a pearl worth $20,000 can be brought in with a duty of 10 per cent.; the addition of a simple gold wire makes it a piece of jewelry, with a duty of 60 per cent. It would seem that an amendment might be made to the tariff by which an importer, whether a private buyer or dealer, could be called upon to pay a 60 per cent. duty on a high valuation of the setting of the ring, brooch, or jewel, such as $20, $25 or $50; while the contents of the ring or ornament, whether a pearl, diamond, emerald, or a collection of stones, should pay a duty of only 10 per cent. This duty would sufficiently protect the jewelry industry, and would at the same time prevent the levying of an unjust and unexpected impost upon a fine pearl or gem of any kind.

It is eminently desirable that those residing in the United States who purchase pearls in foreign countries, should, if possible, consult with the United States consul in the city where they make their purchase, in case they wish to bring the pearls into the United States. In this way a proper declaration can be made, they will be correctly instructed as to the duties upon the pearls, whether unstrung, strung, or set, and they will thus avoid all complications when they reach the United States. Of course, this may not be necessary should the firm with which they are dealing be able to attend to the matter for them.

It must not be forgotten that the duty of 25 per cent. on precious stones, which was imposed during Cleveland's administration, was enacted for the purpose of obtaining an increased revenue for the government, and there is no doubt but that the time was one of great financial stress. Yet even with the duty two and a half times as high as in the previous years, only a small fraction was added to the income of the Government. But one adequate explanation can be given of this remarkable decrease in the recorded imports, more especially when we consider that legitimate dealers could, at that time, buy precious stones in New York City for less than it cost them to purchase them abroad and pay the duty. It seems, therefore, that a 10 per cent. rate is calculated to produce the best and most satisfactory results in every way.

As examples of the difficulties encountered in the attempt to arrive at a proper classification of pearls we cite the following cases which have been the subjects of recent litigation: In 1901, two very valuable collections of pearls were brought to this country. One of these consisted of 45 drilled pearls weighing in all 672⅛ grains and entered at $60,734; the other, of 39 pearls, having an aggregate weight of 678¾ grains and entered at $63,070. At first a duty of 20 per cent. ad valorem was imposed upon these pearls under Section 6 of the Tariff Act, treating them as "unenumerated articles partly manufactured," according to the rule that had been followed since the enactment of the present tariff. This was protested, and the case was brought before the Board of Appraisers.[1] Subsequent to the protest, however, the collector reliquidated the entry of the 45 pearls and imposed upon them a duty of 60 per cent. ad valorem, as pearls set or strung. This was done in view of Judge Lacombe's decision in another notable case which had been taken shortly before to the Circuit

---

[1] General Appraisers 5146 (Treasury Department 23748).

Court of Appeals.[1] This decision was to the effect that pearls in any form not especially covered by paragraphs 434 or 436 of the Tariff Act should be referred to one or the other of those paragraphs, by similitude, according to the provisions of Section 7 of the Act.

The testimony taken before the Board of Appraisers revealed the fact that each of the collections of pearls had been inclosed in a handsome silk-lined morocco case, with a groove running through the center; in this groove the pearls were laid, the largest one in the middle and the others disposed on either side, graduated according to their size; the row or series having the effect of a necklace, although the pearls were unstrung. The importer testified that this arrangement was only made in order to enable him to judge of the size and quality of the pearls, and evidence was given showing that it was necessary to rebore some of them and to ream out the holes before any use could be made of the pearls in jewelry. Nevertheless, the appraisers adhered to their opinion that these gems had been selected especially to form a necklace, and that the time and labor requisite for the assembling of a carefully matched and graduated series of pearls suitable for a necklace constituted the main factor in its production, since the cost of stringing it was trifling; they, therefore, considered that such a series of pearls was dutiable, by similitude, under paragraph 434 of the Tariff Act as jewelry. An application was made to the Circuit Court of the Southern District of New York for a review of the appraisers' ruling,[2] the judge decided against the petitioner,[3] and an appeal was then taken from his decision. On December 12, 1904, the Circuit Court of Appeals decided that the pearls were dutiable, by similitude, at 10 per cent. ad valorem, under Section 7, paragraph 436, and the excess of duty collected was refunded.

Another case has to do with a collection of 37 pearls, entered at $220,000, brought to New York in January, 1906. Duty to the amount of $22,000 (10 per cent. ad valorem) was paid by the importer, but the entry was liquidated at 60 per cent. and $110,000 additional duty demanded. This was paid and a protest was made to the Board of General Appraisers, who decided in favor of the petitioner. The Government appealed and the case[4] was tried in the United States Circuit Court on February 24 of this year (1908). It was shown that the pearls had been worn several times in Paris as a necklace, but the defense held that, as they were loose when imported and were not worth more collectively than separately, this was not material. The judge decided for the Government and an appeal has been taken in June, 1908.

[1] December 6, 1901; 112 Fed. Rep. 672.   [3] Dec. 29, 1903.
[2] Suit No. 3328.   [4] Suit No. 4974.

The proper classification of half-pearls has also been a matter of controversy. This question was brought before the Board of General Appraisers in New York on a protest [1] entered in 1897 against the imposition of a duty of 20 per cent. on several lots of so-called half-pearls imported during that year. This duty was imposed under Section 6 of the Tariff Act, providing for a duty of 20 per cent. on "unenumerated partly manufactured articles." The petitioner claimed that half-pearls were dutiable at 10 per cent. ad valorem, "either directly or by similitude or component of chief value, under paragraph 436, or as precious stones, under paragraph 435 of the Tariff Act." After hearing the testimony of a number of competent and reliable experts connected with some of the leading houses dealing in precious stones and pearls, the appraisers decided that the evidence showed that pearls, being the product of animal secretion, could not properly be denominated stones, and that they were not in fact so designated commercially. At the same time, half-pearls could not be looked upon as "pearls in their natural state," since time and labor had been expended in their production; it was, therefore, evident that paragraph 436 did not apply to them. For this reason the original ruling was reaffirmed.

In 1902 a duty of 60 per cent. was levied on an assorted lot of half-pearls under a new ruling which brought them by similitude under the provisions of paragraph 434 of the Tariff Act, providing a duty of 60 per cent. on "jewelry . . . including . . . pearls set or strung." A protest was entered against this ruling also. [2] In the meanwhile Judge Lacombe had given the opinion to which we have alluded above, and the Board of Appraisers upheld the duty of 60 per cent., basing their decision upon the fact that the material of half-pearls was similar to that of pearls in their natural state or of pearls set or strung, thus satisfying the requirements as to similitude of Section 7 of the Tariff Act. The same section provides that, in case two or more rates of duty shall be applicable to any imported article, it shall pay duty at the highest rate, and therefore the 60-per-cent. rate applying to pearls set or strung was imposed, instead of the 10-per-cent. rate on pearls in their natural state. In both of these cases an application for a review was made to the United States Circuit Court. [3]

DUTIES ON PEARLS IN VARIOUS COUNTRIES, MARCH, 1908

	Basis. Amount in money of the country.	U. S. currency.
Great Britain	Free	
British India	Free	
Australia	Free	
New Zealand	Free	

[1] General Appraisers 4166.   [3] Suits Nos. 2781 and 3324.
[2] General Appraisers 5148.

	Basis.	Amount in money of the country.	U. S. currency.
Canada, precious stones (pearls), polished but not set, pierced, or otherwise manufactured	ad val.	10%	
Austro-Hungary, unset	100 kilogr.	60 kr.	$24.00
Belgium, unenumerated.			
Bulgaria, precious stones (pearls) in the natural state, polished, cut, or engraved, but not mounted	kilogr.	75 lev (francs)	14.25
Denmark, unenumerated.			
France		Free	
Germany, wrought (smoothed, polished, perforated), unset	100 kilogr.	60 marks	14.40
Unset, but strung on textile threads or tape for the purpose of packing and transportation	100 kilogr.	100 marks	24.00
Greece		Free	
Holland, unenumerated.			
Italy, precious stones (pearls) wrought	hectogr.	14 lire	2.66
Montenegro, precious stones (pearls)	ad val.	{ min. 10% { max. 15%	
Norway, precious stones (pearls)	kilogr.	{ min. 2^{50} krone { max. 3 "	.66 .80
Portugal, unenumerated.			
Portuguese S. E. Africa (Quilimane, Chinde and Zambesia) Export Duty	ad val.	6%	
Portuguese India, real pearls or seed-pearls	ad val.	½%	
Rumania	kilogr.	20 lei	3.80
Russia, loose or threaded	funt	10 rubles	5.00
Finland		Free	
Servia, threaded for facilitating their preservation or sale	kilogr.	50 dinars	9.50
Threaded for special uses	kilogr.	70 dinars	15.30
Spain, loose or mounted	hectogr.	25 pesetas	4.75
Sweden, not set		Free	
Switzerland, not mounted	100 kilogr.	50 francs	9.75
Turkey, unset	gramme	3 piasters (gold)	
Egypt (on all imports)	ad val.	8%	
China (on all unenumerated imports)	ad val.	5%	
Japan	ad val.	60%	
Persia, Export Duty	ad val.	5%	
Import Duty, precious stones, rough or cut, including fine pearls	ad val.	25%	
Morocco (on all imports)	ad val.	2½%	

	Basis.	Amount in money of the country	U. S. currency.
Guatemala, unenumerated.			
Salvador, precious stones (pearls) un-mounted	kilogr.	10 pesos, nom. val...	9.60
Nicaragua, precious stones (pearls)	kilogr.	100 pesos, " " ...	96.00
Honduras	½ kilogr.	5 pesos, " " ...	4.80
Costa Rica, unset	kilogr.	100 colones, " " ...	96.00
Panama	ad val.	15%	
Mexico, unset	kilogr.	100 pesos, " " ...	96.00
United States, not strung, not set	ad val.	10%	
Strung, set, or not, and split pearls sorted as to either size, quality, or shape	ad val.	60%	
Philippine, unset	ad val.	15%	
Argentine Republic, precious stones (pearls)	ad val.	5%	
Bolivia	appraisal.	3%	
Brazil (natural)	ad val.	2%	
Chili	ad val.	5%	
Colombia, precious stones (pearls) set in jewelry	ad val.	10%	
Ecuador, precious stones (pearls), set or not set	kilogr.	50 sucres, nom. val..	48.00
Paraguay, unset	ad val.	2%	
Peru, unset	appraisal.	3%	
Uruguay	gramme.	13% on eval of 1 peso	.12
Venezuela	kilogr.	10 bolivars	1.90
Cuba, not set	hectogr.	$7.50 surtax of 25%	
Dominican Republic	ounce	6 pesos, nom. val.	5.76

The only changes from the customs lists as they existed in the tariffs of 1896 are as follows:

	1896	1908
Portugal	3% ad val.	unenumerated
Mexico	50 pesos per carat	100 pesos per kilogram
Nicaragua	5 pesos per libra	100 pesos per kilogram
Haiti	20% ad val.	unenumerated
San Domingo	3.60 pesos per ounce	6 pesos per ounce
Argentina	36 pesos per gram	precious stones 5% ad val.
Austro-Hungary	24 florins per 100 kilogr.	60 kroner per 100 kilogr.

In the Parliament of 1727–1732, the duty on pearls and precious stones was abolished in England. We give facsimiles of the title-page and last leaf of the report of this enactment.

108   Anno Regni sexto Georgii II. Regis.

Diamonds, precious Stones, Jewels, and Pearls of all Sorts, shall pass outwards, without Warrant or Fee; may it therefore please pour most Excellent Majesty, that it may be enacted, and be it enacted by the King's most Excellent Majesty, by and with the Advice and Consent of the Lords Spiritual and Temporal, and Commons, in this present Parliament assembled, and by the Authority of the same, That from and after the Tenth Day of April, which shall be in the Year of our Lord One thousand seven hundred and thirty three, all Diamonds, Pearls, Rubies, Emeralds, and all other precious Stones and Jewels, shall pass inwards without Warrant or Fee, in like Manner as they now pass outwards, and free from the Payment of any Duty granted to His Majesty, His Heirs, or Successors, and it shall and may be lawful for any Person or Persons to import or export the same, in any Ship or Vessel whatsoever; any Law, Custom, or Usage to the contrary in any wise notwithstanding, subject nevertheless to the Proviso herein after contained.

Provided always, That nothing herein contained shall extend to annul or make void the Duty granted to His Majesty for the Use of the united Company of Merchants of England trading to the East Indies, by an Act passed in the Ninth and Tenth Years of the Reign of His late Majesty King William the Third, for such Pearls, Diamonds, and other precious Stones or Jewels, as shall be imported into this Kingdom from any Place within the Limits of the Charter granted to the said Company, or to take away or alter any Privileges, Profits, or Advantages, granted to or now held or enjoyed by the said Company.

After 10 April, 1733, Diamonds and all other precious Stones may be imported or exported free from Duty.

Proviso as to the East India Company.

F I N I S.

Anno Regni
GEORGII II.
REGIS

Magna, Britannia, Francia, & Hibernia,

S E X T O.

At the Parliament Begun and Holden at *Westminster,* the Twenty third Day of *January, Anno Dom.* 1727. In the First Year of the Reign of our Sovereign Lord G E O R G E the Second, by the Grace of God, of *Great Britain, France,* and *Ireland,* King, Defender of the Faith, &c.

And from thence continued by several Prorogations to the Sixteenth Day of *January,* 1732. being the Sixth Session of this present Parliament.

LONDON,
Printed by *John Baskett,* Printer to the King's most Excellent Majesty. 1732.

The total value of diamonds and precious stones imported into the United States during the period from 1867 to 1906 inclusive, was as follows:

Glaziers' (except 1873–83) $2,215,972
Dust   6,407,599
Rough or uncut (included with diamonds and other stones, 1891–96)   74,045,291
Set (not specified before 1897)   36,170
Unset (not specified before 1897)   124,615,662
Diamonds and other stones, not set   207,138,629
Set in gold or other metal   17,799
Pearls (from 1903)   7,809,261

Total $422,286,383

CLASSIFIED STATEMENT OF THE IMPORTS OF PEARLS INTO THE UNITED STATES FROM 1891 TO 1907 INCLUSIVE

Year	Pearls 10%	Pearls, including pearls strung but not set 10%	Pearls in natural state, not strung or set 10%	Pearls split etc. 20%
1891	$11,711			
1892	32,023			
1893	6,926			
1894	12,978			
1895		$283,018		
1896		583,214		
1897		392,867		
1898			$491,060	$205,998
1899			1,412,952	389,899
1900			1,163,382	432,528
1901			929,247	1,173,339
1902			1,896,322	1,314,368
1903			2,835,936	7,220
1904			1,680,615	2,908
1905			1,626,476	
1906			2,072,561	218
1907			1,593,498	
	$63,638	$1,259,099	$15,702,049	$3,526,478

NOTE. Previous to 1891 pearls were classified with "jewelry and precious stones," and it was not until 1895 that most of them were reported separately.

There are several things that are essential in pearl buying, and one of the most important of these is that the light in which the pearls are selected shall be absolutely pure daylight, with no reflections from the side or from above that can enhance or detract from the color of the pearl. This must be carefully considered, as it is not uncommon—more especially in certain parts of Europe—that jewelers have for their selling-offices rooms sumptuously fitted up with hangings of different colors, and sometimes with ground glass windows, provided with heavy silk hangings, so that artificial light becomes a necessity to make the article sold plainly visible. In absolutely pure daylight, more especially with an unclouded sky—on such days as are probably more frequent in the United States than in some of the European countries—it is possible to see the exact tint or color of the pearls; that is, whether it is really a pure white with a tinge of pink or an orient tending to cream-white, or whether it is more or less tinted with what is considered a crude or red color in a pearl. Besides this, in a pure light it is possible to see whether the pearl is brilliant, and to estimate the exact degree of its brilliancy; whether there are any cracks, scratches, or mars on the surface; and, lastly, whether the form is entirely regular. If one should select two necklaces, one absolutely perfect and the other having slight blemishes as to color or brilliancy, or with breaks, marks, or irregularities, these two necklaces would be scarcely distinguishable from each other in artificial light, or in daylight which had been partly confused with artificial light; although the differences between the two would signify that the former was worth two or three times as much as the latter.

At great receptions, large, and apparently magnificent pearls are frequently seen, which are really of inferior quality, and yet, owing to the absence of pure daylight, they can easily be mistaken for perfect specimens by any one not especially familiar with pearls. Indeed, if the royalties of Europe should wear all the pearls belonging to the crown jewels at the same time, in a palace or hall lighted with candles, gas, or even with some types of electric light, they would frequently seem to have a quality which many of them do not and never did possess. It is, therefore, essential for the buyer to use every precaution in reference to the light in which he examines his purchase. And we may add that it is just as essential that he should know the dealer from whom he buys; for, sometimes, after a few weeks or months, cracks or blemishes develop that were not apparent at first, more especially when the pearls have been "improved" for a prospective purchaser.

A test to ascertain the quality of pearls is quaintly expressed in a work published in 1778, as follows:

How to know good pearls. To discover the hidden Defects and Faults of a Pearl and to know whether she is speckled or broken or has any other imperfections, the best way is to make trial of it by the Reverberation of the

Sun-beams; for by this means your eye will penetrate into the very Centre of the Pearl and discover the least defect it has; you will then see whether it be pure, or has any spots or not, and consequently you may the better guess its value.[1]

If you can cause a ray of sunlight or of electric light to fall on a pearl, the light will penetrate it and show any specks, inclosed blemishes or impurities. This can probably best be done by wrapping about the pearl a dark cloth of velvet or other material and having the ray fall slantingly, whereby the defects are much more clearly shown than if the ray be allowed to fall directly upon the gem.

A pearl necklace valued at $200,000, shown at one of our recent great expositions, was to all appearances a remarkably beautiful collection, and it was only when the intending purchaser took them from their velvet bed and held them in his hands that he realized that there was not a perfect pearl in the entire collection. It must have taken more than a week of study for the clever dealer to arrange them so that the best part, sometimes the only good part of each pearl, should be where the eye would fall upon it. After they had been turned in the hands a few seconds, not one perfect specimen was visible.

The demand for pearls has been so great, and the enhancement of value so rapid, that the greatest ingenuity has been employed in presenting the best part of the gems to view, as well as in many other ways. The result is that when pearls are to be used as borders or as a gallery on a comb or brooch, they are pierced in such a way that only the best side shall be outward, so that the general effect produced is that of a perfect row of pearls; but a careful examination may show that two thirds or three fourths of them are irregular, and bear abrasion marks, indentations, or other imperfections.

Following the analogy of the well-known precious stones—the diamond, the ruby, the sapphire, the emerald and those of less importance—the pearl is equally potent in creating great and permanent values for itself in catering to the human love of adornment; and though these large values may be greatly in excess of the original price that it commanded in the native oriental market, yet the increased valuation gives profitable livelihood to hundreds of thousands of persons. These embrace the dealers who sell the original pearls in lots, those who clean and treat them, others who drill and string them, and others again who handle them in setting jewelry of all kinds, and also the large number of dealers throughout the entire world who sell either the jewelry or the unmounted pearls. Directly connected

[1] John and Andrew van Rymsdyk, "Museum Brittanicum," London, 1778, p. 8, note.

with the industry in localities where the fisheries are pursued are a sufficient number of persons to populate a city the size of Boston, and to these we may safely add an equal number as herein noted, aggregating about 1,000,000 people whose livelihood is directly dependent upon the production and traffic of the pearl industry, and who for lack of it would be forced to seek some other employment. Brought thus to a concrete form, one may readily grasp the important bearing which the pearl has in a comprehensive estimate of the complexity of the world's civilization as we know it to-day.

TREATMENT AND CARE OF PEARLS

XIV

TREATMENT AND CARE OF PEARLS

THE pearl is at the height of its perfection when taken from the shell; from that moment it never improves. When it is drawn from the depths of the ocean by the hand of man and given to the charmed gaze of the world, it is as complete and perfect in its way as the most beautiful work of art, and, whether as tiny as the point of a pin or as large as a marble, it is always a perfect, fully formed individual; it is always in its maturity.

Who found the first pearl? When did he discover it, and what were his emotions? Was it found by primitive man? Very likely it was discovered by chance in a mother-of-pearl shell cast up by the sea, or perhaps in a mussel in a brook. If this happened in an oriental country, the native must have already seen many equally remarkable objects, endowed with life, while the pearl could charm him only by its luster and purity. But, besides the impression produced by its beauty, it must have aroused in the soul of the discoverer the sensation of wonder which every new and lovely object excites when seen for the first time. That primitive man appreciated the pearl is evidenced by the fact that it is found in the mounds and graves of the American continent, from the State of Ohio to Peru in South America.

Almost all pearls are in perfect condition for setting when they are found; all that needs to be done is to rub them with a damp or moist cloth or with a powder of finely pulverized small or broken pearls, and they are then ready for the succeeding processes. If there are any blemishes, these can be removed by peeling or "faking," although few fine pearls require any such treatment; and then the gems may be drilled, strung, and set, and all that is necessary for their preservation is due care and attention.

Pearls are frequently injured in opening the shells or in removal of the outer layers around the true pearly nacre. Both the Chinese and the Sulu fishermen are very clever in the art of pearl peeling and pearl improving. This method is called "faking," although it is a perfectly legitimate operation. All it requires is a very sharp knife, a set of files, and a powder obtained by grinding pearls or pearl shells. This powder is placed upon a buffer of leather or cloth to polish such parts of a layer as may not have been entirely removed. The Chinese are unusual adepts in pearl peeling and have been frequently known to sell as true pearls scales that they have removed, after filling these scales or peelings with wax or shellac, and strengthening them by cementing them on a piece of mother-of-pearl. They are then set with the convex side up and the edges carefully covered so as to conceal the deception. The Chinese are also very expert in removing layers of mother-of-pearl from an encysted or buried pearl, taking off layer after layer with the greatest care, and with a delicacy of touch that enables them to realize the moment when the pearl itself has been reached, rarely injuring the latter, although the coating is almost as hard as the inclosed pearl.

Peeling is employed to remove a protuberance or acid stain, to smooth a surface broken by abrasion, or to take off a dead spot produced by careless wearing of the pearls and allowing them to rub against one another. There are many instances where, by careful peeling, a perfect layer and skin have been brought to light, and where irregular or broken pearls, or those with a blemish, have been rendered much more valuable by a good peeler. But in many other cases the pearl has not only been reduced in value, but even rendered altogether worthless, when it had a dead center or was pitted with clay or other impurities.

If a pearl has been injured by coming in contact with the acids frequently used in medicine, the surface may become roughened; or it may be scratched by being rubbed against a stone in case of a fall or other accident. If the surface only is injured, it can be restored to its original beauty with only a slight loss of weight by carefully peeling off the outer layers.

In skinning or peeling a pearl, a magnifying glass, or preferably a fixed lens, such as is used by engravers, is of great assistance, and a sharp knife, or, better still, the sharpened edge of a steel file, is a very essential instrument. Gloves are often worn by the peeler so that no perspiration shall reach the pearl and cause it to slip in the hand while it is being manipulated, and thus have a layer or more injured by the knife.

Streeter mentions a very interesting incident in regard to a genuine black pearl. This pearl, set with diamonds, was shown in a jeweler's window; but after exposure in this way for some time to the sun's rays, the brilliant black luster disappeared and gave place to a dull, grayish hue. When the pearl was removed from its setting, it was seen that the part which had not been exposed to the light was of as good color as when first removed from the shell. It was finally determined

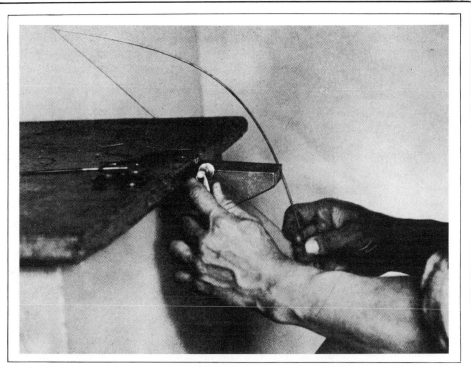

Drilling a pearl by means of the bow-drill

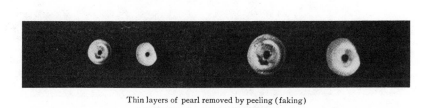

Thin layers of pearl removed by peeling (faking)

Examples of properly and poorly drilled pearls

Side view of same pearls

PEARL DRILLING

Scraping ends of silk threads for stringing pearl necklace

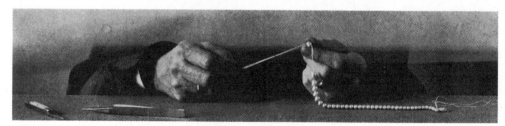

Stringing a pearl collar in sections; cleaning and reaming out a pearl

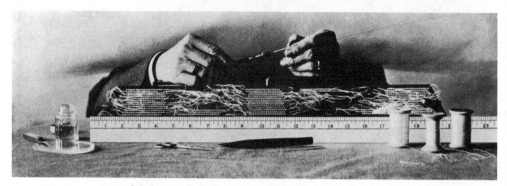

Sliding a pearl along the string in pearl stringing

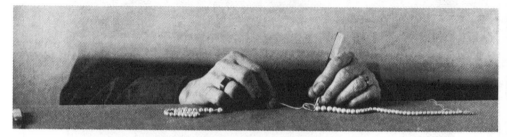

Tying a knot between pearls in pearl stringing

PEARL STRINGING

to skin off the outer layer, an operation which was performed with so much success that the original brilliant black hue was fully restored, proving that the action of the sunlight had only changed the color of the surface. We may add that the pearl, although it was shown in the sun, may never have had a good "skin" or layer exposed; or the layer which was not perfect may have been affected by an exudation of the wearer produced by illness or medicine.

When pearls are of a poor yellow or dull brown tint, unscrupulous dealers sometimes intrust them to an operator who drills them almost entirely through, cracks the skin slightly and impregnates them with a solution of nitrate of silver; this affects the outer layers of the pearls, and, after its decomposition, the metallic silver is deposited, and they become absolutely black. The effect is sometimes hastened by exposing them to the fumes of nitrate of silver. These pearls are then rubbed up or slightly polished and may retain a good appearance for a number of years. The upper layers, however, which have been injured by the chemicals used in the coloring, often scale off, and the poor and unattractive color beneath appears. This is sometimes not detected until years after and when the dealer from whom they were purchased has been forgotten. The breaks or cracks which have been made can readily be detected by means of a pocket lens, if the observer is at all experienced. In many cases the outer layer of the pearl has been colored a good black, although scarcely any crack is visible.

Frequently, when a small knob or protuberance appears in the pearl, or when it has adhered to another pearl or to the shell itself, this protuberance is polished off, and the pearl is drilled at this point. This portion of the surface, however carefully polished, will never have the true orient, but it is placed in the necklace in such a way that it is completely hidden. Often pearls become scratched through rough usage, or by the knife used in opening the shells. These are occasionally polished by means of pearl-powder, or else the entire outer layer is removed, the new skin beneath appearing absolutely bright and perfect. It sometimes happens that a pearl will have a good luster, but a slightly roughened skin. This is at times polished down; but an experienced eye easily detects that it has been tampered with. Yellowish pearls are sometimes bleached by means of strong bleaching substances such as chlorine or other powerful reagents, which, although they may whiten the pearl, cause it to become very friable, as the animal substance becomes more brittle. Pearls treated in this way frequently wear off, layer by layer, until fully one half of the pearl is worn out of the setting. When pearls are stained yellowish from the exudations of the skin, grease, or other impurities, they can be cleaned by putting them in moist caustic magnesia and allowing it to dry on them. When this is removed, the pearls will often be found much purer in color than before.

In various parts of the world certain dubious methods have been used for restoring the beauty of pearls which have grown dim. In India they are rubbed in boiled rice. Some persons have even fed them to a chicken fastened in a coop; after the lapse of an hour or two the chicken is killed, and the pearls rescued from their temporary lodging-place, where they have been somewhat restored by the digestive juices of the fowl.

Some curious tests applied to pearls are given us in a Hindu treatise on gems by Buddhabhaṭṭa. For instance, we read: "If the purchaser conceives a doubt as to the genuineness of a pearl, let him place it during one night in a mixture of water and oil with salt, and heat it. Or let him wrap it in a dry cloth and rub it with grains of rice; if it do not become discolored, it should be regarded as genuine."[1] It is needless to state that these tests would be either useless or injurious.

If the reader is the owner of a pearl or of a pearl necklace and feels that the pearls need treatment, any attempt to follow the directions given by many ancient writers would infallibly result in their injury or destruction.

Pearl drilling is a most delicate operation. It is necessary that the drill points should have the proper shape,—that is, should not be too tapering, but slightly blunt at the end, and turning somewhat in a V-shape,—it is also important that the drill should be revolved with perfect regularity, so as not to jar or jolt the pearl, as this is likely to lead to the cracking of the pearl or to the breaking of the drill. This latter happens not infrequently, and is due either to the structure of the pearl, the clogging of the drill, or to encountering a hard grain of sand inclosed in the pearl. Should the drill break in the pearl, it can best be removed by drilling from a point directly opposite, and slowly forcing the broken drill outward. This process requires great care in the regulation of the speed, and great exactness of direction in order to meet the broken drill accurately.

Pearl drilling was formerly a laborious process, and it was scarcely possible for a driller to perforate more than from forty to fifty pearls per day by means of the bow-drill operated by hand. Now, by the use of a modern machine, 1500 pearls of average size can be drilled without any difficulty in the same time.

Some of the most successful drilling of fine pearls is done by means of the bow- or fiddle-drill. The arm of this is made either of steel or of wood, with a strong cord stretched across it in the style of an archer's bow. The drill is inserted in the end of a brass circular disk

Finot, "Les Lapidaires Indiens," Paris, 1896, p. 24.

with a V-shaped groove on its edge, to admit of the string being passed entirely around it like a pulley, so that when the drill is placed on anything and held at the other side, and the bow is moved up and down, the wheel with the drill end rotates rapidly.

If the pearl is not properly secured, if the drill point is too irregular, if it is not properly centered, or if it is too rapidly rotated at the start, one or more layers of the pearl are likely to be broken, giving an irregular, ragged appearance. If, again, the drill is rotated too rapidly as it is leaving the other side of the pearl, one or more layers are occasionally forced off, and this in turn will produce a break on the pearl. It happens not infrequently that pearls are broken away on the surfaces at both drill holes if the workman is careless.

As pearls have become more valuable, only the most efficient workmen are employed in drilling them. Whereas formerly a drill hole would be half a millimeter in diameter, at present it is much smaller, and such drilling requires the greatest skill in manipulation. The use of these very fine drill holes is due principally to the fact that pearls have become so valuable that the slightest loss, even the fraction of a grain, would amount to a considerable sum in a necklace of large pearls.

When a pearl has been perforated with a very fine drill hole, the hole may be enlarged somewhat by using a slender copper wire, the fineness of the drill hole itself, charged with either diamond-dust, emery, or sand. When the wire thus charged is drawn in and out, the drill hole can be enlarged to any desired size.

A large pearl is held in the hand or secured in a wooden block, or else it is held in a small pair of forceps with a rounded, cup-shaped receptacle at the end, which is usually lined with chamois leather and is pierced with a hole through the center. This hole serves as a guide for the drill, directing it while the pearl is being perforated. Adjustable cups or forceps with cup-like ends of every size are necessary, according to the size of the pearl; and in order that it may be properly seen, it is requisite that the pearl should always be larger than the cup in which it is placed.

The poorest part or spot is selected to form the beginning of the drill hole. The pearl is placed in a pair of calipers with a circular disk, one end of the caliper being placed on the spot to be pierced, the other end naturally touching exactly opposite, the pearl absolutely centering it. As these caliper ends have been rubbed with either rouge, lampblack, or some colored substance that will readily rub off, these two spots of color remain on the pearl and serve as a guide for the driller. The drill end is then placed on the pearl, and the bow moved up and down; and so rapid is this work that five pearls weighing fifteen grains each can be drilled with the greatest care in less than one hour's time. Of small pearls, weighing about one grain, as many as fifty have been drilled in less than one hour by the hand-drill method.

Many of the thinnest and best drills are made out of thin steel needles. These are ground flat by means of a small carborundum wheel, so as to have two flat sides. They are then thin pointed, and with a V-shaped edge. These prevent the drill from clogging up, allowing the fine dust to pass upward and outward readily, and the hard steel almost invariably penetrates the central core of the pearl, no matter how hard or tough this may be. The needle-drill is then secured in a small chuck attached to the brass revolving wheel. Some recommend lubricating a drill with milk when it is employed for piercing a pearl, but a well-made drill, that allows the dust to escape as it is formed, does not require this treatment. The drill should always be made to revolve quite slowly so that no unnecessary heat may be generated by friction to injure the color of the pearl and also to avoid the possibility of the drill becoming clogged by the pearl-dust.

By means of centering calipers or markers, the driller, especially in the drilling of a large pearl, will generally drill first from one end, and then reverse the pearl and drill from the other end, meeting absolutely in the center. This prevents the breaking of the outer layer of the pearl. A skilful workman can, by turning the pearl, so operate the calipers that the true center can be obtained, even if the pearl is not absolutely round, and the drill holes so centered that the irregularity of the pearl is less apparent.

When the pearl has been half drilled through from one side, considerable caution is necessary in drilling from the other, that when the two drill holes are about meeting the drill be not revolved too rapidly, as the clogging is likely to crack the pearl or break the drill. If the pearl is only to be drilled one fourth or one half through, the depth can always be gaged by watching the drill-end, first, by measuring the drill-end itself, and, secondly, by noting to what part of it pearl-powder adheres.

Pearls are more easily manipulated than any other gems. They are also more easily damaged. Still, when properly treated by the workman, there is no material that offers him more satisfactory results than the pearl, if good judgment be used.

Drillers occasionally find that when the drill reaches the center of the pearl, there is a sharp click, the pearl often breaking at this point. This is evidently due to the fact that a harder kernel may exist in the center, such as a tiny grain of sand, which can turn the drill-point; or else the resistance may cause the tiny drill to break.

When a pearl is cracked by a blow or by some accident, it is customary to drill it at the end of the largest crack; this method prevents the crack from extending in that direction. These fissures are sometimes partly filled by means of a solution, and may not be visible at the time when the pearl is bought, but they are liable to appear later.

To illustrate the difference in the care used in drilling, we have selected eight pearls from a paper of poor ones, and reproduce two views of them, one to show the irregularity of the pearls, and the other to show the varying size of the drill holes. Those on the left were drilled by an artist, while those on the right show the work of an inexperienced driller.

At present pendant pearls are never drilled entirely through, and rarely more than half way. But in the Orient, and even in Europe from the fourteenth to the sixteenth century, they were often entirely pierced; even pear-shaped pearls were entirely drilled through, with a metal edge projected below for safety. Frequently old pearls, and more especially oriental pearls, have been entirely drilled through, as are often large oriental rubies, diamonds, and sapphires. When these are set, the holes are either plugged with pearl shell and polished smooth, or a tiny ruby or diamond is set in a metal rim fitting entirely into the drill hole or only slightly projecting. This is well instanced in the portrait of Marguerite of France (1553–1615), in which the artist Delpech shows all the pear-shaped pearls worn by the French queen entirely pierced.

Frequently, where pearls have been drilled by oriental workmen, the drill holes are exceedingly large, five or six times the width of the silk string; in fact often from one to two millimeters in diameter. In the search to supply the great demand, many oriental pearls have been secured which formerly were strung to an oriental jewel by means of a thick wire; it is necessary to close this aperture, as the pearl would lie unevenly on the string. This is done by introducing a mother-of-pearl plug, through which a new drill hole is made. Unless the pearls are unstrung, this is rarely visible; but not infrequently the plug drops out. In other cases the pearl has been drilled not only from end to end, but also from the side, and this third hole is filled with a plug of mother-of-pearl and polished over so as to hide the blemish from the buyer. It is also no uncommon thing for a purchaser to find, after a year, that cracks begin to develop where none apparently existed at the time of his purchase, or they were so minute as to be considered of no consequence.

One of the earliest references to drilling pearls was made by Rugerus, a monk who lived in the eleventh century. He says:

Pearls are found in the sea-shell and shells of other waters; these are perforated with a fine steeled instrument which is fixed in wood, having a small wheel of lead, also another wood in which it may be turned, to which a strap must be placed by which it may be revolved. But should it be necessary that the aperture of any pearl be made larger, a wire may be placed in the opening with a little fine sand, one end of which may be held in the teeth, the other in the left hand, and by the right the pearl is conducted upwards and downwards, and in the meantime sand is applied, that the apertures may become wider. Sea shells are also cut into pieces and are filed as pearls, sufficiently useful upon gold, and they are polished as above.[1]

In "The Toy Cart," a Hindu drama by Sudrake, who lived about the beginning of the Christian era, there is a description of a jeweler's workshop attached to the house of a courtezan. He says: "Some set rubies in gold, some string gold beads on colored thread, some string pearls, some grind lapis lazuli, some cut shells and some grind and pierce coral."[2]

The Chinese and Korean method of drilling pearls differs materially from that of the Occident. A pear-shaped pearl is frequently drilled horizontally and secured by wire or silk, and not drilled perpendicularly, as with us, to have a metal wire or peg fastened into it. If the orientals drill a pearl perpendicularly, the hole is generally carried entirely through it, and a gold knot, which is used as a bead, is placed at the lower end, and sometimes a tiny gem is set in this peg, or else the pearl is secured either by some projection below, or by means of a bit of enamel, or some other object may be attached to the gold or wire below it. Button pearls, especially those of the abalone, are drilled horizontally through the base and secured to the ornament, or to the silk or other material on which they are sewed, by means of a thread or wire; or else they may be drilled from below by means of two sloping holes forming a V, the thread or wire being passed upward until it strikes the angle, and then passed outward again through the other branch of the hole. Many fine, round, and pear-shaped pearls of oriental origin may be seen with this end closed either with a speck of pearl, a diamond, or a ruby.

A most interesting and careful description of the methods of drilling pearls was given by James Cordiner in his valuable volume, "A Description of Ceylon," published in London in 1807, pages 64–66.

[1] "An Essay upon Various Arts, in Three Books by Theophilus, called also Rugerus, Priest and Monk, Forming an Encyclopedia of Christian Art of the Eleventh Century." Translated, with notes, by Robert Hendrie. London, 1847.

[2] "Indian Art," by Sir George C. M. Birdwood (South Kensington Museum Art Books), Pt. II, pp. 188, 248.

The next operation which claims attention is the drilling of the pearls. I neglected to inspect this part of the business; but have been informed that much admiration is excited, both by the dexterity of the artist, and the rude simplicity of the machinery which he employs. A block of wood, of the form of an inverted cone, is raised upon three feet about twelve inches from the ground. Small holes or pits of various sizes are cut in the upper flat surface, for the reception of the pearls. The driller sits on his haunches close to this machine, which is called a vadeagrum. The pearls are driven steady into their sockets by a piece of iron with flat sides, about one inch and a half in length. A well tempered needle is fixed in a reed five inches long, with an iron point at the other end, formed to play in the socket of a cocoa-nut shell, which presses on the forehead of the driller. A bow is formed of a piece of bamboo and a string. The workman brings his right knee in a line with the vadeagrum, and places on it a small cup, formed of part of a cocoa-nut shell, which is filled with water to moderate the heat of friction. He bends his head over the machine, and applying the point of the needle to a pearl sunk in one of the pits, drills with great facility, every now and then dexterously dipping the little finger of his right hand in the water, and applying it to the needle, without impeding the operation. In this manner he bores a pearl in the space of two or three minutes; and in the course of a day perforates three hundred small or six hundred large pearls. The needle is frequently sharpened with oil on a stone slab, and sometimes, before the operation is performed, is heated in the flame of a lamp.

The large pearls are generally drilled first, in order to bring the hand in to work with more ease on those of a smaller size; and pearls less than a grain of mustard-seed are pierced with little difficulty.

After the pearls have been drilled, they must be immediately washed in salt and water, to prevent the stains which would otherwise be occasioned by the perforating instrument.

A quaint description of pearl drilling was given by Anselmus de Boot in 1609.[1]

Since all are not aware of the manner in which pearls are perforated, I wish here to give an account of the method. The handle, A, is held with the left hand, and then the handle, B, of the bow is pushed back and forth with the right hand, so that there is a reciprocal movement of the lance AC. The extreme end, C, has a needle, not so sharp as to come to a point, but slightly blunted. The needle is placed on the pearl which is to be perforated. If the pearls are too small to be held, they are fastened in the case, D, with a small hammer of soft wood, lest they should slip. The board is inclosed on every side by strips of wood so that the water which comes from the pearls shall not flow off. The bow being moved, the needle penetrates and pierces the pearl and it is not corroded by the water.

A mythical story, but a pleasant one, is told of a great pearl collector who had owned a wonderful pear-shaped pearl for many years and

[1] Anselmi de Boodt, "Gemmarum et Lapidum Historia," Hanover, 1609. *Lib*. II, c. 40, "Quomodo margaritae perforuntur," p. 91.

had absolutely failed to find any match for it.   After years of fruitless search he was at last rewarded by finding an absolutely perfect
mate.   He took this to his favorite jeweler in one of the great
capitals of Europe, and ordered the new gem to be pierced to match
the other so that both could be set.   The jeweler called a small German
boy from an adjoining workshop, simply saying, "Jakey, drill this pearl
to match the other."   The collector was dumfounded that no caution
should be given to the boy when so important a piece of work was
intrusted to his care.   Scarcely had the boy left the room when the collector inquired of the jeweler, almost in consternation, "How can you
trust so valuable a pearl to so small a boy without a word of caution?"
To this the dealer replied: "Jakey is the most careful pearl driller I
have ever known.   I know that there will be no failure in the drilling.
I have never cautioned him about such work.   He never has drilled a
pearl wrong.   Had I warned him of the value of the gem or told him
how important a piece of work he was doing, he probably would have
become nervous and, as a result, your pearl would have been cracked.
The conversation had scarcely been completed before Jakey returned
with the pearl as beautifully drilled as the original one which it
matched.

In the Orient and elsewhere, when it is considered desirable to mount
a pearl so that it shall not turn, especially when only one part of
the pearl is perfect and that is to remain outside, the drill hole is sometimes made square, that is to say, drilled round and then reamed
out with a small saw until it becomes square, when a square wire
is inserted; or else the pearl is first drilled with a tiny round hole
and this is then reamed out until it is triangular, when a triangular
wire is introduced.   This method is sometimes used for studs or ring-
settings.

In setting pearls with points or claws on the wire or band of a ring,
the pearls are drilled only half way through.   A gold pin is then inserted, and sometimes a thread is cut into the pearl itself; it is secured
by means of gum mastic or some other strong gum.   Occasionally, to
add greater strength, a side pin is put in, so that the pearl is drilled
with two bits of metal, which penetrate the one side in a perfectly
straight line and the other at an angle of about twenty-five or thirty
degrees (this is called side-pegging).   This gives more strength and
firmness to the pearl itself, and prevents it from twisting or twining
and becoming loose.   Sometimes the pearl hole is drilled so that the
opening is that of a screw-thread, in order to hold it to the earring,
the stud, or the ring.   The gold pin which is inserted to attach the
pearl to the ring or stud has a screw-thread also, and the peg or pin is
screwed on as well as secured.

An ingenious method, termed "keying," for securing the peg in pearls to be set on rings or studs, consists in drilling a hole half through the pearl and then two smaller holes or grooves on each side of the first. Cutting tools of a T-shape are now introduced into the aperture and worked about until the pearl is undercut all around, so that when a peg with a cross-piece is inserted, the latter can be turned within the pearl until it sets at right angles with the widest part of the aperture. In this way the peg is permanently secured and cannot slip out.

The fact that in recent years more pearls have appeared in necklaces that are irregularly bored, that the bore holes are so large that they are plugged with mother-of-pearl, or that one meets with pearls in which a plug has been placed in the side immediately in the center between the two drill holes, is due to the fact that the great demand has resulted in the destruction of many oriental ornaments in which the pearls were drilled in various ways, as well as in the destruction of the different Magyar and other semi-official jewels of eastern Europe.

The most primitive known drills were the flint drills, made by the North American Indians by chipping chert or flint-like minerals to a fine point. With these rude instruments a large, irregular hole was made, which generally measured several times the diameter of the fine drill hole made by a modern pearl driller with an improved drill. The Indians are also said to have used hot copper drills for boring holes.

The earliest, and still a very general and perhaps the best way of drilling pearls, is by means of the bow- or fiddle-drill. This method has been used in a more or less perfected form by all the aboriginal peoples of the New World from Iceland to Tierra del Fuego. But as none of these peoples were familiar with fine, hard steel, they scarcely ever succeeded in making drill holes as fine as those that can be produced by the use of tempered steel. By the latter means, pearls half an inch in diameter are often drilled entirely through with an aperture no larger than a thin bit of straw.

The largest and finest pearls are frequently drilled with the smallest holes, as the slightest loss in weight means a diminution in value. Then, too, a pearl with a small drill hole is not so liable to shift on the string, and thus is less likely to cut the silk thread which holds the pearls together.

It would be difficult to enumerate all the tricks to which some jewelers now resort in order to utilize every fragment of a pearl they can lay their hands on. Some of them are wonderfully clever at reconstruction, but to the woman who loves pearls, nothing can take the place of the soft, beautiful, round gem, with its natural surface.

In sorting pearls for the smaller necklaces, it is customary to open up a number of dozen bunches of the East Indian pearls as they are sent from the East, strung, the ends fastened together in bunches, and then sealed. These pearls are placed on a table and are first arranged according to color and luster on the sorting board. They are then grouped according to size and graduation, the greatest care being exercised in the selection for color, luster, and form. In this way ten necklaces may be re-strung into ten others, the necklaces probably being improved as regards selection, or else better arranged for the uses to which the jeweler wishes to put them.

In the case of the larger necklaces, it frequently requires many years of selection and arrangement before one becomes perfect enough to pass the criticism or suit the fancy of the jeweler.

We have no record as to when the first pearl necklace was strung, nor have we a definite record of the first use of silk for stringing a necklace. The earliest illustration that we have been able to obtain of the use of pearls in the form of a necklace is the one from Susa, in which the pearls were secured with gold. A Syrian necklace, dating about one or two centuries before Christ, was strung by means of a bronze wire. We will endeavor to give a few facts on the interesting process of preparing pearls for wearing.

Pearl stringing is an art, easy as the process may seem, and it is interesting to note the precision, care, and delicacy with which the pearl stringer performs his task. The first step is to grade the pearls according to their size and color, so that they may produce the best possible effect. The largest and finest pearl is placed in the center; alongside of this, on each side, are laid the two pearls next in size which are the most nearly alike in form and hue; and so on to the end of the necklace. This grouping requires both experience and judgment, and is of great importance, since the value of the pearls is often considerably enhanced by a proper arrangement. A skilful stringer is able to grade them so cleverly that only a trifling difference will be found in the weight of the two halves of a necklace.

The stringing process consists in securing the end pearl by a knot to the diamond, pearl, or other clasp which may be used. When a necklace is being strung, the thread is passed through the metal eye, or pearl, or other object that serves as a clasp. It is then tied with one knot, passed through the next pearl, and knotted between that and the second pearl, and sometimes between the second and the third, thus making the joint doubly secure. The other pearls are then strung in their order, a knot being placed after each fifth, fourth, third, or second pearl, or, should there not be enough to give a proper length to the necklace, between each single pearl. The deftness with which the knot is tied so as not to hold the pearl too tightly, and risk the breaking of the thread, and the precision with which forty, fifty, and even sometimes several hundred knots are made on a single string, is a pleasing operation to witness, and requires the greatest care and nicety of touch. If knots are made frequently between the pearls, there is less danger of losing them should the thread break, as only one or two can fall off; sometimes, indeed, when the drill holes are very small, the silk thread, waxed or unwaxed, fits so closely that the pearl does not become detached even when the thread breaks.

The thread used is invariably of silk of the highest standard of purity, strength, and texture, undyed, and not containing any chemicals. Two or three of these threads are held together, then with a knife the edges are very carefully scraped till the combined material of the three threads is less than the thickness of one. Some use a needle to scrape or fray to a sharp point. Then this point is stiffened by means of "white glue," the best material of this kind being pure gum arabic dissolved in water. A little of this is rubbed on the pointed threads. It stiffens in a moment, then the pearls are passed on, one after the other. If the pearls to be strung are already on a necklace, this process is simplified by the unknotting of the end of the necklace to be re-strung; two or three of the pearls are slid on to the new string, the ends or points of the new necklace thread are twisted together with the old ends and the pearls are simply transferred.

Frequently the holes have been drilled so as to leave the rims rather sharp; in this way the thread may be frayed out or even cut. This sharp edge can easily be removed by careful reaming. Silk of pure quality is the best material known for stringing pearls. A series of experiments were made with every available fiber of sufficient durability from every quarter of the globe, but silk alone was found to possess the strength, the flexibility, and the smoothness necessary to permit a very fine set of threads to pass through an opening as small as the drill hole of a pearl. In the case of a long chain or sautoire, more than three hundred pearls will be strung on a single row, one of over eighty inches in length containing over three hundred pearls, and it requires a degree of neatness and patience that few possess to do this in exactly the right way, so that the thread may not be cut, that the pearls may not be too tightly strung, and that the ends shall be carefully attached at the clasp, so that the necklace may hang well and there may be no danger of the ends breaking loose.

According to the frequency with which it is worn, a necklace should be re-strung every three, six, or twelve months. The proper time for re-stringing can generally be determined by the stretching of the thread so that it can be seen either between the pearls or at either end, giving the impression that one or more pearls are missing. A newly strung necklace is taut.

Where a collar is from thirteen to fourteen inches in length, there are frequently twenty-three rows of pearls, kept straight by four jeweled bars, and sometimes from ten to twenty-five pearls in a section between a bar. This would mean that there are more than two thousand pearls in a collar of small pearls. When one considers that at each bar and at the catch and clasp of the collar it is necessary to make a knotting, it is not surprising that it requires from three to four days' time of a very expert pearl stringer to string or re-string such a pearl collar. A splendid example of such a twenty-three-row collar is that belonging to Señora Diaz, wife of the President of the Republic of Mexico.[1]

Frequent stringing may sometimes serve as a protection for pearls, as, if wax is used, the drill hole is likely to become coated with wax from the thread, and this prevents the absorption by the pearl of perspiration or moisture of any kind through the thread. Indeed, the thread itself, when waxed, does not readily absorb moisture, and as the interior of the pearl also becomes waxed, this serves to protect it from the absorption of humidity of any kind.

In making pearl necklets or muff-chains, a piece of gold wire of the proper strength and pliability is taken. This wire is passed through the hole of the pearl and then cleverly bent into a loop on each side and firmly soldered. It is important that the wire should be very slightly smaller than the dimension of the hole in the pearl so that it may fit closely. Sometimes, instead of this method, a ring is soldered to one end of the wire before this is passed through the pearl, the other end being then secured in the manner described above. Still another method is occasionally employed; in this a piece of the wire is bent into a ring, but not quite closed, the aperture being just large enough to admit the wire that has traversed the pearl; in this way the wire can be introduced into the opening in the ring, which grips it tightly, and is then soldered to it. In many cases two small rings are strung on the wire on each side of the pearl before the loops are made, so that they interpose between the latter and the pearl itself. This serves to protect the sides of the pearl, as there is otherwise some danger that the hole may become chipped or ragged; the same result can be obtained if small caps, closely fitting the pearl, are used instead of the rings. This is, however, only possible when the pearl is quite round, and in this case the effect produced is often very attractive.

Many of the pearls set as rings and studs are no longer set in points, but are set upon a peg, or are "pegged," as it is termed. Setting a

[1] See portrait facing page 442.

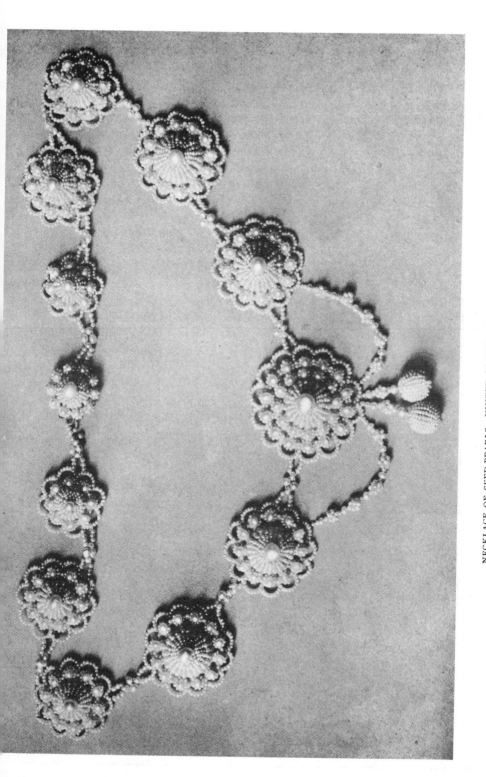

NECKLACE OF SEED-PEARLS. UNITED STATES. CIVIL WAR PERIOD

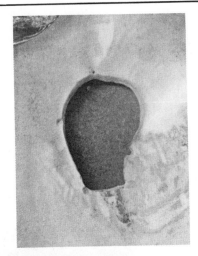

MOTHER-OF-PEARL SHELL
FROM TAHITI

Illustration of a mother-of-pearl shell, showing where a blister has been cut out. In this instance a large pear-shaped pearly blister appeared almost in the center of the shell. A dealer removed this by means of a saw, and was surprised to find that the mother-of-pearl, instead of remaining intact, parted in two pieces. Between these two pieces was a mass of green and white calcareous matter. The two upper figures show the pearly side and the outside of the shell whence the blister was cut. The figures below show the inside and outside of each half of the blister and the earthy matter inclosed.

A is the pearl sawn from the shell.

B is the piece of pearl that parted from the back of this pearly mass.

C and F are two views of the included calcareous matter.

D is the reverse of A, showing the cavity.

E is the reverse of B; originally A rested on B.

There was no indication of any hollow space, or that the mass was not perfect.

pearl in claws generally hides more than one half of the entire sphere. But if the pearl is not properly secured upon the peg, it will occasionally fall off. However, this can be obviated to a great extent by attaching the pearl to a double peg which keeps it from turning and also prevents its falling off. Pearls have occasionally been damaged with the shellac used, or when the gold peg on which the pearl is placed was too hot.

In mounting very small pearls as link chains so as to form a continuous pearly rope without any break in the way of gold links, occasionally V-shaped cavities are drilled into each end of the pearl, and the setting itself is hidden in this V-shaped cavity. This is only done where the pearls are small and not of great value.

The jeweler, in setting pearls, must use the greatest possible care, first, in cutting away the settings, as they are fastened to the pearl, not to scratch or mar it; and then, when he files the settings, not to allow the file to touch the pearl, as both the steel tool and the file would injure it. He must particularly avoid placing the pearl too close to a diamond, ruby, or other precious stone; for, even if the pearl only slightly touches the gem against which it is set, a knock of the hand may mar the pearl's surface. More especially, as pearls are set at present, "pegged" and without points, it is of the greatest importance that they be worn in such a way that they may not touch the unexposed edges of any precious stones, as this also would injure the pearls. For lack of this precaution fine pearls have frequently been harmed.

A large jewelry firm has under consideration the following pearl order: Any workman who in any way mutilates a pearl by filing, imperfect drilling or shaping, or in any way affects the shape of a pearl, without the authority of the foreman, will be called upon to pay for the same.

As pearls are natural objects, any change of the same to fit the setting, or for attachment to any gold object, mutilates the gem and greatly affects its value. If belonging to a customer, this frequently means its replacement, often at a great cost to the jeweler.

Pearl "blisters" frequently have the appearance of being empty; they are generally filled with a fluid, either water or the product of animal and vegetable decomposition. These contents usually emit a peculiar and unpleasant odor. As the exterior of the inclosure gradually wears away and disappears, the contents of the blister are slowly absorbed by the shell itself, and any organic or insoluble substances are deposited on its inner surface.

Thus, when a shell shows any protuberance on this surface, the peeler will cut or scrape away a portion of the decaying shell behind the spot. Should he discover the hole of a borer, he lays the shell aside; but if he finds it to be perfect at this spot, it is evident that the inclusion came from within, and frequently it turns out to be an included pearl. This is removed by breaking the shell, or by cutting around the protuberance very near to its edge, and then breaking away the shell. The pearl is often visible, and layer after layer of the covering mass is removed with the greatest care by the peeler, who is rewarded by bringing to light pearls of various qualities, and frequently those of great value.

An instance in which, by opening a pearl blister, the speculator received a good reward is given by Streeter, who says: "The *Harriet* had the good luck to find, in 1882, a pearl 103 grains in weight, which was inclosed in a huge blister. It was a fine *bouton,* of splendid color in the upper portion, but a trifle chalky below. This was attributed to the admission of salt water into the shell through a hole made by a borer which happened to pierce the shell just where the pearl lay, and had penetrated the latter for almost a quarter of an inch."

Sometimes pearl masses are hollow. Barbot[1] mentions that a French merchant residing in Mexico, having bought one of these pieces from a fisherman at a low price, resolved to satisfy his curiosity by finding out what was inside. He split it in two parts and was agreeably surprised to find a pearl weighing 14¼ carats (57 grains), so round, of such good water, and such fine orient, that he sold it in Paris for nearly 5000 francs ($1000) in 1850.

Seed-pearl work was introduced into the United States, about seventy years ago, by Henry Dubosq, who had studied the methods employed in Europe and has been succeeded in this industry by his son, Augustus Dubosq. The father bought a large quantity of English seed-pearl jewelry, brought it to this country, and hired a number of girls to take it apart carefully and re-string it with white horsehair, to learn how it was made. With no more teaching, he established an industry that has already lasted for three score and ten years.

Seed-pearl jewelry was most in vogue from the year 1840 to 1860. It was generally sold in sets, in a case consisting of a collar, two bracelets, two earrings, a small brooch, and a large spray or corsage ornament. If the object was almost round, occasionally there was a larger central pearl, weighing from one to five grains, usually a button pearl; or, if the ornament was elongated, there were generally three larger pearls. These sometimes possessed a fairly good luster. Seed-pearl jewelry was at one time so popular, and the values were so small in this country, that a $1000 seed-pearl set formed a principal feature of the Tiffany exhibit at the International Exposition held at the Crystal Palace, New York, in 1855.

[1] Charles Barbot, "Traité Complète des Pierres Précieuses," Paris, 1858, pp. 464, 465.

Seed-pearl tiaras sell for from $75 to $200 or $300 each. The work is almost entirely done by girls, either German or of German origin. As labor is higher and pearls have advanced in price, none of the old work could now be duplicated for the amount it cost twenty or thirty years ago. The stringing of the pearls on the English scroll means probably twelve hours of continuous work. An efficient pearl worker receives $3.50 a day, which consists of not more than eight hours, as, owing to the very trying character of the work, clear daylight is necessary to see the holes in the small pearls and in the mother-of-pearl shell.

The foundation of all seed-pearl work is mother-of-pearl. The shell is brought in thin plates, measuring from one and one half to two and one half inches square. One of the most popular and attractive patterns is the English scroll. If a design is to be repeated, a brass figure is made. For the fabrication of a brooch, for instance, a design is first made by drawing on a paper or cardboard; then a brass plate or pattern is cut out, leaving spaces wherever there are to be no pearls. After this a slab of stock mother-of-pearl, nearest the size of the brass plate, is selected, and is sawn out, using the brass plate as a guide for the outlines. The mother-of-pearl is then pierced wherever a pearl is to be secured, and the pearls for its embellishment are chosen, and are strung onto the mother-of-pearl outlines with a special horsehair thread. All the work that remains for the jeweler is the addition of a pin or catch on the back. A representation is given of the designs, the brass plate, the mother-of-pearl, the horsehair, the pearls, and the completed brooch made by this model.

Fine horsehair is used for stringing seed-pearls, because the holes drilled in them are usually too small to admit of the use of silk, and it is very important that what is known as pulled hair, taken from a living horse, should be used, as otherwise the hair is too brittle. This hair, in bunches of from eight to fourteen inches in length, is sold at an average price of $1.50 a pound, and frequently only one ounce is selected for use from the entire pound.

All the pearls used by the seed-pearl workers are purchased in strings and bunches; the finest are those known as the Chinese seed-pearls; they are drilled and strung in bunches, weighing three ounces, and are worth $40 an ounce. They are drilled with so fine an aperture that silk will not pass through the pearl, and only horsehair can be used. The Indian Madras pearls, however, have a larger drill hole and can be strung with silk; they are at present worth from eight to fifteen cents a grain, that is, $48 to $90 per ounce.

Immense quantities of these very minute pearls are also used in bunches or strings, sometimes as many as twenty or thirty strings being grouped together and either bound straight or else twisted into veritable ropes of pearls.

Seed-pearls are sold by the ounce, a single ounce frequently containing as many as 9000,—that is, fifteen pearls to the pearl grain or sixty to the carat,—selling for from $48 to $60 an ounce. Naturally, some of these pearls are even smaller than this, but the average is maintained by those that are a little larger.

Pearls as small as 100 to a diamond carat are drilled and used in seed-pearl work. Diamonds, rubies, and even sapphires, however, are cut in brilliant form when they are as small as 250 to 300 to the carat, or 45,000 to the ounce. The price of these small pearls, however, is only from eight to fifteen cents per carat, whereas diamonds of this size are worth from $200 to $300, their value being three times that of those weighing one sixteenth to one eighth carat each. This is due to the fact that the labor expended in cutting the smaller diamonds is much greater than that bestowed upon the pearls, which simply require drilling and not cutting.

"Half-pearl," as we have mentioned, is the name given to such pearls as are round and spherically domed and are either somewhat flat or almost the shape of one half of a whole pearl of the same diameter. They are produced in two ways: some are cut away as hemispheres from the inner surface of the shell of the pearl-mussel, but more usually they are the better portions of defective whole pearls which are sawn or split by hand into two "halves" with a minute saw, the defective part being rejected altogether or classified as inferior half-pearl, while the better half is classified as a I or II quality half-pearl. Frequently a fine specimen is obtained from an elongated pearl, and sometimes two, three, or even four half-pearls are secured from the various bright parts of a round pearl. In splitting half-pearls, the pearl to be operated upon is held by hand in a kind of grooved vice or pincers and sawn through with a very fine saw; this process is at once simple, rapid, and of insignificant cost.

Only pearls which cannot be cut are filed. In this process the poorer side of the pearl in question is laid upon the file, and the operator takes a piece of ordinary hard wood, so formed that he can grasp it firmly in his hand, presses it down upon the pearl, and rubs the latter on the file, removing all but the good side. In this way a half-pearl is produced.

The smaller half-pearls are from .5 to .75 millimeters in diameter, and an ordinary ounce of half-pearl material will number 18,000. Of the manufactured half-pearls there are, on an average, 20,528 to an ounce.

The half-pearl industry is largely carried on in Idar, on the Nahe

River, and in Oberstein, in the Duchy of Oldenburg, Germany. The pearls are usually purchased from London or Paris houses in lots valued up to $12,000 or more, although some of the firms buy directly from India. In Idar about one hundred people are employed in this industry. Frequently it is pursued in the home of the manufacturer, who may employ from one to a dozen or more workers. These generally include a sorter or arranger, and a marker to indicate the part of the pearl which should be sawn off. There is also a trimmer or one who finally adjusts the pearls.

An unusually clever bit of deception was practised by an American pearl fisher who had found two pearl blisters of almost identical size. Both of these blisters were hollow, and were alike in form. The pearl dealer very cleverly polished down both sides, rounded off the edges, cemented the two backs together, and except for a tiny edge they had all the appearance of a drop pearl that was fairly perfect on both sides. It required but a little heating to separate the parts and show the deception.

In setting half-pearls, they are generally selected from large lots with great care as to their being of uniform size. A circular place for the setting is often drilled with a steel drill, either for several or for a single one. The half-pearl is frequently placed on one or more tiny disks of paper, to give it the exact height in the setting, and the edge of gold is rubbed up against the pearl, which is thus secured in its place; or else tiny edges of gold are left projecting between each pearl. These are pressed down after the pearl is in place. This process requires great delicacy and skill and is frequently employed in the decoration of pearl lockets and watches. In some of the cheaper work, the half-pearls are cemented into the shallow disks that were drilled for them, but frequently they are secured by metal points skilfully raised out of the disks in which the pearls are set, and then pressed down to hold the latter in place. Although apparently fraily set, it is surprising that half-pearl ornaments have been owned for more than a century, scarcely a pearl dropping out; and even if one or two pearls should be lost from the piece of jewelry, the expense of replacing them is not very great. They are often not as safely set when they are mounted with diamonds, rubies, or other stones, more especially in rounded rings or bracelets.

In drilling gold for the setting of half-pearls, where the hole must not be carried right through the metal, a so-called "pearl drill" is used. This is designed to cut a hole with a flat base in comparatively thin layers of metal without disfiguring the opposite side, a task that can easily be accomplished if care be taken not to drill deeper than is strictly necessary for the safe adjustment of the pearl. For the construction of this drill a piece of round steel wire of suitable size is chosen; this is hammered flat at one end and then filed away at each side, leaving a small spike standing in the center, which projects a little beyond the cutting edges and acts as a pivot on which the drill revolves. The steel on both sides of this spike is filed down to a fine edge, care being taken to preserve the horizontal line, so that when the spike is embedded in the metal both cutting edges come into play simultaneously. If the drill is in good condition, it does its work very rapidly, since it is used in an upright drill-stock, whose weight gives a uniform and constant pressure. A good range of sizes of this drill should be kept ready for use, so that one may be found to suit the dimensions of any given pearl. This is essential in order to make an opening just large enough to hold the gem, so that it may fit tightly, without the necesssity of reaming out the hole.

Half-pearls were frequently used with the most pleasing effect in the decoration of antique watches. A number of remarkable examples of this type are among the collection of antique watches of Henry Walters of Baltimore. This collection had been acquired by Tiffany & Co. after the sale of the San Donato Palace, the watches having been withdrawn from the prince's collection by his sister sometime before the sale.

In mounting pearls on gold, a white paste is sometimes employed in half-pearl mounting, which is called by the French jewelers *gouache*. This substance contains white lead, and its use is liable to be injurious to the workmen, cases of lead colic having been recently recognized as thus produced. This subject has lately (1907) been brought forward at the Société Médicale des Hôpitaux in Paris. The cases were at first mistaken for appendicitis, but proved to be well-marked cases of lead poisoning. They had not been reported previously, and are evidently not frequent, those noted being confined to instances in which the employees had carelessly been in the habit of removing an excess of the paste with the tongue.

Pearls that are constantly worn with judicious care do not seem to deteriorate in any way. By judicious care we mean that pearls should not be dropped or thrown down violently or placed on any substance which is likely to act injuriously on the surface of the pearl itself.

Strings of pearls should never be dipped into water or solutions of any kind, because the string which passes through them is likely to absorb and to draw the liquid into the pearl, and as the pearl is made up of many concentric layers, it is quite possible that, through capillary action, some liquid, either pure, or stained with a foreign substance, might be brought into the pearl, which would in this way eventually become discolored. Rings and brooches containing half-pearls fre-

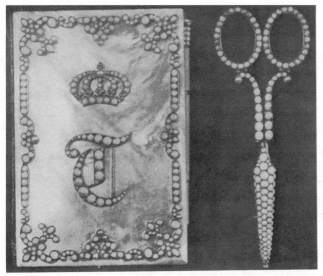

Ladies' sewing case and scissors inlaid with half-pearls
Eighteenth Century

Watch incrusted with half-pearls
Paris Exposition, 1900

Snuff-box, ivory inlaid with fresh-water pearls
Eighteenth Century.   Collection, Metropolitan Museum of Art

Watch incrusted with
half-pearls
Paris Exposition, 1900

Watch incrusted with
half-pearls
Paris Exposition, 1900

Miniature of Catherine Emilie Peake, by Richard Cosway. Gold
frame, surrounded by half-pearls.   Eighteenth Century

EVOLUTION OF A SEED-PEARL BROOCH

Mother-of-pearl plate   Brass model   Pearl brooch completed
Design of brooch   Mother-of-pearl sawn out

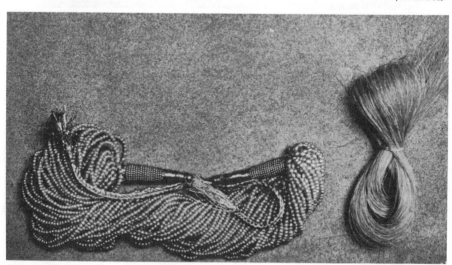

Seed-pearls, Indian strings   White horse-hair for stringing

quently change color from this cause; but contact with the skin, or with lace, or with fabrics which are not stained with certain chemical solutions, seems to have no injurious effect upon pearls.

It is quite possible that in some instances where pearls which have been inherited are thought to have changed and lost their beauty, this belief has been owing to an exaggerated opinion of their quality on the part of those who expected to inherit them and who never had the opportunity to examine them carefully. In other words, in many cases where pearls are believed to have lost their luster, to have died, or partly died, there seems, from the personal observation of the writer, to be little doubt that they never were really fine pearls, and that no change had actually taken place in them. That pearls change but slightly is evidenced by the fact that a splendid necklace belonging to the Empress Eugénie, which was purchased about the year 1860, is in as good condition to-day as when it first passed into the hands of the unfortunate empress of the Second Empire. Many of the pearls in the royal treasury in Vienna that belonged to Maria Theresa, and those that were disposed of at the sale of the French crown jewels in 1886, as well as the pearls that are in the imperial collection at St. Petersburg, do not seem to show any appreciable evidence of age.

The pearl is of a lower hardness than any of the precious or semiprecious stones, and almost as soft as malachite, though not so friable or liable to break as is that mineral; nevertheless, it is in many ways one of the most indestructible of natural objects of the low hardness. Still, pearls, and especially fine pearls, require some care; but, if the same attention is accorded them as would be given to a fine piece of lace, velvet, or other fabric, or to a fine jewel, they will last for a number of generations. If, however, pearls are worn at all times without removal, if they are worn in the bath, if they are thrown on a dressing-table, dropped on the floor, or otherwise ill-treated, if they are worn on dusty automobile rides, in bicycle riding, or during other gymnastic or violent exercise, it is inevitable that their sides will rub together and wear one another away. If they are worn in the bath or in swimming, the silk string which holds them, should it become soaked, may draw some of the water, accompanied perhaps with dust and perspiration, through the drill hole into the center of the pearl, and this is likely to be absorbed in turn by the various layers of the pearl, in some instances undoubtedly affecting the color, changing it to a yellow or a gray. It would be well not to wear pearls under the exceptional conditions above mentioned; and, if they are carefully wiped at times, so as to remove any perspiration or dust, their color is not likely to be affected for a long period of time.

Dr. George Harley writes in the "Proceedings of the Royal Society," March 1, 1888, p. 463:

On one occasion being desirous to crush into powder a split-pea sized pearl, we folded it between two plies of note-paper, turned up the corner of the carpet, and placing it on the hard, bare floor, stood upon it with all our weight. Yet, notwithstanding that we weigh over twelve stone, we failed to make any impression whatever upon the pearl, and even stamping upon it with the heel of our boot did not suffice so much as to fracture it. It was accordingly given to the servant to break with a hammer, and on his return he informed us that on attempting to break it with the hammer against the pantry table, all he succeeded in doing was to make the pearl pierce through the paper and sink into the wooden table, just as if it had been the top part of an iron nail, and that it was not until he had given it a hard blow with the hammer against the bottom of a flat-iron that he succeeded in breaking it.

As the foregoing and other notes had appeared on this subject, the author was led to observe that pearls are possessed of greater durability than is generally supposed. In order to demonstrate this satisfactorily, he took a number of American pearls and placed them upon different kinds of woods, such as white and yellow pine, white oak, teak, ash, cherry, chestnut, and rosewood. He then stood upon them, thus bringing a weight of more than two hundred pounds to bear upon them by means of his heel. The pearls were driven into the different woods, with the single exception of the rosewood, which offered greater resistance so that the pearl only entered partly. In but one instance did a pearl suffer by a slight scaling off. This shows the strength of the many concentric layers, both mineral and vegetable.

This does not signify that pearls should be stepped upon, trodden upon, or thrown about, as it is not unlikely that a pearl would crack if it should fall from some height upon a hardwood or stone floor.

It is believed by many that wrapping pearls in dyed velvets or in fatty woolen materials, and locking them up in safe-deposit vaults, may slightly change them. On the other hand, there is no doubt that sunlight will bleach a pearl, and hence it is that wearing them in the light and air cannot injuriously affect them.

For cleaning pearls, first rub them with a cloth dipped in alcohol diluted with warm (not hot) water, or in a weak solution of soap and water, then dip another cloth in clean water and rub the pearls until they are dry. Be careful not to leave them wet. Either salt, rice, pearl-powder, or some exceedingly soft substance may aid in cleaning them, but no abrasive such as ground pumice, electro-silicon, or any powder that is sold as a polishing powder, should be used.

There are many things that will cause injury to pearls. Occasionally they are affected by the wearer having exudations from the skin induced by some disease or else by acids which pass out through the pores with the perspiration. A smoky atmosphere in which a sulphuric acid is present owing to sulphur in the coal, violent usage such as knocking severely, or dropping—all of these will in time cause more or less injury to a pearl, more especially to one of the whiter varieties; but it is believed that those of a yellowish cast are not so susceptible. Diderot mentioned this as early as 1765.

The "life" of a pearl is said to be fifty, one hundred, and perhaps even one hundred and fifty years; they certainly last for several generations. It has been asserted, without any particular authority, that pearls from the Pacific Ocean and those from Mexico do not last as long as those from the Orient, but this statement is questionable.

If there be any foundation for the belief that it is not well to lock pearls in a safe-deposit box, this is probably owing to the fact that the absolute exclusion from the air may cause the drying out of the organic constituent of the pearl. This may be obviated by putting the pearls in a piece of linen absolutely free from any chemical, at the same time placing with them a bit of blotting-paper or fiber-paper saturated with water; the whole should then be wrapped up in paraffin paper, which will prevent the evaporation of the moisture.

Many sentimental recitals have appeared in the press during the last ten years in regard to the dying of pearls. In connection with this there is a beautiful though mythical story to the effect that Carlotta, wife of the ill-fated Maximilian, Emperor of Mexico, was the possessor of a large collection of pearls which had died, and that these pearls had been placed in a casket and sunk in the depths of the Adriatic, opposite the beautiful but unhappy palace home, Miramar, in the hope that the salt water would revive and restore their original luster. When, however, the time came to bring up the pearls from the sea, it was found that the casket had, in some way, broken loose from the chains, and all trace of it was lost. It is needless to state that there was absolutely no foundation for this romantic tale; indeed, these very pearls were afterward sold. Furthermore, pearls have never lived, and hence they can never die. They do, however, decay, if exposed to influences which destroy either the calcareous or the animal layer of the pearl itself. This is due to many causes: first, overheating, sometimes through the inexperience of a pearl driller; secondly, undue exposure to heat in the washing of a pearl necklace; thirdly, exposure to acids or acid fumes. Apparently there seems to be some foundation for the belief that if they are confined in safe-deposit boxes, probably in contact with wool or with the colored velvets of jewel-cases, the skin of the pearl may be more or less af-

fected. There is no question that in the oriental fisheries so-called
dead pearls have been found in the shell itself, probably owing to some

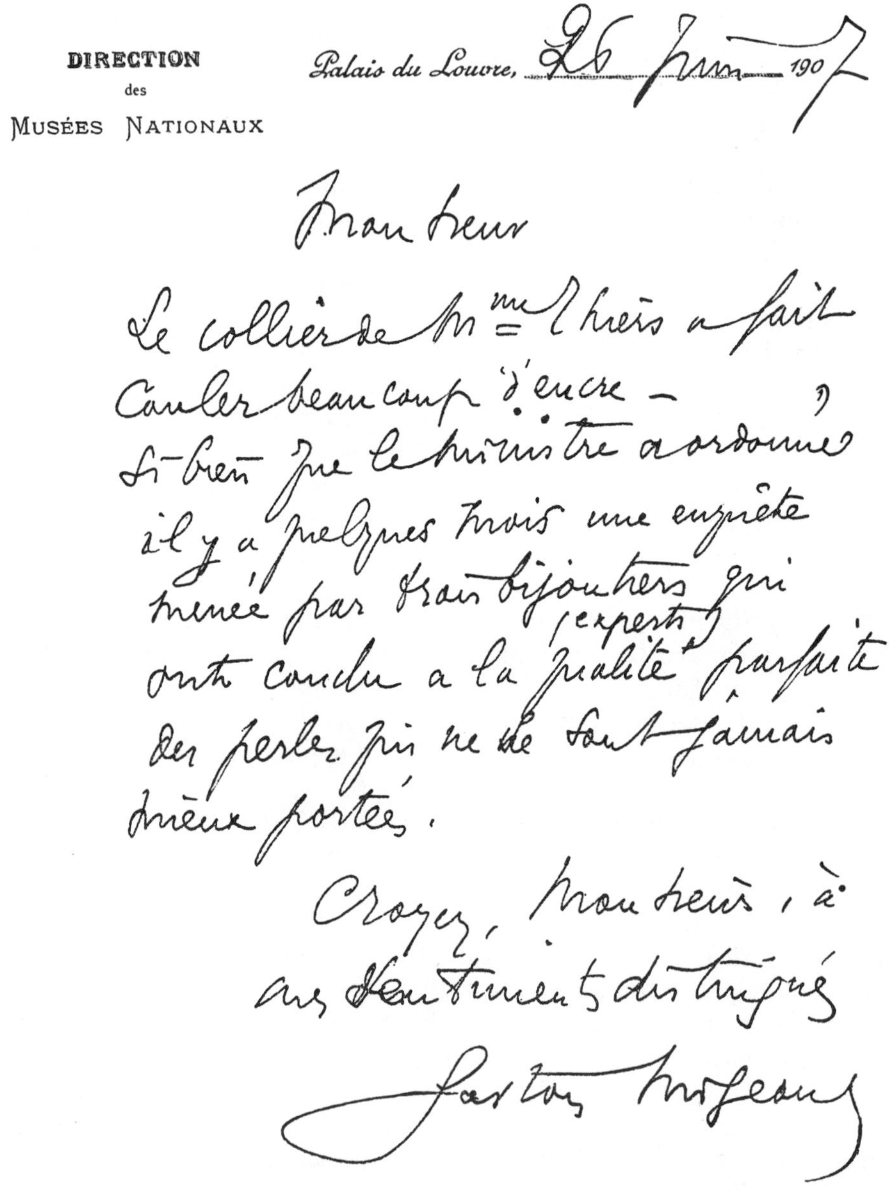

DIRECTION
des
MUSÉES NATIONAUX

Palais du Louvre, 26 Juin 1907

Monsieur

Le collier de Mme Thiers a fait couler beaucoup d'encre — si bien que le ministre a ordonné il y a quelques mois une enquête menée par trois bijoutiers (experts) qui ont conclu a la qualité parfaite des perles qui ne se sont jamais mieux portées.

Croyez, Monsieur, à mes sentiments distingués

Gaston Mogeaud

Facsimile letter of M. Gaston Mogeaud, Director of the Louvre, Paris, stating that the Madame Thiers' pearls are in perfect condition, and have never been in better health.

disease of the pearl-oyster; and they have also appeared in the fresh-water pearl fisheries of the United States, where the pearls have been

MADAME THIERS'S PEARL NECKLACE, BEQUEATHED TO THE LOUVRE MUSEUM, PARIS

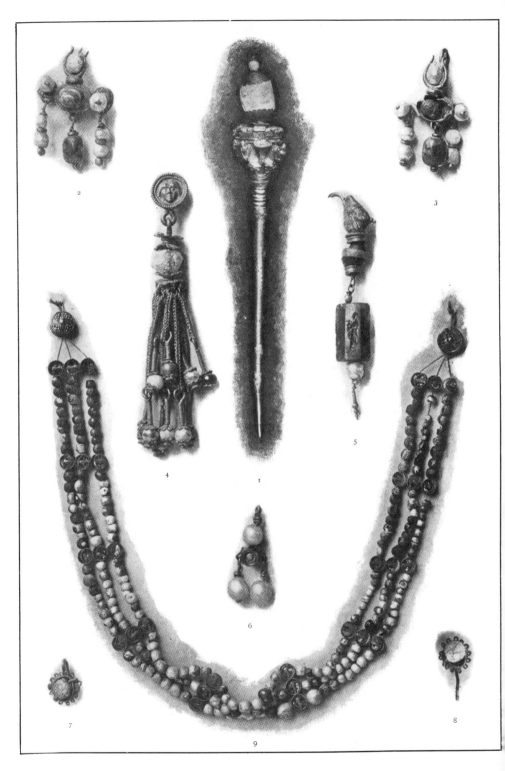

ANTIQUE ORNAMENTS OF PEARLS

No. 1. Gold pin from Paphos, Island of Cyprus, mounted with large marine and small fresh-water pearl, now in British Museum.
Nos. 2, 3, 4, 5, 6, 7, 8. Gold earrings and pins set with pearls, now in the Egyptian Gallery of the Louvre, Paris.
No. 9. Pearl and gold necklace found at Susa, Persia, now in the Louvre, Paris.

too long boiled in the opening of the shell, or where they have been swallowed and have passed through the body of some ruminant, such as a hog, etc.

Probably about no necklace has more been printed than about the famous necklace of Madame Thiers, now in the Louvre Museum of Paris. Article after article has gone over the face of the earth, stating that the pearls in this necklace were dying, and that a record was being kept of the slow death that was overtaking them. Through the courtesy of the director of the museum, M. Gaston Mogeaud, we are permitted to reproduce the following statement from a letter, showing very clearly that there is absolutely no truth in the assertion, and that this necklace has in no way suffered, or is likely to suffer, for many years to come.

"The necklace of Madame Thiers has caused much ink to flow, to such an extent that, a few months ago, the minister ordered an examination to be made by three expert jewelers, who have found that the pearls are in perfect condition, and have never been in better health."

For assuring the safety of jewels there are the primitive methods such as are used in the East Indies, of hiding pearls in out-of-the-way places, where they often escape detection; or else they may be protected by means of an armored room, like the gem-room that contained the wonderful collection of the Duke of Brunswick when he resided in Paris. Decoy necklaces have even been made to represent the original, and so placed that they were taken away by the highwayman or stolen by the burglar under the belief that he was stealing the jewels; while in other cases the pearls have been carried in receptacles that would not be taken for jewel-caskets, a device resorted to by some travelers.

A word in regard to the former system of strong boxes or small safes for the home. These protect from fire and from the ordinary thief, but they have sometimes not proved so invulnerable to the expert cracksman. Quite recently a jewel chest has been devised which can be placed in a trunk and carried from city to city by the owner. It is provided with an exceedingly sensitive electrical apparatus, by means of which a loud burglar alarm is set off should the chest be lifted even one thirty-second of an inch or jarred ever so slightly. This alarm is set automatically when the owner turns the key, and if once started, it will ring for a couple of hours, stopping only when the box is unlocked, thus preventing the carrying away of what is otherwise a portable box.

Lastly, there are the more advanced methods, in use during the past two centuries, such as taking the jewels to a banker and allowing him to place them in his vault, where they are guarded as well as are his own belongings, but not always with the security of the modern safe-deposit vaults, where the gems are absolutely under the control of the owner, and can frequently be obtained at any hour of the day; or as safely kept as they are when deposited in the safe deposit of the jeweler, in whose establishment they can be cleaned, repaired, added to, or changed without risking their removal to another building.

XV

PEARLS AS USED IN ORNAMENTS AND DECORATION

PEARLS AS USED IN ORNAMENTS AND DECORATION

And the necklace,
An India in itself, yet dazzling not.
BYRON, *Marino Faliero.*

THE brilliant diamond and the love of its possession has capti-
vated many to such a degree that it has often been the cause
of intrigue and bloodshed; and national history has been influenced by its acquisition or retention. The pearl, however,
though the most quiet of gems, has, in its own way, found favor in the
sight of emperors and empresses, kings and queens, generals, nobles,
and priests; and even savages have admired its quiet, stately dignity.

The following pages are devoted to a description of the various
ornamental uses of the pearl in different times and countries. Naturally, many of the famous pearls in the following chapter, if considered purely as ornaments, might have found a place here.

The Egyptians of olden times do not appear to have used fine
pearls, although they probably knew of them on account of the prox-
imity of the Red Sea. M. J. de Morgan, the explorer, says: "In the
tombs of Dashour I have never seen any; the only ones that I know of
in Egyptian jewelry belong to the Ptolemaic period and are mounted
in Greek style."[1]

This statement is confirmed by Dr. William F. Petrie, the well-
known Egyptologist, who writes under date of July 26, 1907: "The
pearl was often used in Roman jewelry in Egypt, but I do not know
of any instance of it in pure Egyptian work. The Romans pierced
it and hung it by gold wire on earrings. They also made glass,
pearl-like beads, called *luli* by the modern natives. These beads are
made by silvering glass beads and then flashing over them another
coat of glass."

Among specimens of the late Egyptian work we may note here some
objects in the Louvre:

A pleasing decoration on gold wire is a necklace in the collection of

[1] "Délégation en Perse," Paris, 1905, Vol. VIII, p. 52.

the Egyptian Gallery. In this very small pearls are used as a connective decoration for the points of leaves, and to hold the leaves and ornaments is a gold wire which is secured by bending. This piece comprises 104 pearls, a greater number than is contained in any other object of antiquity found in Egypt.

An Egyptian pendant of unknown origin is also shown in this collection. At the lower end is a bull's head, caparisoned, and the tip of each horn is fitted with a ball like the *embolados toros* of the Spanish bull-fights. The rein is double, and above this there are two rondelles of an unidentified material; then comes a rondelle of lapis lazuli, and after this a rondelle of gold. The whole is strung with twisted gold wire. The center stone is an hexagonal amethyst, evidently a crystal, the two faces of which had been polished and incised. One of these faces represents a priest with a staff of office, and the other a priest holding an incense-burner with the hieroglyph of the altar. With one hand he is offering the two sacrifices, the mineral and the vegetable; in the other he holds a garland of flowers or leaves. Above this is an Oriental pearl somewhat worn and abraded. All these are secured by a twisted gold wire, to which four tiny gold beads of graduated size are affixed at the top of the pendant.

There are six other pendants and earrings in the Egyptian Gallery, all of which contain pearls, and in most instances these pearls have been drilled and suspended by metal wires, unless they are used as an ornament facing outward. In four instances they are secured by a peg of gold.

The Assyrian and Persian bas-reliefs show that the sovereigns and great personages of those countries adorned themselves profusely with pearls. They wore them not only in their jewelry, but also on their garments and even in their beards![1] The coins of the Persian kings also bear testimony to the use of the gem in ancient Persia, since the sovereigns are represented wearing tiaras ornamented with triple rows of pearls.[2] The same may be said of the imperial Roman diadem from the time of Caracalla (188–217 A.D.).

One of the most interesting of all ancient pearl necklaces,[3] containing more pearls than any other that has been found, and in a better state of preservation, is the Susa necklace now in the Persian Gallery of the Louvre Museum. It consists of three rows, each containing 72 pearls, so that there are 216 in all. Ten gold bars, formed of three small disks, each about five millimeters in diameter, divide the necklace into nine equal sections; at each end there is a disk, ten milli-

[1] De Morgan, "Délégation en Perse," Paris, 1905, Vol. VIII, p. 52.
[2] Imhoof-Blumer, "Porträtköpfe auf antiken Münzen," pl. 7, figs. 12 sqq.

[3] See "Délégation en Perse," Vol. VIII. "Recherches Archéologiques." Paris, 1905, third series, pp. 51–2, pl. 5.

meters in diameter, to which the three strands are secured. If there was any other setting, it has evidently disappeared, although it is quite possible that there may only have been a string at each end, as in the East Indian necklaces.

This ornament was found on the site of the ancient Susa or Shushan by M. J. de Morgan, February 10, 1901, in a bronze sarcophagus, which contained the skeleton of a woman, adorned with a great number of gold ornaments set and incrusted with precious stones. M. de Morgan gives *circa* 350 B.C. as the probable date of these objects. The pearls were much deteriorated. About 238 were found, but many of them crumbled away when they were touched. M. de Morgan considers that the necklace was of the type of the "dog-collar" of to-day, and he believes that it originally comprised from 400 to 500 pearls.

According to a personal communication from M. P. Cavvadias, of the Société Archéologique d'Athènes, there are no pearls on the ancient ornaments preserved in the National Museum at Athens. This is hardly surprising in view of the fact that the greater part of these ornaments belong to the archaic period of Greek art; that is to say, to a time when the pearl was evidently unknown to the Greeks.

The fact that we do not find more evidence of the use of pearls in Greece at a later period need cause no surprise, when we consider how many of the treasures of Greek art have disappeared in the course of more than twenty centuries. There can be no question that they were known and used as ornaments at an early time, as we can infer from the description of them by Theophrastus and later Greek authors.

Dr. Edward Robinson of the Metropolitan Museum of Art, and other authorities on Greek art and archæology, maintain that the Arethusa necklace, and other ornaments of that time, depicted on coins, etc., were meant to represent gold ornaments, as it is believed by many that pearls were unknown in Greece at that period.

One of the most interesting specimens showing the use of a pearl in ancient times is a very beautiful pearl pin from Paphos, on the Island of Cyprus, which is mounted with a large marine pearl, probably the largest antique pearl ever found, measuring fourteen millimeters in diameter, and weighing about 70 grains. This, unfortunately, has been very much abraded and worn away, although more than half of the pearl is still present. It is surmounted by a small fresh-water pearl, four millimeters in diameter, weighing about two grains and in a much better state of preservation. This unusually interesting example of prehistoric pearl is in the Greek and Roman department of the British Museum, and we are able to show it by the courtesy of the keeper of that department, Dr. Charles Hercules Read.

In excavations made last spring (1907), in the Hauran district in Syria, Azeez Khayat found a number of loose pearls which had formed a necklace. The tomb in which they were discovered was cut in the rock, and appeared to be of Roman origin. The pearls were still attached to the old bronze wire with which they had been strung. Mr. Khayat also mentions the finding of a pearl pin, and a single earring bearing a pearl, in a rock-tomb at Cæsarea, in Syria. Rock-cut tombs from ten to twelve feet in depth are frequently discovered, and they probably date from the beginning of the Christian era.

The habit was so common of using pearls as a base to throw up the brilliance of other gems, that we may, perhaps, believe even in Caligula's slippers of pearls, with rubies and emeralds set upon them like flowers.

The Roman ladies had a special favor for pearls as earrings, and it was one of their consuming ambitions to possess exceptionally fine specimens for this purpose. They preferred pear-shaped pearls, and often wore two or three of them strung together. They jingled gently as they moved about—a fitting accompaniment, it may be said, to their graceful movements—and from this jingling the name *crotalia,* or "rattles," was applied to them.

The description given by Pliny of the pearl ornaments of Lollia Paulina is the principal claim which the wife of Caligula has on our interest.

I myselfe have seen Lollia Paulina when she was dressed . . . so beset and bedeckt all over with hemeraulds and pearles, disposed in rewes, ranks, and courses one by another; round about the attire of her head, her cawle, her borders, her perruke of hair, her bongrace and chaplet; at her ears pendant, about her neck in a carcanet, upon her wrest in bracelets, & on her fingers in rings; that she glittered and shone againe like the sun as she went. The value of these ornaments she esteemed and rated at forty million Sestertij[1] and offered openly to prove it out of hand by her bookes of accounts and reckonings. Yet were not these jewels the gifts and presents of the prodigall prince her husband, but the goods and ornaments from her owne house, fallen to her by way of inheritance from her grandfather, which he had gotten together even by the robbing and spoiling of whole provinces. See what the issue and end was of those extortions and outrageous exactions of his: this was it. That M. Lollius, slandered and defamed for receiving bribes and presents of the kings in the East; and being out of favor with C. Cæsar, sonne of Augustus, and having lost his amitie, dranke a cup of poison, and prevented his judiciall triall: that forsooth his neece Lollia, all to be hanged with jewels of 400 hundred thousand Sestertij, should be seene glittering, and looked at of every man by candle-light all a supper time.[2]

[1] Equivalent to about 1,250,000 ounces of silver; Hardouin says 7,600,000 francs.

[2] "Naturall Historie," London, 1601, *Lib.* IX, c. 35.

TYSZKIEWICZ BRONZE STATUETTE OF APHRODITE, SHOWING EARRINGS
OF PEARL AND GOLD OF EARLY GREEK PERIOD

Now in Museum of Fine Arts, Boston, Mass.

ANTIQUE PEARL ORNAMENTS

No. 1. Gold earring with turquoise top. Two pearls, two garnets, and two pearls. Found in southern Siberia in 1726; believed to be of the second century, A.D.

No. 2. Brass earring with one pearl and glass beads. Fourth century, A.D.

No. 3. Brass dress pin. Sphere of amber, surmounted by a pearl. Found near village of Mzchet Caucasus. Fourth century, A.D.

No. 4. Carnelian dress pins with pearl tops. Early Christian.

No. 5. Gold earring, hook and eye type. From Olbia, the site of an ancient Greek colony. Fourth century, A.D.

Nos. 1 to 5 are from the collection of the Imperial Hermitage in St. Petersburg.

Nos. 6–8–9. Pearl and gold earrings, Greek, from the Island of Cyprus. Second century, A.D.

No. 7. Roman brooch (pearls and gold), found in the river Thames, England. Ninth century, A.D.

And the taste of the Roman ladies for pearls has perpetuated itself in Italy, though other of the luxurious habits which in their case accompanied it, have long since died out. The women of Florence even now are not content if they do not possess a necklet of pearls, and this generally forms the marriage portion of the middle-class women. It is thought, just as it was in ancient Rome, that this gives an air of respectability, and forms a sure protection from insult in the street or elsewhere.

One of the earliest illustrations showing a pearl earring is the one in the ear of Julia, the daughter of Titus, incised on a splendid aquamarine in the Bibliothèque Nationale. This gem was formerly in the Treasury of St. Denis, and is considered to belong to the Carlovingian period.[1]

So large and heavy were the earrings worn in Rome that there were women known as *auriculæ ornatrices,* special doctresses whose sole occupation was the healing of ear tumors and of injured or infected ears. In a similar way, at the present day, we have the ear piercer, whose vocation, however, is rapidly becoming useless because of the ingenious modern devices for holding the pearls to unpierced ears; and we must consider this eminently desirable when we think of the ear-piercing outfits of the former jeweler, who never disinfected his apparatus, and when we recall the fact that it was always expected that the ear would swell, first, from the crude awl that was used, and, secondly, from the unsterilized instruments.

That the Romans believed in decorating the statues of their goddesses with pearls and dedicating them as offerings, is evidenced by the gift of Cleopatra's pearl, which was cut in halves to make earrings for the Venus of the Pantheon; and by the buckler of British pearls for the statue of Venus Genetrix, given by Julius Cæsar. Quite a number of statues and busts of the Roman period, and some of an earlier time, have the ears pierced for the reception of earrings, and it is highly probable that pearls were used for this decoration. Among these are the busts of Pallas and Juno Lanuvina in the Vatican; that of Eirene, a marble copy of a work of Cephisdotus, in the Glyptothek, Munich, and the Venus de Medici in the Uffizi, Florence.

Pottier[2] mentions several other Greek statues which show that earrings were used for their adornment; as, for example, the winged Victory of Archernos, in Delos; the head of one of the caryatids found at Delphi, a cast of which is in the Louvre; the archaic Aphrodite of the Villa Ludovisi; the Athena from the frieze of the temple at Ægina;

[1] MS. Bibliothèque Nationale, Paris, 2089, XLVII, No. 12.

[2] "Dictionnaire des Antiquites Grecques et Romaines," ed. by Deremberg and Saglio: Art. "Inaures" by Pottier, Paris, 1899, Vol. III, pp. 440–447.

the Venus of Milo, etc. In other instances the ornament was simply painted on the ear as is shown in the Aphrodite in white marble which has been found in Marseilles. This may also have been the case in the frieze at Olympia. The earrings used in these statues were usually metal disks entirely covering the lobe of the ear. We have, however, many representations of pearl earrings in the paintings at Pompeii, and on cameos and coins. These show us several of the types mentioned by Pliny and other authors; still, they are smaller and more unpretentious than we might expect in view of the well-known luxury of the Roman ladies in this respect. The greater part of the earrings represented show a pearl suspended from a single wire; there are

Pearl earrings from Herculaneum and Pompeii

some, however, with three pearls, one above the other,[1] and a few bearing several pearls loosely hung together, answering to the description of the *crotalia*. Others, again, bear pear-shaped pearls or *elenchi*.[2] It is a singular fact that scarcely any of the busts of Roman women are ornamented with earrings, but it is quite possible that the cause for this must be sought in the desire of the artist to dispense with unimportant details which might detract from the general effect he wished to produce. We may note, however, four female figures in the Gallerie des Empereurs in the Louvre Museum, with the ears pierced for the reception of earrings (Nos. 1195, 1202, 1230, and 1269).

Many numismatists, among them Dr. F. Louis Comparette,[3] believe

[1] Babelon, "Cab. des Antiq.," pl. 33, fig. 3.
[2] Duruy, "Hist. des Romains," Vol. I, p. 511.
[3] Custodian of the coin collection of the Philadelphia Mint.

that the necklaces and earrings represented on Greek coins from the fifth century B.C. are intended to represent pearl ornaments, since the personages depicted are in all cases female divinities, goddesses, or nymphs, held in great veneration in the city where the coins were minted, and it is almost certain that the artist intended to portray the choicest and most beautiful of gems as an adornment for the beautiful head of the city's patron.

The Syracusan coins, by Euvenetus, minted in the early part of the fifth century B.C., and bearing the head of Arethusa, seem to be the earliest coins showing a neck and ear ornament. This was later imitated on the Greek and Greco-Roman coins. A coin of Sulla shows a double necklace, one strand consisting of round beads and the other of pendants. The later coins almost always represent the goddesses with neck and ear ornaments. Some of the latter, however, resembling amphoræ, are neither round nor pear-shaped.

In view of the great fondness of the Romans for pearls, it is not surprising that many of these gems have been found in the excavations at Pompeii, Herculaneum, and Capodimonte. The collection of earrings preserved in the Naples Museum is especially noteworthy. Here we can see earrings consisting of a simple golden hoop, from which hangs a wire bearing a single pearl; others in which a cross-bar is attached to the hoop, and at each end of this bar is a loosely hung wire with a pearl at its extremity, this earring suggesting the *crotalia* mentioned by Pliny (see Fig. A); and still others wherein the pearls are strung directly on the hoop. The cross-bars are of various designs, sometimes entirely smooth, and again shaped like a cornice or a pediment; in other cases we have an earring with two pearls on a wire, then a pierced transparent stone, and beneath that, two pearls terminating the large drop. A few of the earrings are more elaborate, as, for example, one represented in Fig. B which was found in Pompeii, March 8, 1870. Here there is an emerald in the center, surrounded by gold rays, between which were set eight pearls, two of which are now missing; above is a small pearl. The single earring shown in Fig. D came from Herculaneum, and bears a circlet of thirteen pearls, alternating with rubies and other stones; beneath there is a link from which depends a pearl about seven and a fifth millimeters in diameter, and weighing nearly twelve grains. The fact that we know the latest date to which these pearls can be assigned, namely, 79 A.D., renders them peculiarly interesting and valuable from a historical point of view. Naturally, many of them are calcined or otherwise damaged, but others are fairly well preserved as to form, although the luster has departed from them. There are twenty-seven earrings in the collection, and the pearls number about one hundred. No great pearls were found.

In the Roman excavations, and in those of other early remains, many objects are found in which there may be a sapphire, an emerald, or several other stones, pierced, and pendant on a gold wire, with a blank space between, showing that something was there originally. This object has apparently decomposed and fallen away. We may reasonably suppose that it was either a pearl or a glass bead, and it is unlikely that glass would be used in connection with the more precious materials. This pearl or glass may have been affected by the organic acids or the acids resulting from the decomposition of the body with which the ornament was buried for a score of centuries.

Among the ancient jewels containing pearls which are preserved in the Hermitage at St. Petersburg, we may mention a broken gold ring with a roughly cut turquoise and two pendants, each set with two pearls separated by a garnet. This object was found in southern Siberia during the reign of Peter the Great, and may belong to the second century before Christ. Also may be noted a pair of gold earrings, with an engraved six-rayed star, in the center of which a pearl is set, while below hang three pendant sticks, two of which have a pearl at the extremity. These earrings were found in 1892 in a tomb situated close to the site of the ancient town of Chersonesus, in the Crimea. As a coin of the Emperor Gordianus III (224–244 A.D.) was discovered in the same tomb, we may assign the earrings to the first half of the third century A.D.

Beside another pair of earrings, one of which is set with a pearl, and two pearl-headed pins, all from the neighborhood of Tiflis, in the Crimea, we may especially refer to an earring made of a plain, thick, golden wire, on which seven pearls are threaded; one of these occupies the center and the others are grouped around it. This earring was purchased in 1903 by the Russian Imperial Archæological Commission from a collector residing at Odessa; it is said to have been found on the site of the ancient Greek colony of Olbia, but we have no definite external or internal evidence to sustain this view.

We may also note the gold necklace and earrings[1] containing pearls found near the site of Olbia during the reign of Napoleon III, and now in the collection of the Roman, Campana. These objects are especially interesting owing to the fact that the pearls are drilled and a gold cap is set on each side.

A pair of pearl earrings were found in a tomb on Mount Mithridates, near Kertch, in the Crimea. These earrings probably belong to the third or fourth century of our era. Of the four pearls which originally adorned the cross-bars, only one has been preserved. Another pair of earrings was discovered in the same place. It is probable that

[1] Imperial Museum of Archæology, St. Petersburg, Russia.

they were ornamented with pearls in a similar way, but the latter have entirely disappeared.

Gabriele Bremond states in his "Viaggi di Egitto," *Lib.* I, c. 30, that it was a Mohammedan custom to embroider baldachins and carpets of precious metals with pearls. This use is especially typified in a baldachin of gold embroidered with pearls which is over the sepulcher of Mohammed at Mecca.[1]

When the Mohammedans captured the Persian city Ctesiphon, in 637, they collected an immense booty. Each of the 60,000 soldiers received the value of 12,000 dirhems ($1560), a total of $93,600,000. Among the treasures sent to Caliph Omar (581–644), in Medina, was a crown, perhaps that of Khusrau I (499–579), which Tabari says was studded with 1000 pearls each as large as a bird's egg.[2] There was also a wonderful carpet 450 feet long and 90 broad, with a border of emeralds, rubies, sapphires, and pearls, representing luxuriant foliage and beautiful flowers. Tabari states that it was called the "Winter Carpet," because "the Persian kings used it in winter when there was no longer verdure or flowers, for whoever was seated on this carpet thought he looked out upon a garden or a green field.[3]

On the occasion of the marriage of the Caliph Al-Mamun (786–833) with the daughter of Hassan Sahal, all the grandees of Al-Mamun received slaves of both sexes as presents from the bride's father. The preliminary negotiations were held at Fomal Saleh, and the road traversed by the bride and bridegroom to reach Bagdad, a distance of one hundred miles, was covered with mats of cloth of gold and silver. We are told that the bride wore on her head-dress a thousand pearls, each of which is said to have been of enormous value.[4]

Describing the birthday festival of Kublai Khan (*circa* 1275 A.D.), Marco Polo says: "The Great Kaan dresses in the best of his robes, all wrought with beaten gold; and full 12,000 Barons and Knights on that day came forth dressed in robes of the same colour, and precisely like those of the Great Kaan, except that they are not so costly; but still they are all of the same colour as his, and are also of silk and gold. Every man so clothed has a girdle of gold; and this as well as the dress is given him by the Sovereign. And I will aver that there are some of these suits decked with so many pearls and precious stones that a single suit shall be worth full 10,000 golden bezants [about $25,000]."[5]

In the Kan period, in China, the dead bodies of the emperors were

[1] "Della Storia Naturale delle Gemme delle Pietre e di tutti i Minerali," Giacinto Gimma, Naples, 1730.
[2] Tabari, "Chronique," translated by Zotenberg, Paris, 1869, Vol. II, p. 304.
[3] *Ibid.*, Vol. III, p. 417.
[4] Alexander, "The History of Women," London, 1782, Vol. II, p. 136.
[5] "The Book of Ser Marco Polo, the Venetian," trans. and ed. by Col. Henry Yule, London, 1871, Vol. I, p. 343.

embalmed and wrapped in a garment ornamented with pearls. They were then inclosed in a case of jade.[1]

Speaking of the jewels of the King of Maabar, or what is now known as the Coromandel Coast, Marco Polo tells us: "It is a fact that the king goes as bare as the rest, only round his loins he has a piece of fine cloth and round his neck he has a necklace entirely of precious stones,—rubies, sapphires, emeralds, and the like, insomuch that this collar is of great value. He wears also hanging in front of his chest from the neck downwards, a fine silk thread strung with 104 large pearls and rubies of great price. The reason why he wears this cord with the 104 great pearls is (according to what they tell) that every day, morning and evening, he has to say 104 prayers to his idols. Such is their religion and custom; and thus did all the kings his ancestors before him, and they bequeathed the string of pearls to him that he should do the like."[2]

A favorite East Indian amulet is known as the "Nao-ratna" or "Nao-ratan," and consists of "nine gems": in former times the pearl, ruby, topaz, diamond, emerald, lapis lazuli, coral, sapphire, and a stone, not identified, called the gomeda. At the present time these stones are generally the coral, topaz, sapphire, ruby, flat diamond, cut diamond, emerald, hyacinth, and carbuncle. This talisman may suggest the Urim and Thummin or sacred oracle of the Jews, which was said to have been taken from Jerusalem in 615 A.D. by Khusrau II, the Sassanian Persian king.

The East Indian custom for persons of quality was to wear a pearl between two colored stones in each ear, that is, either between two rubies or two emeralds; and Tavernier noted, about 1670, that there was no person of any consideration in those regions who did not wear, in each ear, a pearl set between two colored stones. Another favorite ornament for women in India is a girdle elegantly embroidered, bearing a large pendant pearl in front, where it is fastened.[3]

A necklace of twenty-seven pearls bears in India the name of *nakshatra mālā, nakshatras* (originally "stars") being the name of the twenty-seven divisions of the Hindu zodiac.[4]

In the Indian jewels often a small spot of enamel is fastened or melted on to a gold wire, and then one or several pearls are hung upon it; or beads of some gems, as sapphire, ruby, emerald, or even glass, may be added or alternated with pearls. Then the enamel stop-piece is turned down and the other end of the gold wire is twisted on to the

[1] De Mély, "Les Lapidaires Chinois," Paris, 1896, p. 178.
[2] "The Book of Ser Marco Polo, the Venetian," trans. and ed. by Col. Henry Yule, London, 1871, Vol. II, p. 275.
[3] Alexander, "The History of Women," London, 1782, Vol. II, p. 172.
[4] Max Müller, "Rig-Veda Samhita," 1862, Vol. IV, p. 64.

EAST INDIAN NECKLACE OF PEARLS, TABLE DIAMONDS, GLASS BEADS, GOLD AND ENAMEL
Property of an American lady

CROWN OF RECCESVINTHUS AND OTHER GOTHIC
CROWNS OF THE SEVENTH CENTURY

From the treasure of Guarrazar, near Toledo

Musée de Cluny, Paris

setting, loosely, in such a manner as to swing freely. It is the effect of these dozens or even hundreds of swinging drops that add such grace and elegance to East Indian jewelry.

In China, such precious stones as the ruby, sapphire—both blue and yellow—the emerald, and the pink tourmaline, are not facetted, as with us, but are generally polished in conformity to the shape of the bead or other ornament, and never have a lathe-turned or cut appearance; they are either set in cabochon or as beads, rounded, oval, or elongated. All these forms, and the colors used by the Chinese, lend themselves well to combinations with pearls; and hence pearls are often found in Chinese jewelry, especially in those ornaments which are flexible and graceful, in which the pearls and gems are strung on wire and allowed to swing freely with a gentle tinkle when the wearer moves. This is not unlike the setting of such gems in ancient Roman times. An admirable example is shown and described in Bushell's "Chinese Art" (Vol. II, plate 108, page 90). In this head-dress of a Manchu lady, there are combined with the pearls, jadeite, amethyst, amber, and coral, on a gilt silver openwork, with blue kingfisher feathers. This great cap of state is an admirable example of pure Chinese design and workmanship. The pendant strings of pearls are occasionally relieved by a bit of carved jade, carnelian or coral, especially the latter. Another example, the "cap of state" has silver-gilt openwork and immortelles (Taoist symbols), and is much enhanced in beauty by a decoration or inlay of plates of the beautiful blue feathers of the kingfisher, which are used so extensively and effectively in Chinese jewelry. The pearls are scattered at intervals over the cap, and ten strings of them hang from the sides of it. This is believed to be of Manchu origin by Dr. Stephen W. Bushell, the great Chinese scholar, to whom we are indebted for the use of the illustration. We are also told that young ladies in China wear a sort of crown constructed of pasteboard, covered with silk. This is adorned with pearls, diamonds, and other jewels.[1]

The pearls on many Chinese ornaments were generally strung upon silk, often with half a dozen or a dozen seed-pearls above and below the large pearl, to hold the latter in place, and also to add a softness to the whole jewel. The end pendant pearl, even if pear-shaped, was usually pierced entirely through, and a wire that was worked through it was flattened out, and this gold head was again ornamented in some way. A Chinese pendant from the China-Japan war-loot offers an excellent illustration of this kind of pearl-setting. This was preserved in a double box of finely carved gold.

The rosaries containing 104 pearls, which are used to-day, were mentioned centuries ago by Marco Polo, and an excellent pearl string

[1] Alexander, "The History of Women," London, 1782, Vol. II, p. 171.

of this kind has been in the Russian Treasury at Moscow for over two hundred years. Dr. Stewart Culin, the archæologist, who has paid much attention to Chinese customs, informs us that the black and white counters made for use in games by the Chinese are called black and white pearls.

Dr. T. Nishikawa writes us in 1908 that pearls were used in Japan for ornamental purposes more than a thousand years ago. Large abalone pearls are found in images of Buddha made in 300 A.D. Freshwater pearls, usually from Dipsas and Unio, were also used. A beautiful color-print was made by Hoku'ai of the first pearl, called "tide-jewel" by the Japanese.

Most interesting pearls are those in a brooch in the British Museum, which was discovered in 1839 while excavating a sewer opposite Ludgate Hill in Thames Street, at the depth of about nine feet, in a darkcolored artificial stratum of earth, unaccompanied by any remains that could aid in throwing light upon its history. It is four inches and a half in circumference, and is composed of a circular compartment an inch and a quarter in diameter, set with variegated enamel, representing a full-faced head and bust, with a crown on the head, and the drapery of a mantle, formed of threads of gold effectively arranged so as to mark the features of the face and the folds of the drapery; this is inclosed in a border of rich gold filigree-work, set at equal distances with four pearls.[1] Dr. Charles Roach Smith attributes this brooch to the time of King Alfred, and supposes it to have been executed in England by a foreign artist. He only ventures a conjecture that the head might be that of King Alfred.

Crowns, both ancient and modern, are richly ornamented with pearls. We shall treat of the crown of the Holy Roman Empire and of the imperial Austrian crown in the following chapter. One of the most interesting and ancient is the famous crown of Khusrau II (reigned 590–638), made in the latter part of the sixth century, which was brought to light by Shah Abbas after a thousand years of concealment in an obscure fortress among the mountains of Lauristan. It does not contain diamonds among its ornaments, but is incrusted with pearls and rubies.[2]

From the representation given on the cup of Khusrau, the throne of the Sassanian Persian kings appears to have been as large as a couch; it was supported by four winged animals, whose model had been borrowed by the Sassanians from their ancestors, and it was covered with an embroidered stuff thrown over mattresses and cushions. If we

[1] From a letter of Charles Roach Smith, Esq., F. S. A., to John Gage Rokewode, F. R. S. "Archæologia," Vol. XXIX, p. 70.

[2] Augustus C. Hamlin, "Leisure Hours Among the Gems," Boston, 1884, p. 22.

may believe Tabari ("Chronicles," trans. by Zotenberg, Vol. II, p. 304), this throne was of gold, enriched with precious stones, and surmounted by a crown of gold and pearls, so heavy that the sovereign could not wear it, and therefore had it suspended above his head.[1]

One of the crowns in the Hermitage at St. Petersburg was discovered in 1864 in a tumulus near Novo-Tcherkask, with many other valuable objects, all of which had apparently been buried with some important personage. This crown resembles somewhat that of Reccesvinthus in the treasure of Guerrazar, although some portions of it seem to belong to the period of the Roman empire. The conjecture has been made that the crown may have been worn by a queen since it is decorated with a finely executed bust of a woman in amethyst. The crown itself is of pure gold, and was bordered with two rows of pearls, which have disappeared, leaving only the small disks to which they were attached; besides these, it was ornamented with a number of uncut precious stones. The date of this object cannot be exactly determined, although the consensus of opinion is that it belongs to about the third century after Christ. Possibly the bust and some other portions, which appear to be of Greco-Roman workmanship, are of this period, while the rest of the crown was executed one or two centuries later; it is about seven inches in diameter and two in height.[2]

Toward the end of the year 1858 a French officer who lived in Spain, while making some excavations on a property he owned there, discovered fourteen small gold crowns. They were taken to the Spanish mint and are said to have been melted for bullion. New excavations on the same spot brought to light eight other crowns of considerable weight, of the finest workmanship, and incrusted with precious stones, pearls, etc. There is no doubt that these crowns were buried in the early years of the eighth century, when the Arabs, led by Tarik, invaded Spain and forced the Gothic dynasty to take refuge in the north of Europe. The importance of this discovery is very great, since it gives us positive evidence of the development of the goldsmith's art in Spain at that early period. An inscription proves that one of the crowns was dedicated in the second half of the seventh century, and it is one of the few authentic memorials we possess of that epoch. In February, 1859, the eight crowns were purchased by the French government and placed in the Musée de Cluny. Two other crowns found in the same place were added in 1860, and complete the collection.

The largest of these crowns is that of the Gothic king, Reccesvinthus, who was King of Spain from 649 to 672. It is composed of a

---

[1] Dieulafoy, "L'art antique de la Perse," Paris, 1884. Pt. V, p. 137.
[2] See Maskell, "Russian Art" (South Kensington Museum Handbooks), London, 1884, pp. 83, 84; also "La Russie Méridionale," by Reinach-Kondakoff-Tolstoy, pp. 489, 490.

wide band of solid gold, ten centimeters wide and twenty-one centimeters in diameter (about four and eight inches respectively). This band, which opens by means of a hinge, is surrounded by two borders of gold set with the red stones of Caria, called "gemmae alabandenses," and the band itself is studded with thirty large oriental sapphires of the greatest beauty. Thirty fine pearls of appropriate size alternate with the sapphires on a ground incrusted with the red stones above mentioned. From twenty-three small gold chains depend large letters in cloisonné, and also incrusted, forming the sentence: RECCESVINTHUS REX OFFERET. Each letter has a gold pendant with a pearl from which hangs a pear-shaped sapphire.

The crown is suspended from four chains, converging to a double floral ornament of solid gold, adorned with twelve sapphire pendants. This ornament, the leaves of which are open, is surmounted by a capital of rock crystal, then comes a ball of the same material, and the whole is terminated by the gold center to which the four chains are attached.

The cross, which is suspended underneath the crown by a gold chain, is remarkable for its elegance and its richness. It is of solid gold and is inlaid with six very fine sapphires and eight large pearls, each of which is mounted in relief with claws. At the back, the cross still bears the wire by which it was attached to the royal mantle. The inside of the crown is quite smooth; the outside is composed of elegant fleurettes in openwork, the leaves being filled with the same species of red carnelian mentioned above. There are thirty sapphires, all of the finest water, and a few of them show the natural facetted crystallization; the two principal ones, placed in the center of the band, are thirty millimeters in diameter. The pearls are of an exceptional size, and only a few of them have been injured by time. The total number on the crown, cross, and top ornament, is seventy, thirty of which are unusually large. The chains are each composed of five openwork ornaments with an enamel paste inlaid in the gold edge. A close examination of the crown shows that it had been worn before the king presented it to some church.

The royal Hungarian crown given to St. Stephen by the pope in the year 1000 A.D., when Hungary became an empire, is one of the most ancient crowns in existence. It contains 320 pearls and was procured in Byzantium. It was pledged to the emperor, Frederick IV, by Queen Elizabeth of Hungary, probably about 1440.

In the cathedral of Prague (the metropolitan church of St. Vitus) there may be seen the crown which was made by the order of Charles IV (1378) out of four pounds, ten and a quarter ounces of gold. It is adorned with twenty-nine pearls, forty-seven rubies, twenty sapphires, and twenty-five emeralds. The value of the gold and gems was estimated at $10,000 in 1898, which is probably less than it would be worth to-day. The sacred crown worn by St. Wenceslaus was inserted within the crown of Charles IV at the instance of Queen Blanca. The golden scepter and the golden orb are of very beautiful workmanship. The scepter has six rubies, eight sapphires, and thirty-one pearls. There may also be seen in the treasury a gilded monstrance, in the style of the Renaissance, studded with pearls and precious stones, a gift of the princely family of Schwarzenberg. Within the same cathedral, in the tabernacle of the chapel of St. Ludmilla, wife of the first Duke of Bohemia, is the head of that saint, bearing a crown studded with 1800 pearls.[1]

The crown of Vladimir, with its singular and thoroughly Russian form, is preserved in the treasury of the Kremlin at Moscow, and has been used at the coronation of all the Russian emperors. It has borne the name of the crown or cap of Monomachus from the reign of Ivan IV. Although, to judge from this designation, the crown was probably executed in the twelfth or thirteenth century, there is a legend to the effect that it was sent, in 988, from Byzantium by the ruler as a gift to St. Vladimir. It is executed in filigree work, and is surmounted by a plain cross with four pearls at the extremities; between these pearls are set a topaz, a sapphire, and a ruby. The crown itself is ornamented with four emeralds, four rubies, and twenty-five pearls from Ormus, set in gold. The cap has a bordering of sable fur, and is lined with red satin. (See Maskell, "Russian Art," London, 1884, p. 125.)

The imperial state crown of her Majesty Queen Victoria, was made in the year 1838 by Messrs. Rondell and Bridge, with jewels taken from old crowns, and others furnished by command of her Majesty. It consisted of diamonds, pearls, rubies, sapphires, and emeralds, set in silver and gold. It had a crimson velvet cap with ermine border, and was lined with white silk. Its gross weight was thirty-nine ounces five pennyweights troy. The lower part of the band above the ermine border consisted of a row of 129 pearls, and the upper part of a row of 112 pearls; between these rows, in the front of the crown, was a large sapphire (partly drilled) purchased for the crown by his Majesty George IV. In the front of the crown, and in the center of a diamond Maltese cross, was the famous ruby said to have been given to Edward, Prince of Wales (the Black Prince), by Don Pedro, King of Castile, after the battle of Nájera, near Vittoria, 1367 A.D. This ruby was worn in the helmet of Henry V at the battle of Agincourt, 1415 A.D. It was pierced quite through, after the eastern custom, the upper

[1] "Die Domkirche bei St. Veit in Prag," Prague, 1890, pp. 13, 19, 21.

part of the piercing being filled up by a small ruby. From the Maltese cross issued four imperial arches composed of oak leaves and acorns, thirty-two pearls forming the acorns. From the upper part of the arches were suspended four large pendant, pear-shaped pearls with rose diamond cups.[1] Writing in 1850, Barbot, the French jeweler, placed the value of this crown at $600,000.

The crown of St. Edward, the official crown of England, is used at each coronation.[2] The original crown of this name was destroyed by the republicans in 1649, but at the time of the coronation of Charles II, another crown was made to take its place, under the direction of Sir Robert Viner. As far as can be known, this crown was an exact copy of the older one, which was worn by Edward the Confessor, and perhaps even by King Alfred. The crown in use at present is of gold, richly studded with pearls and precious stones of various kinds: diamonds, rubies, emeralds, and sapphires. There is a mound of gold on top, and on this a cross of gold ornamented with very large oval pearls, one attached to the top and the two others pendant from the ends of the cross. The present arrangement of the jewels cannot date back earlier than 1689, as the crown was found to be despoiled of them at the time of the accession of William and Mary. Those now in the crown are acknowledged to be inferior to the former ones.

The orb or mound which is placed in the king's hand immediately after his coronation, is a ball of gold, six inches in diameter, surrounded by a band of the same metal ornamented with roses of diamonds set around other precious stones, and bordered with pearls. It is surmounted by a cross, embellished with four larger pearls at the angles near its center, and three others at the ends. The orb, including the cross, is eleven inches high, and it is figured on the coins of many of the English kings, who are represented holding it in their left hands.

The regalia of Scotland,[3] consisting of the crown, scepter, and sword of state, are preserved in the castle of Edinburgh. It is not certainly known at what time this crown was executed. At the coronation of Robert Bruce (1274–1329) a simple circlet of gold was used; this fell into the hands of the English after the battle of Methven in 1306. In 1307 Edward I issued a pardon at the request of his "beloved Queen Margarate," to a certain Galfredus de Coigniers, who was said to have concealed and kept "a certain coronet of gold with which Robert the Bruce, enemy and rebel of the King, had caused himself to be crowned in our own Kingdom of Scotland."

---

[1] Abridged from a description by Professor Tennant.

[2] Davenport Debrett, "Dictionary of the Coronation," London, p. 52.

[3] Sir Walter Scott, "Description of the Regalia of Scotland," Edinburgh, 1869.

HER MAJESTY, QUEEN ALEXANDRA OF GREAT BRITAIN AND IRELAND, EMPRESS OF INDIA

CROWN OF ST. EDWARD

The official crown of England

Sir Walter Scott, in his account of the regalia, gives it as his opinion that the present crown was probably made for Robert Bruce at a later date, and that it was used at the coronation of his son, David II (1324-1376). The style of workmanship indicates a fourteenth-century origin. The crown was originally open and was arched over by James V (1512-1542). As Scott notes, this was done to many royal crowns in the fifteenth and sixteenth centuries, in order to assimilate them to the type of the old imperial crowns.

The following description is slightly abridged from that given by Sir Walter Scott:

The lower part consists of two circles, the undermost much broader than that which rises over it; both are of the purest gold and the uppermost is surmounted by a range of *fleur-de-lis* interchanged with crosses *fleurées,* and with knobs or pinnacles of gold topped with large pearls; this produces a very rich effect. The under and broader circle is adorned with twenty-two precious stones, betwixt each of which is interposed an oriental pearl. The stones are topazes, amethysts, emeralds, rubies and jacinths; they are not polished by the lapidary, or cut into facets in the more modern fashion, but are set plain, in the ancient style of jewellers' work. The smaller circle is adorned with small diamonds and sapphires alternately. These two circles, thus ornamented, seem to have formed the original Diadem or Crown of Scotland, until the reign of James V, who added two imperial arches rising from the circle, and crossing each other, closing at the top in a mound of gold, which again is surmounted by a large cross *patée* ornamented with pearls and bearing the characters J.R.V. These additional arches are attached to the original crown by tacks of gold, and there is some inferiority in the quality of the metal.

The bonnet or tiara worn under the crown was anciently of purple, but is now of crimson velvet, turned up with ermine—a change first adopted in the year 1695. The tiara is adorned with four superb pearls set in gold, and fastened in the velvet which appears between the arches. The crown measures about nine inches in diameter, twenty-seven in circumference, and about six and a half in height from the bottom of the lower circle to the top of the cross.

The scepter, made by order of James V at the time he added the arches to the crown, is a slender silver rod about thirty-nine inches long. An antique capital of embossed leaves supports three small figures representing the Virgin Mary, St. Andrew, and St. James, above which is a crystal ball, surmounted by an oriental pearl.

The regalia have passed through many vicissitudes. After the execution of Charles I, his son Charles II was crowned King of Scotland at Scone on January 1, 1651. On the advance of the parliamentary army into Scotland, the regalia were placed in the care of the Earl Mareschal who preserved them in his castle of Dunrottar, and here they were kept until the castle was besieged and on the point of falling into the hands of the English. In this extremity, they were rescued by Christian Fletcher, wife of the Rev. James Granger, minister of Kinneff. She obtained permission from the English general to pay a visit to the Lady Mareschal and succeeded in carrying off the regalia. Her husband buried them in the church of Kinneff, just in front of the pulpit. When they were brought to light again after the Restoration, an Act of Parliament was passed which, after reciting Christian Fletcher's services in the matter, stated: "Therefore, the King's Majestie, with advice of his estates in Parliament, doe appoint Two Thousand Merks Scots to be forthwith paid unto her by his Majestie's thresaurer, out of the readiest of his Majestie's rents, as a testimony of their sense of her service."

In 1707, after the union of England and Scotland, it was considered wiser to remove the regalia from public view, since they were calculated to arouse memories of the old Scotch monarchy. These precious objects were therefore inclosed in a chest, which was their usual receptacle, and locked up in the crown-room, a strong vaulted apartment in Edinburgh Castle. There the regalia remained until 1817, when, as doubts had been expressed as to their existence, a commission of investigation was appointed, one of the members being Sir Walter Scott. The chest—which had probably been the jewel-safe of the Stuarts—was forced open, and the regalia were found within, just as they had been deposited in 1707.

An imperial German crown does not exist; a design has been made and accepted, but at the present date, 1907, it has not yet been executed. On festive occasions, when the imperial insignia are necessary, the Prussian insignia are used, especially the Prussian royal crown. This consists of a circlet of gold set with thirteen diamonds. On this are five leaves, each composed of three larger diamonds and a smaller one, and four prongs, each bearing a diamond and above it a large pearl. From the five leaves start the same number of semicircular arches, tapering toward the central point, where they unite. Each of these is set with ten diamonds of decreasing size. On the center rests an imperial globe. It consists of a large Indian-cut sapphire, —the counterpart of the one on the Austrian imperial crown, evidently dating from the time of the Crusades,—and above it rises a chaplet ornamented with diamonds. The crown has a lining of purple velvet reaching to the arches. Between the arches are eight pearl pendants of an average weight of 80 grains; they are 25 millimeters in length, and have a fine, brilliant white color, although they are not perfectly regular in form.

In addition there belongs to the regalia a pearl necklace of three rows; the first consists of thirty-seven pearls averaging 28 grains each; the second of thirty-nine pearls averaging 34 grains, and the third of forty-five pearls averaging 39 grains. There is also a guard chain of 114 pearls, averaging 20 grains, making a total of 2280 grains for the chain. These pearls are also of irregular form.[1]

The crown jewels of the Sultan Abdul-Aziz (1830–1876) were of immense richness and value. At the exhibition in Vienna, 1873, many of these were exhibited in a building created specially for the purposes of display and protection. They were in five compartments, in what might be termed five impregnable fire-proof safes of a peculiar construction. Among other interesting objects was the armor of Sultan Murad I (1319–1389), the founder of the Ottoman empire in Europe. This armor is of the most delicate oriental workmanship. Diamonds, pearls, and rubies are worked broadcast over it with exquisite taste.[2]

In Germany and Austro-Hungary there are many valuable ecclesiastical ornaments, some of which possess great interest for the history of early German art. They also serve to show the appreciation of the pearl even in the Dark Ages and the Early Renaissance period.

One of the most curious productions of early German art is a reliquary in the form of a sack, which is from Enger near Herford, and is exhibited in the Kunstgewerbe Museum in Berlin. It is set with cameos and pearls; several of the latter have dropped out; a few, however, remain in their setting. According to a very probable tradition, this reliquary was given by Charlemagne to the Saxon duke, Wittekind, on the occasion of his baptism in 785. It is of very rude and primitive workmanship and, if we accept the tradition, it is not unlikely that it was executed at Aix-la-Chapelle.[3]

An interesting example of German art, from the time of Archbishop Egbert of Treves (977–993), is a frame now in the Beuth-Schinkel Museum, at Charlottenburg. This was probably the framework of a portable altar. It is decorated with a simple geometrical design in the three primary colors, and has four polished stones and four pearls on the outer border of gold filigree. Another example of the art of Treves at the time of Archbishop Egbert is the Echternacher Codex. The gold-plated cover is a worthy product of the school: ivory, enamel, and mosaic are combined in its decoration with rows of pearls. Among the representations of many saints, appears the figure of the Empress Theophanu, daughter of the Greek emperor, Romanos II, with the inscription "Theophaniu imp." Opposite is a youthful figure,

---

[1] Communicated by Prof. H. Schumacher of Bonn and Johann Wagner & Sohn, Jewelers of the German Court.

[2] W. Jones, "Crowns and Coronations," London, 1883, p. 425.

[3] Otto von Falke and Heinrich Frauberger, "Deutsche Schmelzarbeiten des Mittelalters," Frankfort-on-the-Main, 1904, p. 2.

probably that of her son, Otho III. It seems likely that the work was executed, at the command of the empress, between 983 and 991.[1]

In the cathedral of Treves is the portable altar known as the altar of St. Andrew. This was primarily a reliquary and secondarily an altar. In memory of the relic of the sandal of St. Andrew, which was greatly prized by Archbishop Egbert, this altar bears the representation of a foot executed in wood and covered with plates of gold. The front of the case is divided into three fields; that in the middle containing a Byzantine lion in gold relief, and the others the symbols of the four evangelists in enamel work. The border is formed of rectangular pieces of enamel and smaller ones of gold, and it is set with round stones alternating with half-pearls; the ends are covered with filigree and enamel work wherein are embedded strings of pearls. A coin of Justinian II is set in the middle of the back of the case and is surrounded by a wreath of larger pearls.[2]

A gold cross, the work of Rogkerus Theophilus, is in the Kunstgewerbe Museum in Berlin, and comes from Herford. The frame, which is of wood, is covered with plates of gold; at the extremities and in the center are groups of precious stones surrounded by pearls; at the base is a fine Augustan cameo with a wreath of pearls; the entire cross is covered with filigree work and decorated with pearls in groups of threes. The arrangement of the precious stones, and the enhancement of their beauty by means of the circles of pearls, are highly artistic. As a work of Rogkerus, this cross must have been executed at the very end of the eleventh century and it may be regarded as one of the finest examples of the art of this period.[3]

A very rich collection of ecclesiastical ornaments is contained in the treasury of the cathedral of Gran in Hungary.[4] One of the most interesting objects is a reliquary in the form of a Latin cross, which is of great historical and artistic value. An inventory made after 1528 describes it briefly: "crux aurea continens lignum vitae" (a gold cross containing the wood of life). Although this reliquary probably belongs to the end of the twelfth century, the inventory of 1659 describes it as a gift of King Stephen, and proceeds to say that the kings of Hungary took their coronation oath upon it. This custom has been preserved to the present day, and Emperor Francis Joseph, on the occasion of his coronation as King of Hungary, June 8, 1867, swore, upon this cross, to uphold the constitution and the laws of the land.

[1] Otto von Falke and Heinrich Frauberger, "Deutsche Schmelzarbeiten des Mittelalters," Frankfort-on-the-Main, 1904, pp. 6, 7.

[2] Ibid., p. 9.

[3] Ibid., p. 16.

[4] Josef Dankó, "Aus dem Graner Domschatz," Gran, 1880, pp. 64–66.

The cross is decorated with plates of gold in filigree design, and has four en cabochon cut sapphires and eighteen oriental pearls.

The greatest treasure of the collection is known as the cross of Corvinus, King of Hungary, and is decorated with a great number of pearls.[1] It is a remarkable example of early Italian Renaissance art. The entire structure is about twenty-eight inches high; the pedestal is triangular and ornamented with pearls and precious stones; three sphinxes bearing shields with the arms of Corvinus support a disk from which springs a triangular support sloping outward; on the three sides are mythological figures. Upon this base rests the chapel, a light Gothic structure with the figure of the Saviour bound to a pillar in the center, and the busts of three prophets in the niches outside. Above all is the crucifix, on each side of which are figures of the Blessed Virgin and of St. John. Around the base and about each division of this elaborate design is a row of pearls; the Gothic chapel is surmounted by a close-set row, and each of its six pinnacles terminates in an oval pearl. The cross itself has fifteen large pearls disposed in twos and threes, and many smaller ones. There are at least two hundred pearls on the whole structure.

Another cross, with the arms of the primate, George Szolepchényi, and bearing the date 1667, is of pure design and richly decorated with pearls and precious stones.[2] It is quite possible that this cross, which seems to belong to a better period, was bought by the archbishop, who afterward added his arms. There are thirteen oriental pearls, three at the top, three at the end of each of the arms, and four at the intersection. This cross was used as an "instrumentum pacis," for the kiss of peace, on solemn occasions such as coronations.

We may also note the pendant with the image of the Virgin Mary as patroness of Hungary, which is of gold enamel and has two pendant pearls and a sapphire, and likewise the pectoral cross of the primate, Emerich Losy; this is of gold, decorated with green, blue, and black enamel, and has three pendant pear-shaped pearls, one quite large, as well as thirty-four smaller round pearls.

Among the many valuable and interesting objects in the treasury of the house and chapel of Maria Loretto am Hradschin,[3] at Prague, there is a monstrance of silver-gilt, thirty-seven and a half inches high and fifteen and three quarter inches wide. It dates from the beginning of the eighteenth century, and is not a harmonious whole, but only a combination of different ornaments of precious stones, corals, and several hundred pearls of various sizes. All these are the devo-

---

[1] Josef Dankó, "Aus dem Graner Domschatz," Gran, 1880, pp. 67–70.
[2] *Ibid.*, pp. 74, 75.

[3] "Katalog der Schatzkammer von Maria Loretto am Hradschin zu Prag," Prague, 1891, pp. 34, 40.

tional offerings of now unknown givers, and many of the pieces are of artistic workmanship. This monstrance owes its origin to Josef von Bilin, who was a monk of the Capuchin order and a sacristan of Maria Loretto. On account of the many pearls which adorn it, it is known by the name of the "Pearl Monstrance."

Another monstrance of Arabic gold, of the year 1680, is twenty inches high and is studded with fifty-one pearls, of which twenty-nine surround the disk, while the remainder are on the plate and the base. There are also two crowns of silver-gilt for the statues of the Virgin and of the Infant Jesus. The larger of these crowns has eighteen diamonds, a ruby, and 102 pearls set in two rows; while the smaller has nineteen diamonds and a great number of pearls; both crowns are made up of the offerings of the faithful.

In a historic description of the pearls in the treasury of the Kremlin, Margeret, a Burgundian captain ("Estat de l'empire de Russie," 1649), says that the treasury was "full of all kinds of jewels, principally pearls, for they are worn in Russia more than in the rest of Europe. I have seen fifty changes of raiment for the emperors around each of which there were jewels for a bordering, and the robes were entirely bordered with pearls, some with a border of pearls measuring a foot, half a foot, or four inches in width. I have seen dozens of bedcoverings embroidered with pearls."[1]

In the treasury of the celebrated Troiza Monastery near Moscow, there is an immense collection of ornamental objects for ecclesiastical use, the value of which has been estimated at many millions of rubles. Here may be seen miters and bishops' crooks—many of them of solid gold and set with precious stones—Bibles and missals in golden bindings, priestly vestments, altar-cloths, etc., all literally covered with pearls. There is also a dish filled with large pearls of enormous value.[2]

The use of fresh-water pearls in one of the most interesting ecclesiastical objects of antiquity is shown in the "Shrine of St. Patrick's Gospels," which is in the Dublin Museum. It was purchased by the Irish Royal Academy in 1845 for £300 ($1500). This shrine, known as the "domnach airgid," is of Irish manufacture and was perhaps made in the eleventh or twelfth century. It was found in the neighborhood of Clones, in County Monaghan, and is ornamented with three bosses which contained uncut crystals, and are decorated with figures of grotesque animals and traceries enameled in blue paste; between these may be seen representations of four horsemen. On each of the four corners there was a fresh-water pearl, one of which still remains

---

[1] Maskell, "Russian Art" (South Kensington Museum Handbooks), London, 1884, pp. 119, 120.

[2] Baedeker, "Russland," Leipzig, 1888, p. 317.

in its setting. According to George Petrie, LL.D., in his "Christian Inscriptions in the Irish Language," the shrine bears an inscription to the effect that it was made by John O'Barrdan at the instance of John O'Carbry, Abbot of Clones, who died in 1353.

Dr. R. F. Scharff informs us that there is also in the Dublin Museum a modern Celtic gold brooch, presented to Queen Victoria on the occasion of her visit to Ireland in 1849, and containing a pearl of beautiful luster, discovered in Lough Esk, which is in the western part of Ireland. Dr. Scharff says that this pearl is undoubtedly from the *Margaritifera margaritifera*.

Mr. W. Forbes Howie of Dublin writes that the shrine of O'Donnel, made in 1084, originally contained pearls. It still retains some pieces of amber and coral. Mr. Howie believes that fresh-water pearls were freely used in the decoration of ancient Irish shrines.

The inventories of jewels and ornaments belonging to the kings and queens of France, to the nobility, and to the treasures of the Sainte-Chapelle, in Paris, and of the abbey and church of St. Denis, all mention a large number of objects decorated with pearls.[1] The more important of these are given below.

The following ornaments decorated with pearls are mentioned in the inventory of Louis, Duke of Anjou, which was made *circa* 1360:[2]

A large silver-gilt foot for a vase or chalice, resting upon six lions couchant, and set with groups of four pearls with a garnet in the middle.

A half girdle of gold with a hinge bearing two ornaments, one a balas set between two eagles. Between the ornaments is a gold bar set with eight pearls in two rows. In front is a clasp with a large sapphire in the middle, surrounded by two balases and two sapphires alternating with pearls.

A gold brooch having a balas-ruby in the middle, and at each side four sapphires and four clusters each of five quite large pearls.

A gold brooch of a very pretty design, with five balas-rubies, two sapphires, and eight very round pearls weighing about four carats each. At each end of the brooch is a flat pearl weighing about five carats.

There is in the Bibliothèque Nationale[3] in Paris, the original record of the execution of the testament of the Comte de Montpensier, son of

[1] One of the authors has in his possession twenty manuscript volumes of these inventories. They are careful copies from the originals, most of which are in the Bibliothèque Nationale in Paris. These copies were executed for M. E. Molinier, a conservator of the Louvre Museum, and were disposed of after his death in 1906. The values in the money of the times are usually given, and we have endeavored where possible to indicate the equivalent in United States currency, taking account of the progressive changes in the French monetary standard.

[2] "Inventaire des Joyaux de Louis Duc d'Anjou," In De Laborde's "Emaux," Paris, 1853, Vol. II.

[3] Bibliothèque Nationale. MS. fr. 6542 (suppt. 4622) parchemin original, 13 pp. in folio.

the Duc de Berri. This document was written in 1398, and it mentions that the sale of the jewels and plate of the count produced the sum of "2390 livres tournois 11 sols 3 deniers [about $8265]." In the record we have a description of "a large gold cup, weighing 5 marcs, 7 ounces, 1 gros [nearly 3 lbs.], whereon there is a crown of precious stones." The decoration of the cup comprised thirty large pearls, six balas-rubies, and four sapphires, and we are told that the Duc de Berri retained it for his own use.

An early mention of the use of pearls in rings occurs in the inventory of the Duc de Berri,[1] to whom we have just referred. This inventory, which was made in 1416, notes a gold ring with black enamel, set with a pearl called "the great pearl of Berri."

The inventory of the personal property of Marguerite, Countess of Flanders, the mother of the Duke of Burgundy, was made in 1405.[2] In this inventory we have a list of an immense number of ornamental objects of every sort and kind, and everything, from the ducal crown to the smallest trinket, is garnished with pearls. In most cases the number of pearls is given, and we find that no less than 4494 are enumerated. Evidently the duchess was ever ready to honor the precious gem to which she owed her name, and fully recognized its poetical significance. The following are a few of the more noteworthy ornaments in the inventory:

The circlet of the great crown, composed of eight sections; four of which each comprise sixteen pearls, four diamonds, and four balas-rubies, with a sapphire in the center; the four others contain sixteen pearls, four diamonds, and four sapphires, with a balas-ruby in the center; beside this there are two pearls in each section. Also, eight large fleurons of the great crown, four of which bear each twenty-three pearls, five diamonds, three balas-rubies and a sapphire, and the other four each twenty-three pearls, five diamonds, four sapphires, and a balas-ruby; and eight small fleurons of the said crown garnished each with a pearl, a sapphire, and a balas-ruby. The whole is valued at 8724 florins ($22,682).

A gold cap with ten large ornaments fashioned like brooches, five of which are each of six pearls and a balas-ruby, and the other five each of five pearls and two balas-rubies, and between each ornament there is a balas-ruby. This is appraised at 2159 florins ($5613).

A headdress garnished with balas-rubies and sapphires and tassels of large pearls, each of six pearls, and with a row of larger balas-rubies, larger sapphires and larger pearls. This was estimated at 2030 florins ($5278).

A gold necklace, enameled white and green, garnished with nine rubies, thir-

---

[1] De Laborde, "Emaux," Paris, 1853, Vol. II, p. 437.

[2] "Inventaire des Biens de Marguerite de Flandre," Bibliothèque Nationale coll., Moreau, 1725 (Mouchet 5).

teen diamonds and thirteen pearls, with a clasp of three small rubies, and three large pearls with one large diamond in the center. The worth of this necklace is given as 1923 florins ($5000).

The jewels and ornamental objects in this inventory are appraised at the sum of 56,129 florins,—about $145,000,—equivalent to a much larger sum to-day in consideration of the greater purchasing power of money in the fifteenth century.

In 1480, during the reign of Louis XI, an inventory was made of the objects preserved in the treasury of the Sainte-Chapelle in Paris.[1] We select the following items from this inventory:

A very beautiful cross, covered with gold, bearing on one side a crystal reliquary which contained a piece of the True Cross. On supports attached to the cross were images of the Virgin Mary and of St. John, each holding a reliquary. The cross itself rested on a square silver-gilt base bearing the images of the four evangelists. The ornamentation consisted of fifty large Scotch pearls and 142 small ones, intermixed with garnets and emeralds; there were also many balas-rubies and sapphires of different sizes. The inventory says: "The goldsmith Nicholas Roet declares that the stones are genuine and that the pearls are from Scotland."

Another gold cross, resting on a silver-gilt base which bore the arms of France and Burgundy, was decorated with fourteen sapphires, twenty balas-rubies, and twenty-four Scotch pearls. On the base were the figures of St. Louis and of the queen, kneeling in prayer.

Still another cross, covered with gold and of Venetian workmanship, bore thirty-nine pearls, twenty-seven balas-rubies, and four sapphires. A clasp attached to this cross was set with four large perforated pearls surrounded by small emeralds and sapphires.

A silver-gilt ornament, consisting of a golden image of St. Louis seated on a silver throne and holding in his hand a reliquary decorated with twelve pearls, six emeralds, and six Alexandrian rubies. The crown of the image was set with four large oriental pearls, three balas-rubies, etc.

An ivory image of the Virgin Mary, supported by a silver-gilt base with the arms of France. This base was borne by four lions. On the head of the Virgin was a crown of gold adorned with eight large, round, oriental pearls and four small ones, as well as four emeralds and four balas-rubies. On the breast of the image was a very large, square emerald.

A splendid miter studded with good-sized pearls and decorated with emeralds, rubies, sapphires, and balas-rubies. The pendants were covered with seed-pearls and precious stones.

A fine chasuble of Indian satin lined with crimson taffeta and covered with lilies, birds, unicorns, etc., embroidered in gold and pearls. It was also adorned with small clusters of pearls and with two shields bearing the arms of France and Navarre, quartered.

[1] Bibl. Nat. MS. Latin. 9941 (suppt. 1656), folio, parchment, 40 leaves.

A beautiful copy of the gospels with covers of gold, ornamented with fourteen large sapphires, thirteen balas-rubies, two cameos, and eighty-nine good-sized pearls.

The following items are taken from the inventory of the treasury of the abbey of St. Denis, made in 1534, during the reign of Francis I. This record is in the Bibliothèque Nationale, in Paris:[1]

A crown of gold, with four fleurons, garnished with several balas-rubies, emeralds, sapphires, and pearls; valued at 59,980 crowns (about $135,000).

A golden cross and within it a piece of the True Cross which belonged to "Jeanne d'Evreux, royne de France et de Navarre," valued, with the pearls that decorate it, at 345 crowns ($776).

A wooden chest containing eleven cases in which were many precious stones and large and small pearls, both oriental and Scotch; valued at 1858 crowns ($4180).

A number of priestly vestments embroidered with seed-pearls are inventoried at 1200 crowns ($2700).

A blue satin chasuble bordered with pearls is valued at 350 crowns ($787).

An altar-table, set in the "grand altar," is described as elaborately decorated with "arches and pillars and images of gold" in low relief, and garnished with precious stones and pearls. The value is given as 1203 crowns ($2700).

Another altar-table similarly ornamented is valued at 2645 crowns ($5850). Above this table was a great cross of gold with a silver border, called the "cross of St. Eloysius" (the patron saint of goldsmiths); this was valued at 2291 crowns ($5154).

Over the sarcophagus containing the body of St. Denis, there was "a large tabernacle of wood-work resembling a church, with a lofty nave and low arches." In this nave and in the transepts there were three representations of sarcophagi; the whole was covered with gold, precious stones, and pearls, and was valued at 7275 crowns ($16,368).

The head of St. Denis, incased in gold, was borne by two silver-gilt angels, while a third held a small shrine containing a portion of the jaw-bone of the saint. All these objects were studded with precious stones and pearls, and were valued at 5622 crowns ($12,650).

There were also in the treasury several miters covered with "ounce-pearls" and decorated with gold and silver bands; on this field several larger pearls were set. One of these miters is valued at 964 crowns ($2169) and another at 509 crowns ($1135).

The total value of the articles inventoried is 185,500 crowns (at least $417,375).

Inventories of the property of the dukes of Lorraine, dated 1544, 1552, and 1614, mention a number of pearl ornaments. In the in-

---

[1] Bibliothèque Nationale MS. fr. 18766 (S. Germain fr. 910) 40 ff. Bound in green velvet.

ventory of 1544, made about the time of the accession of Francis I of Lorraine, we read of "a very fine case of silver-gilt around which are thirteen personages in gold, and on the lock three balases and five pearls." The inventory of 1552, made while Charles II was duke, mentions "a cap of crimson velvet whereon there are large pearls," and another cap "entirely covered with pearls." It is, however, in the inventory of 1614, made a few years after the accession of Henry II of Lorraine, that we find the greatest number of items relating to pearls. An estimate of the value of the rings and jewels was "faicte du commandement de son Altèze par jouailliers et Lapidaires et Espertz dudit ars." All these jewels were to remain forever the property of the Duchy of Lorraine. Among the items relating to pearls, the following are worthy of note:

A gold collar with seven settings, each containing one large diamond and two large pearls. The diamond in the center was believed to weigh fifteen carats, and the collar was valued at 35,000 crowns (about $70,000).

Another collar contained seven diamonds and sixteen pearls set in pairs, and was considered to be worth 19,750 crowns (about $40,000).

A collection of one hundred large pearls, some weighing twenty grains, some twenty-four, some twenty-eight, and a few thirty-two grains, were estimated at 12,000 crowns ($24,000).

A large pearl, very nearly pear-shaped and almost as large as a pigeon's egg, was set down at 2000 crowns ($4000).

A very fine pear-shaped pearl weighing forty-eight grains was valued at 800 crowns ($1600).

Another pear-shaped pearl weighing about thirty-two grains was placed at 500 crowns ($1000).

Four other pear-shaped pearls, nearly as large as the one above-mentioned, were estimated at 300 crowns ($600), while a round "pearl of Seville" was valued at only fifty crowns ($100).

Six clusters of pearls, each containing two of fourteen grains, and four of eight grains, were thought to be worth 700 crowns ($1400).

A large chalice was decorated with seven large oriental emeralds and eight clusters, each composed of fourteen fine, round pearls, six of twelve grains and eight of eight grains; the whole valued at 2400 crowns ($4800).

A hat ornament composed of eleven fine rubies and ten large, round pearls, each weighing twelve grains, was estimated at 800 crowns ($1600).

A similar ornament, composed of thirteen rubies and fourteen pearls, partly flat and partly round, was placed at 2000 crowns ($4000).

A collar set with seven fine rubies and the same number of round pearls, each weighing twelve grains, and with seven other pendant pearls, was valued at 550 crowns ($1100).

There was also a bed called the "bed of pearls," which was elaborately decorated with ornamentation in gold and richly studded with pearls.

The inventory made in 1634 of the ornaments, etc., contained in the abbey of St. Denis, offers some new material and a fuller description of a few of the objects mentioned in the inventory of 1534. The most noteworthy entries are given below:

A golden scepter upon a staff of wood. The scepter bears the figure of Charlemagne seated upon a throne; at the corners are two lions and two eagles (one of the latter was lacking in 1634). The figure holds a scepter in its right hand, and a globe surmounted by a cross in its left; on its head is a crown with a large, round, oriental pearl valued at 200 livres ($135). The throne rested on a fleur-de-lys, beneath which was a ball of gold ornamented with eight oriental pearls. Around the throne was the inscription: "Sanctus Carolus Magnus Italia Roma Gallia Germania," and three clusters of three pearls each. The value of this scepter was given at 3300 livres, or about $2200.

The reliquary of the hand of St. Thomas. Two angels, resting on a silver-gilt base, bore the crystal receptacle containing the relic. The ornamentation consisted of eight clusters of four large pearls each, with a small diamond in the center. On the hand was a gold band bearing the inscription: "Hic est manus beati Thomae apti. quam misit in latus domini nostri Jesu Christi." On the hand was a pontifical ring set with a large sapphire. The reliquary also bore the images of St. John the Baptist, of St. Thomas, and of the Virgin Mary. It was valued at 5590 livres, or about $3700.

A vessel made of a porphyry resembling jasper and embellished with forty-six pearls; estimated at 1500 livres ($1000).

A cope given by Anne of Bretagne, Queen of France, and bearing six scenes from the life of the Virgin Mary embroidered in gold and pearls; the whole bordered with pearls and gold of Cyprus. On the cope were the letters A and S, and the words "plutost mourir." There were two ounces of pearls. Valued at 2000 livres ($1350).

A vase of rock crystal, of antique workmanship, with a cover and base of silver-gilt; the top decorated with a band of amethysts, garnets, and sapphires, alternating with Scotch and oriental pearls. On the base are various precious stones and twenty-three Scotch and oriental pearls, and the inscription "Hoc vas sponsa dedit Anor. regi Ludovico." This vase was given by Eleanor of Aquitaine to her husband, Louis VII of France (1137–1180), by whom it was bestowed upon Suger, Abbot of St. Denis (1082–1152). The goldsmith work and decoration belong to the time of Suger. The vase is now in the Louvre.

A chalice of agate, with two handles, and engraved with the figures of men, animals, and birds. It stood on a foot of gold adorned with sixteen sapphires, forty-four pearls, and twenty-two clusters of fourteen pearls each. This chalice rested upon a paten of porphyry decorated with seven fishes inlaid in gold, and with a bordering of pearls and precious stones disposed around the edge. Both together valued at 25,000 livres (about $16,000).

A vase of agate with a foot of silver-gilt, and furnished with a cover and a spout in the form of a serpent, both of silver-gilt. Around the base an inscription: "Dum libare deo gemmis debemus et auro, Hoc ego Sugerus offero

大清國慈禧皇太后

THE EMPRESS DOWAGER OF CHINA

From a portrait painted by Miss Katharine A. Carl

Hungarian Aigret   Earring, Hungary

Earrings, Nijni-Novgorod   Spanish earring

PEARL ORNAMENTS OF THE SEVENTEENTH AND EIGHTEENTH CENTURIES

vas domino" (Since we should pour libations to God out of gems and gold, I, Suger, offer this vessel to the Lord). This vase, which is now in the Louvre and is of sardonyx, was enriched with many precious stones and with nineteen Scotch and oriental pearls. The value given was 1500 livres (about $1000).

A book beginning: "Kyrie Eleison," with covers of wood, one overlaid with gold and the other with silver. On the golden cover was an ivory crucifix, and images, in ivory, of the Virgin Mary and of St. John. The cross was bordered with seed-pearls, as were the diadems of the images. The cover was also decorated with an engraved crysolite, an engraved peridot, and with sapphires, emeralds, and garnets.[1]

A curious item regarding the use of pearls in embroidery is contained in one of the inventories of the dukes of Burgundy, made in 1414; this reads as follows:

The sum of 276 livres 7 sols 6 deniers tournois (about $960), the price of 960 pearls destined to ornament a dress; along the sleeves are embroidered the words of the song "Madame, je suis joyeulx," and the notes are also marked along the sleeves. On each sleeve are 264 pearls which help in forming the notes of the said song, numbering 142; that is to say, a square made of four pearls for each note.[2]

Mention is made in two old French documents of the use of pearls from Compiègne in ornamentation. In the "Inventaire de la royne Clémence," in 1328, we read of "a cock covered with precious stones and bearing a pearl of Compiègne"; and in the "Comptes Royaux," under date of 1353, appears this item: "For four pearls, oriental, Scotch and of Compiègne, for the said arm-chair, 48 crowns." As these pearls could not have been found in Compiègne, we may suppose that there was a market for their sale in that place, which gave rise to the designation.[3]

The English authority and writer on early English silver, F. Alfred Jones, communicated, under date of September, 1907, that pearls were rarely used in old English plate; in fact, any such embellishments were of exceedingly infrequent occurrence. They are, however, frequently mentioned in the inventory of the marvelous collection of gold plate dispersed by Charles I of England, which may have dated from the time of the looting of the churches and monasteries by Henry VIII.

The following items are from the inventories of Philip II of Spain

[1] Bibl. Natl. MS. français, 4611, folio, pp. 433 in parchment.
[2] "Inventaires des Ducs de Bourgogne," De Laborde, "Emaux," Vol. II, p. 438.
[3] See De Laborde. "Emaux," Paris, 1853, Vol. II, p. 437.

and of Margarita, wife of Philip III. The original documents are in the Austrian archives.

A golden cup which came from England. Around the foot was a wreath of fifteen fleurons, each containing pearls, and also four St. Andrew's crosses comprising eighteen pearls each. The interior of the cup showed scenes from the life of St. George and was studded with pearls, while thirty-one pearl pendants hung from the edge. 11,897 reals (about $1700).[1]

Some curious jewels, belonging to Queen Margarita, wife of Philip III of Spain, were entered in an inventory made in 1611.

An imperial eagle, full of diamonds, that came from England, with two pendants of two pearls, which could be unhooked from the said eagle and were worn by her Majesty at two masks as earrings. Valued at 77,000 reals (about $11,000).

Gold earrings, enameled in various colors, with seven diamonds in each one and three pendant pearls, two small ones of equal size and the other shaped like a pear. Valued at 1320 reals ($188).[2]

In the older Spanish jewelry pearls were frequently entirely pierced through, as if they had been worn in necklaces; and if hung as drops of one to three or more, they were strung on a wire, the upper end usually forming an ornament, and they were kept from falling off below by flattening the lower end of the wire, this flattening acting as a stop. These styles have a marked resemblance to the oriental methods elsewhere described, and suggest the derivation of the early Spanish pearl mounting from the Moorish occupation of the country. If they were set singly on any part of the jewel, they were put on a wire peg fastened to it, and then the end of the wire which projected was hammered flat to keep the pearl in place. Excellent examples of these styles are the Spanish earrings in the collection of the Hispano-American Museum of New York. The same method was used in Transylvania in the seventeenth century with remarkably artistic effect.

The pearls of the Virgin of the Rosary in the church of St. Domingo, Lima, were famous. It is believed that they were sold in the war of independence. Those of the monstrance in the sanctuary of the cathedral of Lima were sold during the last war with Chile. The monstrance of the cathedral of Cuzco still shows pearls and emeralds, but they are of small size.

A lady who left a great fortune in pearls to the church of Nazareno

---

[1] "Jahrbuch der kunsthistorischen Samm-   Vienna, 1895, Vol. XIV, Pt. II, p. 52. "In-
lungen des allerhöchsten Kaiserhauses,"   ventories of Philip II" (1598–1607).
                [2] *Ibid.*, Vol. XIX, Pt. II, p. 170.

and the House of the Poor of the church of St. Peter, Lima, was Doña Maria Fernandez de Córdoba, from the family of Borda, grandmother to the minister of Peru in Washington. She was a descendant of Hernan Cortés and of Pizarro by her ancestor Carmen Cortés.

The pearls of Lima figure prominently in the history of the Peruvian families. The war of independence, which ended in 1822, was followed by the suppression of the entailed estates; this forced a division of the family fortunes, and it became necessary to sell the family jewels in Europe. Thither went all the famous pearls of the Peruvian aristocracy, whose luxury is proven by the fact that in 1780 there were in Lima no less than two thousand private carriages.

One of the most remarkable uses of Bohemian pearls was that of a large triptych owned by Count Moritz of Lobkowitz and Duke of Raudnitz. It measured six or more feet in height. The entire borders were ornamented with pearls. The center of the triptych represented the ascension of Christ on a chariot drawn by lambs. In the panel to the right was the Angel Gabriel, and to the left the Virgin Mary praying. The borders and lettering were magnificently embroidered and decorated in Bohemian pearls. This object probably dated from the sixteenth or early part of the seventeenth century. It was estimated by one of the authors to contain at least one hundred thousand pearls.

Madame Zelie Nuttal, the great Maya scholar, personally writes that pearls are not mentioned either as articles of tribute or of decoration in ancient Mexican codices; possibly a lack of fine, hard instruments with which to drill holes in pearls may have caused them to be comparatively little used in personal adornment. Neither do they appear to have been found incrusted in prehistoric objects, and we have no written evidence of their having been used in this way. We do not know of any instances of the wearing of pearls by the Indian women, but the women of the higher classes used to wear them profusely, more especially drop-earrings and pendants. Madame Nuttal also communicates as follows:

Bernadino de Sahagun states: "There are also pearls in New Spain, and they are familiar to everybody. They are named epyollotti,[1] which means the heart of the shell, because they are formed in the shell of the oyster." In Molina's dictionary "seed-pearls" are named "piciltic epyollotti," which means "water-stars," a poetical name, composed of the word a = att = water, and cittallin = star. The latter name leads us to infer the possibility that the "star-skirt, or skirt of, or with stars," the "cittallin icue" of the living image of the goddess "Tlamateculitti" was decorated with *pearls,* although it is only described (Book II, chap. 36) as being "of leather, cut into strips at the bottom (forming a fringe), at the end of each of which hung a small shell named

[1] From *eptli*—shell, and *yollott*—heart, *i.e.,* life.

'cueclitti' which reproduced a sound when she walked." As it is stated that this "star-skirt" was worn over "a white one" it seems as though it must have been of the kind represented in codices and sculptures, made of open-work and netlike, and studded with round objects—possibly pearls—at the crossings or in the centers of the open spaces.

Oil-paintings of the madonnas represent them with robes richly embroidered with pearls, and wearing "ropes of pearls." The Virgin of the Rosario, in the church at Santo Domingo, Mexico, was noted for her pearls, and there is a small oil-painting of this virgin, in which she is depicted with a wealth of pearls.

In the Bohemian National Exposition, held at Prague in 1891, Count Schwarzenberg exhibited four embroideries, each fourteen by eight inches. They were embroidered with Bohemian pearls found on his domains a century or more previous, and contained many thousands of pearls.

In Hungary pearls have always been the favorite jewels, especially among the aristocracy, and they have served to adorn the national costume of both men and women. A century ago nearly every family of distinction owned a necklace, but most of the pearls were small and of indifferent quality. Since that time fine pearls have become more usual, and many wealthy Hungarian families have acquired beautiful pearls of good size and excellent quality, and many splendid necklaces can now be seen in Hungary. The following are some of the finest:

A necklace of three large rows, owned by the Archduke Joseph and valued at one million francs.

A still larger necklace in the possession of Prince Nicholas Esterhazy; this, however, is an entailed heirloom, and may not be parted with without the king's permission.

A very fine necklace of five rows, also an entailed heirloom, owned by Count Maurice Esterhazy.

A large necklace, possessed by Countess Alois Karoly, wife of the late ambassador in London. This is another entailed heirloom; its value is at least a million and a half francs.

An unusually large necklace of four rows, such as one rarely sees, owned by the Countess Wenkheim. The pearls are white, and have a good shape, but not much brilliancy. The average size of these pearls is approximately twenty-four grains.

An equally large necklace consisting of a single row, averaging twenty-six grains, in the possession of Countess Louis Batthyani.

There are a great many other necklaces of fine quality, worth from 300,000 francs down to 100,000 francs, belonging to families such as those of Count Joseph Hunyadi, Countess Festetics-Hamilton, Count Landor Nako, Peer Leo Lanczi, Count Albert Apponyi, Mr. Eugene Dreher, Madame Emma de

Bachrach, etc., etc. Indeed, almost every wealthy family of the better class owns a necklace worth up to 100,000 francs and over.

The portraits of the fifteenth, sixteenth, and seventeenth centuries afford us many interesting evidences regarding the various forms of jewelry in which pearls were used. Indeed, had we no other records, these pictures alone would prove the great popularity of the gem as an ornament.

In the finely executed portrait of the Duchess Anne de France, she wears a coronet with three pearls at each point. It seems to be made up of three large pearls, set on a row of pearls circling the entire top of her head, beneath which is a row of great emeralds, and then another row of pearls. Flaring downward and entirely covering the side of her head near the ear, are two rows of pearls with a row of fine emeralds between them, the rows of pearls deflecting slightly downward until the chin line is reached, and then turning back and slightly upward, meeting at the back. As in the crown cap, the same severe decoration in pearls is the main feature, and is repeated on each side of the robe, the front of which is of ermine. Beginning on a line with the shoulder is a broad band of pearls and emeralds set in gold which extends below her waist. At the top of this are six pearls set in a straight line. Then from the end of this line, dropping straight down, is a row on each side. Between the two rows is a gem, then two great pearls and another gem, then two more pearls, this being repeated to below the waist. The ermine is held at her waist by a trefoil reversed; that is, two pearls above and one below a great gem, and then a trefoil reversed below this. This portrait is dated 1498 and is on a triptych in the cathedral of Moulins.

Quite unique is the pearl decoration in a picture of St. Barbara, painted by an artist of the French school, and dated 1520, which is in the National Museum of Budapest. This artist uses pearls with the utmost severity of taste and richness. Beginning a trifle above the center of her forehead is an emerald ornament, and on each side there extend to the back of her head three rows of pearls, not placed exactly one row above the other, but the rows intertwined with each other. The whole is enriched by a great string of pearls about her neck. The effect produced is extremely artistic and beautiful.

Catharine de' Medici wore two rows of pearls on her bonnet, and a quaint necklace in sections of two rows of four pearls, with a large pearl between; a pear-shaped pendant on a Renaissance jewel; a row of pearls around her low-cut bodice, and a girdle of jewels alternating with pearls, which extended to the lower end of her gown. In addition to all this, she wore a bracelet of jewels with a pearl set between each ornament. This artistic combination is best shown in her portrait in the Uffizi, Florence (No. 726), painted by an unknown artist.

One of the most unique, rich, and chic collections of pearls, and one worn with unusual grace, is that of the Infanta Isabella Clara Eugenia, shown in the fine portrait of her by Coello Sanchez. In this portrait her hat shows the plumes embroidered with slanting rows of three, four, and five pearls. In the center of the hat is an ornament shaped like a flower, with seven large pearl petals surrounding a great pearl center. The hat is tilted to one side showing her hair on the left, while a little to the right of the center of her forehead, and touching it, there hangs from her hair a great pear-shaped pearl, which adds a wonderful amount of character to the jeweling of her head. Around her neck is a high fluted ruff; below which is a collar of large gems relieved by an ornament of two pearls placed between each gem. The same interesting motive is carried out in a girdle of gems which comes down very low to her waist, terminating in a large jeweled heart ornament. The painting shows sixteen remarkable pearls in the collar, and thirty-six pearls in the jeweled girdle.

A very interesting collection of portraits was exhibited last spring (1907) at the Bibliothèque Nationale in Paris. The pictures are contained, in some instances, in old illuminated manuscripts, while in others they are contemporary crayon sketches. Many pearl decorations are represented, and we give a few of the most important.

The portrait of Anne de Bretagne (1476–1514), wife of Louis XII, from the "Heures d'Anne de Bretagne," illuminated by Jean Bourdichon, represents the queen kneeling; she wears a collar ornamented with groups of four pearls alternating with precious stones.

A crayon sketch of Françoise de Foix, Comtesse de Châteaubriant (1490–1537), who became the mistress of Francis I, shows her wearing a hood or coif ornamented with forty oval pearls. She also wears a necklace of sixty fine round pearls.

Diane de Poitiers (1499–1566), granddaughter of Charles VII and Agnes Sorel, is represented with a headdress similar to that worn by the Comtesse de Châteaubriant. It has a border of sixty round pearls. This crayon is of the time of Jean Clouet.

A portrait of Philip Strozzi (1541–1582) who, although an Italian, had the rank of colonel-général in the French army, is interesting as an illustration of the wearing of earrings by the men of this period. The fine round pearl which hangs from his ear strikes us now as a curious ornament for a warrior.

A crayon sketch of Gabrielle d'Estrées (d. 1599), mistress of Henri IV, is attributed to the hand of Daniel Dumonstier. Here may be seen a splendid pearl necklace, which apparently consists of six sections, each comprising three rows of eight round pearls, the sections being connected with each other by a large oval pearl. The necklace, which hangs down over the bosom, is fastened by a clasp in the form of a four-leaved clover, from which depend two other sections similar to those described above, and terminating in an oval pearl pendant.

The portrait of the Duchesse de Montpensier (1627–1693), the "Grande Mademoiselle" of Mme. de Sévigné's letters, is from the "Maximes de nostre salut," dedicated to the duchess by the author, M. de la Serre, and is attributed to Nicolas Jarry. It represents the duchess wearing a beautiful necklace of round pearls and a large pear-shaped pearl earring, while another pear-shaped pearl depends from a clasp which serves to loop up her fichu on the shoulder.

A fine example of the Renaissance style existing in the sixteenth century is that of a gold and enamel necklace of Italian workmanship, embellished with pearls. This necklace was presented to the Louvre Museum by Don A. de Rotschildt. The two-pearl motive is carried out exquisitely, two pearls appearing in a small connecting ornament between two larger enameled and engraved gold plaques, which represent scenes from the life of our Saviour.

At the exposition of 1900 there was shown in the Russian Pavilion a most interesting collection of jewelry of decidedly oriental character, dating from the sixteenth to the eighteenth century. These jewels were said to have belonged to the Emir of Bokhara. They differed slightly from the East Indian in character, and generally consisted of combinations of pearls, rubies, and emeralds, the three colors of these gems predominating. One of the most interesting of the necklaces, acquired by J. Gelatley, Esq., shows an arrangement of the pearls which is peculiarly attractive and decorative.

The heraldic significance of pearls has at times been very important. While in the eighteenth century the crowns of the French nobles were surmounted with silver points, it appears that in the sixteenth century they were provided with pearl points. According to Rudolphus,[1] the dukes wore a leaf crown of eight leaves, with or without as many commingled pearl points; the marquises a crown of four leaves with twelve pearl points, or with four groups of three pearls set one over the other; and the counts, a pearl crown which sometimes had four pearls in each corner, one above the other. The viscounts wore a gold ring set with four pearls, and the barons a gold ring entwined with pearls.

The same is true of the English coronets. Instead of the pearls which they bore at an earlier period, silver balls are now used on those of the English barons, viscounts, earls, and marquises. This change probably owed its origin to the desire on the part of the sovereigns to confine the official use of pearls and other precious stones to themselves. The rules at the coronation of Edward VII forbade the use of

[1] "Heraldic. Curios.," Pars III, c. 8, p. 12.

pearls except as a special royal privilege. The earl's coronet has eight balls raised on points, with gold strawberry leaves between the points. The marquis wears one with four gold strawberry leaves and four silver balls alternating, the latter raised above the rim.[1]

A pearl and gold ring, formerly belonging to Washington, is now in the possession of Vice-Chancellor E. B. Leaming, of Camden, N. J. It bears in the center a lock of Washington's hair under a conical glass, around which is a setting of blue and white enamel with a square of red at each corner. The whole is surrounded by a circle of thirteen pearls. This ring was presented by Washington to Lieutenant Richard Somers prior to the latter's departure on the expedition against the Algerine pirates in Tripoli, in the course of which he lost his life. Before his departure he left the ring with his sister, Sarah Keen. Vice-Chancellor Leaming's paternal grandmother inherited it as heir to Somers's estate, and from her it descended successively to her son and grandson. The lock of Washington's hair is admitted to be one of only three now existing, of the other two, one is at Washington's headquarters at Newburg and the other in the museum at Boston. The ring was exhibited at the Centennial Exposition in 1876.

And what a wealth of pearls was seen at the marriage of the late Emperor Frederick III of Germany with Princess Victoria, in 1858! The wedding gift of the bridegroom consisted of a necklace of thirty-six enormous pearls, three superb ones in the middle, and graduated in size toward the ends. From her mother, Queen Victoria, the bride received a diamond necklace and three massive brooches set with unusually large pearls; and from Prince Albert, a magnificent hair-net of pearls, diamonds, and emeralds. The king and queen of Prussia presented a diadem of brilliants surrounded with a splendid circlet of pearls. On the day of her entry into Berlin, the queen bestowed on the bride a costly brooch of pearls and diamonds, representing a bouquet, the leaves of which consisted of diamonds, while the flowers themselves were of pear-shaped pearls of large size, one weighing 160 grains, and fourteen of them weighing 600 grains together.

One of the most splendid and best known collection of pearls, and one worn with as much grace as any in Europe, consists of those owned by the dowager Queen Margherita of Italy, whose name signifies pearl, and who has always been fond of the ocean jewel. Her husband, King Humbert, made her many presents of this regal gem. A photograph, signed by the queen and sent to us for this volume by her gracious courtesy, shows her wearing her magnificent twelve strings of pearls, a pearl bracelet, and a pearl tiara with pear-shaped pearl tips.

At the coronation of Edward VII and Queen Alexandra, beside her

[1] Debrett, "Dictionary of the Coronation," p. 127.

MARGHERITA, DOWAGER QUEEN OF ITALY

COLLECTION OF BLACK PEARLS BELONGING TO AN AMERICAN LADY

coronation crown, the latter wore many of her richest and most beautiful jewels. These consisted of seven immense rows of pearls, each twenty-four to thirty inches in length, hanging below five large neck circlets of diamonds and a great corsage ornament which covered her entire bodice; and beneath part of this was a splendid ornament of diamonds with large, pear-shaped pearls.

A careful study of the decorations conferred by potentates and governments shows that the pearl is rarely used in the ornamentation of these marks of distinction. A notable exception is that given by the Siamese government. This decoration is known as the nine-jewel Siamese decoration, and bears a large center pearl. It is only conferred on nine members of the royal Siamese family, including his Majesty the King of Siam. The central pearl represents the king and the eight other jewels surrounding it the members of his family. It is strange that Siam should find so much significance in white, as is illustrated by the white elephant, and also by the use of the white pearl for this order.

The Order of Christ, the chief Portuguese order, has a long cross enameled in bright red surcharged with a white cross and bordered with fine pearls. The effect is both striking and beautiful.

The order of the crown of India is a jeweled badge with a device composed of the imperial cipher, E. R. and I., in diamonds, pearls, and turquoises, set within a border of pearls and surmounted by the imperial crown.[1]

A remarkable pearl necklace was recently the subject of litigation in England. It was the property of the late Duchess of Sermonata, an Englishwoman who married an Italian. She was a daughter of the late Lord Howard de Walden, one of the wealthiest of the English nobility. The duchess was in the habit of investing all her spare cash in pearls, and it seems that she chose a very good form of investment, since pearls have increased in value to a greater extent even than diamonds during the same period. Of the ten rows of which this necklace consisted, six were deposited for safekeeping in a London bank and the other four were in Florence at the time of the death of the duchess. She had bequeathed the gems at the bank to her niece, Miss Henrietta Ellis, and had left directions that, if her pearl necklace was in London when she died, it should be sent to her Italian executors. All the jewels are now claimed by these executors, while Miss Ellis contends that it was the intention of the duchess to leave to her the pearls in the hands of the London bankers. The necklace consisted at one time of ten rows; the first, thirteen and a half inches long, comprised forty-one pearls; the second, fourteen inches in length, thirty-

[1] "Illustrated London News," April 13, 1878, p. 347.

nine pearls; the third, fourteen and a half inches, forty-three pearls; the fourth, seventeen inches, forty-seven pearls; the fifth, seventeen and a half inches, forty-nine pearls; the sixth, nineteen and a half inches, fifty-five pearls; the seventh, twenty-one inches, sixty-six pearls; the eighth, twenty-three and a half inches, seventy pearls; the ninth, twenty-six inches, eighty-two pearls; the tenth, twenty-nine and a half inches, ninety-one pearls. The total number of pearls is 583, and the necklace is valued at $150,000.

A widely advertised necklace of large size was shown in the English section of the Paris Exposition of 1900. This necklace consisted of forty-six pearls weighing 1596 grains, and was valued at $450,000. It was loaned by an English gentleman now dead, and was returned to him at the close of the exposition and later dispersed.

In regard to the possession of pearls by families in the United States, we may safely say that there is not a letter in the alphabet under which we cannot find the names of from one to a dozen families, owning single strings or collections from the value of $10,000 to $200,000, or even more. If one is a wearer of jewels, pearls are an absolute necessity; indeed, they are as essential and indispensable for the wealthy as are houses, horses, and automobiles. At no period in the world's history have pearls been more widely distributed; and some of those of to-day are finer in quality and orient, and also more carefully matched, than those in the great collections of the past. Of course there are exceptions, where royal personages have been careful observers and have used good taste, but it is a question whether there have ever been more critical or better buyers, as far as selection is concerned, than are many American men and women who have purchased this gem.

One of the largest pearl necklaces in the United States is in the possession of an American lady. There are perhaps thirty pearls in the necklace, weighing in all about 1400 grains; the largest pearl weighs nearly 120 grains. There is also one of 75 grains and one of 70 grains, the others graduating down to 20 grains.

With increasing wealth, and a demand for rich rather than gaudy or showy jewelry, there is nothing that commends itself so highly as the pearl, which acts as a foil to the diamond, emerald, ruby, and sapphire, and at the same time harmonizes with them and in fact with all the colored stones. The true pearl, as it increases in size and beauty, becomes proportionately more rare and costly; and yet it differs from other jewels in the fact that they are mined in the depths of the earth, and their existing quantity is speculative, while the home of the pearl is much more accessible, and it is possible to make an estimate of the number of pearls in course of growth. Pearls, however, are forming all the time, while other gems are perhaps to-day as they were ages before the advent of man. Nevertheless, even if pearls were cultivated as they should be, and people cared for the mollusks as the oyster-gatherer does for his crop—by planting it, guarding it and gathering it systematically—still, the ever-increasing demand would more than balance the greater supply. As we have said, at no time since pearls were worn have they enjoyed such favor; and while they have always increased in value, this increase has never been so rapid as in the past ten years. They are jewels which can be worn by young or old, and which adapt themselves to every fabric that man or woman can use for attire; whether they are white, gray, or black, they are never obtrusive, but always have a refining effect. Round as the globe upon which we live, they will probably be worn and appreciated as long as life exists upon this sphere.

It is interesting to note the change of taste and the difference of opinion, at various epochs, in regard to the respective merits of pear-shaped and round pearls. In the Roman period the pear-shaped pearls were more highly valued; in the eighteenth century round pearls were esteemed the more valuable, while at the present day they are both on about the same basis.

With the progressive twentieth century taste for independence in fashion, our modern ladies take from every epoch what they think will best suit their superrefined beauty. Therefore we are not surprised to find in their jewel-cases the long earrings and large brooches adorned with seed-pearls, similar to those worn by their grandmothers of the early Victorian period. Although these jewels cannot be considered very beautiful according to the artistic standard of to-day, they, nevertheless, lend to their wearers a certain quaint dignity and piquancy which is very attractive.

As an instance of modern pearl-wearing by a lady of the present century, we may note a portrait in which there is a simple necklace of large pearls; over this a collar of twenty-three rows of pearls with a diamond centerpiece, and to relieve the severity, a sautoir, which is made up of alternate pearls and diamonds, and pearl earrings. No better illustration can be given than the portrait of Señora Carmen Romero Rubio de Diaz, wife of President Porfirio Diaz of Mexico, which, by her courtesy, we are able to figure.

The gathering of a great necklace is not the work of a day; it often requires many years. Such necklaces are frequently held for a long time by dealers or by a number of people who are interested in their sale, and whenever one or more pearls can be purchased which form a better graduation or which are of better color or more perfect, they are usually purchased to improve the necklace if the price is a proper one.

In the early sixties, when most American women aspired to owning a pair of diamond earrings, it was not uncommon for ladies to start with a hundred, two hundred, five hundred, or a thousand-dollar pair, and, for a dozen years to come, to add an annual sum of one hundred, two hundred or five hundred dollars to increasing the size of these by exchanging them with the dealer at the cost price and paying the difference between the value of the pair that had been purchased and that of the new pair. In this way ladies who never would have thought of spending five thousand dollars for a pair of earrings, virtually made a savings-bank of the jewels. This is frequently done with pearls. A small necklace or a few pearls will be purchased; these are added to annually or at such times as the owner may have spare savings or gifts to invest. It is not uncommon for a family to buy a pearl for a daughter on her first birthday, and each succeeding year add one pearl to this, so that she may first wear one pearl, then two, then three, and by the time the young lady makes her début in society, a good start has been made toward a pearl necklace. It was the custom of King Humbert of Italy to present his queen, Margherita, with one fine pearl every year, and with this succession of annual gifts she possessed one of the finest collections in Europe.

In the portraits of the four daughters of the present Czarina of Russia, the Grand Duchesses Tatiana, Olga, Maria, and Anastasia, we can see that their pearl necklaces were built up gradually, as that of the eldest daughter is notably longer than those of her younger sisters. These pearls were annual gifts from the Czar and Czarina and from others of the imperial family.

There are few ornaments worn by man or woman that have not at one time or another been bepearled, either with large or small pearls, with one pearl or many pearls, with pearls of high or low degree, and no object is ever made the less rich by the addition of the peerless gem of the ocean depths.

As the prices of pearls have increased, naturally the single objects containing them have also become more costly. It is not unusual to see rings with pearls each costing from $5000 to $10,000, $20,000, and even $30,000 and over, the pearls not infrequently being in button form.

Rings are occasionally made up of one white and one black oriental pearl, and if a pink one is combined with these, it is either a freshwater or a conch pearl. Such rings sell for $5000, $8000, $10,000 and $15,000 each.

Pendant pearls, either round, ovate, drop, or pear-shaped, sell from $5000 to $10,000, $20,000, $50,000, and even $100,000.

The prices for one or two choice pearls worn for the adornment of a man's shirt-front are $2000, $3000, $5000, and even $10,000.

SEÑORA CARMEN ROMERO RUBIO DE DIAZ, WIFE OF
PRESIDENT PORFIRIO DIAZ OF MEXICO

Jade jar inlaid with pearls set with fine gold
Heber R. Bishop Collection, Metropolitan Museum of Art

Japanese decoration set with pearls
Order of the crown of the First Class. Metropolitan Museum of Art

In link buttons, slightly ovate, button or round pearls are used, the link being made up of one white and one black pearl, costing $2000, $3000, $5000, and even $20,000 a set.

It is not unusual for a man to wear a scarf-pin set with a round, ovate, or pear-shaped pearl costing $2000, $5000, $10,000, $15,000, and even $30,000.

For men's scarf-pins, a variety of colors are frequently selected, such as a white oriental, a pink American, a pink conch, or a gray and black oriental pearl.

Single pearl necklaces sell for $1000, $2000, $5000, $8000, $10,000, $15,000, $20,000, $50,000, $100,000, $250,000, and $500,000 necklaces are not unknown.

Tiaras sell for $10,000, $20,000, $50,000, and $100,000.

Waistcoat buttons, sometimes made up of baroque pearls, cost from $200 to $500; sometimes, however, when fine pearls are used, the price paid for a set of five or six buttons is as high as $10,000.

It is scarcely possible to mention all the various forms in which the pearl has been worn: whether as a spray of many small pearls or a few large ones, either round, ovate, or pear-shaped for aigrets; in points on crowns, used either for ornamental or heraldic purposes; for the decoration of the orbs, scepters, and crowns of kings and emperors; for forming an edging on bonnets, caps, fillets, or diadems; in pendant form, usually consisting of one, although sometimes of three or four pendants in rows and lines to ornament the side of the face; or one, two, three, or a bunch together to adorn the ear; as a single pearl on a wire or a group of them, as worn in the nose of the East Indian beauty; as a single, two, three, or many-rowed necklace to grace the lady, the queen, and the empress; or else in six, ten, to twenty or more rows with a tiny gold jeweled bar, or a large diamond center, in the form of collars; as a long chain from four to ten feet long to hang from the neck to the waist, or else to be worn once, twice, or thrice around the neck, hanging down and then encircling the waist in the form of a sautoir; either as a single drop, consisting of an ovate or pear-shaped pearl or a number of them together in the form of a pendant combined with diamonds; as a single pearl surrounded with pearls or diamonds for buttons to adorn my lady's crown; in rows, or combined with jewels and enamel, in the form of a bow-knot with long bunches of pearls, for shoulder bars; either as one pearl alone or alternating with gold wire, with jewels, or with many pearls, in endless forms, as bracelets; either as a single row, two rows or alternate rows in infinite variety on bodices, as worn in the past more than at the present; in a single row on ornamenting metal, enamel, or jewelwork in the form of girdles; in five hundred forms for rings; as an embroidery or in rows pendant on slippers; and, finally, as a stole. These are only a few of the uses to which a lady can put pearls.

By men, pearls are worn to adorn the shirt, to wear in the scarf, as link buttons, as waistcoat buttons, or as a fob. The pearls used in this way are sometimes quite as expensive as any of those worn by the ladies.

Ecclesiastics, for more than two thousand years, have appreciated the richness of bepearling. In Russia we find pearls decorating crosses, missal covers, vestments, bindings on books, chalices and crook-tops; they are employed as borders to ikon frames, or for the decoration about the Madonna and Child. In Persia we find pearl-embroidered rugs, pillows, and bolsters. Half-pearls are used in quaint decorations for watches, snuff-boxes, miniatures, and portrait frames. Even saddles and horse-trappings in the East do not escape the charm and beauty of the pearl. Even the English coronation spoon is known for the pearls which ornament it.

Of the many forms of earrings that have come down to us, none is simpler or daintier than a single pearl worn as an ear-screw, or partly or entirely strung on a thin gold wire. Another dainty style is three pearls, worn one below the other as in ancient Rome, known as a triclum; or the round pearl with a pear-shaped pendant or bunches of pearls known as crotalia, also worn in ancient times.

A pearl necklace is usually clasped either by a round or ovate pearl, drilled so that the catch and snap are contained within the pearl itself, or else by a pearl surrounded by diamonds, rubies, or other gems. Such a clasp frequently serves to bind from two to fifteen rows of pearls, the first or smaller row encircling the neck, and each row in turn being larger until the fifteenth row reaches to the bosom or even to the waist.

Pearl collars are usually made up of four, six, ten, twenty, and even twenty-five rows; often of very small pearls, generally fitting closely to the neck. The pearls are held in position either by four gold, diamond, or jeweled bars, or frequently the entire front of the collar is occupied by a large diamond ornament.

In ancient times, pearls were a favorite decoration of crosses; frequently an entire cross was made up of pearls, either of a single or a double row. Many portraits dating from the fifteenth to the seventeenth century show the cross used in connection with a necklace, this either starting from the top of the cross or from each side at the end of each arm. Sometimes from below the arms and the lower part of the cross there hung pendant either round or pear-shaped pearls. We have other instances where at the top, the cross was attached to a pearl necklace, while below each of the two arms there hung a pearl, and from the lower part of the cross a double necklace again reached to the back of the neck. Frequently a festoon collar will be made up of five rows of pearls, each of a graduated length, and pendant on each a diamond. Recently pearls have been drilled and invisibly joined by fine platinum links, so as to form a continuous ribbon or even a collar two inches wide; occasionally, a Greek border or some other design, of larger pearls or of diamonds, rubies, sapphires or other gems, is interwoven. This constitutes a veritable, smooth pearl cloth, or pearl mesh, very beautiful and also comfortable to wear. Indeed, a purse, measuring five by six inches, has been made of this cloth of pearls.

Dust pearls, too minute to drill, and numbering over 100,000 to the ounce, were used, in the latter part of the eighteenth and the early part of the nineteenth centuries, for the embellishment of the hair-work then so much in favor and which was placed under glass. Where foliage was represented the leaves were made of the most minute seed-pearls, graduated in size and set on an outline of enamel or white paint, the pearls being cemented to the outline. This added a softness to the hair-work and other decoration.

As long as the pearl has been known, there has been a desire to obtain possession of one in some of its degrees of perfection, and for this reason many attempts have been made to prepare something that might pass for a pearl or even suggest a pearl. Sometimes the mother-of-pearl shell has, naturally, a protuberance, either round or pear-shaped, which, if cut off and highly polished may resemble an imperfect pearl; and this operation is often so cleverly performed that, at the first glance, this object may pass for a true pearl. In Russia, and especially in Bohemia, they have gone farther than this. They have cut out a bit of mother-of-pearl shell, leaving a piece of the natural shell for the top, or the part that will be visible, and rounding off the rest of the surface so as to give it a pearly effect. These objects are of trifling value and are used in necklaces and earrings, and in the ornamentation of icons and miniature frames and even as beads. Glass with either an exterior or interior coating of a nacreous substance is sometimes made absolutely round, while at other times it is made with many imperfections so as to resemble either a marine baroque or a fresh-water irregular pearl. The North American Indian, as described elsewhere, has coated little balls of clay with a powder made from a pearl-bearing fresh-water mussel and then baked them.

XVI

FAMOUS PEARLS AND COLLECTIONS

XVI

FAMOUS PEARLS AND COLLECTIONS

The kingdom of heaven is like unto a merchant man, seeking goodly pearls: who, when he had found one pearl of great price, went and sold all that he had and bought it.

St. Matthew, XIII, 45, 46.

IN the course of twenty centuries many pearls and pearl collections have become famous, either because of their intrinsic value or else through historic associations. An attempt is made here to list briefly the more important of these. While we have purposely omitted any mention of the pearl collections in private hands at the present time, some of which are more valuable than many of those noted in the following pages, we have, nevertheless, given the principal sales of pearls at auction during the past twenty years. Many specimens of remarkable size and beauty have changed hands in this way, more especially in England.

CLEOPATRA PEARLS. Next to that "pearl of great price," mentioned by Christ, probably the most famous of all pearls were the two which Pliny records as having been worn in the ears of Cleopatra, "the singular and onely jewels of the world and even Nature's wonder." This writer does not note their size, but estimates their value at sixty million sestertii. We have already quoted the passage in which Pliny relates how one of these pearls was dissolved and swallowed by Cleopatra in order to win a wager she had made with Antony. After the death of that queen the other pearl "was cut in twaine, that in memoriall of that one halfe supper of theirs, it should remaine unto posterite, hanging at both the eares of Venus at Rome in the temple of Pantheon."[1] Budé estimated the value of the pearl dedicated to Venus at 250,000 *escus* of gold.[2]

Another famous pearl mentioned by Pliny was the one which Julius Cæsar presented to Servilia, mother of Brutus, the value of which he notes as six million sestertii.[3]

[1] Pliny, "Naturall Historie," London, 1601, *Lib.* IX, c. 35.
[2] Budé, "De Asse," Paris, 1514.
[3] Pliny, "Historia Naturalis," *Lib.* IX, c. 35.

PEROZ PEARL. The historian Procopius,[1] of the sixth century, tells of a magnificent pearl which belonged to Peroz, or Firuz (459–484), one of the Sassanian kings of Persia. In the course of his disastrous battle with the White Huns, in which both he and his sons perished, Peroz, having a presentiment of the misfortune about to befall him, took the pearl from his right ear and cast it away, lest any one should wear it after him. This pearl is described as being "such as no king had ever worn up to that time." Procopius, however, thinks it more probable that the ear of Peroz was cut off in the combat, and he states that the emperor (Zeno, 426–491) was very anxious to buy the gem from the Huns, but that all search for it was in vain. Nevertheless, a rumor was current that it was recovered later, but that another pearl was substituted for it and sold to Kobad, a successor of Peroz.

A different version is given by Panciroli,[2] who quotes Zonaras, a Byzantine historian of the twelfth century, as his authority. According to this version Justinian the Great, who succeeded to the throne forty-three years after the death of Peroz, offered one hundred pounds of gold (about $25,000) for the pearl, but the barbarians refused to part with it, preferring to keep it as a memorial of Persian folly. On the coins of Peroz he is represented wearing an earring with three pendants, one of which may have been this wonderful pearl.

CHARLES THE BOLD. One of the greatest jewels of the fifteenth century was that belonging to Charles the Bold, Duke of Burgundy (1433–1477). According to notes and drawings[3] made in 1555 by J. J. Fugger of Nuremberg, who was the banker jeweler of his generation, this consisted of a large pyramid diamond five eighths of an inch square at the base, with the apex cut as a four-rayed star in relief; surrounding this were three rectangular pyramid-shaped rubies and three magnificent pear-shaped pearls, and a large ovate pearl was suspended from the lowest ruby. The pear pearls are described as measuring half an inch in diameter and must have weighed about sixty grains each. This magnificent jewel was probably the most celebrated in Europe during the fifteenth century. According to Comines, on the defeat of the Grand Duke and the plundering of his baggage by the Swiss at Granson in 1476, the ornament was found by a careless soldier who tossed it away, but retained the gold box containing it. On second thought, he searched for and recovered the jewel and sold it to a priest for one florin, and the ecclesiastic sold it to a Bernese govern-

---

[1] "Historia," *Lib.* I, c. 4, ed. Niebuhr, Bonnae, 1833.

[2] Panciroli, "Rerum Memorabilium, libri duo," Frankfort, 1660, Pt. I, p. 44. We have been unable to find this statement in the An-

nals of Zonaras; it was possibly derived from some gloss or annotation.

[3] Published by Lambeccius in "Bibliotheca Cæsarea," Vol. II, p. 516.

ment official for the sum of three florins. Some years later this jewel, together with the ducal cap of Charles the Bold, which was covered with pearls, and bore a plume case, set with diamonds (points), alternating with pearls and balas-rubies, was sold by the Bernese government to Jacob Fugger, as related by J. J. Fugger in the manuscript above noted, "for no more than 47,000 florins." In the vain hope that it would be purchased by Emperor Charles V, grandson of Charles the Bold, Fugger held the jewel for many years, but he broke up the cap and reset the stones in it for Maximilian II. The brooch was finally sold to Henry VIII of England just before his death, and it passed to his daughter and successor, Bloody Mary, who presented it to her Spanish bridegroom, Philip. Thus, after seventy-six years, the jewel was restored to a descendant of the original owner. This history has been given at some length owing to its illustration of the manner in which great pearls were easily lost on battle-fields and were passed about from one country to another.

TARAREQUI PEARLS. The early American fisheries yielded several magnificent pearls, many of which eventually became part of the imperial Spanish jewels. Prominent among these was the *Huerfana* or *Sola*. According to Gomara, this was secured in 1515 from the Indians at Tararequi, in the Gulf of Panama, in a large collection which weighed 880 ounces. It was pear-shaped and weighed thirty-one carats. Gomara states that this pearl was purchased from Gaspar de Morales, leader of the Spanish expedition, by a merchant, for the sum of 12,000 castilians. "The purchaser could not sleep that night for thinking on the fact that he had given so much money for one stone, and sold it the very next day to Pedrarias de Avila, for his wife Donna Isabel de Bovadilla"; and afterward it passed to Isabella, wife of Emperor Charles V (1500–1558). It was remarkable for its luster, color, and clearness, as well as for its size. Another large pearl in this collection weighed twenty-six carats.

OVIEDO PEARL. As already noted on page 237, in his "Historia natural y general de las Indias," published at Toledo in 1526, Gonzalo de Oviedo wrote of having purchased at Panama a pearl weighing twenty-six carats for which he paid 650 times its weight in fine gold, and which he claimed was the "greatest, fairest and roundest" that had ever been seen at Panama. Probably this was the twenty-six-carat pearl obtained at Tararequi by Gaspar de Morales in 1515. At 650 times its weight in gold the value of this pearl would be $2294.54; representing a base of $.2124 per grain; but at a base of $5 per grain the same pearl would be worth $54,080, equaling 15,320 times its weight in gold.

TEMPLE OF TALOMECO. Among great collections of pearls, some writers would place that described by Garcilasso de la Vega as having been found by De Soto and his followers in 1540 in the Temple of Talomeco near the Savannah River in America.[1] According to Garcilasso, the quantity of pearls there was so great that 300 horses and 900 men would not have sufficed for its transportation, vastly excelling every other if not all other collections in the history of the world. Unfortunately the accuracy of this account has not been unquestioned.

LA PEREGRINA. Most celebrated among the early American pearls was La Peregrina (the incomparable), or the Philip II pearl, which weighed 134 grains. According to Garcilasso de la Vega, who says that he saw it at Seville in 1597,[2] this was found at Panama in 1560 by a negro who was rewarded with his liberty, and his owner with the office of alcalde of Panama. Other authorities note that it came from the Venezuelan fisheries in 1574. It was carried to Spain by Don Diego de Temes, who presented it to Philip II (1527–1598). Jacques de Treco, court jeweler to the king, is credited with saying that it might be worth 30,000, 50,000 or 100,000 ducats, as one might choose to estimate, for in fact it was so remarkable as to be beyond any standard valuation. If we can credit Garcilasso, at one time this pearl decorated the crown of the Blessed Virgin in the church of Guadeloupe, which was resplendent with gems.[3] A contemporaneous account[4] notes that it was worn at Madrid by Queen Margarita, wife of Philip III, at the fêtes given in celebration of the treaty of peace between that country and England in 1605.

CHARLES II PEARL. Somewhat similar to the foregoing was the pearl of Charles II of Spain (1661–1700), which was presented to that monarch by Don Pedro de Aponte, Conde del Palmer, a native of the Canaries. This gem was found in 1691, or more than a century after La Peregrina. These two pearls were nearly equal in size, and for many years they were worn as earrings by the successive queens of Spain. It is reported that they were destroyed in 1734, when a large portion of the old palace at Madrid was burned.[5]

The jewels of the Spanish crown have passed through so many vicissitudes that it is not surprising that but few of them remain in the

[1] See p. 254 for Garcilasso's description.
[2] Garcilasso, "Historie des Incas, Rois du Pérou," Amsterdam, 1704, Vol. II, p. 352.
[3] Ibid., p. 351.
[4] Miscel. Academ. Nat. Curios, Dec. 1, Ann. II, obs. 288.
[5] "Hawkins' Voyages," Hakluyt Society, 1878, p. 315 note.

QUEEN ELIZABETH OF ENGLAND

ELIZABETH OF FRANCE

MARY, QUEEN OF SCOTS

The property of the Earl of Leven and Melville.   About 1559-1560

Spanish treasury. After the overthrow of the Spanish monarchy by the French in 1808, Ferdinand VII, during the time of his exile, disposed of many of these jewels. It is asserted that, after the deposition of Queen Isabella, in 1868, the crown jewels were divided between herself and her sister, the pious Duchesse de Montpensier, and a considerable portion was eventually distributed among the numerous descendants of the latter. It is also stated that there is no mention of the Spanish crown jewels during the reign of King Amadeus, the first sovereign of the restored monarchy. There are, however, great quantities of pearls and other gems belonging to the various madonnas in the Spanish churches, as, for example, Nuestra Señora de Atocha, Cavadonga and others.

PEARLS OF MARY STUART. The pearls owned by the unfortunate Mary Queen of Scots (1542–1587) were among the most beautiful in Europe. Inventories of these[1] show great *bordures de tour* of large pear pearls with *entredeux* of round pearls, long ropes of pearls strung like beads on a rosary, carcans or broad belts set with pearls, and a large number of loose pearls. Many of these appear in the portraits of this popular queen; but probably the most remarkable exhibition of them is in the portrait now owned by the Earl of Leven and Melville,[2] which appears to agree fairly well with the inventories of her jewels, although this portrait is not wholly free from impeachment as to its accuracy and contemporaneousness.

After the downfall of the queen, most of her jewels were sold, pawned, or lost by theft. A number of them passed into the possession of Queen Elizabeth in 1568, in a manner not wholly satisfactory to lovers of justice. Some of these were described in a letter dated May 8, 1568, and addressed to Catharine de' Medici by Bodutel de la Forest, the French ambassador at the English court, as "six cordons of large pearls, strung as paternosters; but there are about twenty-five separate from the others much larger and more beautiful than those which are strung. They were first shown to three or four jewelers and lapidaries of this city, who estimated them at three thousand pounds sterling, and who offered to give that sum; certain Italian merchants who viewed them afterwards valued them at 12,000 *escus,* which is the price, as I am told, this queen [Elizabeth] will take them at. There is a Genevese who saw them after the others and estimated them as worth 16,000 *escus* [$24,000]."[3]

Catharine de' Medici, who was a mother-in-law of Mary Stuart, was

[1] See Robertson, "Inventaires de la Royne d'Ecosse," Bannatyne Club, 1863.
[2] See Lang, "Portraits and Jewels of Mary Stuart," Edinburgh, 1906.
[3] Teulet, "Relations politiques de la France et de l'Espagne avec l'Ecosse," Vol. II, p. 352.

very anxious to obtain these pearls; but the ambassador wrote on May 15, 1568, that he had found it impossible to purchase them; for, as he had told her from the first, they were intended for the gratification of the Queen of England, who had purchased them at her own price, and was even then in possession of them.[1]

QUEEN ELIZABETH'S PEARLS. Although in her youth she is said to have had a distaste for personal decorations, in her later years Queen Elizabeth entertained an extravagant fondness for pearls. In speaking of her portraits, Horace Walpole says: "A pale Roman nose, a head of hair loaded with crowns and powdered with diamonds, a vast ruff, a vaster fardingale, and a bushel of pearls, are features by which everybody knows at once the pictures of Queen Elizabeth."[2] And to the end, her love for them was unabated, for in the last tragi-comic scene of her life, to meet the Angel of Death himself, she was dressed up in her most splendid jewels with great pearl necklaces and earrings and pendants, as Paul Delaroche so successfully pictured in his remarkable painting in the Louvre.

The faded waxwork effigy of her, long preserved in Westminster Abbey in that curious collection of effigies[3]—the "Ragged Regiment," as Walpole called them—has a coronet of large spherical pearls in wax, long necklaces of them, a great pearl-ornamented stomacher, pearl earrings with large pear-shaped pendants, and even broad, pearl medallions on the shoe-bows. In accordance with that singular custom which prevailed from the time of Henry V (1422), to that of Queen Anne (1714),[4] this effigy lay on her coffin at the funeral and caused, says Stow in his Chronicle, "such a general sighing, groning, and weeping, as the like hath not beene seene or knowne in the memory of man." A contemporaneous poet wrote that when the corpse with the effigy passed down the Thames to lie in state at Whitehall,

> Fish wept their eyes of pearl quite out,
> And swam blind after.

GRESHAM PEARL. During the reign of Queen Elizabeth, Sir Thomas Gresham, the merchant prince, was credited with possessing a pearl valued at £15,000, which he reduced to powder and drank in a glass of wine to the health of the queen, in order to astonish the Spanish ambassador, with whom he had laid a wager that he would give a more costly

[1] Teulet, "Relations," etc., p. 364.
[2] Walpole, "Anecdotes of Painting in England," London, 1849, Vol. I, p. 151.
[3] An interesting account of this collection was given in a little book, now quite rare, published in London in 1793 by John Roberts, entitled "A View of the Waxen Figures in Henry VII's Chapel."
[4] Bolton, "Curious Relics of English Funerals," Boston, 1894, p. 233.

dinner than could the Spaniards.[1] No other information regarding this pearl seems available. The valuation certainly appears excessive when compared with that of some other pearls of that period.

We quote an item from Burgon,[2] taken from the manuscript journal kept by Edward VI:

25 [April, 1551]. A bargaine made with the Fulcare for about 60,000 l. that in May and August should be paid, for the deferring of it. First, that the Foulcare should put it off for ten in the hundred. Secondly, that I should buy 12,000 marks weight at 6 shilinges the ounce to be delivered at Antwerpe, and so conveyed over. Thirdly, I should pay 100,000 crowns for a very faire juel of his, four rubies marvelous big, one orient and great diamount, and one great pearle.

RUDOLPH II PEARLS. The scientific, art-loving, but eccentric Rudolph II (1552–1612), Emperor of the Holy Roman Empire, gathered about him at Prague a great collection of jewels and wealth of all sorts. The values of his pearls and precious stones, of the gold and silver articles, was estimated by the archæologist, Jules Cæsar Boulenger, at seventeen millions of gold florins, which was a very considerable sum at that time, as appears when we consider that one hundred gold florins annually was deemed a good salary for an official at the emperor's court. De Boot mentions a pearl belonging to Rudolph II which weighed "thirty carats and cost as many thousands of gold pieces." It is quite likely that this was the one noted by Gomara as coming from the Gulf of Panama,[3] and which Rudolph probably inherited from his grandfather, Emperor Charles V. The pearl bought by Oviedo in Panama, prior to 1526, may be one of the principal decorations of the imperial crown of Austria.

We read in that curious and interesting book, "The Generall Historie of the Turkes," by Richard Knolles,[4] that Abbas the Great, Shah of Persia (1557–1628), after having defeated the Turks in many battles, desired to form an alliance with Emperor Rudolph II, and to induce him to break his engagements with the Turks. To this end Shah Abbas, in 1610 sent an embassy to Prague, with many valuable gifts for the emperor, among which were "three orientall pearles exceeding big." It has been conjectured, and it is also claimed, that these may be three of the eight pear-shaped pearls which are now to be seen in the crown of Rudolph II. One of the largest pearls in the Austrian crown, as we have stated, is most probably the Oviedo pearl.

[1] Lawson, "History of Banking," London, 1750, pp. 24, 25.
[2] Burgon, "The Life and Times of Sir Thomas Gresham," London, 1839, Vol. I, p. 69.
[3] See p. 451.
[4] London, 1631, p. 1297.

CHARLES I PEARL. Admirers of Vandyke's pictures of Charles I (1600–1649) readily recall the pearl pendant from his right ear, which appears in nearly all of his portraits by that artist. Janin wrote: "This pearl in the ear of his majesty was greatly coveted, and as soon as his head had fallen, the witnesses of the dreadful scene rushed forward, ready to imbue their hands in his blood in order to secure the royal jewel." It seems more probable that the martyr king would have left this gem in the hands of a trusty friend for his family than to the risk of injury by the ax and to be torn from his mutilated head by a scrambling mob.

OWING to their control of the great fisheries, the most valuable collections of pearls have been held by eastern monarchs, and particularly by those of India and Persia. It has been estimated that one third of the portable wealth of these countries is in jewels. Most Orientals are as suspicious of interest in their jewels as they are of inquiry regarding their harems, imagining, doubtless, that the interest conceals a sentiment of cupidity, hence it is not practicable to give a minute description of them. However, several travelers have recorded glowing accounts of collections which they have examined, which read much like a description of Aladdin's palace in the Arabian Nights. Among these, some of the greatest are the

PEARLS DESCRIBED BY TAVERNIER. For accounts of remarkable pearls in eastern countries in the seventeenth century, we are indebted to that well-informed old French jeweler, Tavernier, one of the most remarkable gem dealers the world has ever known. He made numerous journeys to Persia, Turkey, Central Asia, and the East Indies, gaining the confidence of the highest officials and trading in gems of the greatest value. After amassing a large fortune and purchasing a barony near Lake Geneva, he died at Moscow in 1689 while on a mercantile trip to the Orient, at the age of eighty-four years. His "Voyages," published in 1676–1679, reveal a critical knowledge of gems, a remarkable insight into human nature, and the absence of any intention to impart misleading information.

In the first English edition of his travels, published in 1678, Tavernier gave sketches of five of the principal pearls which came under his careful observation.

Figure 1 of Tavernier's diagram shows what he considered "the largest and most perfect pearl ever discovered, and without the least defect." The weight of this pear-shaped gem does not appear to have been noted, but from the sketch it may be estimated at about 500 grains. Tavernier states that the bloodthirsty Shah Sofi, King of

Persia, purchased it in 1633 from an Arab who had just received it from the fisheries at El Katif. "It cost him 32,000 tomans, or 1,400,-000 livres of our money, at the rate of 46 livres and 6 deniers per toman ($552,000)."[1]

Very much smaller but more beautiful than this great pearl, was the one which Tavernier saw in 1670 at Ormus in the possession of the Imam of Muscat, who had recently recovered the Muscat peninsula from the Portuguese. The jeweler stated that although this weighed only twelve and one sixteenth carats (forty-eight and a quarter grains),[2] and was not perfectly round, it surpassed in beauty all others in the world at that time. It was so clear and lustrous as to appear translucent. At the conclusion of a grand entertainment given by the Khan of Ormus, at which Tavernier was present, the Prince of Muscat drew this gem from a small purse suspended about his neck, and exhibited it to the company. The Khan of Ormus offered 2000 tomans (about $34,500) for it, but the owner would not part with his treasure. Tavernier states that later the prince refused an offer of 40,000 escus ($45,000) from Aurangzeb, the Great Mogul of India.[3]

Figure 3 in the diagram represents a pear-shaped pearl of fifty-five carats (220 grains) which Tavernier sold to Shaista Khan, uncle of the Grand Mogul. Although of large size and good shape, this was deficient in luster. According to the jeweler, this pearl was from the Island of Margarita on the Venezuelan coast, and was the largest ever carried from Occident to Orient.

Tavernier listed among the Great Mogul's jewels a large olive-shaped pearl, perfect in form and luster. The weight was not noted, but from the sketch which he gave (see Fig. 4) it may be estimated at about 125 grains. It formed the central ornament of a chain of emeralds and rubies, which the Mogul sometimes wore about his neck. He also listed a round pearl of perfect form (see Fig. 5). The weight of this also is not noted, but from the sketch it may be estimated at 110 grains. This was the largest perfectly spherical pearl known to Tavernier. Its equal had never been found, and for that reason it was kept with the unmounted jewels.

Among the other pearl treasures of the Great Mogul, Tavernier noted the following:

(a) Two grand, pear-shaped pearls, one weighing about seventy ratis,[4] a little flattened on both sides, and of beautiful water and good form. (b) A button-shaped pearl, weighing from fifty-five to sixty ratis, of good form and good water. (c) A round pearl of great per-

---

[1] Tavernier, "Travels in India," London, 1889, Vol. II, p. 130.

[2] Tavernier used the Florentine carat, which equaled 3.04 grains troy.

[3] Tavernier, "Travels in India," London, 1889, Vol. II, p. 110.

[4] One *rati* equaled seven eighths of the Florentine carat, or 2.66 grains troy.

fection, a little flat on one side and weighing fifty-six ratis; this had been presented to the Great Mogul by Shah Abbas II, King of Persia. (d) Three round yellowish pearls weighing from twenty-five to twenty-eight ratis each. (e) A perfectly round pearl, thirty-five and a half ratis, white and perfect in all respects. This was the only jewel purchased by the Great Mogul himself, the others being inherited or coming to him as presents. (f) Two pearls perfectly shaped and equal, each weighing twenty-five and a quarter ratis. (g) Also two chains, one of pearls and rubies of different shapes pierced like the pearls; the other of pearls and emeralds, round and bored. All of these pearls were round and ranged in weight from ten to twelve ratis each.[1]

PEACOCK THRONE. The famous Takht-i-Tâ'ûs, or "Peacock Throne," at Delhi doubtless contained the greatest accumulation of gems in the seventeenth century. It was completed, in the eighth year of his reign (1044 A.H., 1634 A.D.) by Shah Jehan, greatest of Mogul sovereigns, who likewise built the Taj Mahal at Agra, one of the most beautiful edifices ever designed by man. Abd-al-Hamid, of Lahore, in his Pâd-shâh-nâmah, "Book of the King," composed prior to 1654, writes as follows:[2]

In the course of years many valuable gems had come into the imperial jewel-house, each one of which might serve as an ear-drop for Venus or as an adornment for the girdle of the Sun. Upon the accession of the emperor, it occurred to him that, in the opinion of far-seeing men the acquisition of such rare jewels and the keeping of such wonderful brilliants could render but one service, that of adorning the throne of the empire. They ought, therefore, to be put to such a use that beholders might benefit by their splendour and that majesty might shine with increased brilliancy.

As described by Tavernier in 1676, great quantities of pearls were used in the ornamentation of this throne, the arched roof, the supporting pillars, the adjacent sun-umbrellas, being well covered with these gems, many of them of great value. The choicest one was pear-shaped, yellowish in color, and weighed about fifty carats (200 grains);[3] this was suspended from a great ruby which ornamented the breast of the peacock. "But that which in my opinion is the most costly thing about this magnificent throne is that the twelve columns supporting the canopy are surrounded with beautiful rows of pearls, which are round and of fine water, and weigh from 6 to 10 carats each [24 to 40 grains]."[4] The total value of the jewels entering into the

[1] Tavernier, "Travels in India," London, 1889, Vol. I, pp. 397–399.
[2] Sir Henry Miers Elliot, "The Mohammedan Period as described by Its Own Historians," Vol. V. of "The History of India,"
ed. by A. V. W. Jackson, New York, 1907, p. 324.
[3] See Fig. 2 of Tavernier's diagram.
[4] Tavernier, "Travels in India," trans. by V. Ball, London, 1889, Vol. II, p. 384.

ornamentation was estimated at 160,500,000 livres or $60,187,500; and the present value of the throne as it stands in the shah's palace at Teheran, whither it was carried by Nadir Shah from the sack of Delhi in 1739, even though divested of many of its most valuable gems, is estimated at $13,000,000.[1] The designer of the Peacock Throne was Austin de Bordeaux, who also planned the magnificent Taj Mahal. He was named by Shah Jehan, "Jewel-Handed," and received a salary of two thousand rupees a month.

SHAH'S "TIPPET." Sir Harford Jones Brydges' description of the jewels of the Shah of Persia at Teheran is of particular value, since he had formerly dealt in jewels and was an expert in such matters. He says:

I was particularly struck with the king's tippet, a covering for part of his back, his shoulders and his arms, which is only used on the very highest occasions. It is a piece of pearl work of the most beautiful pattern; the pearls are worked on velvet, but they stand so close together that little, if any, of the velvet is visible. It took me a good hour to examine this single article, which I have no fear of saying can not be matched in the world. There was not a single pearl employed in forming this most gorgeous trapping less in size than the largest marrow-fat pea I ever saw raised in England, and many—I should suppose from 150 to 200—the size of a wild plum, and throughout the whole of these pearls, it would puzzle the best jeweler who should examine them most critically to discover in more than 4 or 5 a serious fault. The tassel is formed of pearls of the most uncommon size and beauty; and the emerald which forms the top of the tassel is perhaps the largest perfect one in the world. . . . For some days after I had seen these jewels, I attempted to make an estimate of their value, but I got so confused in the recollection of their weight and the allowance to be made in some of them for their perfection in water and color, that I gave it up as impossible. I cannot, however, think I shall much mislead if I say that on a moderate, perhaps a low calculation, their value cannot be less than fifteen millions [sterling?] of our money.[2]

SHAH'S PEARLS IN 1820. Nearly a century ago the elaborate state costume of the Shah of Persia was described by the English artist, Sir Robert Ker Porter. In this description he mentioned particularly the pearls in the tiara, the pear-shaped pearls of immense size with which the plumes were tipped, the two strings of pearls—"probably the largest in the world"—which crossed the king's shoulders, and the

[1] Benjamin, "Persia," p. 73.
[2] Brydges, "An Account of the Transactions of His Majesty's Mission to the Court of Persia, in the Years 1807-1811," London, 1834, p. 383.

large cushion incased in a network of beautiful pearls against which he reclined.[1]

PEARLS OF THE GAIKWAR OF BARODA. Among the greatest jeweled treasures of India are those of the present Mahratta Gaikwar of Baroda, who has precedence over all the rulers in India at all functions, and is one of the most prominent and enlightened of the Indian princes. He governs a province of about 8225 square miles and 2,415,396 inhabitants in the northwestern part of India, 248 miles north of Bombay. Most of these treasures, whose value is estimated at a dozen million dollars, were collected by his predecessor, Mahratta Khandarao, who lived in barbaric splendor, and they are rarely worn by the present gaikwar. These treasures include a sash of one hundred rows of pearls, terminating in a great tassel of pearls and emeralds; seven rows of superb pearls whose value is estimated at half a million dollars; a litter set with seed-pearls, quantities of unstrung pearls, and more remarkable yet, a shawl or carpet of pearls, which closely resembles the "tippet" at Teheran described by Brydge. This carpet is said to be ten and one half feet long by six feet wide, and to be made up of strings of pearls, except that a border, eleven inches wide, and also center ornaments, are worked out in diamonds. Some writers assert that this costly ornament was originally intended by the late Mahratta Khandarao as a covering for the tomb of Mohammed. Others state that it was designed as a present for a woman of whom he was enamoured, but that the British resident interfered, claiming that the wealth of Baroda was not sufficient to warrant such an expensive gift on the part of the ruler. This ornament is now retained among the regalia at Baroda, and is probably the most costly pearl ornament in the world, its value being estimated at several million dollars.

SUMMER PALACE IN 1860. Many superb pearls were among the precious objects in the Yuen-Min-Yuen or Summer Palace in Pekin at the time of its capture by the European forces in 1860. Numbers of these were lost in the confusion of the sacking and plunder, when the soldiers' pockets were filled and the floors were strewn with jewels, beautiful objects of gold and silver, rich silks and furs, carved jade, lapis lazuli, etc. Some of the pearls found their way to Europe, and especially to France and England. They were of good size and luster and were mostly yellowish in color. Unfortunately, many were crudely drilled with large holes, and had been strung on gold wires by which they were attached to the idols they decorated at the time they

[1]Porter, "Travels in Georgia, Armenia, Ancient Babylon," etc., London, 1821, Vol. I, p. 325.

GAIKWAR OF BARODA, 1908

LA PELLEGRINA. For nearly a century there has been in Russia one of the most lovely pearls in the world; this is La Pellegrina, formerly owned by the Zozima brothers of Moscow, who were antiquarians of note in St. Petersburg. In 1818 a small book of forty-eight pages was written about this beautiful gem by G. Fischer de Waldheim, vice-president of the Imperial Medico-Chirurgical Academy, probably the only book ever devoted to a single pearl. According to this writer, La Pellegrina was purchased at Leghorn by one of the Zozima brothers from an English admiral who had just returned from India. It combines all the requisites of perfection: it is absolutely spherical and has never been pierced; its luster, its silvery sheen, make it appear almost transparent, and for a pearl of this high grade, it is of remarkable size, weighing 111½ grains.

The Zozima brothers retained it in a sea-urchin shell mounted in gold and with a convex lens as cover; this was contained in a silver box, and this in turn in another box studded with gems. Although the lens enlarged the appearance of the pearl, it detracted from its beautiful form, giving it an oval shape. But when removed from the triple inclosure, it rolled about like a globule of quicksilver, and surpassed that metal in whiteness and brilliancy.

Everything that is beautiful and perfect takes such possession of the beholder that words become insufficient to express his feelings; and that is what happened to me in the case of La Pellegrina of Zozima. One must have seen an object of this kind in order to appreciate the impression it makes. As an evidence of this, I shall note the last visit which I made to the owner in company with several distinguished persons.

After having examined many curious medals and coins, and also some pearls which exceeded in size the one of which I treat, and after they had received their due meed of admiration, La Pellegrina appeared, rolled upon a sheet of paper by the owner's little finger. Attention and admiration was depicted on every face; a perfect silence reigned. It was only when the pearl had been removed very carefully lest it should slip away, and was again triply enclosed, that we recovered the power of speech and could unanimously express our admiration.[1]

As it had been stated that this pearl was in the possession of the Princess Youssoupoff, Mr. Henry W. Hiller of New York, who was in St. Petersburg, courteously made inquiries and was successful in obtaining a view of the two splendid pear-shaped pearls. These are almost exactly alike, but neither of them can well be La Pellegrina, since this is a round pearl; possibly the one on the right may be La Peregrina.[2]

[1] De Waldheim, "Essai sur la Pellegrina," p. 48.   [2] See p. 452.

HER GRACE, THE DUCHESS OF MARLBOROUGH

THE HOPE PEARL. WEIGHS 1800 GRAINS

Actual size

The owner of La Pellegrina in 1818, Z. P. Zozima, died in Moscow at a great age, in 1827. He was a Greek dealer in curiosities and gems, who had resided in Moscow for a long time, and had many clients among the nobility of Russia. It is stated that a few months before his death the best pieces of his collection, including La Pellegrina, were stolen from him by a compatriot.

Moscow Pearl, 1840. The German traveler, Johann Georg Kohl (1808–1878), in the account of his travels in Russia, relates an interesting incident connected with a beautiful pearl in the Imperial Treasury. Shortly previous to 1840, a rich Moscow merchant died in a convent, whither he had retreated after the manner of the wealthy pious ones of his nation. Feeling the approach of age, he had given up the toils of business to his sons. His wife was dead, and the only beloved object which even in the cloister was not separated from him was a large, beautiful, oriental pearl. This precious object had been purchased for him by some Persian or Arabian friend at a high price, and, enchanted by its water, magnificent size, and color, its perfect shape and luster, he would never part with it, however enormous the sum offered. He himself inhabited an ordinary cell in the convent; but this object of his love reposed on silk in a golden casket. It was shown to few persons, and favorable circumstances and strong recommendations were necessary to obtain such a favor. A Moscow resident reported the style and manner of the ceremony. On the appointed day he went with his friends to the convent, and found the old gentleman awaiting his guests in his holiday clothes. Their reception had something of solemnity about it. The old man went into his cell and brought out the casket in its rich covering. He spread white satin on the table, and, unlocking the casket, let the precious pearl roll out before the enchanted eyes of the spectators. No one ventured to touch it, but all burst into acclamations, and the old man's eyes gleamed like his pearl. After a short time it was returned to the casket. During his last illness, the old gentleman never let the pearl out of his hand, and after his death it was with difficulty taken from his stiffened fingers.

There seems to be a great similarity between the description of this pearl and that of La Pellegrina, although we have been unable to verify our surmise as to their identity.

The Hope Pearl. In the first half of the last century, Henry Philip Hope, a London banker, brought together a great collection of gems, among which were many pearls. The most famous of these was the often-described Hope pearl, one of the largest known; the value of which, however, is not in proportion to its size, owing to its irregular formation. As described in the catalogue of the Hope collection, published in 1839, this oriental pearl is of an irregular pear-shape, weighs 1800 grains, or three ounces, measures two inches in length, and in circumference four and one half inches at the broadest and three and one fourth inches at the narrowest end. The color at the larger end is of a bronze or a dark green copper tint, this gradually clearing into a fine white luster for within one and one half inches of the smaller end. This baroque pearl was firmly attached to the shell, and it yet shows the point of attachment, which has been polished so as to correspond to the remaining portion. It is attractively mounted, the smaller end being capped with an arched crown of red enameled gold set with diamonds, rubies, and emeralds.[1] After remaining in the Hope jewel collection at the South Kensington Museum for many years, it was sold at auction, at Christie's, in 1886, when that collection was placed on the market. This pearl is now held by Messrs. Garrard & Company of London, at the price of £9000.

The Hope collection also contained about 148 pearls of good form. Notable among these were the following: (a) a conical pearl weighing 151 grains, cream-white in color, from Polynesia; (b) a bouton pearl of 124 grains, bluish-white at the top and encircled by a dark bronze color; (c) an oval cream-colored pearl, weighing ninety-four grains, from the South Seas; (d) an eighty-nine-grain, roughly spherical pearl, one side bluish and the other of a light bronze; (e) an eighty-five-grain, acorn-shaped, bluish-white pearl, with a band of opaque white near the base; (f) an oval conch pearl, pink in general color and somewhat whitish at the ends, weighing eighty-two and one fourth grains; (g) another conch pearl, seventy-seven and one half grains, button shaped, yellowish-white with a slight shade of pink; (h) a seventy-six-and-one-half-grain drop-shaped pearl of a *chatoyant* aspect, of white color shaded with red, purple, and green; and (i) a pear-shaped Scottish pearl of thirty-four and three fourths grains, of a milky bluish caste, slightly tinged with pink.

VAN BUREN PEARLS. Among the collections of the United States National Museum are two pendant pearls each weighing about thirty grains, and a necklace containing 148 pearls with an aggregate weight of 700 grains. These were presented in 1840 to President Van Buren by the Imam of Muscat. They were deposited in the vaults of the Treasury Department, where they remained until a few years ago, when, by the order of the Secretary of the Treasury,

[1] Hertz, "Catalogue of the Collection of Pearls and Precious Stones Formed by Henry Philip Hope, Esq.," London, 1839.

they were transferred to the custody of the National Museum where they now are.

THIERS NECKLACE. In the galleries of the Louvre at Paris may be seen a pearl necklace formerly owned by the wife of President Thiers (1797–1887), consisting of 145 pearls in three rows. The weights of the three largest individual pearls are fifty-one, thirty-nine, and thirty-six grains, respectively. The aggregate weight is 2079 grains, and the value at the time of their deposit was estimated at 300,000 francs. This is on a base of $2.02; at a higher valuation the figures would be:

$$\$148,947 = \$71.64 \text{ per grain; base, } \$5$$
$$238,315 = 114.63 \quad " \quad " \quad " \quad 8$$

the last being very probably nearer to the correct value of the necklace at the present time.

TIFFANY QUEEN PEARL. Doubtless the most famous pearl ever found within the limits of the United States, and likewise one of the choicest, is the well-known "Queen Pearl," found in Notch Brook near Paterson, New Jersey, in 1857.[1] In form it is a perfect sphere, and weighs ninety-three grains. The history of the discovery and of the sale of this beautiful gem is set forth on page 260.

THE BAPST PEARLS. Very practical is the account given by Streeter of a pair of magnificent spherical pearls exhibited at the Paris Exhibition in 1878 by Messrs. Bapst of Paris. One of these pearls—then weighing 116 grains—was purchased by Mr. Streeter in 1877, and by him sold to a leading merchant of London, who skilfully removed a blemish on it, reducing it to 113¾ grains in weight. After holding it for some months, it occurred to him that it would match a pearl sold by Hunt and Roskell to Dhuleep Singh about fifteen years previously. On comparison, the two were found to match perfectly, one weighing 113¾ and the other 113¼ grains. The two were eventually sold early in 1878 "for £4800, which was even then much below their value, and to-day they would be worth £10,000. They were exhibited in the great Paris Exposition in 1878, where they attracted universal attention, and were pronounced by connoisseurs to be the most extraordinary pair of pearls ever seen in Europe. They were sold from the exhibition to an individual for a very large sum."[2]

[1] "Gems and Precious Stones of North America," New York, 1892, Pl. VIII, p. 229.   [2] Streeter, "Pearls and Pearling Life," London, 1886, pp. 295, 296.

The "Southern Cross." The "Southern Cross" is an unusual pearl or rather cluster of pearls which attracted much attention twenty years ago. It consists of nine attached pearls forming a Roman cross about one and one half inches in length, seven pearls constituting the shaft or standard, while the arms are formed by one pearl on each side of the second one from the upper end. The luster is good, but the individual pearls are not perfect spheres, being mutually compressed at the point of juncture and considerably flattened at the back. If separated, the aggregate value of the individual pearls would be small, and the celebrity of the ornament is due almost exclusively to its form. This striking formation was exhibited at the Colonial and Indian Exhibition at London in 1886, and later at the Paris Exhibition in 1889, where it was the center of interest, and obtained a gold medal for the exhibitors. It is reported that an effort was made to bring about its sale at £10,000, the owners suggesting that it was especially appropriate for presentation to Leo XIII, on the occasion of his jubilee in 1896. The writers have been unable to obtain information as to its present location.

Much information relative to the "Southern Cross" was volunteered by Henry Taunton in the very interesting account of his wanderings in Australia. He presents apparently reliable statements showing that it was found on March 26, 1883, off Baldwin Creek in Lat. 17° S. and Long. 122° E., by a boy named Clark, in the employ of James W. S. Kelly, a master pearler. When delivered to Kelly, it was in three distinct pieces, but the boy reported that it was in one piece when he found it a few hours before. Kelly sold it in the three pieces in which he received it for £10 to a fellow pearler named Roy; Roy sold it for £40 to a man named Craig, and he sold it to an Australian syndicate.

However, according to Taunton's positive statement, there were only eight pearls in the cluster when it was sold by Kelly in 1883, and to make it resemble a well-proportioned cross—the right arm being absent—another pearl of suitable size and shape was subsequently secured at Cossack and attached in the proper place to the others, which, in the meantime, had been refastened together by diamond cement, thus making three artificial joints in the present cluster. "As if to assist in the deception, nature had fashioned a hollow in the side of the central pearl just where the added pearl would have to be fitted; and—the whole pearling fleet with their pearls and shells coming into Cossack about this time—it was no difficult matter to select a pearl of the right size and with the convexity required. The holder paid some ten or twelve pounds for the option of selecting a pearl within given limits; and then once more, with the aid of diamond cement and that of a skilful 'faker,' this celebrated gem was transformed into a perfect cross."[1]

MORGAN-TIFFANY COLLECTION. Probably the most interesting assortment of American pearls is the Morgan-Tiffany Collection in the American Museum of Natural History, New York. The excellence of this collection lies, not in the high cost of any individual pearl, but in its illustrating in a comprehensive manner the great variety, colors, and forms of American pearls. Not only are the many varieties of fresh-water pearls represented, but likewise abalone pearls from the Pacific coast, conch pearls from the Gulf of Mexico, and a good assortment of pearly concretions from edible oysters and clams of the Atlantic coast.

This collection contains 557 species of white and colored Unio pearls, four multicolored, five mallet-shapes, 166 baroques, thirty-nine hinge-pearls, twenty pearlaceous masses, thirty-four clam (Venus) pearls, fifteen abalone pearls, eleven conch pearls, and twelve oyster (Ostrea) pearls. The collection was exhibited in two parts, the first at the Paris Exposition of 1889, and the second at the Paris Exposition of 1900. On each occasion a gold medal was awarded.

COUNT BATTHYANI'S PEARL. A curious history is connected with a beautiful black pearl[2] which was at one time in the possession of Count Louis Batthyani, the premier of the revolutionary government of Hungary. The count was shot in 1849, by the orders of a court-martial, and on the eve of his execution he gave the pearl, which he had worn mounted on a scarf-pin, to his trusty and faithful valet. The latter left it to his son, who, when in straightened circumstances, sought to raise money upon the pearl. The pawnbroker of the small town was distrustful of its value and took it to Budapest for appraisal. There the suspicions of the authorities were aroused, an investigation was ordered, and it was finally discovered that the pearl had been stolen one hundred and fifty years before from the English crown. The English government redeemed it for the sum of £2500 ($12,500). How it came into the possession of Count Batthyani is a mystery; probably he purchased it from some antiquarian.

In 1900 there was shown in Paris one of the most important black pearls of any time, a pear-shaped pearl of forty-nine grains, of a most wonderful black color with a green sheen, as perfectly formed as though it had been turned out of a lathe; it did not terminate in a point at the small end, but was slightly flattened. It was so beautiful an object that it almost seemed it should never be drilled for

[1] Taunton, "Australind," London, 1900, p. 231.   [2] Austrian Court Journal, 1899.

mounting. This pearl ultimately sold for more than $30,000, and it is probably the finest black pearl that has ever reached the European markets.

According to a personal communication from E. Z. Steever, governor of the District of Sulu, the largest pearl that he has seen in the islands belongs to the sultan, and is now in the possession of Hadji Butu, former prime minister. It is an oblate spheroid, there being a trifling difference between the two diameters. The upper hemisphere is very beautiful; the lower one has a few minute, black specks which are superficial and could be easily removed, the pearl not having been treated since it was taken from the oyster. This pearl measures five eighths of an inch at its greatest diameter and is said to weigh twelve carats (forty-eight grains). Hadji Butu informed Governor Steever that the sultan had refused $25,000 for the pearl in Singapore.

The Nordica pearl is the finest abalone of which we have any record. It weighs 175 grains, is a drop pearl of a greenish hue, with brilliant red fire-like flashes, and serves as a pendant to the famous collection of colored pearls belonging to the beloved and admired American prima donna, Madam Nordica.

At the International Exposition in Paris in 1889, Mr. Alphonse Falco, president of the Chambre Syndicale, exhibited a round pearl, white and lustrous, weighing seventy grains, and valued at 50,000 francs.

Augusto Castellani, the well-known Italian jeweler of Rome, in the year 1868, during the Papal régime, executed a crown for King Victor Emmanuel II. This crown was destined for the Church of the Holy Sepulchre, in Jerusalem, and on it is a pearl which, although slightly irregular, is as large as the famous Gogibus pearl.

A remarkable golden-yellow pearl from Shark's Bay, West Australia, is in the possession of a New York lady; it weighs thirty and one half grains, is perfectly round, and is without a flaw or blemish.

CROWN JEWELS OF FRANCE. The collection of gems known as the Crown Jewels of France owed its origin to Francis I (1494–1547). While in Bordeaux, on his way to meet his bride, Eleanor of Austria, sister of Emperor Charles V, Francis created by letter patent the Treasure of the Crown Jewels, giving to the state a number of his most valued diamonds, under the condition that at each change of sovereign a careful inventory should be made. The original collection consisted of six pieces of jewelry valued at 272,242 *"écus soleil,"* or about $700,000. The crown jewels have passed through many vicissitudes in the course of time. A number of the gems were at various

THE GREAT SÉVIGNÉ OF THE FRENCH CROWN JEWELS

Containing "The Regent Pearl," weighing 337 grains, and four pear-shaped pearls of 100 grains each; also 100 carats of diamonds.

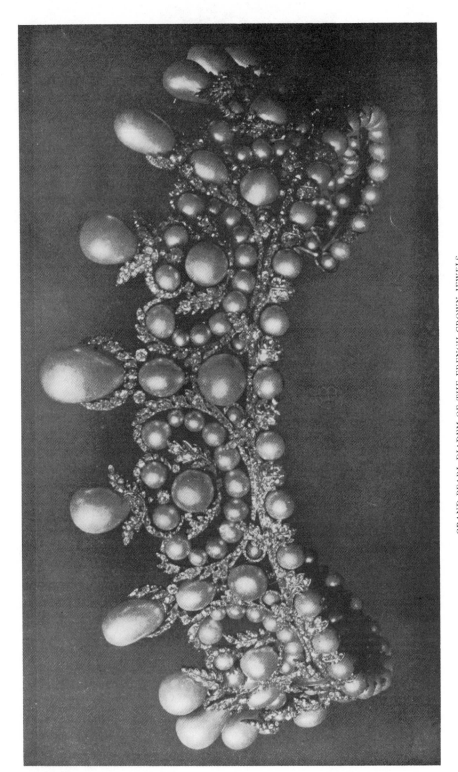

GRAND PEARL DIADEM OF THE FRENCH CROWN JEWELS   Worn by the Empress Eugénie

Containing 212 pearls weighing 2452 grains, and 1990 diamonds weighing 74 27-32 carats.

times pledged as security for loans made in France and Italy, and it is said that in 1588, during the reign of Henry III, all the jewels disappeared from the royal treasury. Henry IV strove to regather the scattered ornaments, but it was only in the reign of Louis XIV that the collection became really important. At the time of the French Revolution, in 1791, an inventory was made by the order of the National Assembly.

The jewels were then deposited in the Garde-Meuble, where they were exposed to public view. Either they were very carelessly guarded, or the guardians were in collusion with a band of thieves, for the room wherein they were kept was entered on five successive nights, and when the theft was finally discovered only about 500,000 francs' ($100,000) worth of the gems remained. Many of the most valuable objects were, however, traced and recovered. Napoleon I, when he became emperor, made every effort to enrich the treasure, and purchased gems to the value of 6,000,000 francs ($1,200,000), and subsequent rulers added to the collection on various occasions.

At the time of the official inventory in 1791 the entire collection of pearls was estimated at about 1,000,000 francs ($200,000). The finest specimen in the collection was a splendid round pearl weighing 109¼ grains[1] and estimated at 200,000 francs ($40,000), or $366 per grain, on a base of $3.35. Then came two pear-shaped pearls of a fine orient and well-matched, weighing respectively 117¾ and 113 grains, and valued at 300,000 francs ($60,000) or $260 per grain, on a base of $2.25. In addition to the above there were twenty-five separate round pearls which had constituted the necklace of the queen; they ranged in weight from 36 to 165½ grains, and were valued at about 90,000 francs ($18,000). Beside the pear-shaped pearls to which we have alluded, there were two other pairs, each valued at 32,000 francs ($6400) for the two pearls; they averaged about 100 grains in weight. In addition to these there were two weighing respectively 175½ and 205¼ grains, each valued at 20,000 francs ($4000), and seven others ranging in weight from 92½ to 167 grains and valued at from 10,000 to 15,000 francs ($2000 to $3000). The best oval pearl was one weighing seventy-six and one half grains and estimated at 20,000 francs ($4000); there were two others, one of ninety-three grains, valued at 12,000 francs ($2400), and one of 121 grains, valued at 10,000 francs ($2000). We may also mention an egg-shaped pearl weighing 145¼ grains, estimated as worth 10,000 francs ($2000), and a button pearl of 198 grains entered at 15,000 francs ($3000). Beside these separate pearls there were eleven strings comprising 310 pearls, weighing in all 6778 grains and valued

[1] See p. 461.

at but 29,400 francs (about $6000). The average per pearl was 95 francs ($19), less than one dollar a grain.

These pearls, according to their beauty, would now be worth from four to six times the valuation here given, so that the two large pear-shaped pearls of the French crown may be worth to-day $200,000 and the great round pearl from $100,000 to $250,000.

Many of these pearls were a century old. They were collected at a time when not as much attention was paid to their absolute perfection and beauty as at the present time, for there probably never has been a period when rare and perfect pearls, diamonds, or rubies have been appreciated so much more highly than those of mediocre quality.

RECAPITULATION

	No.	Weight in grains Average	Total	Francs	Value U. S. currency
Round Pearls	I		109¼	200,000	$40,000
	3	79	238½	29,000	5,800
	11	77½	804½	37,300	7,460
	7	64½	450½	19,400	3,880
	14	53¾	753¼	23,100	4,620
	43	34½	1488½	16,100	3,220
Pear-shaped	2	115⅜	230¾	300,000	60,000
	4	99¼	397½	64,000	12,800
	6	163⅛	978¾	92,000	18,400
	8	114¼	914¼	55,000	11,000
	47	42¼	1989¾	24,600	4,920
Oval	3	27	290½	42,000	8,400
	9	72½	654¼	20,100	4,020
	11	43	473¾	5,000	1,000
Egg-shaped	1		145¼	10,000	2,000
Irregular	12	39½	475¼	7,300	1,460
Button	1		198	15,000	3,000
	6	66¼	398	4,900	980
Baroque	4	37½	150¾	1,500	300
Strings	310	21⅞	6778	29,400	5,880
	503	35⅜	17,919¼	995,700	$199,140

After the downfall of Napoleon III and the proclamation of the French Republic, the jewels were inventoried, and, by a law passed December 10, 1886, it was decreed that a large part of the treasure should be sold at public auction. The sale was held in the Pavillon de Flore, a part of the Palace of the Tuileries, on May 12, 1887, and, very naturally, all the principal gem-dealers and collectors were represented. A number of remarkable pearl ornaments were among the objects offered at this sale, one of the most beautiful being a diadem of an exceptionally artistic openwork design, adorned with large, round pearls and surmounted by a row of magnificent pear-shaped pearls. The total number of pearls in this diadem was 212, and their weight 2452 grains. It was sold for the sum of 78,000 francs ($15,600). The coronet which accompanied the diadem comprised 274 pearls, weighing 984 grains; the design was similar to that of the diadem, but the points consisted of a round and a pear-shaped pearl in alternation. This ornament realized the sum of 30,000 francs ($6000). A large brooch of very elaborate and beautiful design, beside a number of smaller pearls, comprised four fine, pear-shaped pendants, weighing 100 grains each, and two choice bouton pearls, and had in the center the famous pearl known as "La Régente," which was purchased in 1811 for 40,000 francs ($8000). This splendid ornament brought the sum of 176,000 francs ($35,200). Four other brooches each contained seven pearls and many brilliants, the twenty-eight pearls having a total weight of 1496 grains, an average of more than fifty-three grains. Each brooch had two pearls surrounded with brilliants, and five large, pear-shaped pearls set as pendants. They were sold to different purchasers at prices ranging from 18,500 francs ($3700) to 43,000 francs ($8600), the four together realizing 113,-500 francs ($22,700).

Six pearl necklaces were also offered. One of forty-seven pearls weighing 698 grains was sold for 34,600 francs ($6920), and two others, each consisting of fifty-eight pearls, with a total weight of 524 and 400 grains respectively, brought the sum of 22,300 francs ($4460) and 15,000 francs ($3000). Another necklace composed of thirty-eight round pearls and nine pear-shaped pendants, the total weight being 1612 grains, sold for 74,300 francs ($14,860). The two finest necklaces were broken up into a number of separate lots. One of them, consisting of 362 pearls and weighing in all 5808 grains,—an average of a trifle over sixteen grains,—was offered in four lots which together brought 295,800 francs ($59,160). The other necklace comprised 542 pearls weighing 6752 grains, and was disposed of in eight lots, realizing in all 331,800 francs ($66,360). Two bracelets adorned with 202 pearls and a number of small brilliants were purchased for the sum of 90,200 francs ($18,040). The total amount realized for the pearl ornaments was 1,261,500 francs ($252,300). There are several American ladies who own single strings of pearls which are of more value than the whole pearl parure of the Empress Eugénie.

Pearl parure of the crown jewels of France, worn by the Empress Eugénie, and sold at the Tuileries, May 12, 1887.

A diadem containing 212 pearls, weighing 2452 grains.

A coronet with 274 pearls, weighing 984 grains.

Four brooches, each containing four large pearls, two round, two pear-shaped, and three smaller ones, weighing in all 1496 grains.

A larger brooch comprising four large pear-shaped pearls, each weighing 100 grains. In the center is the pearl called "La Régente."

Two necklaces each consisting of forty-seven pearls, with an aggregate weight of 698 and 1612 grains, respectively.

Two bracelets with 202 pearls, weighing 2000 grains.

Five buttons, three with nine and two with ten pearls.

A necklace of 542 pearls, weighing 6752 grains.

Another necklace of 362 pearls, weighing 5808 grains.

Two other necklaces, each containing fifty-eight pearls, the total weight being 400 and 524 grains, respectively.

SUMMARY OF THE PRINCIPAL PEARL ORNAMENTS FORMING THE PEARL PARURE OF THE FRENCH CROWN JEWELS SOLD AT PALACE OF THE TUILERIES IN MAY, 1887

Designation	No. of pearls	Aggregate weight grains	Amount rec'd francs
1 necklace	362	5,808	295,800
1 "	542	6,752	331,800
1 "	47	698	34,600
1 "	58	524	22,300
1 "	58	400	15,000
1 "	47	1,612	74,300
2 bracelets	202	2,000	90,200
1 large diadem .	212	2,452	78,000
1 coronet	274	984	30,000
1 brooch	45	1,200	176,000
4 brooches	28	1,496	113,500
Total	1875	23,926	1,261,500

IMPERIAL AUSTRIAN SCHATZKAMMER. The weights and values of the great gathering of pearls of the imperial Austrian Schatzkammer were carefully estimated by one of the authors and by his friends, and it is the first attempted inventory ever published.[1]

The imperial crown of the Holy Roman Empire, preserved in the treasury of the imperial Burg at Vienna, and known as the crown of Charlemagne, has in front seventeen pearls weighing 424 grains, of which two weigh fifty-six grains each. The remaining fifteen pearls average 20.8 grains. The values of these pearls are as follows:

	$2.50	Base $5.00	$7.50
15 pearls, 20.8 grains....	$16,224	$32,448	$48,672
2 56-grain pearls.......	15,680	31,360	47,040
Total	$31,904	$63,808	$95,712

[1] The senior author was permitted to handle these treasures in 1899.

THE IMPERIAL AUSTRIAN CROWN

Made by order of Emperor Rudolph II, in 1604

MRS. GEORGE J. GOULD

At the back of the crown there are eighteen pearls, weighing 180 grains. One of these has a weight of twenty-six grains; the remaining seventeen average 9.058 grains. The values estimated are as follows:

	$2.50	Base $5.00	$7.50
17 pearls, 9.058 grains...	$3,487.55	$6,975.10	$10,462.65
1 pearl of 26 grains	1,690.00	3,380.00	5,070.00
Total	$5,177.55	$10,355.10	$15,532.65

The pearls in the cross surmounting the crown have a weight of thirty-six grains.

The imperial cross is profusely ornamented with pearls in front, while the back consists simply of silver-gilt. There are three strings of pearls in the front running in each direction. The total weight of the pearls is 4092 grains; one weighing sixty-four grains, and the smallest two grains.

The crucifix of the Golden Fleece is ornamented only in front with pearls; these have a weight of but 136 grains.

The imperial Austrian regalia, dating from the time of Emperor Rudolph II, are also in the imperial Burg; some of the emperor's jewels were sold at auction in Prague in 1728.

The crown is adorned with two rows of pearls, weighing respectively 960 and 840 grains; between these rows are pearls having a total weight of 440 grains, while the ornaments and eight large drops weigh 2052 grains. The largest pearl on this crown weighs 104 grains. It is drop-shaped and belonged to Rudolph II; it is slightly uneven and the color, although white, is not that of a new pearl, but this pearl has a positive history of three hundred and six years, and at the present time is probably the oldest known unchanged pearl with a direct and authentic record.

The imperial orb is studded with pearls weighing in all 1560 grains. Four of these weigh forty grains each, the others are of lesser size, the smallest weighing ten grains. The scepter is adorned with pearls to the weight of 300 grains.

In addition to these insignia and regalia there are in the treasury two magnificent pearl necklaces, deposited by Empress Maria Theresa in 1765. The first consists of a single string of 114 large-sized pearls with the "Baden Solitaire," a diamond of 30 carats, as a clasp. Three of these pearls weigh from 92 to 100 grains each, and the whole string has a total weight of 3400 grains. This would give us the following values, the pearls averaging 29.82 grains.

Base		
$2.50	....	$253,432.12
5.00	....	506,864.24

The other necklace contains 121 pearls of a total weight of 3788 grains, arranged in three rows; these pearls average 31.3 grains, the largest weighing forty grains and the smallest ten grains. The necklace has a diamond clasp of 14½₃₂ carats. The pearls are worth:

Base	
$2.50	$296,450
5.00	592,900
7.50	889,350

Two bracelets with brilliant clasps, belonging to the same set, and consisting of 240 medium-sized pearls weighing 2800 grains and averaging 11⅔ grains, have the following values:

Base	
$2.50	$81,658.80
5.00	163,317.60
7.50	244,976.40

There is also a pendant of six pearls, weighing 300 grains and averaging fifty grains; these pearls are worth:

Base	
$2.50	$37,500
5.00	75,000
7.50	112,500

Another necklace, bequeathed to the treasury by the late Empress Caroline Augusta, consists of eighty-six pearls disposed in two rows, the largest pearl weighing seventy-two grains and the smallest eight grains. The total weight is 2600 grains and the average, 30.2. This necklace is worth:

Base	
$2.50	$196,088.60
5.00	392,177.20
7.50	588,265.80

Still another necklace is composed altogether of black pearls, of which there are thirty, the largest weighing forty-eight grains and the smallest ten. The total weight of this necklace is 1040 grains, an average of 34.66 grains for each pearl. On a base of $5 this necklace is worth $180,150.

The diamond crown of the empress bears pearls to the total weight of 2000 grains; among them are four weighing 100 grains each. These pearls alone, on a $5 base, would be worth $200,000.

The total weight of the pearls in all these ornaments is 35,816 grains, equal to four and a half pounds, avoirdupois, and they are worth from $2,000,000 to $4,000,000.

The so-called crown of Charlemagne bears the inscription: "Chuonradus Rex Dei gratia Romanorum Imperator Aug." It is believed to be a work of the twelfth century and originally the royal crown of Conrad III, king of the Germans (1093–1152), the first Hohenstaufen.[1] The arch is said to have been added to adapt this royal crown for use in the expected coronation of Conrad as emperor. He died, however, while making preparations for his journey to Rome.

The imperial vestments used in the coronation ceremonies of the Holy Roman Empire, were produced in the celebrated Hôtel de Tirâz, at Palermo. Roger II, King of Sicily (1096–1154), after a victorious campaign in Greece, brought back with him to Sicily a number of skilled silk-weavers and embroiderers, whom he established at Palermo. The imperial mantle is of a brilliant purple and bears an inscription, embroidered in gold and pearls, stating that the garment was made in the royal manufactory in the year 1133. Two pearl-embroidered representations of a lion, who has stricken down a camel and is about to tear it to pieces, also appear on this mantle. This symbol of royal power was used frequently by the Saracens, and it is said that Richard Cœur de Lion had this design embroidered on his saddle-cloths.[2]

A fine collection of large baroque pearls is preserved in the Grüne Gewölbe (Green Vaults) in the palace at Dresden, which contains the treasures of the royal family of Saxony. Most of these were mounted during the eighteenth century by J. M. Dinglinger (1665–1731), the famous goldsmith to the Saxon Elector, King Augustus II of Poland, and who is sometimes called the German Cellini. A splendid specimen of his work is a vase of Egyptian jasper in the shape of a shell, bearing a representation of Hercules fighting with the Nemæan lion; this bears reference to the immense personal strength and power of Augustus II, whose portrait is painted in enamel on a mirror at the back. The pedestal is adorned with a great many precious stones, pearls, and enamel paintings in the shape of cameos representing the twelve labors of Hercules. A dragon is studded with emeralds and its back is formed of pearls, with a large sardonyx in the middle. Hercules and the lion are in enamel.

In the same collection may be seen the figure of a dwarf made by Ferbecq, who was one of the goldsmiths of King Augustus. The body of the dwarf is formed of a baroque pearl, which is studded with small diamonds. His sleeves and trousers are in black and green enamel; his hat is also of green enamel and on it is a string

[1] Winckler, "Die Reichskleinodien," Berlin, 1872, p. 17.   [2] *Ibid.*, p. 9.

of diamonds. In his right hand the dwarf holds a spit and in his left a roast. On his left shoulder he bears a goose, the upper part of whose body is formed of a pearl; at his right side hangs a bottle also formed of a pearl. The gilded pedestal is ornamented with white enamel-work on a pink ground. Above and below, it is set with white and yellow diamonds. Another figure, similarly formed of a large baroque pearl ornamented with gold and diamonds, shows a drunken vintager and his dog; and an exceedingly grotesque, ornamented baroque is said to bear a striking resemblance to Señor Pepe, the court dwarf of Charles II of Spain.

Exhibited at the Palace of Rosenberg at Copenhagen, are similar designs in which large baroques form the principal part of fish, birds, dragons, mermaids, etc. Prominent among them is the figure of a skater, executed by the jeweler, Diederichsen; it is said that this was made for Frederick VII, who died before it was accepted, and in 1895 it was presented to the museum.

A beautiful and costly figure of this nature was completed a year or two ago by the court jeweler, Alfred Dragsen, of Copenhagen. This is nearly four inches in height, and represents a female snake-charmer. A very long baroque pearl forms the body from the shoulders to the knees, and the head, arms, and the legs below the knees are of gold. The figure is ornamented with a diamond-studded garland, ruby necklace and earrings, and garters set with similar gems, a red enameled girdle ornamented with pearls, and golden anklets with black pearls. With a flute she charms a serpent twined about her body and grasped with the other hand.

What is said to be the finest collection of black pearls in all Europe is that belonging to the Duchess of Anhalt Dessau, Germany. It consists of three large caskets of black pearls that have taken a century to collect. It is traditional in the family that these pearls are never to be sold except as their last possession, since they know they will always find a purchaser.

A collection of pearls had been kept for many years in the Monte de Piedad of Mexico City, which it was claimed had been pawned by a friend of the Empress Carlotta, in order to provide her with money at the time of the assassination of Emperor Maximilian. These pearls were contained in a necklace and a pearl and diamond tiara, which were sewn upon cardboard covered with black velvet, and had the appearance of not having been disturbed for many years. The necklace consisted of old pearls, both of the so-called Madras and Panama varieties; in the center was a large diamond medallion from which pear-shaped pearls were suspended. It is believed that these pearls were part of Empress Carlotta's marriage portion, and that they

MADAM NORDICA

THE MADAM NORDICA COLLECTION OF COLORED PEARLS

The Nordica drop pearl weighs 175 grains

came from the Austrian crown jewel collection. None of them possessed much quality with the exception of one, a large pear-shaped pearl which was set at the base of the necklace and weighed eighty-four grains. The drilling of this pearl was of a very old style, being of that type in which a tube is inserted in the drill hole, through which a gold wire passes to hold the pearl; a diamond is then set at the base of the tube to disguise the drill mark. It is, however, possible that the pearl came from the East Indies, where large drill holes are usual.

The pearls were sent in bond to the United States as a collection, and then to Europe, where they were sold separately, the pear-shaped pearl appearing again in the New York market in 1906. There has been some doubt as to these really being Carlotta's pearls, but the Mexican account is fairly consistent, and it satisfactorily disposes of the newspaper romance in which it was claimed that Carlotta had taken these pearls with her to Europe and that they had been buried in a casket in the Adriatic Sea.

RECENT AUCTION SALES. In England and in France, more than in the United States, great auction sales of jewels are common. They are held in London, principally at Christie's, originally a coffee house, established in 1880; and in Paris, at the Hôtel Drouot. Good prices are generally realized, as the buyers of the entire continent attend these great sales. The purchases are usually made by dealers who frequently do not neglect each other's interests if private buyers are present. No matter how great may be the amount involved, no matter whether the collections consist of paintings, furniture, or jewels, there are always buyers, to a much greater extent than in the United States. The price for fine jewels may naturally vary a trifle in the different markets, according to the conditions of payment. It must evidently make an appreciable difference whether almost the entire amount is paid in cash or whether a credit of thirty days is extended, or one for a much longer period, in some countries for as long as one, two, or even three years.

While in the United States such sales of valuable jewels are very unusual, it has been a custom in England and in France for many years, in the settlement of estates, to sell not only the furniture, bric-à-brac, etc., but also the jewels. Sales of this kind are naturally calculated to attract not only the dealers, but also many rich collectors and connoisseurs, and as they are frequently widely advertised, and London and Paris are, at the most, but one to two days' journey from all parts of Europe, many people attend, most of the private buyers being represented by their agents. By means of these sales many heirlooms, which have been handed down from generation to generation, often pass into the possession of strangers. In the matter of jewels, it has been frequently noticed that dealers are in the majority of cases the ultimate buyers, and it has also been inferred that when an outsider participates in the bidding, the prices are advanced to such an extent that it does not often appear profitable for him to buy in the face of such competition. All manner of people have had their estates disposed of in this way, and the list of these sales during the past twenty years is a striking one: royalty, nobility, merchants, and people in many other conditions of life find a place in it.

It is not an infrequent custom in London for solicitors to advance money on jewels, and when the payments are not forthcoming these jewels are sold. Hence, many sales appear at the larger auction rooms in which no name is given, the owners frequently being people of high degree.

On July 19, 1892, a necklace containing eighty-five graduated pearls of unusual size and quality, the property of the late H. W. F. Bolchow, M.P., was sold in London for the sum of £2500 ($12,500). Another necklace of 146 fine graduated pearls disposed in two rows, brought £2400 ($12,000); a single-row necklace of eighty-five pearls realized £1600 ($8000), and one of 118 pearls in two rows £1660 ($8300).

An exceptionally fine pearl necklace which belonged to her Grace the late Caroline, Duchess of Montrose, mother of the present Duke of Montrose, was sold at Christie's on April 30, 1895. The necklace comprised 362 graduated pearls, arranged in seven rows of forty-four, forty-seven, forty-eight, forty-nine, fifty-two, fifty-eight, and sixty-four pearls, respectively. The amount realized for this ornament was £11,500 ($57,500).

On July 9, 1901, a pearl necklace, advertised as the property of "a French lady of rank," and several other valuable pearl ornaments, were offered at Christie's. It is supposed by many that these jewels belonged to Madame Humbert. The necklace was composed of six rows of graduated pearls consisting of fifty-five, sixty-one, sixty-seven, seventy-three, seventy-nine, and eighty-nine pearls, respectively, a total of 424, all of good color and luster. A London dealer considers that they owed their value mainly to skilful matching and fineness of color; they are perhaps a century old. As may be supposed, there were many bidders who competed eagerly for possession of this fine ornament, and it was at last adjudged for the sum of £20,000 ($100,000). While this was, up to that date, a record price in an auction room, it was by no means an exceptional figure for private sales; indeed, at about the time this necklace was sold, a London dealer disposed of another for £34,000 ($170,000).

A necklace, the property of the late Lady Matheson of the Lews, was sold at Christie's, March 5, 1902. Well-matched and graduated round pearls, to the number of 233, were disposed in four rows, and strung with seed-pearls between. This necklace, which had been presented to Lady Matheson at the time of her marriage in 1843 by Sir Jamsetjee Jejeebhoy, Bart., brought the sum of £6800 ($34,000).

A pearl necklace, containing fifty-three fine and graduated pearls, was sold in London, June 9, 1902, for the sum of £2250 ($11,250). Another necklace of sixty-eight fine round pearls, one of which formed the clasp, brought £1580 ($7900). A beautiful pearl and brilliant pendant of open scroll design, with a large, round white pearl in the center, and a large, pear-shaped black pearl as drop, realized £800 ($4000), and a pearl collar of ten rows of round pearls brought £820 ($4100). A noteworthy offering at this sale was a rope of 135 pearls, an heirloom sold under the will of Lady Marianna Augusta Hamilton. These pearls had been given to Lady Augusta Anne Cockburn in 1769 on the occasion of her marriage to Sir James Cockburn, Bart., by her godmother, Augusta, Duchess of Brunswick, sister to George III. The rope was sold for £900 ($4500).

The pearls of Lady Dudley were sold at Christie's on July 4, 1902. Among them was a magnificent necklace of forty-seven slightly graduated round pearls, of large size and unusually brilliant orient; their gross weight was 1090 grains. This necklace brought the sum of £22,200 ($111,000). A single pear-shaped pearl of the finest orient mounted with a diamond cap, as a pendant, and weighing 209 grains, was sold for £13,500 ($67,500). A rope of 222 graduated round pearls of the highest quality, weighing 2320 grains was purchased for £16,700 ($83,500), and a pearl and brilliant tiara brought £10,300 ($51,500). The entire casket of thirty-one lots realized £89,526 ($447,630).

At the sale of the jewels of Mlle. Wanda de Boncza, at the Hôtel Drouot, Paris, December 6, 1902, a fine necklace was disposed of for the sum of 150,000 francs ($30,000), and a rope of 100 small pearls realized 38,100 francs ($7620); the proceeds of the entire sale of these jewels were 1,249,578 francs ($249,915).

Among the Aqualia jewels, sold in London in 1903, was a pearl necklace that brought £4480 ($22,400).

The jewels of the late Marquis of Anglesey, an enthusiastic jewel and art collector, were disposed of at Christie's on May 4 and 5, 1904. At the time of his death, the marquis was supposed to be a bankrupt, but the value of the gems which he had purchased had increased so rapidly that the sale realized the sum of £22,988 10s. ($114,942), more than enough to cover all the obligations of the estate. Of this amount a magnificent drop-pearl, mounted as a scarf-pin, brought £4000 ($20,000); another drop-pearl of the finest orient, weighing 105½ grains, but slightly cracked, was sold for £3700 ($18,500). Four other drop-shaped pearls, mounted as scarf-pins, were sold for £5220 ($26,100), one of them bringing £1720 ($8600). A single bouton pearl, used as a coat fastener, realized £980 ($4900), and a pearl trefoil was purchased for £580 ($2900). One fine large bouton pearl, set as a stud, was disposed of for £3000 ($15,000), and another somewhat smaller bouton, also set as a stud, brought £1600 ($8000).

A splendid necklace comprising forty-nine well-matched and graduated pearls of fine quality, weighing 563½ grains, was sold in London on June 29, 1905, for the sum of £4700 ($23,500). At the same sale a necklace of thirty-two graduated pearls, weighing about 890 grains, brought £2600 ($13,000).

On July 20, 1905, a pearl necklace comprising forty-five graduated pearls of fine orient, with a cabochon ruby clasp, the gross weight being 832 grains, was sold in London for £3150 ($15,750).

A fine pearl and brilliant pendant was disposed of at the sale in London, February 21, 1906, of the stock of Mr. E. M. Marcoso. This pendant was composed of one large white brilliant, weighing 18½₃₂ carats, and a drop-shaped pearl weighing 75¾ grains. The ornament brought the sum of £2050 ($10,250).

A pearl necklace composed of 285 well-matched and graduated pearls disposed in five rows was sold in London on June 13, 1906, for the sum of £10,000 ($50,000). At the same sale a three-row necklace, with 213 graduated and matched pearls of fine orient, brought £3200 ($16,000), and a rope of 237 fine pearls realized £2800 ($14,000).

Among the jewels disposed of at a sale in London on July 11, 1906, may be mentioned a five-row pearl necklace of 445 graduated oriental pearls which was sold for £2500 ($12,500). Three other necklaces were offered at the same sale; one of fifty-five matched and graduated pearls of fine quality bringing £3400 ($17,000); one of fifty-seven pearls, £2700 ($13,500), and the other of 219 well-matched and graduated pearls realizing £2350 ($11,750). Still another necklace of 417 matched and graduated pearls arranged in five rows was sold for £4800 ($24,000). A splendid pearl drop, of the finest orient, brought £1650 ($8250), and a pearl rope of 191 oriental pearls, arranged so as to form three single-row necklaces, realized the sum of £3700 ($18,500), three fine black pearls, mounted as studs, were sold for £1000 ($5000). The most important necklace was reserved for the end of the sale; this was composed of forty-seven large oriental pearls, and was purchased for the sum of £10,000 ($50,000).

At the sale of the Massey-Mainwaring collection at Christie's on March 18, 1907, a five-row pearl necklace consisting of 471 graduated pearls, with a bouton pearl in the center, was sold for £4600 ($23,000).

Another collection, sold at Christie's, April 15, 1907, was the property of the late Mrs. Lewis-Hill, and the proceeds of the first day's sale reached the heretofore unapproached total of £94,805 ($474,025), thus exceeding by $26,395 the amount obtained in one day by the sale of Lady Dudley's jewels. Among the valuable pearls in this collection, we may mention a pair of large bouton pearl earrings, with small diamond tops, which brought £1180 ($5900). The enthusiasm and interest of the assembly were aroused by a necklace of forty-five large, graduated pearls of fine orient, with a bouton pearl and brilliant cluster snap; after spirited bidding this was bought for £6100 ($30,500). The greatest event of the day, however, was the appearance of a splendid rope of 229 pearls of very good form, well-matched and graduated. The opening bid was £10,000 ($50,000) and after a warm contest the pearls were finally acquired for the sum of £16,700 ($83,500). A necklace consisting of fifteen graduated drops, each formed of one bouton pearl, one brilliant, and one pear-shaped pearl drop, depending from a narrow band of small diamonds, drew forth a bid of £5000 ($25,000) and was finally awarded for the sum of £12,200 ($61,000). A pearl rope of 183 graduated and well-matched pearls realized £7200 ($36,000), and a fine pair of pearls set as earrings brought £3400 ($17,000). The crowded auction room, the keen competition among the bidders, and the amount obtained for these jewels are good indications of the firmness of the market at the present time.

At the sale of the jewels of Lady Henry Gordon-Lennox, held at Christie's on May 12, 1907, a splendid necklace was offered. It comprised 287 graduated pearls of the finest orient, disposed in five rows, with a large circular pearl. This magnificent ornament was sold for £25,500 ($127,500).

On July 11, 1907, a splendid necklace of forty-nine graduated pearls, of fine luster and carefully matched, was sold at auction by Debenham and Storr of London, for the sum of £5600 ($28,000). At the same sale a single-row necklace of forty-five pearls brought £5300 ($26,500), and a rope of oriental pearls realized £4500 ($22,500).

At the auction sale of the collection of the late Bishop Bubics of Hungary, among other objects, a very handsome saber pouch was offered, of the style worn by the Hungarian hussars. It was of green silk and richly embroidered with hundreds of pearls of varying sizes. After a spirited competition this pouch was sold to Prince Esterhazy for 13,500 crowns ($2700). Some time after the sale a letter from the late bishop was found, containing the statement that he had borrowed the ornament from the jewel-room of the Princess Esterhazy. Naturally, Prince Esterhazy was not called upon to pay the amount of his bid. It is a gratification to know that at least one of the remarkable Magyar jeweled ornaments has escaped the cupidity of enterprising jewelers who have broken up so many of these ornaments for the gems which they contained.

A BRIEF SUMMARY OF THE PRINCIPAL GREAT PEARLS OF HISTORY

	Weight
Charles the Bold's Pearls. Three, each about	60 grains
Gomara Pearl, 31 carats	124 "
Oviedo Pearl, 26 carats (probably the Morales or Pizarro Pearl)	104 "
La Peregrina	134 "
Charles II's Pearl, (nearly equal to La Peregrina)	
Morales and Pizarro Pearl, 26 carats	104 "
Rudolph II's Pearl, 30 carats	120 "
Tavernier's Pearls:	
Shah Sofi's Pearl (estimated)	500 "
Imam of Muscat's Pearl, 12¹⁄₁₆ carats	48¼ "
Shaista Khan Pearl, 55 carats	220 "
Great Mogul's Pearls:	
Peacock Throne Pearl	200 "
Two pear-shaped, one about 70 ratis	186.2 "

Tavernier's Pearls—*Continued:* Weight

Olive-shaped pearl (estimated)	125 grains
One button-shaped, 55–60 ratis	146.3–159.6 "
One round pearl, 56 ratis (gift of the Shah Abbas II)	148.9 "
Round pearl (estimated)[1]	110 "
Three yellowish pearls, 25–28 ratis	66.5– 74.5 "
One perfectly round pearl, 35½ ratis	94½ "
Two perfectly shaped and equal, each 25¼ ratis	67.1 "
La Reine des Perles	109¼ "
Pearls of Maria Theresa's Necklace (three)	92–100 "
La Régente (now owned by Princess Yousoupoff)	337 "
La Pellegrina	111½ "
The Ynaffit, pear-shaped	143⅛ "
Hope Pearl, drop-shaped baroque	1800 "

Also in the Hope Collection:

Conical pearl	151 "
Bouton pearl	124 "
Oval pearl	94 "
Roughly spherical pearl	89 "
Acorn-shaped pearl	85 "
Oval conch pearl	82¼ "
Button-shaped conch pearl	77½ "
Drop-shaped pearl	76½ "
Pear-shaped Scotch pearl	34¾ "

Van Buren Pearls:

Two, each about	30 "
Also necklace 148 pearls	700 "
Tiffany Queen Pearl, American	93 "
Black bouton earring-pearl	88 "
White bouton earring-pearl	93 "
Bapst Pearls, two	113¼ and 113¾ "
Round pearl of Paris Exposition of 1889	70 "
Mme. Nordica's Pearl (abalone)	175 "
Great Bahama Conch Pearl	138¼ "
The Queen Conch Pearl	90 "
W. H. Moore's Pearl (Arkansas pearl, brown)	122½ "
Shark's Bay Pearl, golden yellow	30½ "
Rudolph II Crown Pearl, 26 carats	104 "
Carlotta's Pearl, pear-shaped	84 "
Marquis of Anglesey's Pearl, drop-shaped	105½ "
Black pear-shaped pearl (Lower California)	49 "

[1] As this pearl was brought from the East later on, it may be the same as the Reine des Perles, stolen from the French crown jewels in 1791. It is evidently the same as the La Pellegrina of the Zozima brothers (1814) and later stolen from them, reappearing as the pearl described by Kohl, in 1840, first in the possession of a Russian merchant and then later in the Russian Treasury.

XVII

THE ABORIGINAL USE OF PEARLS, AND THEIR DISCOVERY IN MOUNDS AND GRAVES

XVII

THE ABORIGINAL USE OF PEARLS, AND THEIR DISCOVERY IN MOUNDS AND GRAVES

THE use of pearls by the aborigines of the territory now comprised in the United States is proven by their appearance in the mounds and certain graves of pre-Columbian date. This is of great interest in view of the unique system of burial and the great variety of objects buried with the pearls. It is evident from the quantities discovered in some of the mounds that a very great number of pearls, many of large size, must have been owned by these aborigines, and they were evidently quite expert in the art of drilling them. Pearls must have been freely used for ornamental purposes, and it is clear that many rivers in this region must have produced them in great numbers, when we consider that in all probability the mussels were taken only as they were required for food or for bait in fishing, and had probably reached their full growth.

It is not unlikely that pearls were used on this continent for a long period, and they may have been in use centuries before any employment was made of them in Europe. In the age of the mound-builders there were as many pearls in the possession of a single tribe of Indians as existed in any European court. We have no means of ascertaining the precise date of any of these burials, and there are no historical records relating to this region, such as were kept in Mexico as well as in Europe and Asia. No trace has been found of the employment of pearls, either for decoration or ornament, by the aborigines of Europe or Asia; either they did not use them or else the pearls have entirely passed away in the course of twenty or more centuries. We do know, however, that neither pearls nor Unio shells were used by any of the lake-dwellers of Switzerland or the adjacent countries.

Many eminent archæologists have investigated the finding and history of the pearls of the mound-builders of Ohio and Alabama, especially Squier and Davis, F. W. Putnam, Warren K. Moorehead, C. C. Jones, W. C. Mills, and Clarence B. Moore. The discoveries made up to 1890 were fully treated by one of the writers in several pamphlets (one of them, "Gems and Precious Stones of North America").

It is not unlikely that the Indians of the Atlantic coast may have known of pearls from the common clam as well as from the edible oyster. The former may have often contained pearls weighing from fifty to one hundred grains each, as at that period the mollusks were permitted to attain their full growth, and perhaps were not eaten except when they were as small as little-neck clams; the larger ones were sought for the purple spot which held the muscle, and was used for wampum. We have no record of the finding of pearls in any graves north of Virginia, as the many graves opened in the past century have failed to reveal them, nor has the use of pearls been mentioned by any of the early writers. They may have been worn, but if so they have passed away or may have been mistaken for ashes if they had decrepitated.

The first English settlers found the Indians of the tidewater region of what now constitutes the Middle States using pearls quite freely and esteeming them among their favorite treasures and ornaments. Captain John Smith, and all the early chroniclers of the Virginia colony, have given many accounts of this aboriginal use of pearls.

In view of the general interest awakened by the tercentenary of the founding of Jamestown, and the exposition in commemoration thereof, the "American Anthropologist" devoted its first number for 1907 principally to topics relating to the Virginia Indians.[1] Among these articles is one of much interest by Mr. Charles C. Willoughby, of the Peabody Museum at Cambridge, Massachusetts, dealing with the tribes occupying tidewater Virginia at the time of the first colonization, their habits and customs, their distribution, and their subsequent history of diminution and almost of extinction. These were a branch of the Algonquian stock, and extended as far south as the Neuse River in North Carolina. To the south and west they were hemmed in by tribes of Iroquoian and Siouan race, and on the north they were separated from other hostile Indians by the Potomac River and Chesapeake Bay. The powerful confederacy under Powhatan comprised some thirty tribes or "provinces," covering most of the tidewater region of Virginia proper. To the greater chiefs, John Smith states that tribute was paid, consisting of "skinnes, beads, copper, pearle, deere, turkies, wild beasts and corne."[2] Many other references in this article confirm and illustrate this general statement, especially regarding pearls, both as to their use by the living and their deposit with the remains of the dead.

In the account given of the native clothing, the outer mantles are

---

[1] "American Anthropologist," Lancaster, Pa., Vol. IX, No. 1, Jan.-March, 1907, pp. 57-86.

[2] "True Travels," Richmond edition, 1819, p. 144.

described, made usually of deerskin with the hair removed, and bordered with a fringe. These were often "couloured with some pretty work, . . . beasts, fowle, tortayses, or such like imagery,"[1] or adorned with shells, white beads, copper ornaments, pearls, or the teeth of animals.[2] Strachey describes a wonderful cloak made of featherwork, belonging to an Indian princess, the wife of a deposed chief, Pipisco; with it she wore "pendants of great but imperfect couloured and worse drilled pearles, which she put into her eares," besides a long necklace made of copper links.[3]

With regard to such ornaments, Mr. Willoughby says (p. 71) that "the ears of both sexes were pierced with great holes, the women commonly having three in each ear, in which were hung strings of bones, shell, and copper beads, copper pendants, and other ornaments. Captain Amidas met the wife of a chief who wore in her ears strings of pearl beads as large as 'great pease' which hung down to her middle.[4] The husband of this woman wore five or six copper pendants in each ear. It was a common custom for the men to wear a claw of a hawk, eagle, turkey, or bear, or even a live snake as an ear ornament."

"Bracelets and neck ornaments of various kinds of beads were common. Beads of copper seem to have been most highly valued in the early colonial period. These were made of 'shreeds of copper, beaten thinne and bright, and wound up hollowe,' and were sometimes strung alternately with pearls which were occasionally stained to render them more attractive.[5] Beads of polished bone or shell were strung into necklaces either alone or with perforated pearls or copper beads. Some of these chains were long enough to pass several times around the neck. Necklaces of such construction as to be easily identified were worn by messengers as a proof of good faith. Powhatan gave Sir Thomas Dale a pearl necklace, and requested that any messenger sent by Dale to him should wear it as a guaranty that the message was authentic."[6]

"Pearls of various shapes and sizes were comparatively common, but symmetrical pearls of uniform size were more rare. Strachey writes of having seen 'manie chaynes and braceletts (of pearls) worne by the people, and wee have found plentie of them in the sepulchers of their kings, though discoloured by burning the oysters in the fier, and deformed by grosse boring.' One of Hariot's companions obtained

---

[1] Strachey, "Historie of Travaile into Virginia Britannia," Hakluyt Society, London, 1849, p. 65.
[2] Smith, *op. cit.*, p. 130.
[3] Strachey, *op. cit.*, p. 57.
[4] Smith, *op. cit.*, p. 83.
[5] Strachey, *op. cit.*, p. 67. "The 'blue' or

'violet-colored' pearls shown in White's original drawings are probably stained pearls." These were most probably the dark purple pearls of the round clam or quohog of the coast, although it is possible that they were only glass beads.
[6] Smith, *op. cit.*, Pt. II, p. 19.

from the Indians about five thousand pearls, from which a sufficient number of good quality and of uniform size were obtained to make a 'fayre chaine, which for their likenesse and uniformitie in roundnesse, orientness and pidenesse of many excellent colours, with equalitie in greatnesse, were verie fayre and rare.'[1]

"Those who have examined the thousands of pearls from the Ohio mounds, to be mentioned later, can readily understand these conditions. The pearl beads from the mounds vary in diameter from about an eighth of an inch to nearly an inch, the great majority being small and irregular, although there are many among them of good form and value. It is probable that most of the Virginia pearls were obtained from the fresh-water mussel (Unio)"; not unlikely from the common marine clam (*Venus mercenaria*), or the common oyster (*Ostrea virginica*).

As regards the burial of pearls with the dead and their use in religious rites, curious and quite full accounts are given by Strachey, Smith, Hariot, and Beverley.[2] There was a "temple," also occupied as a residence by one or more priests, in the territory of every chief. This building was usually some eighteen or twenty feet wide, and varied in length from thirty to one hundred feet, with an entrance at the eastern end, and the western portion partitioned off with mats to form a sort of sanctuary or "chancel." Within this were kept the dried bodies of deceased chiefs, and an image of the god, called Okee, made in the shape of a man, "all black, dressed with chaynes of perle." Full descriptions of these idols and their manufacture are given by Hariot and Beverley, also of the process of preserving the remains of the chiefs.[3] After the body had been disemboweled, the skin was laid back and the flesh was cut away from the bones. When this operation was completed, the skeleton, held together by its ligaments, was again inclosed in the skin, and stuffed with white sand, or with "pearle, copper, beads, and such trash sowed in a skynne."[4] It was then dressed in fine skins and adorned with all sorts of valuables, including strings of pearls and beads. The same kinds of treasures were also deposited in a basket at the feet of the mummy.

Captain Smith describes the temple of Powhatan, at Uttamussack, which was in charge of seven priests, and was held in great awe by "the salvages." At a place called Orapaks, was also his treasure-house, fifty or sixty yards long, frequented only by priests, where he kept a great amount of skins, beads, pearls, and copper, stored up

---

[1] Thomas Hariot, "A Brief and True Report of the New Found Land of Virginia," Holbein edition, p. 11.

[2] Willoughby, "American Anthropologist," Lancaster, Pa., Vol. IX, No. 1, January, 1907, pp. 61, 62

[3] Beverley, "History of Virginia," 1722, pp. 167, 186.

[4] Strachey, *op. cit.*, p. 89.

against the time of his death and burial. A vivid account is given of the four grotesque images that stood guard at the corners of this building, all made "evill favouredly according to their best workmanship."[1]

The use of pearls as ornaments, and their deposit with the remains of chiefs and persons of distinction, have already been described as familiar among the Indian tribes of tidewater Virginia, in the notes above cited from early explorers and colonists. It is a curious circumstance, however, that this habit does not appear to have extended in that part of the country much beyond the dominions of Powhatan, as no pearls have been noted in the Indian graves in Maryland. This statement, in reply to a letter of special inquiry, is made by Dr. P. R. Uhler, of the Peabody Institute of Baltimore, who has been making very careful studies of all aboriginal remains in that region, for the Maryland Academy of Sciences.

It would seem from this and other evidence, that the use and appreciation of pearls must have been in some way a tribal matter, familiar to some and not to others, of the Indian peoples. In the Mississippi Valley, the ancient population known as the mound-builders, by some regarded as a distinct and earlier race, and by others as of true Indian stock, although much more advanced in arts and culture, have left in their mounds most remarkable quantities of pearls. But here again, the same feature appears, that these treasures are not found wherever there are mounds, but only in certain regions. Of these, by far the most celebrated is that of the Scioto and Miami valleys, in Ohio. Outside of these, no large amounts have been found, and only at a few localities are they met with at all.

The valleys of the Miami and Scioto rivers and their tributaries contain many remarkable mounds and "earthworks," which have attracted much attention, and have been more or less explored at different times, with increasing care and thoroughness as archæological science has advanced. It may be well to give a brief, general account of these investigations and some leading features of the mounds as a whole, before going into particulars as to the occurrence of pearls.

The first important and scientific study of these remarkable structures was that conducted in the early forties by Dr. Edwin H. Davis and Mr. E. George Squier, and published in their celebrated and standard work entitled "Ancient Monuments of the Mississippi Valley," issued by the Smithsonian Institution in 1848. This book and the "Correspondence" in regard to the mounds by the same writers, published in 1847, were the first works issued by the Smithsonian Institution.

---

[1] Smith, *op. cit.*, p. 143.

According to Squier and Davis,[1] two quarts of pearls were originally deposited in one of these mounds. The writers consider that the pearls were probably derived from the fisheries in the southern waters, and they regard their presence in the Ohio mounds as a proof of "an extensive communication with southern and tropical regions and a migration from that direction."

A number of pearls or pearl beads from the Ohio mounds and which formerly belonged to the Squier and Davis collection, are now in the Blackmore Museum at Salisbury, England. According to a communication from Dr. H. P. Blackmore, director of the museum, these pearls, which originally formed five necklaces, have been much injured by the action of fire at the time the bodies of those interred in the mounds were burned. Mr. Blackmore considers that the greater part of the pearl beads are of mother-of-pearl cut from some large shell, made into a round shape and perforated, but, after very careful examination, he is of the opinion that about ten may be classed as natural pearls. Their present color is a dull, leaden gray, rather lighter than the "black pearl" of commerce. The size of these pearls or beads varies from four millimeters to twenty millimeters in diameter. One of the necklaces consists of thirty-three beads well graduated, but of a dead white color from the action of the earth.

A quarter of a century later, when the Centennial Exposition was in preparation, the Smithsonian Institution undertook the formation of a public exhibit illustrating American archæology, and engaged Prof. F. W. Putnam, of Cambridge, Massachusetts, to open and examine some of the most remarkable of the mounds described by Squier and Davis. These explorations were continued for some years, partly for the government and partly for the Peabody Museum of Archæology at Cambridge, and their results were exhibited at the Columbian Exposition in 1893. The mounds explored were chiefly in the valley of the Little Miami, and particularly those known as the Turner group.

A very important series of explorations was also carried on by Mr. Warren K. Moorehead, covering the years from 1887 to 1893, largely in preparation for the Columbian Exposition. These investigations were mainly in the Scioto valley, in the counties of Ross, Franklin, and Pickaway, Ohio. Among the most important results then obtained were those from the mounds of the "Porter" and "Hopewell" groups, in Ross County.

Since that time, much valuable work has been done by Mr. Moorehead and others, and particularly under the auspices of the Ohio State Archæological and Historical Society. The latest and most complete investigation was made for this society in 1903, by its curator, Prof.

[1] Squier and Davis, Smithsonian "Contributions to Knowledge," Vol. I, 1848, p. 283.

William C. Mills, in the Harness mound, seven miles north of Chillicothe, Ohio, near the Scioto River, in Ross County. This locality had been previously explored in part, by Professor Putnam in 1885, and Mr. Moorehead in 1896; it was now systematically examined down to the original surface at every point.

Squier and Davis divided these ancient monuments into four classes: (1) Altar mounds, which contain what appear to be altars, and are also called hearths, of stone or hardened clay; (2) Burial mounds, containing human bones; (3) Temple mounds, with neither altars nor bones, but seeming to have had some special religious significance; and (4) Anomalous mounds, including "mounds of observation" and others of mixed or uncertain character. The burials are found to be of two kinds, simple interment and cremation; and these are sometimes met with in the same mound.

This classification has been generally followed in describing these ancient structures, although the whole subject is obscure and difficult, from our ignorance of the purposes and conditions of their formation. In many of the mounds of the first two classes especially, not only have pearls been found, but quantities of interesting and remarkable objects, many of which have been brought from distant points, and prove clearly the existence of an extensive intertribal commerce at a remote period. Galena from Illinois and Wisconsin, mica from North Carolina, obsidian from beyond the Rocky Mountains, and seashells from the Gulf coast, are among these objects, and particularly native copper from Lake Superior, from which many articles were fashioned by hammering. Pearls are extremely abundant, and were at first supposed to have been brought from the coast, and may have been the pearls of the common clam and the common oyster, the pearls being found in opening the mollusks for food; but the recent development of pearl hunting in the western rivers, where the freshwater mussels (Unios) are so abundant and produce such beautiful pearls, shows that these treasures were undoubtedly gathered, partly, if not wholly, in the region where the mounds exist. The enormous numbers found are, indeed, no source of surprise, as such quantities of pearls have been obtained, for over twenty years past, from the same regions. The mollusks are still abundant in all the streams of the Mississippi Valley, except where they have been reduced or exterminated by the reckless methods of pearl hunting employed where the "pearl fever" has prevailed.

It is quite possible that the fresh-water Unios were not sought for their pearls alone, but were also used as food, and perhaps as bait for fishing. They were evidently gathered in great quantities, as is shown by the old heaps of shells found along the banks of streams at many points; and doubtless there are multitudes of such heaps that have never been observed. They are known as far north as Idaho, as communicated by Dr. Robert N. Bell, State mineralogist, and they extend still farther north, as noted by Dr. Harlan I. Smith, in his "Preliminary Notes on the Archæology of the Yakima Valley."[1] He says: "Small heaps of fresh-water clam-shells were examined, but these being only about five feet in diameter and as many inches in depth, are hardly to be compared to the immense shell-heaps of the coast."

These Unio shell-heaps are frequent in the South, and some of the Spanish chroniclers of De Soto's expedition in 1540–1541, describe the gathering and cooking of the mussels, and the finding of occasional pearls therein. The same writers also give glowing accounts of the pearls possessed by the natives. Some of these accounts may be exaggerated, but they cannot be wholly so. It would seem that some of the pearls may have come from marine shells, and others from those of the rivers and streams; but there are few pearl-producing shells on our own coasts, and it is not very likely that there was any trade or intercourse with the West Indian Islands, where marine pearls occur freely.

Albert H. Pickett, in his "History of Alabama," refers to the accounts of De Soto's historian, Garcilasso de la Vega, and holds that the pearls which he noted were evidently from the Unios of Alabama. "Heaps of mussel shells," he says, "are now to be seen on our river banks wherever Indians used to live. They were much used by the ancient Indians for some purpose, and old warriors have informed me that their ancestors once used the shells to temper the clay with which they made their vessels. But as thousands of the shells lie banked up, some deep in the ground, we may also suppose that the Indians in De Soto's time, everywhere in Alabama, obtained pearls from them. There can be no doubt about the quantity of pearls found in this State and Georgia in 1540, but they were of a coarser and more valueless kind than the Spaniards supposed. The Indians used to perforate them and string them around their necks and arms like beads."[2]

The use of fragments of these shells in tempering the clay for pottery, alluded to in the preceding paragraph, is well known. Prof. Daniel S. Martin describes an old village site in South Carolina, near the Congaree River, a few miles south of the city of Columbia, where the ground had been plowed, and along the furrows the soil was gleaming with brilliant pearly fragments of Unio shells, intermingled with bits of pottery.

Mr. Clarence B. Moore discovered pearls pierced for stringing in

[1] "Science," April 6, 1906, Vol. XXIII, No. 588.

[2] "History of Alabama," Charlestown, 1851, Vol. I, p. 12.

several of the mounds at Moundville, Alabama. He also found a sheet-copper pendant, elongated oval in outline, with an excised repoussé decoration, embracing a swastika within a circle, and a triangle. This pendant, which lay near the skull of burial No. 132, bears a perforated pearl nearly seven millimeters in diameter and weighing about nine grains; it is fastened to the pendant by a piece of vegetable fiber that passes through the pearl. With another burial (No. 162), the skeleton of an adult, was an elliptical gorget of sheet-copper decorated with a pearl.[1] In a personal communication Mr. Moore states that all the pearls found by him in the mounds were very much disintegrated by the lapse of time; he also writes that he has never found any shells immediately with the pearls, although masses of Unio shells were often met with in the mounds. He believes the shell-fish had been used for food.

Unio shell-heaps exist likewise on the shores of the inland lakes of Florida, and in middle Georgia and Alabama; and several of them on the banks of the Savannah River, above Augusta, are fully described by Colonel Charles C. Jones.[2] He says: "In these relic-beds no two parts of the same shell are, as a general rule, found in juxtaposition. The hinge is broken, and the valves of the shell, after having been artificially torn asunder, seem to have been carelessly cast aside and allowed to accumulate."

Thus, in addition to the historical evidence, physical proof is abundant of the pearl fisheries of the aboriginal tribes of the South. In order to ascertain the precise varieties of shells from which the southern Indians obtained their pearls, Colonel Jones invited an expression of opinion from a number of scientists whose studies rendered them familiar with the conchology of the United States. Their responses throw considerable light upon this inquiry, though with some curious variation.

Prof. William S. Jones, of the University of Georgia, says that he has seen small pearls in many of the Unios found in that State.

Prof. Jeffries Wyman, on the other hand, after a careful and extensive series of excavations in the shell-heaps of Florida, failed to find a single pearl. "It is hardly probable," he remarks, "that the Spaniards could have been mistaken as to the fact of the ornaments of the Indians being pearls, but in view of their frequent exaggerations, I am almost compelled to the belief that there was some mistake; and possibly they may not have distinguished between the pearls and the shell

[1] "Moundville Revisited," Reprint from the Journal of the Academy of Natural Sciences of Philadelphia, Philadelphia, 1907, Vol. XIII, pp. 398-403.

[2] "Antiquities of the Southern Indians," New York, 1873, p. 483; also, "Monumental Remains of Georgia," Savannah, 1861, p. 14.

beads, some of which would correspond with the size and shape of the pearls mentioned by the Spaniards."

Prof. Joseph Jones, whose investigations throw much valuable light upon the contents of the ancient tumuli of Tennessee, says: "I do not remember finding a genuine pearl in the many mounds which I have opened in the valleys of the Tennessee, the Cumberland, the Harpeth, and elsewhere. Many of the pearls described by the Spaniards were probably little else than polished beads cut out of large sea-shells and from the thicker portions of fresh-water mussels, and prepared so as to resemble pearls. I have examined thousands of these, and they all present a laminated structure, as if carved out of thick shells and sea conchs." This point will be referred to again.

Dr. Charles Rau[1] writes: "I learned from Dr. Samuel G. Bristow, who was a surgeon in the Army of the Cumberland during the Civil War, that mussels of the Tennessee River were occasionally eaten 'as a change' by the soldiers of that corps, and pronounced no bad article of diet. Shells of the Unio are sometimes found in Indian graves, where they had been deposited with the dead, to serve as food during the journey to the land of spirits."

Dr. Brinton saw on the Tennessee River and its tributaries numerous shell-heaps consisting almost exclusively of the *Unio virginianus* (Lamarck). In every instance he found shell-heaps close to the water-courses, on the rich alluvial bottom-land. He says: "The mollusks had evidently been opened by placing them on a fire. The Tennessee mussel is margaritiferous, and there is no doubt but that it was from this species that the early tribes obtained the hoards of pearls which the historian of De Soto's exploration estimated by bushels, and which were so much prized as ornaments."[2]

A source has recently been pointed out whence small pearls, and perhaps some fine specimens, could have been obtained by the Indians of Florida, and in considerable quantities. In the Unios of some of the fresh-water lakes of that State, there were found not less than 3000 pearls, most of them small, but many large enough to be perforated and worn as beads. From one Unio there were taken eighty-four seed-pearls; from another, fifty; from a third, twenty, and from several, ten or twelve each. The examinations were chiefly confined to Lake Griffin and its vicinity. It is said that upon one of the isles in Lake Okeechobee are the remains of an old pearl fishery, and it is proposed to open the shells of this lake, which are large, in hopes of finding pearls of superior size and quality.

---

[1] "Ancient Aboriginal Trade in North America," Report of the Smithsonian Institution for 1872, p. 38 of the author's reprint.

[2] See "Artificial Shell Deposits in the United States," in the Report of the Smithsonian Institution for 1866, p. 357.

The use of the pearl as an ornament by the southern Indians, and the quantities of shells opened by them in various localities, make it seem strange that it is not more frequently met with in the relic-beds and sepulchral tumuli of that region; but, after exploring many shell- and earth-mounds, Colonel Charles C. Jones failed, except in a few instances, to find pearls.[1] A few were obtained in an extensive relic-bed on the Savannah River, above Augusta, the largest being four tenths of an inch in diameter, but all of them blackened by fire. Many of the smaller mounds on the coast of Georgia do not contain pearls, because at the period of their construction the custom of burning the dead appears to have prevailed very generally; hence, it may be that the pearls were either immediately consumed or so seriously injured as to crumble out of sight.

This absence of pearls tends somewhat to confirm the opinion that beads made from the thicker portions of shells that were carved, perforated, and brilliant with nacre, were regarded by the imaginative Spaniards as pearls. More minute investigation, however, will doubtless reveal the existence of pearls in localities where the pearl-bearing shells were collected. Perforated pearls have been found in an ancient burying-ground located near the bank of the Ogeechee River, in Bryan County, Georgia; and many years ago, after a heavy freshet on the Oconee River, which laid bare many Indian graves in the neighborhood of the large mounds on Poullain's plantation, fully a hundred pearls of considerable size were gathered.

It seems quite clear that many of the pearls reported by the early Spanish voyagers were really such, although it is well known also that shell beads have been found in mounds in connection with pearls; but the numbers found in Ohio, by Professor Putnam, Mr. Moorehead, and others, leave no room for doubt in this matter. That the Indians of the South also had these pearls, both drilled and undrilled, is beyond question.

The same fact comes to view, however, in these various accounts, that has been alluded to already, viz., that the use of pearls among the aborigines appears to have been local, and probably tribal. All the fresh waters of North America contain Unios, especially in the Mississippi basin and in the South, and all the Unios are more or less pearl-bearing; but it is only at certain points that pearls are found deposited in ancient graves, sometimes, however, in extraordinary quantities.

Father Louis Hennepin relates that the Indians along the Mississippi wore bracelets and earrings of fine pearls, which they spoilt, having nothing to bore them with but fire. He adds: "They gave us

---

[1] "Antiquities of Southern Indians," p. 490.

to understand that they received them in exchange for their calumets from nations inhabiting the coast of the great lake to the southward, which I take to be the Gulph of Florida."[1]

The statement here made, that the Indians perforated their pearls only "with fire," evidently refers to the use of a heated copper wire, or point, as mentioned by Pickett and others of the early explorers. This point is of importance, as apparently indicating a marked difference between the Indians met with by the first European visitors, and the mound-building people of an earlier time, among whom the perforation was made with small stone drills. On this point, a recent letter from Prof. Wm. C. Mills, who has conducted the very full exploration of the Harness mound in Ohio, is of interest. He describes the small and carefully-wrought flint drills, which he found, and believes to have been made and used for this purpose. In size and form they answer all requirements; they are delicate little implements, somewhat T-shaped or gimlet-shaped, an inch and a quarter long; the narrow boring part is about an inch in length and tapers from one eighth of an inch to quite a fine point; the wider upper end is abruptly expanded into the transverse handle, which is about a quarter of an inch thick, *i.e.,* lengthwise of the instrument, and half an inch in span, *i.e.,* across, so as to give a good hold for the fingers to rotate the drill, just as in an ordinary gimlet.

Passing now to the actual discoveries of pearls in the mounds of the Mississippi Valley, these will be reviewed in the order of the successive explorations in which they were made known. As already stated, the only region where any large amounts have been encountered, is that of the Scioto and Miami valleys in Ohio. Even here, pearls are found only at certain points, and though the numbers are great, the graves which contain them are few. They were apparently buried only with the remains of individuals of especial distinction, probably either chiefs or eminent medicine-men. The accounts of recent explorations in these mounds bring to mind very forcibly the statement before cited from Captain John Smith, as to Powhatan's treasure-house, where all his most valued articles, including pearls, were collected and kept, in preparation for his death and burial. Pearls appear also to have been used only by the more cultured tribes, and were kept in the larger and more prosperous communities exclusively. They are confined to the great "mound groups," and are not found in isolated mounds. The tumuli of northern Ohio, the hill mounds, and the village sites along the smaller streams, have yielded practically none.

According to the manner of burial, the pearls vary greatly in their

[1] Transactions of the Philosophic Society for 1693.

present condition. Where they have been placed with cremated bodies, they are, of course, much damaged, being blackened and largely decomposed. Otherwise, although injured in color and luster, the mere fact of burial in the ground has not entirely ruined them. They are generally perforated, so as to be strung or attached to garments, and traces of both these methods of use are sometimes clearly shown.

The term "pearl beads," often employed by writers, is uncertain in meaning; as it may refer either to actual pearls, bored so as to be strung, or to imitations thereof made from pearly shell. With regard to this point, although such quantities have been obtained, there seems to have been very little close examination as to their structure, which would at once indicate the facts, according as the minute layers of the pearly material are concentric or not. The only distinct testimony is that we have cited above from Prof. Joseph Jones,[1] who states that he has examined large numbers, and found them to be apparently cut from shells. He makes the suggestion that they may have been carved from the thicker portions of the fresh-water Unios. This is not only probable, but would go far to solve the mystery of the enormous numbers found, as compared with anything known of the yield of genuine pearls by these mollusks, even with all the pearl hunting of recent years. An interesting fact bearing directly on this question is the discovery in the Taylor mound, at Oregonia, Warren County, Ohio, of several Unio shells in which had been made a circular hole, two thirds of an inch in diameter, either for some ornamental use of the shell or to extract pieces to be shaped into beads. These may have been made in either of two ways. Firstly, by breaking pieces of the shell from one of the valves, as a lapidary "roughs out" a piece of gem material before he begins to grind it into shape; or, secondly, by cutting out a circular disk of shell by means of a hollow copper drill or a hollow reed, just as they perforated hard pieces of quartz or granite for pipes, or as they trephined circular disks from the skulls. Decorated disks of Unio shell were also found in the same mound. If the ancient people made beads in this manner, there is little difficulty in accounting for the quantities described, especially in connection with the evident gathering of Unios on a large scale, as shown by the widely distributed shell-heaps already described. They certainly did make beads from various marine shells, and these are found with the pearl beads in many of the mounds, as particularly noted by Professor Jones, cited above, and by others.

In the recent exploration of the Harness mound, by Professor Mills, a very curious discovery was made of imitation pearls of a kind never before met with; these were made of clay, modeled apparently after

[1] See p. 494.

the larger natural pearls associated with them, and after being baked hard, had been "covered with a flexible mica," so as to resemble pearls.[1] The mica was a silvery mica that may have been burned and would pulverize into a gray powder with a pearly luster, as almost all micas are too resilient to be attached in any other way.

Taking up now the history of pearl discovery in the mounds, the first definite record goes back to about 1844, when perforated pearls were found by Dr. Edwin H. Davis[2] on the hearths of five distinct groups of mounds in Ohio, and sometimes in such abundance that they could be gathered by the hundred. They were generally of irregular form, mostly pear-shaped, though perfectly round ones were also found among them. The smaller specimens measured about one fourth of an inch in diameter, but the largest had a diameter of three quarters of an inch.

The next great discovery of these Unio pearls was in the Porter group of mounds, in the Little Miami Valley, explored by Prof. Frederick W. Putnam, and Dr. Charles L. Metz, who procured over 60,000 pearls, nearly two bushels, drilled and undrilled, undoubtedly of Ohio origin; all of them, however, decayed or much altered, and of no commercial value. In 1884 these scientists examined the Marriott mound, where they found nearly one hundred Unio shells, and among other objects of special interest six canine teeth of bears, that were perforated by a lateral hole near the edge at the point of greatest curvature of the root, so that by passing a cord through this, the tooth could be fastened to any object or worn as an ornament. Two of these teeth had a hole bored through near the end of the root on the side opposite the lateral perforation, and the hole countersunk in order to receive a large spherical pearl, about three eighths of an inch in diameter. When the teeth were found, the pearls were in place, although chalky from decay. Upward of 250 pearl beads were found here, concerning which they say: "The pearl beads found in the several positions mentioned are natural pearls, probably obtained from the several species of Unios in the Ohio rivers. In size they vary from one tenth of an inch to over half an inch in diameter, and many are spherical. They are neatly drilled, and the larger from opposite sides. These pearls are now chalky, and crumble on handling, but when fresh they would have formed brilliant necklaces and pendants."[3]

It is easy to see, even at a glance, that most of those in this great deposit of 60,000 are true pearls. Many are very irregular in form, and quite a number are the elongated, somewhat feather-shaped,

[1] "Exploration of the Edwin Harness Mound," Columbus, O., press of F. J. Heer, 1907, p. 76.
[2] "Ancient Monuments of the Mississippi Valley," Squier & Davis, Washington, 1848, p. 232.

[3] "Explorations in Ohio," from the Eighteenth Report of the Peabody Museum, Cambridge, 1886, p. 462.

Necklace of fresh-water pearls and cut shell beads, from Mound No. 25

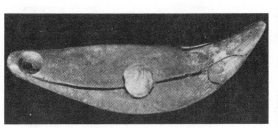

Bear-tooth inlaid with fresh-water pearl from the neck of skeleton No. 209, Mound 23

Perforation in charred, cut fresh-water pearl; weight, 5569 grams

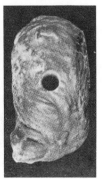

Perforated fresh-water pearl; weight, 22,955 grams

FRESH-WATER PEARLS FROM HOPEWELL GROUP OF MOUNDS, ROSS COUNTY, OHIO

Group of charred, cut fresh-water pearls; more than 100,000 found in mounds

Finger-shaped piece of lignite inlaid with fresh-water pearl

Copper bird, 15⅞ inches long with eye
of fresh-water pearl

FRESH-WATER PEARLS FROM HOPEWELL GROUP OF MOUNDS, ROSS COUNTY, OHIO

"hinge-pearls," that are found in the region of the hinge-teeth of Unios. A large and interesting exhibit of these is shown in the Field Museum of Natural History, Chicago. But thousands of spherical pearls were also obtained, from the "altars" or "hearths" of mounds belonging to the first division of Squier and Davis's classification, above noted. From the Turner group, in Clermont County, in the Little Miami Valley, Professor Putnam obtained for the Peabody Museum as much as half a bushel of pearls of this character. As these had been exposed to fire, nearly all were blackened, some cracked, and all greatly impaired.[1]

The next great series of explorations were those conducted by Mr. W. K. Moorehead in the Scioto Valley, in the counties of Ross, Franklin and Pickaway, Ohio. He opened and examined a number of mounds, and found pearls or pearl beads in ten or twelve of them, but the larger deposits were confined to certain limited districts, which seem to have been occupied by tribes more advanced in culture and in traffic than the rest. In these, the pearls and also objects of other kinds brought from a distance, are principally found. The scattered mounds, not associated with any village or community sites, have few of these valuable objects.

But even where they are found freely, pearls were apparently used or possessed by only a few individuals. Mr. Moorehead investigated in all 117 burial mounds, containing about 1400 skeletons. Pearls were met with in only seven of these mounds, and in connection with but twenty-two skeletons. These, however, yielded a total of 2600 pearls, apparently from Unios, the numbers found with single skeletons varying from 18 to 602, an average of 118. It thus appears that in Mr. Moorehead's researches, pearls were found in about one mound out of seventeen, and in these, with about one skeleton out of eight.

From "altar mounds," pearls have been in some cases taken in vast numbers. Professor Putnam's discoveries are mentioned above; and Mr. Moorehead obtained tens of thousands from two altars or hearths in the Hopewell group, which will be described hereafter.

When found in the burial mounds with skeletons, pearls are generally seen to have been placed at the wrists or ankles, or about the neck, or in the mouth. Sometimes they are found on copper plates, and occasionally they show evidence of having been sewn or attached to a garment. Particulars on these points will be given further on. Mr. Moorehead has also found bears' teeth, set with pearls, as Putnam and Metz did in the Marriott mound, lying with or near skeletons.

In the case of the altar mounds, there seems to have been a different procedure, not a burial, but a great funeral sacrifice in honor of some

[1] Collection of Peabody Museum of Archæology, Cambridge, Mass.

very distinguished person, in which treasures of every kind, including great stores of pearls, were consumed, or meant to be. Of this, Mr. Moorehead says, in a letter to the author: "In the case of all altar offerings, a fire had been kindled . . . and all these things were heaped upon it. They were utterly ruined, save a few; . . . those at the top were not so much affected as those at the bottom."

Mr. Moorehead's investigations already mentioned were in the years 1888 to 1891 inclusive; he next took up especially the remarkable Hopewell groups of mounds, in 1891–1892, and explored these extensively for the archæological exhibit at the World's Columbian Exposition of 1893, at Chicago.[1] This was his most important and elaborate investigation, and will be described in some detail. In 1896, he made a partial exploration of the Harness mound near Chillicothe, which has been fully completed more recently by Prof. William C. Mills, and will also be described further on.

The investigations made in the Hopewell group of mounds were recorded by Mr. Moorehead in a series of articles in the "Antiquarian."[2] He gives a general account of the remarkable region of ancient remains in Ross County, Ohio. The State archæological map shows the "mound belt," as a strip of country some fifteen miles wide and one hundred miles long, extending through the Scioto Valley, from about Columbus to Portsmouth. The ancient works noted on this map, though not all that exist there, yet number over 900 mounds, 24 village sites, 36 circles of earth and stone, 87 other inclosures and works of similar character, and 31 sites of gravel or kame burials. Five groups of mounds in particular exist in Ross County, all of them showing a "high culture" state. "All of the lower Scioto Valley," says Mr. Moorehead, "was occupied by a mound-building tribe ranking higher in intelligence and numerically stronger than that of any other section of the whole Ohio region." Among the many remarkable ancient works in that part of the country, the five groups in Ross County are the most important, and among these, the Hopewell group is preëminent. The first published notice of them, which appeared in 1820, was by Mr. Caleb Atwater.[3] Squier and Davis examined and described them in the years 1844–1846, and obtained large and notable collections from them which are now in England, in the Blackmore Museum at Salisbury, as not enough interest in such matters then existed in America to induce the purchase and retention of these valuable treasures. From that time until 1891, when Mr. Moorehead began his explorations there, no one had paid much attention to these mounds, all

---

[1] Now in the Field Museum of Natural History, Chicago, Ill.

[2] "American Archæologist," May, 1897, to May, 1898.

[3] "Archæologia Americana," 1820, p. 182.

published accounts being derived from those of Squier and Davis. They described them under the name of Clark's works, from the owner of the farm within which they lie; but the property has since passed into the possession of Mr. M. C. Hopewell. From this fact, yet more from his kind and intelligent interest in the work of exploration, his name has been given to the group.

The Hopewell works are situated on the north fork of Paint Creek, about one third of a mile from the stream. The intervening space is low bottom-land, and the works stand upon a terrace about twenty feet high, from which again there is a rather steep rise of thirty or forty feet more, to the general level of the country. They consist of a nearly quadrangular inclosure, about half a mile in length (strictly 2800 feet), and half as much in width, occupying the entire breadth of the terrace. At its eastern end, this large inclosure opens into a second and smaller one, an exact square of 850 feet. Within the main inclosure are one or more village sites, a number of separate mounds, and especially a group of several connected elevations, together known as the Effigy mound, these being much the highest and most conspicuous, and themselves surrounded by a semicircular inclosure. The whole suggests a defensive work, or "walled town"; but the wall, although strongly and carefully built, partly of stones and partly of hard clay, is so low—only from four to six feet in height—that it could not have been a very formidable obstacle to a vigorous assault; and, moreover, the whole is overlooked and "commanded" from the bluff above it. The mounds, as Squier and Davis examined them, were pronounced to be mainly of the sacrificial or "altar" type. Since their very full and accurate account was published, time and the hand of man have reduced and almost obliterated portions of the wall and some of the smaller mounds, while the creek has slightly shifted its course. When they wrote their description, it was a little nearer than it is now; and they then expressed the belief that it had formerly washed the base of the terrace where the works are located.

Mr. Moorehead's exploring party, aided by Dr. H. T. Cresson, began operations at this notable group of mounds in August, 1891, and continued them through about seven months, without interruption, much of the time in severe winter weather. The work was carried on under authority of the Anthropological Department of the Columbian Exposition of 1893, at Chicago. All the most interesting and important of the very extensive body of relics obtained was displayed there; and the whole remains as a permanent exhibit in the Field (Columbian) Museum of Natural History.

The Hopewell group comprises in all some twenty larger and

smaller mounds within the general inclosure, besides a few unimportant ones outside of it, and the main connected group in the special inclosure near the center. These latter form together what is known as the Effigy mound, a name based upon its general resemblance to a reclining human figure; but it is not constructed on a human or animal design, as are the effigy mounds properly so called. After working for a time upon some of the others, and finding much interesting material, Mr. Moorehead set his men to work upon the Effigy mound, and spent most of his time and effort upon that remarkable structure, of which he made a very thorough and systematic exploration.

The Effigy mound is about 500 feet long and 220 feet wide, and rises 23 feet above the general surface at its highest point. It proves to belong to the fourth class of Squier and Davis, those of mixed character, with both altars and burials, as it contained three large altars and as many as 175 skeletons, nearly all of adults.

Reviewing now the entire exploration of the Hopewell group, the first mound opened, known as No. 17, was of considerable size, nearly ninety feet in diameter, and was notable for a layer of mica—some 3000 sheets—that extended almost entirely through it. It contained a rude altar, with ashes and bones, some copper implements, bone needles, sharks' teeth, and nearly 200 pounds of bright galena. The next examined, No. 18, contained several decayed skeletons, and a good example of an "altar," together with ornaments cut from human skulls. The next, No. 19, had an altar of earth, partially hardened by heat, which was taken out entire and boxed. It was roughly cubical, about three feet each way. In the "bowl," or concavity, on the top of it, were various minor implements, with some galena and mica, etc. The next attacked was a large mound, No. 2, which had been partly opened by Squier and Davis, nearly fifty years before. It is remarkable for its immense store of roughly chipped flint disks, over 8000 in number, of which 600 were taken out by Squier and Davis, and most of the remainder by Mr. Moorehead. It would seem to have been a place of storage for partly worked material of this kind, to preserve it from the hardening effect of long exposure to the air.

Several other mounds yielded little of importance, save that from the soil on the site of No. 1, which had been obliterated, were taken a number of fragments of bone, curiously ornamented with finely carved patterns. Two others, Nos. 4 and 5, had peculiarly constructed altars, of which an extended account is given.

The first discovery of pearls by Squier and Davis was made in their mound No. 9, now obliterated by a railroad. With the pearls, they report as found on the top of a small altar, broken instruments of obsidian, cut patterns of mica, vestiges of cloth, etc.

Mr. Moorehead's first discovery of pearls was in a small but interesting mound, No. 20, about forty feet in diameter. It had been reduced by plowing to only some two feet in height; and its contents would ere long have been broken into and scattered by the same process. This was strictly a burial mound, and soon yielded five skeletons, one of them being that of a child, nine or ten years of age. With these bones were numerous objects: two large shells made into cups for drinking, several copper articles and ornaments, among them a broad copper bracelet encircling the right wrist, and several hundred pearl and shell beads and small shells. The same mound yielded later some other children's remains, but with no important objects. A finely polished pipe and two bear's teeth coated with copper were also found.

Mr. Moorehead points out the evidences of a long occupation of this site by a cultured tribe, who had commerce with the South and West more than with the North or East.

Work was then begun, in the latter part of September, on a large and important mound known as the Oblong (No. 23), 155 feet long by 100 feet wide, with an elevation at present of 14 feet, and originally of perhaps 20 feet. This mound yielded thirty-nine skeletons, lying at depths varying from eight and three fourths to eleven feet below the present surface, nearly on the base-line of the mound. Some of these were surrounded by boulders, others were much charred, and a good deal of variety exists in their condition, all of which Mr. Moorehead describes particularly. All manner of relics and objects were obtained, including pearl beads and a splendid copper ax of seventeen pounds' weight, of course entirely too large for any practical use, and hence plainly a ceremonial object or badge of some high distinction. Among the most remarkable of the many interesting objects discovered here were the large canine teeth of bears,[1] which had not only been drilled through near the base of the root for suspension, like many others, but had also been partly drilled at the middle of one side, and a large pearl inserted into the cavity. These singular ornaments were found at the neck of a skeleton, and had evidently been worn as pendants. It will be remembered that almost identical specimens were found by Professor Putnam in the Marriott mound in the Miami Valley.[2] The one here figured is now in the Field Museum of Natural History, Chicago, with most of the other Hopewell material.

Another somewhat similar example of the taste and art of the same

[1] See p. 499.   [2] See p. 498.

people, also preserved in the Field Museum, came from the mound known as No. 25. This consisted of a large figure of a bird, in hammered copper, fifteen and seven eighths inches long, with a pearl inserted to form the eye. The head is quite expressive, and the tail-feathers well represented, although the wings and the general proportions are rude. This is shown about one third of the actual length.

The Effigy mound was next examined. The first trial shafts proved it to be evidently of human construction, and not of glacial origin, as some had supposed. One or two open cuts were then begun, using teams with a large shovel until indications of burials were found, when the further work would be carried on by hand, with extreme care.

After about two weeks, in which time several skeletons were unearthed, with some shells, beads, and copper ornaments, a burial of extraordinary character was reached on November 14. Here was lying a skeleton which the newspapers soon reported as "The King of the Mound-Builders." It was much decayed, but was covered and surrounded with a wealth of relics. The skull was surmounted by a tall cap or helmet of copper, from which extended a wonderful pair of antlers, exactly imitating those of a deer, but made of wood and covered with copper. The whole skeleton, to quote the words of Mr. Moorehead, "glittered with mica, pearl, shell, and copper." Plates of the latter were above, beneath, and around it, with bears' and panthers' teeth, etc., and over 1000 beads, many of them of pearl. The succeeding month, during which the last cut was finished down to the base-line, and a third one much advanced, revealed numerous skeletons, with abundant objects of the same general kind, including a remarkable separate deposit of copper articles of curious workmanship, ornaments of cut mica, and one of cannel coal, fragments of meteoric iron and celts made therefrom, and "many thousand pearl and shell beads." The latest trophy here unearthed was another enormous ax of copper, nearly two feet in length, unparalleled in the world.

The first altar was next reached; it was about four by five feet, and some six inches deep, and had an immense variety of objects upon it and around it, nearly all entirely ruined by the fire. Among them were pearl beads.

The largest altar had been not only heaped with all sorts of valuables, but they had been piled around it so as to form a sloping mass of twelve feet or more in diameter at the base. Among these was a layer of mica plates of extraordinary size, eighteen or twenty inches in diameter. It is not easy even now to obtain sheets of mica of such dimensions, in any quantity. Carvings and effigies in bone and slate, rock-crystal arrow-heads, obsidian knives, etc., etc., damaged and broken by heat, were cemented together by half-melted copper. The pearl and shell beads taken out amid the ashes are estimated at not less than 100,000.

The Effigy mound, "a place for ceremony, for sacrifice, for burial," as Mr. Moorehead calls it, thus combining the character of the first three classes distinguished by Squier and Davis, is seen not to have been constructed at one time, but to have developed gradually through perhaps a long period. The several altars, the more important burials, the store of copper objects, each was surmounted by a small and separate mound. "These may have been built on the level dance or ceremonial floors, from time to time. When the entire floor was covered, the people brought large quantities of earth and gravel, heaped it on top of the irregular contour of the small mounds, and this formed the present Effigy."

The population that occupied the main inclosure was apparently not very large, as compared with some other of the important earthworks, such as Fort Ancient, or Madisonville. From the distribution of village-site debris, Mr. Moorehead estimates that there could have been only from two hundred to three hundred lodges, even if these were all occupied at the same time. But the indications of traffic and of art show that it must have been a community advanced in culture beyond most of its neighbors. Mr. Moorehead believes it to have been a sort of capital among a body of allied or affiliated tribes who made and occupied the similar earthwork towns of the "mound belt,"—a center of production and distribution of art objects, and a place for the holding of great religious ceremonials. It may be noted, however, that the art was developed in certain directions and not in others wherein it might be expected. In hammered copper-work and in drilling, it was most remarkable, in the latter extending even to the perforation of quartz crystals, but of pottery there is little, and that not very choice—a striking contrast to the abundant and elaborately ornamental potter's art of the tribes in the Southwest.

Tonti, the historian of La Salle's expedition, in the eighteenth century, states that La Salle actually saw mound-dwellers among southern tribes of Indians, living very much as the Ohio mound-builders must have done, and quite untouched as yet by any contact with the whites. Tonti describes the dwellings, made of sun-dried mud and with dome-shaped roofs of cane; two of them were larger and better constructed than the rest, one the chief's house and the other a temple, both about forty feet square. The latter held the bones of deceased chieftains, and was surmounted by three rude, wooden eagles. In the center was apparently "a kind of altar," where was maintained a perpetual fire of logs, watched by two aged men. A recess, to which strangers were not admitted, contained the treasures of the tribe, especially pearls from the Gulf, as he was told. The chief returned the visit of La Salle, coming in great state, with attendants, one of whom bore a disk of copper, supposed to represent the sun, the chief's great ancestor.[1] The wooden eagles recall the large copper bird taken from mound No. 25 at Hopewell; and the copper disk carried before the chief suggests a similar use for some of the large objects of the same metal. The whole account is extremely interesting in its resemblance to the Ohio remains.

The most complete study of these ancient structures is that of the Harness mound, not far distant from the Hopewell, conducted under the direction of the Ohio State Archæological and Historical Society, in 1905, by their curator and librarian, Prof. William C. Mills.[2]

The Harness group contains within and about it fourteen mounds; the works as a whole were described by Squier and Davis, on page 56 of their great report ("Ancient Monuments of the Mississippi Valley," 1848), and have been frequently mentioned and pictured for their striking form,—a large and perfect circle, opening at one side into a smaller circle and also into an exact square. They are located, like the Hopewell, in Ross County, and stand on a terrace of the Scioto River, nearly a mile from its eastern bank, and about eight miles south of Chillicothe.

The square inclosure measures 1080 feet on each side, and the diameters of the two circles are about 1600 feet for the larger and 650 feet for the smaller. In general character, this group closely resembles the Hopewell: there is the same low wall or embankment, some four feet high, though without any ditch as at Hopewell, and the same problem as to its object. A number of small mounds are placed here and there, and one large and important one recalls the Effigy, though it is somewhat less in size and much more regular in form. In 1846, when Squier and Davis examined it, unfortunately most of the ground was covered with woods; but these are gone, and the works have since been much reduced by tillage and partly obliterated by railroad and other constructions.

The one large mound is named for the recent owner of the property, Mr. Edwin Harness; the present owner, his son, Mr. John M. Harness, aided and facilitated the explorations in every way. This fact, as also in the case of Mr. Hopewell, stands in pleasing and honorable contrast to the narrow policy of some land-owners, who refuse permission for any such work, even when the structures are upon unused and valueless ground.

[1] "La Salle and the Discovery of the Great West," Parkman, p. 281.

[2] William C. Mills, "Explorations of the Edwin Harness Mound, Columbus, O."; press of Fred. J. Heer, 1907. "Ohio Archæological and Historical Quarterly," Vol. XVI, No. 2.

The large mound is an almost perfect oval in form, 160 feet long and some 80 feet across at its widest point, which is about one third of the way from the northern end; in height it is nearly 20 feet, or was before its recent removal. It was partly explored by Squier and Davis in 1846, and quite extensively by Professor Putnam in 1885, and, unlike the Effigy mound, had been repeatedly opened and examined in a small way by both official and unofficial explorers. In 1896, Mr. W. K. Moorehead took up the work where Professor Putnam had stopped, and carried it considerably further, under the auspices of the Ohio Archæological and Historical Society; and the same body, in 1905, commissioned Mr. Mills to resume and complete the examination, removing the entire structure down to its base.

The Harness mound, unlike the Effigy, was for burial purposes only. There must have been nearly two hundred. Squier and Davis found one of these, and possibly another which they mistook for an "altar"; and they state their belief that the mound probably contained other burials which their two pits had not revealed. Professor Putnam encountered 12 burials, Mr. Moorehead 27, and the final exploration 133, making a total of 174. Besides these, an unknown number have been disturbed and removed by occasional explorers. Of the 174 recorded, only ten had been buried without being burned; the rest were all cremated, some where they were laid, but most of them elsewhere, and the ashes brought and placed in the grave. This was in all cases carefully prepared, within a small inclosure of logs, the decayed and charred remains of which are clearly traceable. The entire mound itself had been outlined with posts set in the ground, the holes and impressions remaining as evidence of the fact.

Mr. Mills outlines the history of this mound, in a way that recalls Mr. Moorehead's views as to the gradual growth of the Effigy. It began as a place for the holding of funeral rites and the deposit of the dead, marked out by lines of posts, which show that it was from time to time enlarged. Finally, when the place was substantially filled, earth and gravel were deposited over the whole, and slabs of stone (particularly noted by Squier and Davis) were laid around it, upon the lower part of the slope.

Much description is given of the separate graves or burial chambers, which are of several types, and of the various details of the cremated and uncremated interments. The mound is rich in relics, although none of the profuse sacrificial accumulations of the "altars" were encountered, this being a mound of burial only. The relics are of the same kind, in general, as those found in the Hopewell group, and to specify them in detail would be only repetition. From the 133 graves opened in Mr. Mills's final investigation, no less than 1200 specimens were obtained for the museum of the Archæological Society at Columbus. Among these were artefacts of Lake Superior copper (and some pieces of native silver), large shells from the Gulf, galena, obsidian, and much mica, both in "blocks" and cut into ornaments, all showing the same range of aboriginal commerce as already described at Hopewell. In reference to pearls, the following are the principal observations:

Beads made from Unio pearls were very abundant everywhere in the Harness mound, as also beads of shell. They are found in such position as to show that they were strung and worn around the neck or wrists. One burial (No. 100) had some 2100 pearl beads, all rather small, and some of them perfectly round. Several hundred were obtained, however, that ranged from one quarter to one half an inch in diameter. A number of these are shown of natural size. The larger pearls, instead of being bored through for beads, are frequently somewhat flattened by grinding, and then pierced with two holes so as to attach them to a fabric. Very large ones were sometimes set in copper,—a style of work never observed before. Mr. Mills says of this: "Large and select pearls were flattened upon one side by grinding, and then placed upon a circular disk of copper a little larger than the pearl. The edges were then turned (up) around the pearl, holding it in place. Not only were pearls set in this way, but various pieces of shell cut in a circular form." Fine examples of this unique style of jewelry, of natural size, and another copper setting of like character, from which the pearl has been lost, are shown in plates facing pages 499 and 510.

More curious still is the discovery of imitation pearls, made of clay, and apparently modeled from real ones as they reproduce all the irregularities of form of the true pearls. They could easily have been made more nearly spherical, as the beads cut from shell are so regular as to look as though made by machinery. These somewhat irregular clay imitations, found with the genuine pearls, were first coated with a pulverent mica and then burned so as to preserve a pearly appearance.

Other forms of art work were abundantly represented in the Harness mound, such as carvings and decorations in stone and bone; a variety of textile fabrics, of which remnants are preserved when they were in contact with plates of copper,. the salts of the metal having penetrated the fabric and prevented its entire decay; very skilful work in copper, and to some extent in native silver and meteoric iron; and numerous fragments of pottery, more or less ornamental with simple impressed patterns. The "culture," as a whole, appears to have been equal, and very similar, to that of the Hopewell community, and these are regarded as having been the most advanced among the Ohio mound-builders; while the term "Fort Ancient culture" is applied to a somewhat lower grade in the matter of arts, which has its chief illustration among the builders and occupants of that celebrated work. By such researches, thus minutely and systematically conducted, there is now beginning to be possible something like a classification of these ancient unknown tribes, which will doubtless be developed more fully, as investigation shall be extended and its results combined and compared.

As to pearls in the mounds of Illinois, we are informed by the veteran archæologist, Dr. J. F. Snyder, that in 1889 he found the skeletons of three adult Indians at the base of a small mound on the bluffs of the Sangomon River in Cass County. These skeletons were in a squatting posture; artefacts—such as greenstone celts, a bicave stone and a heavy pipe—had only been deposited with one of them. Around each wrist and ankle of this skeleton were perforated beads made from *Marginella* shells, and resting on the sternum was a solitary pearl which had evidently formed the center of a necklace of the same small marine shells. Although much decayed, it still retained something of its original luster. It was spherical, measured approximately seven eighths of an inch in diameter, and was perforated through the middle. Dr. Snyder also states that at the base of one of the large mounds he opened in 1895, in Brown County, on the west side of the Illinois River, he discovered a number of the large canine teeth of the bear, perforated at the roots, so as to be used for necklaces. On the convex side of each tooth were from two to four pits about one third of an inch in diameter, and the same in depth, in which gems had been inserted. Two small pearls were still in place. Near by were the remains of another necklace composed of alternate pearls and bone beads; the latter were oblong and perforated lengthwise. Eight of the pearls were recovered, ranging in diameter from one half to one third of an inch, and pierced through the center, but all were very badly injured by the action of fire.

Mr. David I. Bushnell, who has excavated the McEvers mound in Montezuma, Pike County, Illinois, for the Missouri Historical Society, found in this mound a cyst containing a skeleton six feet in height and also a skull reposing on a bundle of bones near which lay forty-five pearls, one of them weighing fifty-two grains and still showing a beautiful luster. Almost all the objects discovered in the mound will be presented to the Missouri Historical Society. The large pearl would be worth from $12,000 to $15,000 if it were in perfect condition.

We learn from Mr. Richard Herrmann, founder of the Herrmann

Museum of Natural History, Dubuque, Iowa, that on the top of the high cliff from Eagle's Point to its end at McKnight's Spring, there were formerly a great many mounds which were long ago examined by government experts. Many ancient ornaments were found in these mounds, among them a string of pearls, greatly damaged from having been buried for a long period.[1] Mr. Herrmann believes that these pearls were taken from the Mississippi River by the mound-builders.

Enough has been said, in this general sketch, to give some idea of the extent to which pearls, largely those from the fresh-water Unios, were gathered and used by the native tribes of North America, from the ancient mound-builders of the Ohio Valley to the Indians encountered by the explorers and colonists of the fifteenth and sixteenth centuries.

The love of pearls shown by the Indians was as noteworthy as was their devotion to their dead and the superstitious mystery which enshrouds their funeral rites; for, when the human sacrifice was consummated, the act was performed in as earnest a spirit of devotion as was shown by Abraham in his readiness to sacrifice Isaac, and the Indians evidenced an almost pathetic sentiment either of reverence, duty, or supernatural dread.

Dr. J. Walter Fewkes writes that in none of his excavations has he ever noted pearls. Haliotis shells, conch shells, and fragments of the same have been found in the great ruins at Casa Grande, Arizona.

Dr. Charles Hercules Read, director of the Department of Archæology of the British Museum, states that the Mexican mosaic masks in the Christy collection, which are pre-Columbian in origin, and probably date hundreds of years in advance of the conquest, prove of special interest from the fact that five of them contain an inlay of mother-of-pearl shell. The first of these is a plain mask in which the eyes are of mother-of-pearl; the second is a dagger having the details of feather-work in mother-of-pearl; the third, a circular shield center having the eyes, teeth, fingers, and toes of the figures in mother-of-pearl; the fourth, a helmet with small pieces of pearl-shell representing collars around the necks of rattlesnakes; and the fifth is a jaguar in the side of which are similar inlays. These masks are described by Dr. Read in "Archæologia," Society of Antiquaries, London, Vol. LIV, p. 383; in this volume the objects are shown in color. Dr. Read communicates that the pearl jaguar seems to be of more recent execution, but he believes the first four to be original. He is not entirely sure that these objects contain the true mother-of-pearl, the substance having changed so much as to make a decision doubtful even if it were

[1] Herrmann, "Mound-builders of the Mississippi Valley," pp. 92, 93.

extracted. He states, however, that it is a pearly, nacreous shell, resembling that of the ordinary pearl-oyster. In these masks are also other shells, among them a red shell, probably a spondylus, almost as red as coral. The mother-of-pearl is of special interest as it is quite possible that the shell itself was known, and it may be that pearls also formed part of a commerce that existed between the coast and the interior.

We are informed by Mr. E. P. Dieseldorf, of Coban, Republic of Guatemala, that he has never observed pearls in the pre-Columbian graves in Guatemala; he had, however, frequently found marine shells, whole, and elaborated in connection with jadeite beads.

In a personal communication, Mr. Thomas Gann, of Yucatan, states that, in excavating a mound at San Antonio, near the mouth of the Rio Hondo, in Yucatan, he uncovered a small stone cyst or chamber, containing two perforated, pear-like ornaments of considerable size, together with portions of a human skeleton, painted pottery, etc. He also states that ornaments such as beads, gorgets, and ear-pendants, made from the pearly shell of both the oyster and the conch, are of common occurrence in many sepulchral mounds in British Honduras and in Yucatan, and he notes the fact that pink conch pearls are found in considerable numbers at the present day along the coast of British Honduras. There is no especial fishing for pearls, and they are found only incidentally in conchs which have been gathered for food. These pearls are sold by fishermen in Balize at prices varying from two or three dollars to twenty or thirty apiece. In size they range from that of a large pin's head to that of a small pea.

Mrs. Marie Robinson Wright informs us that she has never found pearls in the Bolivian graves, although they are quite plentiful in Bolivia to-day, and hundreds of them are offered in the markets. The pretty girls wear them as earrings and in their *topos*.

There is no doubt that pearls existed long before the advent of man, both in the fresh-water and in the marine form. This is more clearly evidenced by Sir Charles Lyell, who calls attention to the fact that the fresh-water mussel (*Unio littoralis Gray*), formerly found in abundance at Grays Thurrock, Essex, no longer exists in England, but occurs in France, showing that not only had this mollusk been unseen by any Englishman, but that the form had become extinct in an entire country. Thus, both the pearl shell of the ocean and the pearl-mussel of the river, for many centuries produced pearls, which passed away with the shell itself.

A great number of fossil Unios were collected by Barnum Brown from the Laramie clays, 130 miles northwest of Miles City, Montana.

The shells were found in a bed situated about 180 feet above the Fort Pierre shales and, therefore, well above the recognized cretaceous strata. These shells were in fairly good condition and retained the nacreous coloring to a considerable extent. As some of them resemble the modern species, it seems that the same designations might be applied to them.

Prof. R. P. Whitfield, one of our greatest palæontologists, who has carefully examined these fossil shells, suggests that they are probably the progenitors of the species of Unios and fresh-water mussels that now inhabit the Mississippi and Wisconsin rivers and their tributaries, and he proposes the following names for some of them, indicating at the same time the living species with which he compares them: *Unio biæsopoides, Unio æsopoides* and *Unio æsopiformis,* all resembling *U. æsopus* Green; *Unio letsoni* = *U. cornutus* Barnes; *Unio cylindricoides* = *U. cylindricus* Say; *Unio gibbosoides* = *U. gibbosus* Barnes; *Unio pyramidatoides* = *U. pyramidatus* Lea; *Unio retusoides* = *U. retusus* Lam.; *Unio verucosiformis* = *U. verrucosus* Barnes.

Although it is almost certain that these ancient Unios were pearl-bearing, Professor Whitfield informs us that, in a period of fifty years of palæontological research, he has never found a fossil pearl.

We are informed by Sophus Müller, Director of the Royal Danish Museum of Antiquities at Copenhagen, that no Danish ornaments containing pearls have been found dating from an earlier period than 1000 B.C.; he also states that no fresh-water pearls have ever been discovered in the Danish graves.

Dr. H. Ulmann, director of the great Swiss Landesmuseum at Zurich, and Dr. Otto Leiner, director of the Rosengarten Museum at Constance, personally communicated to us that no pearls exist in either of the collections of these great museums, nor to their knowledge have any been discovered in the lake-dwellings or the prehistoric graves of either Switzerland or Baden. This may either be due to conditions favorable to the dissolution of the pearl by the action of the ooze on the lake bottom, or else to the entire absence of knowledge of them on the part of a people who were familiar with many materials, since the museum collections even show jade implements of a number of types.

Dr. Leiner, whose father was curator of the Rosengarten Museum before him, informs us that at Bodman on Lake Constance there were found a large number of bored cylinders, from one fourth of an inch to one inch in length, made out of limestone. They were used for necklaces, somewhat in the style of our Indian wampum, and were either worn alone or in connection with bored cylinders made of the tuff-rock and also of encrinite stems.

Dr. Leiner also asserts that he has never seen *Unio margaritifera* in Lake Constance; nor was there any evidence of shells, broken or otherwise, observed by him in the excavations in the lake-dwellings.

The curator of the Rhodesia Museum, Bulawayo, South Africa, states that in Rhodesia, in the vicinity of Bulawayo, beads made out of the shell of the common Unio or fresh-water mussel (*Unio verreauxi*) have been observed in the graves, although pearls themselves have never been found with them in any burials.

ADDENDA

ONE of the authors used every endeavor in 1893 and 1894 to have a bill passed by Congress for the regulation of pearl-fishing in the United States. These efforts were frustrated by the influence of the local pearl-fishers. An attempt has now been made to preserve the industry in Illinois, where the legislature has this spring passed a bill for its regulation.

The first section of the bill provides:

It shall be unlawful for any person or persons to take or catch, by any means whatever, in any of the navigable waters within the jurisdiction of this State, any mussel, fresh-water clam or shell-fish from the first day of October to the first day of April (both dates inclusive) of each succeeding year.

The bill imposes upon any one who violates these provisions a fine of not less than $25, nor more than $100, or imprisonment in the county jail for a term not exceeding one year, or else both fine and imprisonment at the discretion of the court.

Another section provides that any one not a resident of Illinois, who takes clams, shell-fish, or mussels, without procuring a license, shall be subject to a fine of not less than $50, nor more than $100, or to imprisonment for one year, or to both penalties. The licenses may be procured on application and payment of $50 for each vessel to be employed, and they expire on the first day of October following their issuance. The amount received for these licenses is to be turned over to the State Treasurer at the end of each month and placed to the credit of the State Fish Protective Fund. No boat having more than two bars, each not exceeding sixteen feet in length, shall be used for this fishery, and the space separating the hooks on these bars is not to be less than eight inches.

Miss Carl, the artist who painted the portraits of the Empress and that of the Dowager Empress of China, states that she wears a diamond ring. When she shows this she apologizes for wearing it, stating that it had been given to her by the Viceroy, Li Hung Chang, saying that she, herself, sees no beauty in the sparkle of the diamond; for her there is more beauty in the soft, quiet tones of the pearl than in the brilliancy of the diamond.

During the Boxer War in China, the looting was carried on to so great an extent, that a French hotel-keeper is said to have obtained a basket of pearls, which he bought for a trifle, and which are said to have netted him very nearly $1,000,000.

BIBLIOGRAPHY

BIBLIOGRAPHY[1]

ALBERTUS MAGNUS, B*h* *hop of Ratisbon*
Le libro de le pietre preciose e de le loro u(ir)tude (Old Italian version of the De lapidibus). 14th century MS., vellum, 44ff. 4to. In the library of George F. Kunz.
The Secrets of Albertus Magnus of the Vertues of Hearbs, Stones and Certaine Beasts. *London,* 1617. 12mo.

ANGLERIUS (PETRUS MARTYR)
De orbe nove decades (8). *Compluti,* 1530. Fol.
The Historie of the West Indies. Hakluyt's Voyages. VOL. V. *London,* 1809. 4to.
The Decades of the New Worlde. Translated by Richard Eden. Edited by Edward Arber. *Birmingham,* 1885. 4to.

ANSON (GEORGE), *Baron*
A Voyage Round the World, in the Years MDCCXL, I, II, III, IV. *London,* 1748. 4to.

ATHENÆUS, *Naucratita*
Deipnosophistarum libri quindecim. *Gr.* and *Lat.* 14 vols. *Argentorati,* 1801–07. 8vo.
The Deipnosophists, or Banquet of the Learned, of Athenæus. 3 vols. *London,* 1854. 8vo.

BACCI (ANDREA)
De gemmis et lapidibus pretiosis. *Francofurti,* 1543. 8vo.

BALFOUR (EDWARD)
Article "Pearls." The Cyclopædia of India. Third edition. *London,* 1885. 8vo.

BARBOSA (DUARTE)
A Description of the Coasts of East Africa and Malabar in the Beginning of the 16th Century. *London,* Hakluyt Society, 1866. 8vo.

BARRERA (A. DE), *Madame*
Gems and Jewels: Their History, Geography, Chemistry, and Ana. *London,* 1860. 8vo.

BARTHEMA (LODOVICO DI)
The Travels of Lodovico di Varthema in Egypt, Syria, Arabia Deserta and Arabia Felix, in Persia, India, and Ethiopia, A.D. 1503 to 1508. *London,* Hakluyt Society, 1863. 8vo.

BAUER (MAX)
Precious Stones; a Popular Account of Their Characteristics and Applications, With an Appendix on Pearls and Coral. Translated from the German with additions by L. J. Spencer. *London,* 1904. 8vo.

BAUGNIET (HENRI DE)
Description historique et scientifique de la Collection de Pierres précieuses. *Bruxelles,* 1847. 12mo.

BECKMANN (JOHANN)
Beyträge zur Geschichte der Erfindungen. 2 vols. *Leipzig,* 1782–1805. 8vo.

BELLEAU (REMI)
Les amours et nouveaux éschanges des pierres précieuses: vertus et proprietez d'icelles. Discours de la vanité, pris de l'Ecclésiaste de Salomon. *Paris,* 1576. 4to.

BENJAMIN (BEN JONAH) *of Tudela*
The Itinerary of Rabbi Benjamin of Tudela; Translated and Edited by A. Asher. 2 vols. *London,* 1840–41. 12mo.

BENT (THEODORE AND MRS. THEODORE)
Southern Arabia. *London,* 1900. 8vo.

BENZONI (GIROLAMO)
Novæ novi orbis historiæ; id est rerum ab Hispanis in India occidentali hactenus gestarum . . . libri tres, U.

---

[1] A large number of these works are in the library of George F. Kunz.

Calvetonis opera ex Italicis Hieronymi Benzonis. *Geneva,* 1578. 8vo.
History of the New World; Showing his Travels in America, from A.D. 1541 to 1556. *London,* Hakluyt Society, 1857. 8vo.

BERQUEN (ROBERT DE)
Les Merveilles des Indes Orientales et Occidentales, ou nouveau. Traité des Pierres précieuses, et des Perles. *Paris,* 1661. 8vo.

BERTHELOT (M.)
Traitement des perles. Collection des anciens Alchimistes Grecs, pp. 352–356. *Paris,* 1888.

BEUMENBERGER (JOHANN GOTTLIEB)
Der Vollkommene Juwelier. *Weimar,* 1828. 8vo.

BIEDERMANN (WILHELM)
Untersuchungen über Bau und Entstehung der Molluskenschalen, Jenaische Zeitschrift für Naturwissenschaften, VOL. XXXVI, pp. 1–164. *Jena,* 1901.

BION (JEAN MARIE)
Inventaire des Diamans de la Couronne, Perles, Pierreries, Tableaux, . . . existans au Garde-Meuble . . . fait . . . par ses commissaires MM. Bion, Christin et Delattre. *Paris,* 1791. 8vo.

BIRON (C. C.)
Curiositez de la Nature et de l'Art, aportées dans deux Voyages des Indes. *Paris,* 1703. 12mo.

BLACKSTONE (A. G.)
British Pearls. "Belgravia Magazine," VOL. IX, pp. 343–350. *London,* September, 1869.
Also in "Eclectic Magazine," VOL. X, pp. 604–608. *London,* November, 1869.

BLAKE (HENRY A.), *Sir*
Leasing of the Pearl Fisheries of Ceylon. Sessional Papers, pp. 327–336. *Colombo,* 1906. Fol.

BLANCO (GUZMAN), *President*
Apuntes Estadísticos del Estado Nueva Esparta. *Caracas,* 1876. 8vo.

BOCHART (SAMUEL)
Hierozoicon, sive Bipartitum opus de animalibus Sacræ Scripturæ. *Londini,* 1663. Fol.

BOËTHIUS (ANSELMUS)
Gemmarum et lapidum historia. *Hanoviæ,* 1609. 4to. Recensuit et commentariis illustravit Adrianus Tollius. *Leyden,* 1636. 4to.
Tollius' 3rd edition (1647) appends the Greek text of Theophrastus and a short work, "De Gemmis et Lapidibus," by Johannis de Laet of Antwerp.

BOHADSCH (JOANNES BAPTISTA)
De quibusdam animalibus marinis. *Dresdæ,* 1761. 4to.

BOHLEN (PETER VON)
Das alte Indien, mit besonderer Rücksicht auf Aegypten. *2* vols. *Königsberg,* 1830. 8vo.

BONNEMERE (LIONEL)
Les Perles fines de l'Ouest de la France. Revue des Sciences Naturelles de l'Ouest. VOL. III, pp. 97–99. *Nantes,* 1893.
Les Mollusques des eaux douces de France et leurs perles. Institut International de Bibliographie Scientifique. *Paris,* 1901. 8vo.

BOUCHON-BRANDELEY (G.)
La Pêche et la Culture des Huîtres Perlières à Tahiti; Pêcheries de l'Archipel Tuamotu. Journal Officiel, June 23, 25, 26, and 27. *Paris,* 1885. Translated in Report U. S. Fish Commission, 1885, pp. 353–377.

BOUTAN (LOUIS)
Production artificielle des perles chez les *Haliotis.* Comptes Rendus de l'Académie des Sciences, VOL. CXXVII, pp. 828–830. *Paris,* 1898.
L'Origine réelle des perles fines. Comptes Rendus de l'Académie des Sciences, VOL. CXXXVII, pp. 1073–1075. *Paris,* 1903.

BOWERBANK (JAMES SCOTT)
On the Structure of the Shells of Molluscous and Conchiferous Animals. Transactions of the Microscopic Society, VOL. I, pp. 123–154. *London,* 1844.
Also in Froriep, Notizen aus dem

Gebiete der Natur und Heilkunde, VOL. XXV, 1843.

BREWSTER (DAVID), *Sir*
On the New Properties of Light, Exhibited in the Optical Phenomena of Mother-of-Pearl. Philosophical Transactions of the Royal Society of London, 1814, VOL. II, pp. 397–418. Also in Journal de Physique, VOL. LXXXI, pp. 181–188, 471–473, 1815.
A Treatise on Optics. *London,* 1831. 8vo.

BRODIE (JAMES)
The Pearls of the Ythan, Aberdeenshire. Proceedings of the Royal Physical Society, VOL. III, pp. 394–396. *Edinburgh,* 1867.

BRUCE (JAMES)
Select Specimens of Natural History, Collected in Travels to Discover the Source of the Nile, in Egypt, Arabia, Abyssinia, and Nubia. *Dublin,* 1790. 8vo.

BRYDGES (HARFORD JONES), *Sir*
Mission to the Court of Persia, in the years 1807–11. 2 vols. *London,* 1834. 8vo.

BUCKLAND (FRANCIS TREVELYAN)
Log-book of a Fisherman and Zoologist. *London,* 1875. 8vo.

BUFFUM (E. GOULD)
Pearl Diving [Gulf of California]. "The Pioneer, or California Monthly Magazine," VOL. I, pp. 35–39. *San Francisco,* January, 1854.

BURNHAM (S. M.)
Precious Stones in Nature, Art and Literature. *Boston,* 1886. 8vo.

C. (D.)
Some Account of the British Pearl Fishery now Existing on the Conway. "Loudon's Magazine of Natural History," VOL. III, pp. 132–134. *London,* 1830.

CAIRE (A.)
La Science des pierres précieuses, appliquée aux arts. *Paris,* 1826. 8vo.

CALVERT (ALBERT FREDERICK)
Pearls: their Origin and Formation. *London,* 1892. 8vo.

CAMDEN (WILLIAM)
Britannia; or A Chorographical Description of Great Britain and Ireland, Together With the Adjacent Islands. 2 vols. *London,* 1722. Fol.

CARPENTER (WILLIAM B.)
On the Microscopic Structure of Shells. Report of the British Association for the Advancement of Science, September, 1844, pp. 1–28. *London,* 1845.
The Microscope and its Revelations. Sixth edition. *London,* 1881. 8vo.

CATALOGUE
Diamants, Perles et Pierreries provenant de la collection dite des Joyaux de la Couronne. *Paris,* Imprimerie National, 1887. Fol.

CATTELLE (W. R.)
Precious Stones: a Book of Reference for Jewelers. *Philadelphia,* 1903. 8vo.
The Pearl. *Philadelphia,* 1907. 8vo.

CHAMBERS'S JOURNAL
Pearls and Pearl Fisheries in the Persian Gulf. "Chambers's Journal," VOL. XIX, pp. 157–160. *Edinburgh,* March 5, 1853.
Fishing for Pearls. "Chambers's Journal," VOL. LV, pp. 87–90, February 9, 1878.

CHAPPUZEAU (LE SIEUR)
Histoire des Joyaux, et des principales Richesses de l'Orient et de l'Occident. *Genève,* 1665. 12mo.
The History of Jewels and of the Principal Riches of the East and West. *London,* 1671. 12mo.

CHARDIN (JEAN)
Voyages en Perse, et autres lieux de l'Orient. 10 vols. *Amsterdam,* 1711. 12mo.

CHARTIER (H. LE)
Tahiti et les Colonies Françaises de la Polynésie. *Paris,* 1887. 8vo.

CHURCH (ARTHUR HERBERT)
Precious Stones Considered in Their Scientific and Artistic Relations. South Kensington Museum Handbook. *London,* 1883. 8vo.

CLAVE (ÉTIENNE DE)
Paradoxes, ou traittéz Philosophiques des pierres et pierreries, contre l'opinion vulgaire. *Paris,* 1635. 8vo.

CLAVIO (D. SERVATIO)
Piscatura margaritarum. Miscellanea Curiosa sive Ephemeridum Medico-Physicarum Germanicarum Academiæ, pp. 417–419. *Norimbergæ,* 1685.

CLOQUET (J.) AND MOQUIN-TANDON (M.)
Observations sur les perles des bivalves d'eau douce. Bulletin de la Société d'Acclimatation, VOL. v, pp. 452–461. *Paris,* 1858.

COATE (H. E. ACRAMAN)
Pearl Fishing [Ceylon]. "Time Magazine," VOL. XXI, pp. 646–653. *London,* December, 1889.

COLLETT (OLIVER)
Pearl-oysters and Pearl Fisheries. Journal of the Ceylon Branch of the Royal Asiatic Society, VOL. XVI, pp. 165–197. *Colombo,* 1901. 8vo.

COMBA (B.)
La Madreperla. *Torino,* 1898. 8vo.

COMBER (E.)
Economic Uses of Shells. Journal of the Bombay Natural History Society, VOL. XVI, pp. 462–472. *Bombay,* 1905.

COMYN (TOMÁS DE)
State of the Philippine Islands. *London,* 1820. 8vo.

CORDINER (JAMES)
A Description of Ceylon, Containing an Account of the Country, Inhabitants and Natural Productions. 2 vols. *London,* 1807. 4to.

CORNHILL MAGAZINE
The Pearl Harvest. "Cornhill Magazine," VOL. XIV, pp. 161–173. *London,* August, 1866.

COUTANCE (A.)
La Perle. Diamant et Pierres Précieuses; . . . au point de vue de leur histoire et de leur travail . . . Par E. Jannettaz, . . . A. Coutance. *Paris,* 1881. 8vo.

COXE (DANIEL)
A Description of the English Province of Carolana, by the Spaniards call'd Florida, and by the French La Louisiana, as also of the great and famous river Meschacebe or Missisipi. *London,* 1722. 8vo.

CURZON (GEORGE NATHANIEL), *Baron*
Persia and the Persian Question. 2 vols. *London,* 1892. 8vo.

D. (M. L. M. D. S.)
Le dénombrement, facultéz et origine des pierres précieuses. *Paris,* 1667. 8vo.

DALL (W. H.)
Pearls and Pearl Fisheries. "American Naturalist," VOL. XVII, pp. 579–586 and 731–745. *Philadelphia,* June, 1883, and July, 1883.

DALLAS (W. S.)
On the Natural History of the Cingalese Pearl-oyster and on the Production of Pearls. Annals and Magazine of Natural History, Ser. 3, VOL. I, pp. 81–90. *London,* February, 1858.

DAMPIER (WILLIAM)
A New Voyage Round the World. 2 vols. *London,* 1699–1703. 8vo.

DARBOUX (G.)
L'Industrie des Pêches aux Colonies. Les Produits de la Pêche. *Marseille,* 1906. 4to.

DASTRE (A.)
Les Perles fines; Production naturelle et production artificielle. Revue des Deux Mondes, VOL. CLI, pp. 671–690. *Paris,* 1899.

DAVENPORT (CYRIL JAMES)
The English Regalia. *London,* 1897. 4to.

DELONDRE (AUGUSTIN)
Nacroculture et ostréiculture perlière aux îles Pomotu (Océanie). Bulletin de la Société National d'Acclimatation, VOL. III, pp. 389–390. *Paris,* 1876.

DENNYS (N. B.)
Breeding Pearls. Journal of the Straits Branch of the Royal Asiatic Society, VOL. I, pp. 31–37. *Singapore,* July, 1878.

DICKENS (CHARLES), *Editor*
Pearl Fishery [Ceylon, 1836]. "Household Words," VOL. III, pp. 75–80. *London*, April 12, 1851.
Pearls of Price. "All the Year Round," VOL. XVII, pp. 534–537. *London*, June 1, 1867.
Scotch pearls. "All the Year Round," VOL. XXI, pp. 125–127, January 9, 1869. Also in "Every Saturday," VOL. VII, pp. 157–158. *Boston*, January 30, 1869.

DIEULAFAIT (LOUIS)
Diamonds and Precious Stones; a Popular Account of Gems. Translated from the French . . . by F. Sanford. *Cambridge, U. S. A.*, 1874. 8vo.

DIGUET (LÉON)
Pêche de l'huître perlière dans le Golfe de Californie. Bulletin de la Société Centrale d'Aquiculture. VOL. VII, pp. 1–18. *Paris*, 1895.
Sur la formation de la perle fine cnez la *Meleagrina margaritifera*. Comptes Rendus de l'Académie des Sciences, VOL. CXXVIII, pp. 1589–91. *Paris*, 1899.
Exploitation de l'huître perlière dans le Golfe de Californie. Bulletin de la Société Centrale d'Aquiculture, VOL. IX, pp. 221–235. *Paris*, 1899.

DOBRZENSKY (JACOBUS JOANNES WEN-CESLAUS)
Perlarum maturationis historia. Miscellanea Curiosa Medico-physica, obs. 183, pp. 281–282. *Jenæ*, 1671.

DONNAN (JAMES)
Report on the Inspection of Pearl Banks, 1875. Sessional Papers. *Colombo*, 1875. Fol.
Report on the Inspection of the Pearl Oyster Banks, 1876. Sessional Papers. *Colombo*, 1876. Fol.
Report on the Inspection of the Pearl Oyster Banks, 1878. Sessional Papers. *Colombo*, 1878. Fol.
Report on a Recent Inspection of the Pearl Banks. Sessional Papers. *Colombo*, 1880. Fol.
Report on the Pearl Fishery off Chi-law. Sessional Papers. *Colombo*, 1884. Fol.
Report of an Experimental Cultivation of Pearl-Oysters. Sessional Papers. *Colombo*, 1885. Fol.
Report on a Recent Inspection of the Pearl Banks. Sessional Papers. *Colombo*, 1885. Fol.
Report of a Recent Inspection of the Pearl Banks. Sessional Papers. *Colombo*, 1886. Fol.
Report of an Inspection of the Pearl Banks, 1887. Sessional Papers *Colombo*, 1888. Fol.
Report on the Failure of Oysters on the Cheval Paar, 1887. Sessional Papers. *Colombo*, 1888. Fol.

DONNAN (JAMES) AND TWYNAM (W. C.)
Reports by the Superintendent of the Fishery and the Inspector of the Pearl Banks. Sessional Papers. *Colombo*, 1887. Fol.

DOUMERT (A.)
Nos Parures. Le Jais et les Perles Fausses. *Paris*, 1890. 12mo.

DUBOIS (RAPHAËL)
Sur le mécanisme de la formation des perles fines dans le *Mytilus edulis*. Comptes Rendus de l'Académie des Sciences, VOL. CXXXIII, pp. 603–605. *Paris*, 1901.
Sur la nature et la formation des perles fines naturelles. Mémoires et Comptes Rendus des Séances du Congrès International d'Aquiculture et de Pêche. *Paris*, 1903. 8vo.
Sur l'acclimatation et la culture des *Pintadines*, ou huîtres perlières vraies, sur les côtes de France, et sur la production forcée des perles fines. Comptes Rendus de l'Académie des Sciences, VOL. CXXXVII, pp. 611–613. *Paris*, 1903.
Application des rayons X à la recherche des perles fines. Comptes Rendus de l'Académie des Sciences, VOL. CXXXVIII, pp. 301–302. *Paris*, 1904.
Sur les perles de nacre. Comptes Rendus de l'Académie des Sciences, VOL. CXXXVIII, pp. 583–584. *Paris*, 1904.

Sur le mécanisme sécrétoire pro-
ducteur des perles. Comptes Rendus
de l'Académie des Sciences, Vol.
CXXXVIII, pp. 710–712. *Paris, 1904.*

Dutens (Louis)
Des pierres précieuses et des perles
fines, avec les moyens de les con-
noître et de les évaluer. *Londres,
1776.* 12mo.

Eberhard (Johann Peter)
Abhandlung von dem Ursprung der
Perle, worin deren Zeugung, Wachsthum und Beschaffenheit erklärt, und
eine Nachricht von verschiedenen
Perlenfischereien gegeben wird. *Halle,
1751.* 8vo.

Edouard-Petit (Alix)
Aux îles des perles. Journal le Cor-
respondent, pp. 977–996. *Paris,*
March 10, 1906.

Elgin, *Earl of*
Leasing of the Pearl Fisheries of
Ceylon. Sessional Papers, pp. 650–
651. *Colombo, 1906.*

Emanuel (Harry)
Diamonds and precious stones. *London, 1865.* 8vo.

Entrecolles (F. X. d')
Manière de faire des perles artifi-
cielles. Lettre à Pekin, 4 Nov. 1734.
Lettres édifiantes et curieuses écrites
des Missions Étrangères, Vol. XXII,
pp. 425–437. *Paris, 1736.*

Esteva (José Maria)
Memoria sobre la pesca de la perla en
la Baja California. Boletin de la
Sociedad Mexicana de Geografía y
Estadística, Vol. X, pp. 673–697.
Mexico, 1865.

Farrington (Oliver Cummings)
Gems and Gem Minerals. *Chicago,*
1903. 8vo.

Ferguson (Alastair Makenzie and
John)
All about Gold, Gems, and Pearls in
Ceylon and Southern India . . . Sec-
ond edition. *Colombo, 1888.* 8vo.

Fernandes de Queiros (Pedro)
The Voyages of Pedro Fernandez de

Quiros, 1595–1606. Translated and
edited by Sir Clements Markham. 2
vols. *London,* Hakluyt Society, 1904.
8vo.

Feuchtwanger (Lewis)
A Popular Treatise on Gems, in Ref-
erence to Their Scientific Value.
New York, 1859. 8vo.

Fichtner (Conrad Heinrich)
Einige Nachrichten von dem Rehau-
ischen Perlen Bach. Frankische
Sammlungen von Anmerkungen aus
der Naturlehre Arzneigelehrtheit.
Nurnberg, 1768. 8vo.

Fields (J. T.)
Underbrush. *Boston, 1881.* 8vo.

Filippi (Filippo de)
Sull' origine delle perle. Il Cimento
revista di Scienza, Lettere, ed Arti,
Vol. I, pp. 429–439. *Torino, 1852.*
Translated into German by Dr. Küch-
enmeister in Müllers' Archiv für
Anatomie, Physiologie, und wissen-
schaftliche Medicin, 1856, pp. 251–
269. *Berlin, 1856.*
Mémoire pour servir à l'histoire géné-
tique des Trématodes. Memorie della
Reale Accademia delle Scienze di
Torino, Vol. XV, pp. 331–358, 1855;
Vol. XVI, pp. 419–442. *Torino,*
1857.
Also in Annales des Sciences Natu-
relles, Vol. II (Zool.), pp. 254–284.
Paris, 1854. Nouvelles observations
sur le développement des Trématodes.
Annales des Sciences Naturelles, III
(Zool.), pp. 111–113. *Paris, 1855.*
Encore un mot sur la formation des
perles. Müllers' Archiv für Anato-
mie, Physiologie, und wissenschaft-
liche Medicin, 1856, pp. 490–493.
Berlin, 1856.
Troisième mémoire pour servir à l'his-
toire génétique des Trématodes. Memorie della Reale Accademia delle
Scienze di Torino. Vol. XVIII, pp.
201–232. *Torino, 1859.*

Fischer (Gotthelf)
Essai sur la Pellegrina, ou la Perle
incomparable des Frères Zozima.
Moscou, 1818. 8vo.

FISCHER (P.)
Production artificielle des perles. Journal de Conchyliologie, VOL. XIII, pp. 64–65. *Paris, 1865.*

FRANK LESLIE'S MAGAZINE
Pearl-mussel Fishery of New Jersey. "Frank Leslie's Magazine," VOL. III, pp. 384–386. *New York,* May 23, 1857.

FRÉDÉ (PIERRE)
La Pêche aux Perles. *Paris, 1887.* 8vo.
La Pêche aux Perles en Perse et à Ceylon. *Paris, 1890.* 8vo.

FREDERICK (CÆSAR)
The Fishing for Pearls [in Ceylon, 1563–81]. Translated out of Italian by Thomas Hickoke. Hakluyt's Voyages, VOL. V. *Glasgow, 1904.* 8vo.

FRIEDLAENDER (S.)
Perlen der Juwelierausstellung. Internationale Fischerei-Austellung zu Berlin, 1880, pp. 75–83. *Berlin,* 1881. 8vo.

FRISWELL (J. H.)
Pearls and Oysters. "Once a Week," VOL. III, pp. 78–81. *London,* July 14, 1860.

FRYER (JOHN)
A new Account of East-India and Persia. *London, 1698.* Fol.

GANONG (W. F.)
Bibliography of the Fresh-water Pearl Fishery in New Brunswick. Bulletin of the Natural History Society of New Brunswick, No. XVII, pp. 134–136. *St. John, N. B.,* 1899.

GARNER (ROBERT)
On the Pearls of the Conway River, North Wales. British Association for the Advancement of Science, Report for 1856, PART II, pp. 92–93. *London, 1857.*
Notes on Anatomy and Pearls of *Alasmodon margaritifera.* Proceedings of the Zoological Society of London, VOL. XI, pp. 426–428, 1872.
On the Formation of British Pearls and Their Possible Improvement.

Journal of the Linnean Society (Zool.), VOL. XI, pp. 426–428. *London, 1873.*

GARRAN (ANDREW)
Australasia Illustrated. 3 vols. *Sydney, 1892.* 4to.

GEIGER (MALACHIAS)
Margaritologia, sive dissertatio de Margaritis, in qua, post varia ad Margaritas pertinentia, demonstratur, Margaritas bavaricas, in usu medicinali, viribus et effectibus æquivalere orientalibus et occidentalibus. *Monachii, 1637.* 12mo.

GEMELLI-CARERI (GIOVANNI FRANCESCO)
Giro del mondo. Nuova edizione, accresciuta, ricorretta, e divisa in nove volumi. *Venezia, 1719.* 8vo.

GIARD (ALFRED)
L'épithélium sécréteur des perles. Comptes Rendus de la Société de Biologie, séance du 29 decembre, 1903, VOL. LV, pp. 1222–1225. *Paris,* 1903.

GIBBINS (HERBERT JAMES)
Curiosities of Pearls. "Gentleman's Magazine," VOL. CCLXXVII, pp. 306–315. *London,* September, 1894.

GILL (WILLIAM WYATT)
Life in the Southern Isles; or Scenes and Incidents in the South Pacific and New Guinea. *London, 1876.* 8vo.

GILLMAN (HERBERT W.)
Valuation of Pearls in Ceylon. Journal of the Ceylon Branch of the Royal Asiatic Society, 1887, VOL. X, pp. 32–40. *Colombo, 1888.*

GIMMA (GIACINTO)
Della storia naturale delle gemme, delle pietre, e di tutti i minerali, ovvero della fisica sotterranea. 2 vols. *Napoli, 1730.* 4to.

GODRON (D.-A.)
Les perles de la Vologne, et le Château-sur-perle. Mémoires de l'Académie de Stanislas, 1869, pp. 10–30. *Nancy, 1870.*

GRAND (S.)
Méthode de culture de l'huître perlière

dans les lagons de Tahiti. Revue Maritime et Coloniale, VOL. CXXV, pp. 575–590. *Paris*, May, 1895.

GRAY (JOHN EDWARD)
On the Structure of Pearls, and on the Chinese mode of producing them of a large size and regular form. Thomson's Annals of Philosophy, New Series, VOL. IX, pp. 27–29. *London*, January, 1825; VOL. X, pp. 389–390, November, 1825.

GRIFFIN (G. W.)
The Pearl-shell Fisheries of Queensland. U. S. Consular Reports, No. LV. *Washington*, August, 1885. Also published in Bulletin U. S. Fish Commission, VOL. VI, pp. 433–435. *Washington*, 1887.

GRILL (JOHAN ABRAHAM)
Bericht wie die Chinesen ächte Perlen nachmachen. Abhandlungen der Königlichen Schwedischen Akademie der Wissenschaften for 1772, VOL. XXXIV, pp. 88–90. *Leipzig*, 1776.

GRIMM (HERMANNI NICOLAI)
De piscatura margaritarum apud Insulam Manaar, non procul à Ceylon sitam. Miscellanea Curiosa Medicophysicarum, obs. 36, pp. 99–107. *Norimbergæ*, 1685.

GROSS (J. G.)
De Margaritis, earumque Virtute Medica. *Wirceburgi*, 1744. 4to.

GRYLLS (JAMES WILLYAMS)
The Pearl Fishery. "The New Monthly Magazine," VOL. LXXXII, pp. 70–79. *London*, January, 1848. The Out-station; or Jaunts in the Jungle. *London*, 1848. 16mo.

GUIDO (JOANNIS)
De Mineralibus tractatus absolutissimus. *Francofurti*, 1627. 8vo.

GUILLEMAND (FRANCIS HENRY HILL)
Malaysia and the Pacific Archipelago. *London*, 1894. 8vo.

GÜNTHER (ALBERT)
A Small Fish of the Genus *Fierasfer* Imbedded in the Shell of *Margarita margaritifera*. Proceedings of the Zoölogical Society of London, 1886, pp. 318–320.

HAGUE (W. F.)
On the Natural and Artificial Production of Pearls in China. Journal of the Royal Asiatic Society of Great Britain and Ireland, VOL. XVI, pp. 280–284, 1856.

HALE (SARAH J.)
History of Pearls, Natural and Artificial. "Godey's Magazine," VOL. XLVIII, pp. 533–537. *Philadelphia*, 1854.

HALL (ANNA MARIA)
Pearls and Pearl Divers of the Gulf of California. "St. James's Magazine," VOL. II, pp. 289–295. *London*, October, 1861.

HALL (FRANCIS)
Colombia: its present state. *Philadelphia*, 1825. 12mo.

HAMILTON (ALEXANDER), *Captain*
A New Account of the East Indies. 2 vols. *Edinburgh*, 1727. 8vo.

HAMONVILLE (L. D')
Les moules perlières de Billiers (*Mytilus edulis*). Bulletin de la Société Zoologique de France, VOL. XIX, pp. 140–142. *Paris*, 1894.

HARDY (ROBERT WILLIAM HALL), *Lieut.*
Travels in the Interior of Mexico, in 1825, 1826, 1827, and 1828. *London*, 1829. 8vo.

HARLEY (GEORGE)
Microscopic Examination of Pearls. "The Cheltenham Ladies' College Magazine," No. XVII, pp. 37–42. *London*, 1888.

HARLEY (GEORGE AND HAROLD S.)
The Chemical Composition of Pearls. Proceedings of the Royal Society of London, VOL. XLIII, pp. 461–465, 1888. The Structural Arrangement of the Mineral Matters in Sedimentary and Crystalline Pearls. Proceedings of the Royal Society of London, VOL. XLV, pp. 612–614, 1889.

HARPER'S MAGAZINE
Pearls and Gems. "Harper's New Monthly Magazine," VOL. XXI, pp. 764–780. *New York*, November, 1860. Treasures of the Deep. "Harper's New Monthly Magazine," VOL. LVIII,

pp. 321–336. *New York*, February, 1879.

HARTING (PIETER)
Sur la production artificielle de quelques-unes des principales formations calcaires organiques. Comptes Rendus de l'Académie des Sciences, VOL. LXXIII, pp. 361–362. *Paris, 1871.* Also in Quarterly Journal Microscopic Science, VOL. XII, pp. 118–123. *London, 1872.*

HARTMANN (PHILIPP JACOB)
Exercitatio de generatione mineralium, vegetabilum, et animalium in aere. *Königsberg(?), 1689.* 4to.

ḤASAN IBN YAZĪD (ABŪ ZAID) *Al-Sīrāfī*
Ancient Accounts of India and China, by two Mohammedan travellers who went to those parts in the 9th Century. *London, 1733.* 8vo.

HAWKINS (RICHARD), *Sir*
The Observations of Sir Richard Hawkins in his Voyage into the South Sea, in the Year 1593. Reprinted from the edition of 1622. *London,* Hakluyt Society, 1847. 8vo.

HEEREN (ARNOLD HERMANN LUDWIG)
Historical Researches into the Politics, Intercourse and Trade of the Principal Nations of Antiquity. 3 vols. *Oxford, 1833.* 8vo.

HERDMAN (WILLIAM ABBOTT)
The Pearl Fisheries of Ceylon. Abstract of a discourse delivered at the Royal Institution on March 27, 1903. "Popular Science Monthly," VOL. LXIII, pp. 229–238, July, 1903. Also in Smithsonian Report for 1904, pp. 485–493. Presidential Address, 1905. Proceedings of the Linnean Society of London, 117th Session, pp. 20–30, October, 1905. Report to the Government of Ceylon on the Pearl-Oyster Fisheries of the Gulf of Manar. 5 vols. *London,* The Royal Society, 1903–1906. 4to. The Pearl Fisheries of Ceylon. Proceedings of the Royal Institution of Great Britain, VOL. XVII, pp. 279–287. *London,* March, 1905.

HERTZ (BRAM)
A Catalogue of the Collection of Pearls and Precious Stones formed by H. P. Hope, Esq. *London, 1839.* Fol.

HESSE (P.)
Die Perlfishcherei im Roten Meere. Zoologischer Garten, VOL. XXXIX, pp. 382–385. *Frankfurt-a.-M.,* December, 1898.

HESSLING (THEODOR VON)
Ueber Perlen und ihre Entstehung. Westermann's Illustrirten Monatsheften, 1857. Ueber die Ursachen der Perlbildung bei *Unio margaritifer.* Siebold und Kölliker, Zeitschrift für Wissenschaftliche Zoologie, VOL. IX, pp. 543–546. *Leipzig, 1858.* Die Perlmuscheln und ihre Perlen. *Leipzig, 1859.* 8vo. Die Verbreitung der Seeperlenmuschel und der Perlfischerei. Zeitschrift für die Gesammten Naturwissenschaften, VOL. XIV, pp. 17–32. *Halle,* 1859. Ueber die Befruchtung der Flussperlenmuschel. Siebold und Kölliker, Zeitschrift für Wissenschaftliche Zoologie, VOL. X, pp. 358–363. *Leipzig, 1860.*

HILL (JOHN)
Theophrastus's History of Stones, with the Greek Text and an English Version, and Notes Critical and Philosophical, Including the Modern History of Gems Described by that Author. *London, 1746.* 8vo.

HOLDSWORTH (EDMUND WILLIAM HUNT)
Report on the Conditions and Prospects of the Pearl Oyster Banks, 1868. *Colombo, 1868.* Fol.

HOME (EVERARD), *Sir, Bart.*
On the Production and Formation of Pearls. Philosophical Transactions of the Royal Society of London, VOL. CXVI, pp. 338–341. *London,* 1826.

HORNELL (JAMES)
Biological Results of the Ceylon

Pearl Fishery of 1904, with Notes on Divers and Their Occupation. *Colombo*, 1905. 4to.

Report on the *Placuna placenta* Pearl Fishery of Lake Tampalakamam. Ceylon Marine Biological Reports, PART II, VOL. I, pp. 41–54. *Colombo,* 1906.

Report on the Operations on the Pearl Banks During the Fishery of 1905. Ceylon Marine Biological Reports, PART II, VOL. I, pp. 55–79. *Colombo,* 1906.

HUMBOLDT (FRIEDRICH HEINRICH ALEXANDER VON), *Baron*

Personal Narrative of Travels to the Equinoctial Regions of the New Continent During the Years 1799–1804. 7 vols. *London,* 1814–29. 8vo.

IBN BATUTA

The Travels of Ibn Batuta. Translated from the original Arabic manuscript copies . . . by Samuel Lee. *London,* 1829. 4to.

IRVING (THEODORE)

The Conquest of Florida, under Hernando de Soto. 2 vols. *London,* 1835. 12mo.

ISSEL (ARTURO)

I Molluschi Commestibili, le Applicazioni delle Conchiglie, le Perle e i Coralli. Annali dell' Industria e del Commercio, 1880, num. 28. *Roma,* 1881. 8vo.

JAHN (J. G.)

Die Perlenfischerei im Voigtlande. *Oelsnitz,* 1854. 8vo.

JAMESON (H. LYSTER)

On the Identity and Distribution of the Mother-of-pearl Oysters. Proceedings of the Zoological Society of London for 1901, VOL. I, pp. 372–394.

On the Origin of Pearls. Proceedings of the Zoological Society of London for 1902, VOL. I, pp. 140–166.

The Formation of Pearls. "Nature," VOL. LXVII, pp. 280–282. *London,* January 22, 1903.

JARDINE (FRANK L.)

Report Relating to the Mergui Pearl and Pearl Shell Fisheries. *Rangoon,* 1894. Fol.

JEFFREYS (JOHN GWYN)

British Conchology, or an Account of the Mollusca Which now Inhabit the British Isles and the Surrounding Seas. 5 vols. *London,* 1862–69. 12mo.

JEFFRIES (DAVID)

Treatise on Diamonds and Pearls and the True Method of Manufacturing Diamonds. *London,* 1750. 8vo.

A Treatise on Diamonds and Pearls. Fourth edition, corrected to the present time. *London,* 1871. 12mo.

JONES (CHARLES COLCOCK)

Antiquities of the Southern Indians, Particularly of the Georgia Tribes. *New York,* 1873. 8vo.

JONES (WILLIAM)

History and Mystery of Precious Stones. *London,* 1880. 8vo.

JOPP (ALEXANDER H.)

Days with Industries: Adventures and Experiences Among Curious Industries. *London,* 1889. 8vo.

JORDANUS (CATALANI), *Bishop*

Mirabilia Descripta. The Wonders of the East, by Friar Jordanus (*circa* 1330). Translated by Col. Sir Henry Yule. *London,* Hakluyt Society, 1863. 8vo.

KÆMPFER (ENGELBERT)

The History of Japan. 2 vols. *London,* 1727. Fol.

KAWALL (J. H.)

La pêche des perles en Livonie. Annales de la Société Malacologique de Belgique, VOL. VII, pp. 38–46. *Brussels,* 1872.

KELAART (EDWARD FREDERICK)

Introductory Report on the Natural History of the Pearl-oyster of Ceylon. Proceedings of the Royal Society of Edinburgh, VOL. I, pp. 399–405, 1857.

Also in Madras Journal of Literature and Science, VOL. III, pp. 89–104, 1858.

Report on the Tablegam Pearl-oyster Fishery. Madras Journal of Literature and Science, VOL. III, pp. 105–110, 1858.
Introductory Report on the Natural History of the Pearl-oyster of Ceylon, 1858–59. *Trincomalee, 1859.*
Filaria in Oysters [Ceylon]. Proceedings of the Royal Physical Society, VOL. II, pp. 101–102. *Edinburgh, 1863.*

KELAART (EDWARD FREDERICK) AND MÖBIUS (KARL)
On the Natural History of the Cingalese Pearl-oyster and on the Production of Pearls. Annals of Natural History, 3rd series, VOL. I, pp. 81–100. *London, 1858.*

KELLY (H. M.)
A Statistical Study of the Parasites of the Unionidæ. Illinois State Laboratory of Natural History, VOL. V, pp. 399–418. *Springfield, Ill., 1899.*

KENT (WILLIAM SAVILLE)
Report on the Pearl and Pearl Shell Fisheries of North Queensland. *Brisbane, 1890.* Fol.
On the Experimental Cultivation of the Mother-of-pearl Shell in Queensland. Report Australian Association, VOL. II, pp. 541–548. *Sydney, 1891.*
The Great Barrier Reef of Australia; Its Products and Potentialities. *London, 1893.* Fol.

KING (CHARLES WILLIAM)
The Natural History of Precious Stones and of the Precious Metals. *London, 1870.* 8vo.

KINGSMILL (WALTER)
The Pearling Industry [of Western Australia]. Illustrated Handbook of Western Australia. Paris International Exhibition, 1900. *Perth, W. A., 1900.* 8vo.

KLEIN (JACOB THEODOR)
Tentamen Methodi Ostracologicæ sive Dispositio Naturalis Cochlidum et Concharum. *Lugduni Batavorum, 1753.* 4to.

KLEIN (JOHANN)
De Jure circa Margaritas vulgo von dem Perlen-recht. *Rostochii, 1700.* 8vo.

KNAUTHE (KARL)
Perle in *Unio rostratus.* Zoologischer Garten, p. 155. *Berlin, 1896.*

KOGEL (J.)
Tripang- und Perlenfischerei in niederländischen Australien. *Ausland,* 1857. 8vo.

KÜCHENMEISTER (FRIEDRICH)
Uebersetzung der Arbeit: "Sull' origine delle Perle, del Dottore F. de Filippi." Nebst auf Untersuchungen gegründeten Anmerkungen. Müller's Archiv für Anatomie, Physiologie, und wissenschaftliche Medicin, pp. 251–268. *Berlin, 1856.*
Ueber eine der häufigsten Ursachen der Elsterperlen und das Verfahren, welches zur künstlichen Vermehrung der Perlen dem Königl.-Sächsischen Ministerium der Finanzen vorgeschlagen wurde. Müller's Archiv für Anatomie, Physiologie, und wissenschaftliche Medicin, pp. 269–281. *Berlin, 1856.*

KUNZ (GEORGE FREDERICK)
American Pearls. Proceedings of the American Association for the Advancement of Science, 33rd meeting, pp. 665–674. *Salem, 1885.*
Precious Stones in the United States. "Harper's Magazine," VOL. LXXVI, pp. 97–106. *New York,* December, 1887.
Gems and Precious Stones of North America. A Popular Description of Their Occurrence, Value, History, Archæology, and of the Collections in which they Exist, also a Chapter on Pearls and on Remarkable Foreign Gems Owned in the United States. *New York, 1890.* 4to.
On the Occurrence of Pearls in the United States. Transactions of the American Fishery Society, pp. 16–32. *New York, 1893.*
A Brief History of the Gathering of Fresh-water Pearls in the United States. Bulletin U. S. Fish Commission, VOL. XIII, pp. 321–330. *Washington, 1894.*

On Pearls and the Application and Utilization of the Shells, etc., as Shown at the World's Columbian Exposition. Bulletin U. S. Fish Commission, VOL. XIII, pp. 439–457. *Washington,* 1894.

The Fresh-water Pearls and Pearl Fisheries of the United States. Bulletin U. S. Fish Commission, VOL. XVII, pp. 373–426. *Washington,* 1898.

Fresh-water Pearls of America. "Nature," VOL. LX, pp. 150–152. *New York,* June 15, 1899.

Precious Stones of California. *Sacramento,* 1905. 8vo.

KUSNETZOW (I. D.)

Fischerei und Thiererbeutung in den Gewässern Russlands. *St. Petersburg,* 1898. 8vo.

LAFERRIÈRE (J.)

De Paris à Guatémala. Notes de Voyages au Centre-Amérique, 1866–1875. *Paris,* 1877. 8vo.

LANDAETA ROSALES (MANUEL)

Gran Recopilacion geografica, estadistica e historica de Venezuela. 2 vols. *Carácas,* 1889. Fol.

LANGERUS (JOHANNES)

De dulcissimo margaritas nomine de Margaritæ. Scriptorum Publice propositorum a Gubernatoribus. *Vitebergæ.* 1555. Fol.

LASTEYRIE (FERDINAND DE)

Description du Trésor de Guarrazar. *Paris,* 1860. 4to.

LEACH (WILLIAM ELFORD)

The Zoologists Miscellany; being descriptions of new or interesting animals. 3 vols. *London,* 1814–17. 8vo.

LE BECK (HENRY J.)

An Account of the Pearl Fishery in the Gulph of Manar in March and April, 1797. Asiatic Researches, VOL. V, pp. 393 *et seq. London,* 1798.

LE COMTE (LOUIS)

Memoirs and Remarks . . . Made in Above Ten Years' Travel Through the Empire of China: Particularly Upon Their Pottery, Varnishing,

Silk, and Pearl-fishing. *London,* 1737. 8vo.

LEDELIUS (SAMUEL)

De perlis Lusato-silesiacis. Miscellanea Curiosa, observatio 150, pp. 327–328. *Norimbergæ,* 1690.

LEONARDUS (CAMILLUS)

Speculum lapidum. *Venice,* 1502. 4to.

LEUWENHOECK (M.)

Microscopical Observations on the Salts of Pearls, Oyster-shells, etc. Philosophical Transactions of the Royal Society of London for 1707, No. 311, pp. 2416–24.

LEWIS (JAMES)

On the Coloring Matter of Pearl-shells. Proceedings of the Academy of Natural Sciences of Philadelphia for 1860, pp. 88–89.

LEWIS (J. P.) AND HORNELL (JAMES)

Reports on the Pearl Fishery of 1904. Sessional Papers. *Colombo,* 1904. Fol.

Reports on the Pearl Fishery of 1905. Sessional Papers. *Colombo,* 1905. Fol.

LINSCHOTEN (JOHN HUYGHEN VAN)

The Voyage of John Huyghen van Linschoten to the East Indies. 2 vols. *London,* Hakluyt Society, 1884. 8vo.

LISTER (MARTIN)

A Journey to Paris in the Year 1698. *London,* 1699. 8vo.

LONDON MAGAZINE

Pearls. "London Magazine," VOL. XI, pp. 715–722. *London,* January, 1904.

LOPES DE CASTANHEDA (FERNHAM)

The First Booke of the Historie of the Discoverie and Conquest of the East Indies, Enterprized by the Portingales. *London,* 1582. 4to.

LOPEZ DE GOMARA (FRANCISCO)

Historia general de las Indias con todo el descubrimiento y cosas notables que han acaecido desde que se ganaron hasta el año de 1551. *Çaragoça,* 1552. Fol.

LOVELL (ROBERT)

Panmineralogicon, or an Universal

History of Minerals, containing the summe of all Authors. *Oxford, 1661.* 8vo.

LOVETZKY
Notice sur les perles du gouvernement de Viatka. Bulletin de la Société des Naturalistes. *Moscou, 1830.*

LOW (FRANZ)
Die Flussperlenmuscheln-fischerei in der Moldau in Böhmen. Verhandlungen der Zoologisch-Botanische Gesellschaft, pp. 333–336. *Wien, 1859.*

LUCATT
Rovings in the Pacific from 1837 to 1849. 2 vols. *London, 1851.* 8vo.

M.
L'Histoire Naturelle éclaircie dans deux de ses parties principales, La Lithologie et la Conchyliogie dont l'une traite des pierres et l'autre des Coquillages . . . par M. . . . de la Société Royale des Sciences de Montpellier. *Paris, 1742.* 4to.

McCULLOCH (JOHN RAMSAY)
Article "Pearls." Dictionary of Commerce. *London, 1882.* 8vo.

MACDONALD (ALEXANDER C.)
In Search of El Dorado: a Wanderer's Experiences. *London, 1905.* 8vo.

MACGOWAN (D. T.)
Pearls and Pearl-making in China. Journal of the Society of Arts, VOL. II, pp. 72–75. *London,* Dec. 16, 1853.

McINTOSH (WILLIAM CARMICHAEL)
The Story of a Pearl. "The Zoologist," No. 752, VOL. LXII, pp. 41–56. *London,* February, 1904.

MALCOLM (JOHN), *Sir*
Sketches of Persia, from the Journals of a Traveller in the East. 2 vols. *London, 1827.* 8vo.

MANDEVILLE (JOHN), *Sir*
Le Grand Lapidaire, où sont déclarez les noms de Pierres orientales, avec les Vertues et Propriétés d'icelles, aussi les isles et pays où elles croissent. *Paris, 1561.* 12mo.

MARBODUS, *Bishop of Rennes*
Marbodei, Galli poetae vetustissimi de lapidibus pretiosis encheridion cum scholiis Pictorii Villingensis. *Paris, 1531.* 8vo.
Liber Lapidum seu de Gemmis varietate lectionis et perpetua annotatione illustratus a Johanne Beckmanno. *Gottingæ, 1799.* 8vo.

MARIOT
La reproduction des huîtres perlières aux îles Tuamotu. Bulletin de la Société d'Acclimatation, VOL. I, pp. 341–342. *Paris, 1874.*

MARKHAM (CLEMENTS ROBERTS)
Tinnevelly Pearl Banks. "Intelligent Observer," VOL. IV, pp. 418–426. *London,* January, 1864.
Also in "Electric Magazine," VOL. CXI, pp. 496–501. *London,* April, 1864.
The Tinnevelly Pearl Fishery. Journal of the Society of Arts, VOL. XV, pp. 256–262. *London,* March 1, 1867.

MARTENS (GEORG VON)
Purpur und Perlen. *Berlin, 1874.* 8vo.

MAUNDER (SAMUEL)
The Treasury of Natural History. *London, 1878.* 12mo.

MECKEL VON HEMSBACH (JOHANN HEINRICH)
Mikrogeologie. Ueber die Concremente im thierischen Organismus. Nach des Verfassers Tode herausgegeben und bevorwortet von T. Billroth. *Berlin, 1856.* 8vo.

MÉRY (M.)
Remarques faites sur la Moule des Estangs. Histoires de l'Académie Royale des Sciences, November 22, 1710, pp. 408–426. *Paris, 1732.* 4to.

MEXIA (PEDRO) AND SANSOVINO (M. FRANCESCO)
Treasurie of Auncient and Moderne Times. 2 vols. *London, 1619.* Fol.

MIDDENDORF (ALEXANDER VON)
Reise in den äussersten Norden und Osten Siberiens während der Jahre 1843 und 1844, VOL. II, Zoologie. *St. Petersburg, 1851.* 4to.

MILBURN (WILLIAM)
Oriental Commerce; Containing a

Geographical Description of the Principal Places in the East Indies, China and Japan, with Their Produce. 2 vols. *London,* 1813. 4to.

MITSUKURI (K.)
The Cultivation of Marine and Freshwater Animals in Japan. Bulletin of the Bureau of Fisheries for 1884, VOL. XXIV, pp. 257–289. *Washington,* 1905.

MÖBIUS (KARL)
Die echten Perlen. Ein Beitrag zur Luxus-, Handels- und Naturgeschichte derselben. *Hamburg,* 1858. 4to.

MONARDES (NICOLAS)
Delle cose che vengono portate dall' Indie Occidentali, pertinenti all' uso della medicina. *Venetia,* 1575. 12mo.

MOQUIN-TANDON (M.) AND CLOQUET (J. S.)
Sur la production artificielle des perles, rapport fait à la Société d'Acclimatation, 1858. Journal de Conchyliologie, VOL. X, pp. 87–88. *Paris,* 1858.

MORESBY (JOHN)
New Guinea and Polynesia. Discoveries and Surveys in New Guinea and the D'Entrecasteaux Islands: a Cruise in Polynesia and Visits to the Pearl-shelling Stations in Torres Straits, of H. M. S. Basilisk. *London,* 1876. 8vo.

MORIER (JAMES P.)
A Journey Through Persia, Armenia, and Asia Minor, to Constantinople, in the Years 1808 and 1809. *London,* 1812. 8vo.

MOYNIER DE VILLEPOIX (R.)
Recherches sur la formation et l'accroissement de la coquille des mollusques. Journal de l'Anatomie et de la Physiologie, VOL. XXVIII, pp. 461–518. *Paris,* 1892.

MÜLLER (FELIX)
Über die Schalenbildung bei Lamellibranchiaten. Schneider's Zoologische Beiträge, VOL. I, pp. 206–246. *Breslau,* 1885.

NALDI (PIO)
Delle Gemme, e delle regole per valutarle. 2 vols. *Bologna,* 1791. 8vo.

NATHUSIUS-KÖNIGSBORN (W. VON)
Über die Gestaltungsursachen der Haare, der Eischalen, der Molluskenschalen und der Harting'schen Körperchen. Archiv für Entwickelungsmechanik der Organismen, VOL. VI, pp. 365–393. *Leipzig,* 1898.

NICHOLS (THOMAS)
A Lapidary; or, the History of Pretious Stones; with Cautions for the Undeceiving of all those that deal with Pretious Stones. *Cambridge, England,* 1652. 4to.

NIEBUHR (CARSTEN)
Beschreibung von Arabien. *Kopenhagen,* 1772. 4to.

NILSSON (SVEN)
Historia molluscorum Sveciae terrestrium et fluviatilium breviter delineata. *Lundæ,* 1822. 8vo.

NISBET (JOHN)
Burma under English Rule and Before. 2 vols. *Westminster,* 1901. 8vo.

NITSCHE (HINRICH)
See- und Süsswasser-Perlen. Internationale Fischerei-Ausstellung zu Berlin, 1880, pp. 83–95. *Berlin,* 1881. 8vo.
Die Süsswasserperlen auf der internationalen Fischerei-Ausstellung zu Berlin, 1880. Nachrichtsblatt der Deutschen Malakozoologischen Gesellschaft, pp. 49–64. *Frankfurt,* 1882.

NOÉ (LOUIS PANTALÉON JUDE AMÉDÉE DE), *Count*
Mémoires relatifs à l'expedition anglaise partie du Bengale en 1800 pour aller combattre en Égypte l'armée d'Orient. *Paris,* 1826. 8vo.

NUÑEZ CABECA DE VACA (ALVAR)
La relacion que dio Alvar Nunez Cabeca de Vaca de lo acaescido enlas Indias enla armada donde yua por gouernador Paphilo de Narbaez. *Valladolid,* 1542. 4to. See SMITH (BUCKINGHAM).

OLAUS (MAGNUS)
Historia de Gentium Septentrionalium variis conditionibus. *Basileæ*, 1567. Fol.

OSBORN (HENRY LESLIE)
Observations on the Parasitism of *Anodonta plana*, etc. Zoological Bulletin, pp. 301–310. *New York*, 1898.

OTTÉ (E. C.)
The Oriental Pearl. "Macmillan's Magazine," VOL. IV, pp. 229–237. *London*, July, 1861.

OUSELEY (WILLIAM), *Sir*
Travels in Various Countries of the East; More Particularly Persia, etc. 3 vols. *London*, 1819–23. 4to.

OVIEDO Y VALDES (GONZALO FERNANDEZ DE)
Historia natural y general de las Indias. *Toledo*, 1526. Fol.

PAGENSTECHER (HERMANN ALEXANDER)
Trematodenlarven und Trematoden. Helminthologischer Beitrag. *Heidelberg*, 1857. Fol.
Ueber Perlenbildung. Siebold und Kölliker Zeitschrift für wissenschaftliche Zoologie, VOL. IX, pp. 496–505. *Leipzig*, 1858.
Also in Verhandlungen des Naturhistorisch-medicinischen Vereins, pp. 157–158. *Heidelberg*, 1859.

PALGRAVE (WILLIAM GIFFORD)
Narrative of a Year's Journey Through Central and Eastern Arabia (1862–63). *London*, 1865. 8vo.

PANCIROLI (GUIDO)
Rerum memorabilium jam olim deperditarum. *Ambergæ*, 1599. 8vo.

PARAZZOLI (A.)
La Pesca nel Mar Rosso. Bollettino della Società di Esplorazioni commerciali di Africa, pp. 177–190. *Milano*, June, 1898.

PENFIELD (FREDERIC COURTLAND)
The Lure of the Pearl. "The Century Magazine," VOL. LXXIII, pp. 61–77. *New York*, November, 1906.

PENNANT (THOMAS)
British Zoology. 4 vols. *London*, 1768–70. 8vo.

A Tour in Scotland in 1769. *Chester*, 1771. 8vo.

PERCIVAL (ROBERT)
An Account of the Island of Ceylon. *London*, 1803. 4to.

PEREZ Y HERNANDEZ (JOSÉ MARIA)
Estadistica de la Republica Mejicana. *Guadalajara*, 1862. 4to.

PERLE ET ALGIOFAR
Tesoro delle Gioie Tratto Curioso, . . . come Perle, Gemme, Auori, Unicorni, Bezzari, Cocco, Malacca, Balsami, Contraherba, Muschio, Ambra, Zelieto. *Padoua*, 1630. 12mo.

PERRIER (EDMOND) AND FALCO (ALPHONSE)
Rapports du Jury International. Classe 53.—Engins, instruments et produits de la pêche.—Aquiculture. *Paris*, Imprimerie Nationale, 1901. Fol.

PETAU
Observations sur les moules d'étang dans lesquelles on a trouvé des Perles. Mémoire de l'Académie des Sciences, pp. 23–24. *Paris*, 1769.

PFEIFFER (IDA LAURA)
Eine Frauenfahrt um die Welt. 3 vols. *Wien*, 1850. 8vo.

PFIZMAIER (AUGUST)
Beiträge zur Geschichte der Perlen. Sitzungsberichte der philosophisch-historischen Klasse der kaiserlichen Akademie der Wissenschaften, VOL. LVII, pp. 617–654. *Wien*, 1868.

PHILOSTRATUS
Philostratorum quæ supersunt omnia . . . accessere Apollonii Tyanensis Epistolæ, Eusebii liber adversus Hieroclem, Callistrati descript. *Lipsiæ*, 1709. Fol.

PLAT (HUGH), *Sir*
The Jewel House of Art and Nature . . . Whereunto is added a rare and excellent Discourse of Minerals, Stones, Gums, and Resins. *London*, 1653. 4to.

PLINIUS SECUNDUS (CAIUS)
C. Plinii Naturalis Historia. D. Detlefsen recensuit. 6 vols. *Berolini*, 1866–82. 8vo.

The Historie of the World, Commonly called the Naturall Historie of C. Plinius Secundus. Translated into English by Philemon Holland, Doctor of Physick. 2 vols. *London,* 1601. Fol.

POLO (MARCO)
La Description géographique des provinces et villes plus fameuses de l'Inde Orientale. *Paris,* 1556. 4to.
The Book of Ser Marco Polo, the Venetian, concerning the kingdoms and marvels of the East. Newly translated and edited, with notes, by H. Yule. 2 vols. *London,* 1871. 8vo.

PORTER (ROBERT KER), *Sir*
Travels in Georgia, Persia, Armenia, Ancient Babylonia, . . . during the Years 1817, 1818, 1819 and 1820. *London,* 1821–22. 4to.

POUGET (JEAN HENRI PROSPER)
Traité des Pierres précieuses, et de la manière de les employer en parure. 2 vols. *Paris,* 1762. 4to.

PRATT (CHARLES STUART)
Pearls and Mother-of-pearl. "Popular Science Monthly," VOL. XLIX, pp. 390–398. *New York,* July, 1896.

PRIDHAM (CHARLES)
An Historical, Political and Statistical Account of Ceylon and its Dependencies. 2 vols. *London,* 1849. 8vo.

PROCOPIUS (OF CÆSAREA)
The History of the Warres of the Emperor Justinian . . . Written in Greek by Procopius . . . and Englished by H. Holcroft, Knight. *London,* 1653. Fol.

PUJOL (T. F.)
Estudio Biológico sobre la ostra *Avicula margaritiferus.* Boletin de la Sociedad Mexicana de Geografía y Estadística, VOL. II, pp. 139–150. *Mexico,* 1859.

PULTENEY (RICHARD)
A General View of the Writings of Linnæus. Second edition. *London,* 1805. 4to.

PUTNAM (F. W.)
Notes on *Fierasfer dubius* from Panama Pearl-oysters. Proceedings of the Boston Society of Natural History, VOL. XVI, pp. 343–346. *Boston,* 1874.

PUTON (ERNEST)
Mollusques terrestres et fluviales des Vosges: Le Département des Vosges, statistique, historique, et administrative, par Henri Lepaye et Ch. Charton. 2 vols. *Nancy,* 1845. 8vo.

QUEENSLAND REPORT
Departmental Commission on Pearl-shell and Bêche-de-mer Fisheries. Report, Together With the Minutes of Evidence and Proceedings of the Commission Appointed to Inquire into the General Working of the Laws Regulating the Pearl-shell and Bêche-de-mer Fisheries of the Colony. *Brisbane,* 1897. Fol.

QUIÉVREUX (H.), *Consul de France*
La pêche des perles au Venezuela. Revue Maritime, VOL. CXLVI, pp. 444–448. *Paris,* August, 1900.

RAINERI (ANTONIO)
Fior di pensieri sulle Pietre Preziose di Ahmad Teifascite. *Firenze,* 1818. 8vo.

RAU (SEBOLD FULCO JAN)
Specimen Arabicum, continens descriptionem et excerpta libri Achmedis Teifaschii "De Gemmis et Lapidibus Pretiosis." *Trajecti ad Rhenum,* 1784. 4to.

RÉAUMUR (RENÉ ANTOINE FERCHAULT DE)
Sur la Matière qui colore les Perles fausses, et sur quelques autres Matières animales d'une semblable couleur; à l'occasion de quoi on essaye d'expliquer la formation des Ecailles des Poissons. Histoire de l'Académie Royale des Sciences, 1716, pp. 229–244. *Paris,* 1716.
Observations sur le coquillage appellé *Pinne marine* ou Nacre de perle, à l'occasion dequel on explique la formation des perles. Mémoires de l'Académie des Sciences, pp. 177–194. *Paris,* 1717.

Betrachtungen von der Seemuschel (*Pinna marina*), oder Perlmutter genannt. Königliche Akademie der Wissenschaften in Paris. *Breslau,* 1753. 8vo.

REDDING (ROBERT), *Sir*
Structure, Color, etc., of Irish pearls. Philosophical Transactions of the Royal Society of London, VOL. XVIII. p. 659, October 13, 1688.
A Letter Concerning Pearl-fishing in the North of Ireland. Philosophical Transactions of the Royal Society of London, VOL. XVIII, pp. 659–663. *London,* 1693.

REINAUD (JOSEPH TOUISSAINT)
Fragments Arabes et Persans inédits relatifs à l'Inde antérieurment au xiᵉ siècle de l'ère chrétienne. *Paris,* 1845. 8vo.
Mémoire géographique, historique et scientifique sur l'Inde antérieurement au milieu du xiᵉ siècle de l'ère chrétienne d'après les écrivains arabes, persans et chinois. *Paris,* 1849. 4to.

RENAUDOT, *Abbé*
Ancient Accounts of India and China by two Mohammedan Travellers. *London,* 1733. 8vo.

RIBEIRO (JOÃO), *Capitano*
Histoire de l'Isle de Ceylon . . . traduite du Portugais par Monsr. l'Abbé Le Grand. *Amsterdam,* 1701. 12mo.
History of Ceylon. *Ceylon,* 1847. 8vo.

RONDELETIUS (GULIELMUS)
Universæ Aquitilium Historiæ Pars Altera. *Lugduni,* 1554. Fol.

ROSENBERG (H. VON)
Perlenfischerei auf den Aru-Inseln. Nachrichtsblatt der Deutschen Malakozoologischen Gesellschaft, pp. 29–40. *Frankfurt,* 1884.

ROSNEL (PIERRE DE)
Le Mercure Indien, ou le Trésor des Indes . . . Seconde partie, dans laquelle est traitté des pierres précieuses et des perles, . . . avec un traitté sommaire des autres pierres moins précieuses. *Paris,* 1672. 4to.

ROTHSCHILD (M. D.)
A Hand-book of Precious Stones. *New York* and *London,* 1890. 12mo.

ROUGEMONT (LOUIS DE)
The Adventures of Louis de Rougemont, as Told by Himself. *Philadelphia,* 1900. 8vo.

RUEUS (FRANCISCUS)
De Gemmis aliquot, iis praesertim quarum divus Joannes Apostolus in sua Apocalypsi meminit. *Parisiis,* 1547. 8vo.

RUMPH (GEORG EVERHARD)
Thesaurus imaginum piscium, testaceorum, ut et cochlearum, accedunt conchylia, conchæ univalviæ et bivalviæ denique mineralia. *Lugduni Batavarum,* 1711. Fol.

RUSCHENBERGER (WILLIAM S. W.)
A Voyage Round the World; Including an Embassy to Muscat and Siam. *Philadelphia,* 1838. 8vo.

RZACZYNSKI (GABRIEL)
Historia naturalis regni Poloniæ, Magni Ducatus Lituaniæ, annexarumque provinciarum. *Sandomiriæ,* 1721. 4to.

SAINT LAURENT (JOANNON DE)
Description abrégée du fameux Cabinet de M. le Chevalier de Baillou, pour servir à l'histoire naturelle des pierres précieuses, métaux, minéraux, et autres fossiles. *Luques,* 1746. 4to.

SANDIUS (CHRISTOPHER)
On the Origin of Pearls. Philosophical Transactions of the Royal Society of London, 1674. No. 101, p. 11. *London,* 1674.

SANGER (J. P.), *General*
Census of the Philippine Islands, Taken Under the Direction of the Philippine Commission in 1903, VOL. IV, pp. 533–536. *Washington,* 1905.

SCHMID (JOACHIMUS)
De Margaritis. *Wittebergæ,* 1667. 8vo.

SCHOLTZ (H.)
Schlesien's Land- und Süsswasser-Mollusken systematisch geordnet und beschrieben. *Breslau,* 1843. 8vo.

SEEMAN (BARTHOLD)
Narrative of the Voyage of H. M. S.

Herald, During the Years 1845–51, Under the Command of Capt. Henry Kellet. 2 vols. *London,* 1853. 8vo.

SEPTALA (LUDOVICO)
Ludovici Septalii de Margaritis nuper ad nos allatis, judicium, etc. *Mediolani,* 1618. 4to.

SERVANTUS (CLAVIUS)
De piscatura margaritarum. Miscellanea Curiosa Medico-physica, p. 417. *Francofurti,* 1684.

SEURAT (L. G.)
L'Huître Perlière: Nacre et Perles. *Paris* (1900). 12mo.
Observations sur l'évolution de l'Huître perlière des Tuamotu et des Gambier. *Rikitea,* 1904. 8vo.
The Pearl-forming Properties of the Parasite *Tylocephalum margaritiferæ.* Comptes Rendus de l'Académie des Sciences. *Paris,* March 26, 1906.
Le Nacre et la Perle en Océanie. Pêche, origine et mode de formation des Perles. Bulletin du Musée Océangraphique de Monaco, No. LXXV, May 20, 1906. *Monaco,* 1906. 8vo.

SHIPLEY (ARTHUR E.)
Pearls and Parasites. "The Quarterly Review," VOL. CCII, pp. 485–496. *London,* 1905.

SHIPLEY (ARTHUR E.) AND HORNELL (JAMES)
The Parasites of the Pearl-oyster. Supplementary Report, Herdman's Pearl Oyster Fisheries of the Gulf of Manaar, VOL. II. *London,* 1904.
Further Report on Parasites Found in Connection with the Pearl-oyster Fishery at Ceylon. Herdman's Report on the Pearl Oyster Fisheries of the Gulf of Manaar, PART III, pp. 49–56. *London,* 1905. 4to.

SHIPP (BARNARD)
History of Hernando de Soto and Florida. *Philadelphia,* 1881. 8vo.

SIEBOLD (CARL THEODOR ERNST VON)
Ueber die Perlenbildung chinesischer Süsswasser-muscheln, als Zusatz zu dem vorhergehenden Aufsatze. Siebold und Kölliker, Zeitschrift für wissenschaftliche Zoologie, VOL. VIII, pp. 445–454. *Leipzig,* 1857.
Die Süsswasserfische von Mitteleuropa. Zeitschrift für die gesammten Naturwissenschaften, VOL. XXII, pp. 468–477. *Halle,* 1863.

SIMMONDS (PETER LUND)
On the Pearl, Coral, and Amber Fisheries. Journal of the Society of Arts, VOL. XVIII, pp. 173–200. *London,* 1870.
The Commercial Products of the Sea; or, Marine Contributions to Food, Industry and Art. *London,* 1879. 8vo.

SIMMONDS (VANE)
Fresh-water Pearls. *Charles City, Iowa,* 1899. 8vo.

SIMPSON (CHARLES TORREY)
The Classification and Geographical Distribution of the Pearly Fresh-water Mussels. Proceedings of the U. S. National Museum, VOL. XVIII, pp. 295–343. *Washington,* 1896.
The Pearly Fresh-water Mussels of the United States; Their Habits, Enemies, and Diseases, with Suggestions for Their Protection. Bulletin U. S. Fish Commission, VOL. XVIII, pp. 279–288. *Washington,* 1899.
Synopsis of the Naiades, or Pearly Fresh-water Mussels. Proceedings of the U. S. National Museum, VOL. XXII, pp. 501–1044. *Washington,* 1900.

SMITH (BUCKINGHAM)
Narratives of the Career of Hernando de Soto in the Conquest of Florida, as Told by a Knight of Elvas and in Relation by Luys Hernandez de Biedma, Factor of the Expedition. Translated by Buckingham Smith. *New York,* Bradford Club, 1866. 8vo.

SMITH (HUGH McCORMICK)
The Mussel Fishery and Pearl-button Industry of the Mississippi River. Bulletin U. S. Fish Commission, VOL. XVIII, pp. 289–314. *Washington,* 1899.

SMITH (SIDNEY)
Large Pearl in *Unio margaritiferus.*

"The Naturalist," p. 133. *London,* 1866.

SOMERSET (SOMERS)
The Pearl Fishery of Ceylon. "The Nineteenth Century," VOL. CCCLXIII, pp. 843–851. *London,* May, 1907.

SOTO (FERDINANDO DE)
The Discovery and Conquest of Terra Florida by Don. Ferdinando de Soto and Six Hundred Spaniards his Followers. Written by a Gentleman of Elvas. *London,* Hakluyt Society, 1851. 8vo.

SOUBEYRAN (J. LÉON) AND DELONDRE (AUG.)
De la pêcherie d'Huîtres perlières de Tinnevelly et de la culture artificielle des Huîtres perlières de la même localité. Bulletin de la Société d'Acclimatation, VOL. IV, pp. 398–415. *Paris,* 1867.

SOURÎNDRO MOHUN TAGORE
Mani-málá, or a Treatise on Gems. 2 vols. *Calcutta,* 1881. 8vo.

SPRAT (THOMAS)
The History of the Royal Society of London for the Improvement of Natural Knowledge. *London,* 1667. 4to.

SQUIER (EPHRAIM GEORGE)
Observations on the Aboriginal Monuments of the Mississippi Valley. Transactions of the American Ethnological Association, VOL. II. *New York,* 1847.
Nicaragua: its people, scenery, monuments, etc. 2 vols. *New York,* 1852. 8vo.

SQUIER (E. G.) AND DAVIS (E. H.)
Ancient Monuments of the Mississippi Valley. Smithsonian Contributions to Knowledge, VOL. I. *Washington,* 1848. 4to.

STEARNS (ROBERT EDWARD CARTER)
On Certain Parasites, Commensals, and Domiciliares in the Pearl-oysters, *Meleagrinæ.* Report of the Smithsonian Institution for 1886, pp. 339–344. *Washington,* 1887.

STEUART (JAMES)
An Account of the Pearl Fisheries of Ceylon, with an Appendix. *Ceylon,* 1843. 4to.
Notes on Ceylon . . . To Which are Added Some Observations . . . on the Pearl Fishery. *London,* 1862. 8vo.

STREETER (EDWIN WILLIAM)
Precious Stones and Gems, Their History and Distinguishing Characteristics. *London,* 1877. 8vo.
Pearls and Pearling Life. *London,* 1866, 8vo, pp. xvi, 329.

SWEET (FRANK H.)
Pearl Seeking. "Lippincott's Magazine," VOL. LXI, pp. 375–378. *Philadelphia,* March, 1898.

T. (W.)
Experiments on the Growth of Pearls, with observations on their structure and color. Edinburgh Philosophical Journal, VOL. XI, pp. 39–45, July, 1824.

TASSIN (WIRT)
Descriptive Catalogue of the Collections of Gems in the United States National Museum. Report of the U. S. National Museum for 1900, pp. 473–670. *Washington,* 1902.

TAUNTON (HENRY)
Australind: Wanderings in Western Australia and the Malay East. *London,* 1903. 8vo.

TAVERNIER (JEAN BAPTISTE)
Les Six Voyages de J. B. Tavernier en Turquie, en Perse et aux Indes. *Paris,* 1676. 4to.
Travels in India, 1640–1667 . . . Translated from the original French edition of 1676, with a biographical sketch . . . by V. Ball. 2 vols. *London,* 1889. 8vo.

TAYLOR (W. H.)
Taylor's Submarine Pearl Fishing Company. *New York,* 1848. 8vo.

TEIXEIRA (PEDRO)
The Travels of Pedro Teixeira; with his "Kings of Hormuz" (1608) and Extracts from his "Kings of Persia." *London,* Hakluyt Society, 1902. 8vo.

TENEBRE (JEHAN)
La pêche des perles. A Travers le Monde, pp. 232–234. *Paris, 1905.*

TENNENT (JAMES EMERSON), *Sir*
Ceylon: an Account of the Island. 2 vols. *London, 1859.* 8vo.

THEOPHRASTUS
Theophrastus's History of Stones, With the Greek Text and an English Version, and Notes Critical and Philosophical, Including the Modern History of Gems, Described by that Author, by John Hill. *London, 1746.* 8vo.
Theophrastus von den Steinen, aus dem Griechischen. (By A. H. Baumgärtner.) *Nurnberg, 1770.* 8vo.

THOMAS (HENRY SULLIVAN)
A Report on Pearl Fisheries and Chank Fisheries [of Tuticorin], 1884. *Madras, 1884.* Fol.
The Pearl-oyster of the Gulf of Manaar. Madras Journal of Literature and Science, pp. 89–115. *Madras, 1887.*

THOMPSON (LINDSAY G.)
History of the Fisheries of New South Wales; with a Sketch of the Laws by Which They Have Been Regulated. *Sydney, 1893.* 8vo.

THURM (EVERARD IM)
Sketch of the Ceylon Pearl Fishery of 1903. Spolia Zeylonica, VOL. I, pp. 56–65. *Colombo, 1904.*

THURSTON (EDGAR)
Notes on the Pearl and Chank Fisheries and Marine Fauna of the Gulf of Manaar. *Madras, 1890.* 8vo.
Pearl and Chank Fisheries of the Gulf of Manaar. Madras Government Museum, Bulletin No. 1, *Madras, 1894.* 8vo.

TIFFANY & Co.
Collection of Pearls and the Shells in Which They Are Found in the Brooks, Rivers, Lakes and on the Coasts of the United States. U. S. Section Palais des forêts, chasse et pêche. Expositione Universelle, Paris, 1900. *New York, 1900.* 8vo.

TOLLIUS (ADRIANUS)
Gemmarium et Lapidum Historia. Recensuit ... Adrianus Tollius. *Lugduni Batavorum, 1636.* 8vo.

TOLLIUS (JACOBUS)
Epistolæ itinerariæ, ex auctoris schedis postumis recensitæ, suppletæ et digestæ; annotationibus et figuris adornatæ, cura et studio. *Amstelædami, 1700.* 4to.

TOWNSEND (CHARLES HASKINS)
Report Upon the Pearl Fishery of the Gulf of California. Bulletin U. S. Fish Commission, VOL. IX, pp. 91–94. *Washington, 1891.*

TULLBERG (TYCHO)
Studien über den Bau und das Wachsthum des Hummerpanzers und der Molluskenschalen. Kongliga Svenska Vetenskaps-Akademiens Handlingar, VOL. XIX, No. 3, pp. 1–57. *Stockholm, 1882.*

TWYNAM (WILLIAM C.)
Report on the Pearl Fishery of 1877. Sessional Papers. *Colombo, 1877.* Fol.
Report on the Pearl Fishery of 1879. Sessional Papers. *Colombo, 1879.* Fol.
Report on the Pearl Fishery of 1880. Sessional Papers. *Colombo, 1880.* Fol.
Report on the Pearl Fishery of 1881. Sessional Papers. *Colombo, 1881.* Fol.
Report on the Pearl Fishery held at Marichchikkaddi during 1890. Sessional Papers. *Colombo, 1890.* Fol.
Report on the Pearl Fishery of 1891. Sessional Papers. *Colombo, 1891.* Fol.
Report on the Ceylon Pearl Fisheries. Sessional Papers. *Ceylon, 1899.* Fol.
Report on the Ceylon Pearl Fisheries. Sessional Papers. *Colombo, 1900.* Fol.

TWYNAM (WILLIAM C.) AND DONNAN (JAMES)
Reports by the Superintendent of the Fishery and the Inspector of the

Pearl Banks, 1887. Sessional Papers. *Colombo, 1887.* Fol.

ULLOA (ANTONIO DE), *Admiral*
Relación histórica del viage á la América meridional. 5 vols. *Madrid, 1748.* 4to.

UNGEWITTER (FRIEDRICH H.)
Geschichte des Handels, der Industrie und Schifffahrt von den ältesten Zeiten an bis auf die Gegenwart. *Leipzig* und *Meissen* (1844). 8vo.

VANE (GEORGE)
Ceylon Pearl Fisheries. Report to Governor Sir Henry Ward, February 28, 1863. *Ceylon, 1863.*
The Pearl Fisheries of Ceylon. Journal of the Ceylon Branch of the Royal Asiatic Society, VOL. x, No. 34, pp. 14–32. *Colombo, 1888.*

VASCO DA GAMA
The Three Voyages of Vasco da Gama. *London,* Hakluyt Society, 1869. 8vo.

VAUGHAN (RICE)
A Discourse of Coin and Coinage . . . as also Tables of the Value of All Sorts of Pearls, Diamonds, Gold, Silver, and Other Metals. *London, 1675.* 12mo.

VEGA (GARCILASSO DE LA)
Historia de la Florida. 4 vols. *Madrid, 1803.* 24mo.

VENEGAS (MIGUEL)
A Natural and Civil History of California Containing an Accurate Description of That Country. 2 vols. *London, 1759.* 8vo.

VENETTE (NICHOLAS)
Traité des Pierres qui s'engendrent dans les terres et dans les animaux, où l'on parle des causes qui les forment dans les hommes. *Amsterdam, 1701.* 12mo.

VERE (SCHELE DE)
Jewels of the Deep: Pearls. "Putnam's Magazine," VOL. xi, pp. 278–288. *New York,* March, 1868.

VERNATTI (PHILIBERTO), *Sir*
Pearl-divers in the East Indies, Their Time Under Water. Philosophical

Transactions of the Royal Society of London, for 1669, No. 43, p. 863.

VILLA (ANTONIO)
Sull' origine delle perle e sulla possibilita di produrle artificialmente. Atti del Ateneo Milano, VOL. xv, 1859–60. pp. 165–173. Il Politecnico, VOL. viii, pp. 568–584. *Milano, 1860.*
Des Perles; de leur origine et de leur production artificielle. Traduit de l'Italien par Timothée Coutet. *Paris, 1863.* 12mo.

VINCENT (WILLIAM)
The Commerce and Navigation of the Ancients in the Indian Ocean. 2 vols. *London, 1807.* 4to.

W. (S.)
Recollections of Ceylon: its Forests and its Pearl Fishery. "Fraser's Magazine," VOL. LXII, pp. 753–767. *London,* December, 1860.

WASHBURN (HOWARD E.)
American Pearls. *Ann Arbor, Mich.,* 8vo, 48 pages, paper, 5 plates.

WATKINS (M. G.)
Scotch Pearls and Pearl Hunting. "Gentleman's Magazine," VOL. CCLXXX, pp. 626–629. *London,* June, 1896.

WATTS (MRS. PHILIP)
A Visit to the Ceylon Pearl Fisheries. "The Graphic," VOL. LXXI, pp. 583–597. *London,* May 20, 1905.

WEBER (M.)
Om Perler og Perlefiskerierne. Norske Fiskeritidende, pp. 252–263. *Bergen,* October, 1886. 8vo.
A translation in English by Herman Jacobson is given in Bulletin U. S. Fish Commission, VOL. vi, pp. 321–328. *Washington, 1886.*

WELLSTED (JAMES RAYMOND)
Travels in Arabia. 2 vols. *London,* 1838. 8vo.
Travels to the City of the Caliphs Along the Shores of the Persian Gulf and the Mediterranean. 2 vols. *London, 1840.* 8vo.

WESTERLUND (CARL AGARDH)
Fauna Molluscorum terrestrium et

fluriatilium Sveciæ, Norvegiæ et Daniæ. 2 vols. *Stockholm*, 1871–73. 8vo.

WESTROPP (HODDER M.)
A Manual of Precious Stones and Antique Gems. *London*, 1874. 8vo.

WHIGHAM (H. J.)
The Persian Problem. *London*, 1903. 8vo.

WHITMARSH (HUBERT PHELPS)
Fishing for Pearls in Australia. "The Century Magazine," VOL. XXI, pp. 905–911. *New York*, April, 1892.
Pearl-diving and Its Perils. "The Cosmopolitan," VOL. XVIII, pp. 564–572. *New York*, March, 1895.
The World's Rough Hand: Toil and Adventure at the Antipodes. *New York*, 1899. 12mo.
Working Under Water: the Story of an Amateur Pearl Fisher. "The Outlook," VOL. LXI, pp. 124–129. *New York*, January 14, 1899.

WILLIAMS (CHARLES)
Silvershell; or the Adventures of an Oyster. *London*, 1856. 8vo.

WILSON (D.), *Colonel*
Pearl Fisheries in the Persian Gulf.

Journal of the Royal Geographic Society of London, VOL. III, pp. 283–286. *London*, 1834.

WOHLBEREDT (O.)
Nachtrag zur Molluskenfauna des Königreiches Sachsen. Nachrichtsblatt der deutschen Malakozoologischen Gesellschaft, pp. 97–104. *Frankfurt*, 1899.

WOLF (JOHANN CHRISTOPH)
Reise nach Zeilan. Nebst einem Berichte von der holländischen Regierung zu Jaffanapatnam. 2 vols. *Berlin* und *Stettin*, 1782–84. 8vo.

WOODWARD (HENRY)
Parasitical Animals in *Meleagrina margaritifera* of Australia. Proceedings of the Zoological Society of London, 1886, pp. 176–177.

ZIMMERN (HELEN)
Stories in Precious Stones. *London*, 1873. 8vo.

ZWEMER (S. M.)
Arabia: the Cradle of Islam. *New York*, 1900. 8vo.

INDEX

INDEX